WORKING WITH TECHNOLOGY
LAW AND PRACTICE

AUSTRALIA
LBC Information Services
Sydney

CANADA and the USA
Carswell
Toronto

NEW ZEALAND
Brooker's
Auckland

SINGAPORE and MALAYSIA
Sweet & Maxwell (S.E. Asia)
Singapore

WORKING WITH TECHNOLOGY LAW AND PRACTICE

Edited by

Iain Purvis

Barrister, 11, South Square, Gray's Inn

Jennifer Pierce

Solicitor, Charles Russell

LONDON · SWEET & MAXWELL · 2001

Published in 2001 by Sweet & Maxwell Limited of
100 Avenue Road
London NW3 3PF

Typeset by J&L Composition Ltd, Filey, North Yorkshire
Printed in England by MPG Books Ltd, Bodmin, Cornwall

No natural forests were destroyed to make this product; only farmed
timber was used and replanted

ISBN 0–421 59810–7

A CIP catalogue record for this book is available
from the British Library

ACKNOWLEDGEMENTS

In addition to the individual authors, a number of people were generous in their assistance to the editors. In particular, **Ashley Carr** and **Frances Smith** from 11 South Square, and **Mark Simpkin** from Brick Court Chambers, provided invaluable assistance with co-ordination and **Jane Juby** compiled the final manuscript for the publishers. We are especially indebted to **Derek Chandler** for his help with the proofs, to **Gary Rinck** General Counsel of Pearson Plc, who assisted with the United States material on intellectual property, to **Dr Paul Cozens** of Mathys & Squire who commented on Chapter 9, to **Vivien Irish**, who assisted with part of Chapter 8, to **Professor Adrian Shipwright** of Pump Court Tax Chambers who commented on the references to taxation, to **Nigel Burroughs** of 11 Old Square, who reviewed Chapter 21, to **Mark West** of 11 Old Square, who assisted with a query on Chapter 22, and to **Philippe Courtois** of Courtois Lebel & Associés who dealt with a query on French law. We would also like to thank **John O'Reilly** of Unilever who assisted us in our initial search for an author to cover United Kingdom Government funding.

ACKNOWLEDGEMENTS

PREFACE

All of those who work with rights in technology in some capacity, be it advisory, commercial, or managerial, need to understand, or to master, to some degree, a wide variety of legal disciplines, such as intellectual property, contract and European law. In some cases it is also a requirement to understand an increasingly complex commercial backdrop. This makes the work stimulating, and even occasionally enjoyable. However, the variety and complexity of the issues that arise are an obvious source of traps for the unwary. This, together with the rapid pace of change in many of the disciplines involved makes it an area of work that may be particularly inaccessible on first acquaintance.

Whilst we cannot pretend to cover all the potential subject matter, we have assembled a group of authors who are specialists in some of the more important legal disciplines associated with the field, together with those who are actively involved in funding and commercialisation of technology. In the case of United Kingdom government funding we deviated from this strategy on the basis that the topic is so diverse that many of those with more practical experience would not have a sufficient overview of it.

As with any book of this kind, decisions had to be made as to what to leave out. We have aimed in general to include material that is specific to technology, at the expense of entirely general material that is already covered in many other publications. For example, we have included a chapter on patent prosecution, but have omitted basic contract law. Similarly, we have not included information on conventional banking transactions, partly because loan finance is a less common source of funding for technology in the early stages (when the funding of a business based on technology will differ from other businesses) and partly because banking has been the subject of several other books.

We have given as much detail as we could, bearing in mind constraints of space. However, it is in the nature of a wide-ranging book of this kind that it cannot hope to deal with every aspect of the subject in the same degree of detail as a practitioner's book dealing with one subject only. We have tried to avoid matters of purely historical interest. In those few cases where such material has been included (such as United Kingdom and E.C. funding), this has been judged useful in explaining the general principles.

Both the authors and publishers hope that in addition to the anticipated legal readership, this book will be accessible to and will be read by professionals other than lawyers, and by others whose work relates to some aspect of the creation or management of intellectual property. So far as the legal profession is concerned, we hope that it will be useful not only to those who work with technology every day, but also to those whose practice occasionally brings them into contact with its subject matter. With this in mind, we have included introductory chapters to each legal section, which may be read by those who are new to the subject, and ignored by those with greater knowledge. We have also assembled some statutory, E.C. and governmental materials on a disk, that is slotted into the back cover, so that readers who do not have ready access to these materials can follow up on some of the references in the main text.

As we have said, one of the main challenges faced by those who work with technology is the rapid speed of change in all areas, especially the law. Even in the short time that will elapse between submission of proofs and publication, some parts of the text may start to become out of date. Readers should therefore check that material has not been overtaken or affected by some very recent development.

Iain Purvis and Jennifer Pierce
November 2000

TABLE OF CONTENTS

Chapter 3

Venture Capital

Chapter 4

European Commission Funding

Chapter 5

The University Perspective

Chapter 6

The Industry Perspective

Chapter 7

Introduction to Intellectual Property

Chapter 8

The Law of Patents

Chapter 9

Patenting Procedure and Practice

Chapter 10

The Law of Confidence

Chapter 11

The Law of Copyright

Chapter 12

The Law of Designs

Chapter 13

The Law of Trade Marks and Passing Off

Chapter 14

Ownership of Intellectual Property

Chapter 17

Competition Law in the United Kingdom

Chapter 18

United States Antitrust Law

Chapter 19

Introduction to Chapters on Agreements

Chapter 20

Research and Development Agreements

Chapter 21

Companies and other Business Structures

Chapter 22

Assignments

Chapter 23

Licensing

Chapter 24

Remedies and How to Get Them

TABLE OF CASES

TABLE OF STATUTES

TABLE OF STATUTORY INSTRUMENTS

TABLE OF EUROPEAN LEGISLATION

TABLE OF INTERNATIONAL LEGISLATION

TABLE OF INTERNATIONAL TREATIES/ CONVENTIONS

CHAPTER 1

INTRODUCTION TO COMMERCIAL AND FUNDING CHAPTERS

Jennifer Pierce, Charles Russell

GENERAL

1–01   When people first start work commercialising technology they are often confronted with complex legal and factual issues, together with a great deal of jargon and folklore. Unless they are in a large company with substantial resources, they are also likely to need to seek sources of finance for inventions. We therefore decided to begin this book with a series of short chapters that would introduce readers to the commercial aspects of transactions, and which would acquaint them with some of the terms used. In particular, these chapters should provide both advisers and researchers with an overview of some of the more common sources of finance for technology.

FUNDING

Obtaining funding

1–02   The funding available for research and development is insufficient to meet the current demand, so promoters and inventors of technology may need to consider all the available options, particularly in the early stages. Their advisors need to follow suit if they are to advise on the legal or commercial aspects of any transaction. They may not actively seek finance for their client, but they need to appreciate the commercial context in which funding documents are generated, in order to understand their client's bargaining strength. We have been forced to be highly selective in our choice of material for this section. However, we believe that what we have provided will assist readers to understand important aspects of some of the major sources of funding.

Funding systems and terms

1–03   Even if advisors are not directly involved in funding, they may need to be aware of the sort of terms on which funding may be provided, as it may have an impact on the commercialisation of the property at a later stage. Funding documents may, for example, contain provisions for compulsory licensing of the results of the funded project. It has not been possible to describe all funding programmes in the same depth, due to the number of programmes available. Graham Stroud has produced a detailed account of European Union funding, which explains both the funding process and some of the conditions attached. In contrast, Dr Paul Cunningham, has explained the overall structure of the complex United Kingdom system of state funding, so that readers will understand the provenance of the documentation. Dr Paul Castle has produced a detailed, and yet succinct, chapter on the availability of venture capital funding, and the process of obtaining it.

Insufficient debt finance

1–04   There is insufficient debt finance available for many businesses that commercialise intellectual property, and debt finance may be viewed less favourably than other forms of finance by these businesses. This applies more particularly to businesses that are "early stage", or are in the development phase.[1] Early stage "high technology" businesses are viewed as particularly high credit risks, as the failure rate of new technologies is considerable, and in the initial phases they are far less likely to produce sufficient financial return to service the debt. This is exacerbated by the fact that the major assets of such companies are often intellectual property, which is not viewed as the most desirable security. Lenders tend to prefer tangible assets as security for a variety of reasons.[2] Intellectual property is generally more difficult to value and to dispose of than other assets. It also appears that intellectual property based businesses tend to prefer raising equity finance as a means of sharing risk and minimising current financial obligations.[3] Whilst debt finance is available, it is not the most important source of finance, and there are many publications devoted to it, so it has been omitted.

Other sources of funding

1–05   In many cases, promoters of technology based businesses are obliged to seek some form of government or European Commission funding, or equity investment.[4] In some industrial sectors, larger companies may be prepared to participate, through research contracts and strategic alliances. In certain industries, more especially those relating to healthcare, charities have substantial funds for research, and may provide some form of assistance, provided that the proposed method of participation is compatible with their charitable status. The charity sector would, regrettably, have been too difficult to cover in a publication of this size, as there are many potential sources of funding, and they tend to be specific to particular areas of research, particularly healthcare.

COLLABORATION

Collaboration generally

1–06   Collaboration and funding are often linked. Collaboration may, for example, be a requirement of funding, as is the case with European Union schemes. It may also be a means of funding, insofar as one of the participants in the collaboration may receive payments as part of their motivation for undertaking the research, as is the case with much externally funded university research. A substantial amount of research and development is funded in this manner. The terms on which collaborative research is undertaken are likely to affect the exploitation of the results of the research. They can vary considerably, depending on the circumstances, even when they are negotiated by the same organisation. There can also be a marked difference between the stance of similar organisations within the same sector, and terms may depend on the nature of the research. The commercial sector is experimenting with new business models as well as the public sector.

Different viewpoints

1–07   It can be difficult to ascertain the terms and conditions on which companies and universities will do business. In addition to the considerable variation, the com-

---

[1] See Arthur Andersen's report on "The use of intellectual property as debt finance", produced in November 1998 in conjunction with the Intellectual Property Institute .
[2] *ibid.*
[3] *ibid.*
[4] This would be through a sale of shares in a company.

mercial parameters are shifting constantly, and there can be consequential changes in terms. Accounts of the current terms can vary greatly, even amongst supposedly unbiased commentators. There is undoubtedly a university funding gap of serious proportions, but the way in which the research community will deal with this has yet to be determined. The only way to proceed in this climate is to pool all the available information, from a variety of sources, and to use it to come to the best possible agreement in the circumstances. Chapters 5 and 6 describe the commercial *perspectives* of Dr Campbell Wilson, an executive in AstraZeneca, a major pharmaceutical company, and of Dr Roger Holdom and Mr James Evans, of University College, London and the University of Manchester, respectively. In this context it is crucial that readers consider Chapter 5 on university funding together with Chapter 6 on collaboration with industry.

SHIFTING SANDS

1–08 Unfortunately, funding schemes are revised regularly by successive governments and other administrations. This leads to substantial changes in the nature of funds available, and to the conditions attached to the funding. Chapter 2 contains a description of some of the more recent changes in the United Kingdom. In addition, the popularity of technology stocks can vary considerably from time to time, affecting one of the major external sources of finance for the sector. Whilst few newly formed companies will proceed quickly to a listing, the financing of the more established companies may increase the collaborative projects undertaken within the sector. Government funding of universities has an impact on funding in general, as universities have a greater need of revenue from collaboration with the private sector, and from technology transfer, if government funding is reduced. Also, in view of the importance of technology to the economy, successive governments have introduced various policies to try to stimulate certain aspects of the sector.

CHAPTER 2

PUBLIC SECTOR FUNDING IN THE UNITED KINGDOM

Dr Paul Cunningham, Senior Research Fellow, PREST, Manchester University

INTRODUCTION

2–01 This Chapter outlines the current arrangements for the provision of public funding in the United Kingdom (U.K.). It begins with a brief account of the development of the present day system. Such an account is necessary to provide the context within which the current system operates and the major issues which impinge upon public sector R&D policy. The Chapter then examines the level and flow of research funds, before dealing with the major public sector funders of research. Details of the mission and major programmes of relevant Government departments and agencies are provided, together with information on the mechanisms in place for the support of innovation. It should be recognised that Government policies (and even structures) can be subject to rapid change. For these reasons, all figures provided are for indicative purposes only. Furthermore, as the funding programmes of individual ministries, departments, agencies and research councils are also subject to change, only broader subject area budgets are presented. Finally, the wide range and large number of funding schemes precludes a detailed description of how each may be accessed. To compensate for this, a list of addresses and web-sites is provided on the disk for those seeking more detailed information.

HISTORICAL BACKGROUND

2–02 The establishment of the Royal Greenwich Observatory in 1675 is generally considered to mark the start of Government support for science in the U.K. However, few further significant developments took place until the Geological Survey and the forerunner of the Laboratory of the Government Chemist were set up in 1832 and 1843 respectively. In the 1850s the Government set up a Department of Science and Art, and although a Royal Commission on Scientific Instruction and the Advancement of Science proposed the establishment of state-run laboratories and a Ministry of Science and Education, no further Government action ensued until the founding of the National Physical Laboratory in 1899. While this marked a milestone in the provision of substantial state support for scientific research, the real roots of the current governmental arrangements for science and technology lie in the changes instigated in the early 1900s.

First world war

2–03 The First World War brought with it an alarming realisation of Britain's dependency upon the products of the German industrial system. In response, the Government set up an Advisory Council on Scientific and Industrial Research and a Department of Scientific and Industrial Research (DSIR). This was intended to mobilise British science in the support of the national war effort, through its own laboratories, through its responsibilities for the industrial research associations, and through grants to scientists working in universities. The Government adopted an arm's length approach to the direction of scientific research, an arrangement which was endorsed by the report of the Haldane Committee[1] in 1918. This recognised the importance of scientific research to government, and distinguished between research relevant to particular departmental portfolios and what could nowadays be termed basic research. This distinction led to the conduct of what is now termed "mission-oriented" research by departmental laboratories and to the creation of a series of "research councils" to complement DSIR — thus laying the basis for the U.K.'s research council system.

Second world war

2–04 By the Second World War the importance of scientific and technological research was fully recognised and the 1950s witnessed further growth in Government support for science and technology. Beyond the growth of the DSIR and the other research councils, the university sector underwent an expansion, new organisations such as the United Kingdom Atomic Energy Authority were created and the Ministry of Defence (MoD) and its supply ministries began to consume increasingly larger amounts of funding. In 1959 the first attempt was made to co–ordinate decision making for science and technology, hitherto dispersed among the relevant Government departments, by the creation of an Office of the Minister of Science. This Office had oversight of the DSIR and other research councils, atomic energy and space research, and all general matters of civil science policy outside departmental research programmes.

1960s and 1970s

2–05 A succession of Governments through the 1960s and 1970s undertook several reorganisations of the components of the science and technology system. The first half of the 1960s saw the creation of the Natural Environment Research Council (NERC), the Science Research Council (SRC) and the Social Science Research Council (SSRC) together with a Department of Education and Science (with financial responsibility for the research councils), and a Ministry for technology. Perhaps the most far-reaching effects, however, emerged as a result of the review of Government R&D, conducted by Lord Rothschild in 1971.[2] Not least amongst its recommendations was the proposed observance of the "customer-contractor principle" whereby departments would "commission" research from their own laboratories. Debate on how best to centrally co-ordinate science and technology at government level led to the reorganisation or establishment of a number of policy/advisory bodies, such as the Advisory Board for the Research councils (ABRC) and the Advisory Council for Applied Research and Development (ACARD). Also during this period a new set of concepts — mission orientation, technology policy, social relevance — assumed greater prominence in policy debate.

The 1980s

2–06 Political philosophy regarding the funding of research throughout government, the public sector, the universities and industry underwent a fundamental change in the early 1980s — "market forces" were to regulate the overall level and distribution of research expenditure, rather than government direction, and wherever possible, public

---

[1] *Report on the Machinery of Government*, Cm. 9230, December 1918.
[2] "The Organisation and Management of Government R&D", published in *A Framework for Government Research and Development*, Cm. 4814, November 1971, HMSO.

funding was to be substituted by private funding. The Alvey Programme of pre-competitive collaborative R&D in advanced Information Technology formed, perhaps, the sole exception to this ideology. Over its five year lifetime it attracted £200 million of Government money and a further £150 million from industry. However, its successor programmes received declining levels of support and the general shift in Government attitude continued. This ideology has since led to the privatisation of a number of Government research establishments, and placed pressure on universities to obtain a greater proportion of their funding from private sector sources.[3] The period was also marked by debate over issues such as prioritisation and selectivity and the recognition of "strategic research".

The 1990s

2–07   The early 1990s witnessed a number of substantial changes to the U.K.'s science and technology infrastructure. The Office of Science and Technology (OST) — a cabinet-level body headed by the Chief Scientific Adviser replaced the ABRC. The OST was intended to focus science and technology policy matters, and engage in dialogue with other Departments and Ministries on a more equal footing. The post of Science Minister was re-established as head of a new Office of Public Service and Science.

Realising our potential

2–08   A major landmark in U.K. science policy was the Government's publication in 1993 of the first major review of science for over twenty years. The White Paper, *Realising Our Potential,* set out the Government's strategy for science, engineering and technology (SET). Its aims were to be achieved through a number of measures: a Technology Foresight Programme to promote communication, interaction and understanding between industry, scientific communities and Government departments and to provide policy advice to a new Council on Science and Technology; a clear statement of central and departmental government expenditure and planning in an annually published *Forward Look of Government Funded Science Engineering and Technology*[4]; initiatives to forge closer links between universities, industry and government in order to promote the transfer of technology; and greater innovation support to firms, particularly small and medium-sized enterprises (SMEs).

Reorganisation of the research councils

2–09   An extensive reorganisation of the research councils was also undertaken. Each was required to produce a mission statement making explicit commitment to their contribution to wealth creation and the management structure was modified, each being headed by a Chief Executive and a Chairman, the latter generally drawn from industry. The Government's commitment to the Rothschild "customer-contractor principle" was reaffirmed and a further scrutiny of the public sector laboratories was announced. Finally, the OST was given the lead role in co-ordinating cross-departmental science and technology issues.

---

[3] More recent Government policy has, however, sought to reward research collaboration between the universities and industry, through a variety of schemes.
[4] The *Forward Look* was to make the Government's strategy more explicit to both the research community and industrialists. It reviews trends in the level of public and private funding for SET, sets out the Government's three-year spending plans, provides information on current research and technology development programmes, and identifies future challenges and opportunities for publicly funded SET over the next five to ten years.

1995 Competitiveness White Paper

2–10 In 1995, the Government published a Competitiveness White Paper which tied in many SET issues into the Government's strategy for national competitiveness. It was clear that the DTI had shifted away from an interventionist policy towards a focus on small targeted programmes aimed primarily at improving industrial/academic linkages and networking behaviour within industry for the spread of best practice, particularly in management. With the exception of some downsized efforts into civil aviation and space,[5] the era of large-scale technology generation and R&D support programmes was over. As a postscript, in 1995 the Government moved the OST from the Cabinet Office into the DTI. This prompted concern within the science and technology communities that the science and engineering base would be marginalised to directly serving the interests of industry.

RECENT DEVELOPMENTS AND CURRENT POLICY

2–11 Upon achieving office in 1997, the Labour administration's first action was to undertake a year long Comprehensive Spending Review (CSR), to refocus public spending on its priorities (in particular, health and education), while making savings where possible in other areas. The CSR resulted in £1.1 billion of new spending for the science and engineering base over three years (to 2001). Labour's first major innovation policy statement came in the form of the 1998 Competitiveness White Paper *Our Competitive Future: Building the Knowledge Driven Economy* — essentially, the Government's blueprint for innovation policy. High on the agenda was the widely adopted concept of the "knowledge driven economy" and the understanding that "the United Kingdom's distinctive capabilities are . . . knowledge, skills and creativity". Using the White Paper's terminology, there are three areas in which the Government sees itself being able to contribute to this aim: by strengthening capabilities, by promoting collaboration, and by enhancing competition.

Specific measures to strengthen capabilities included: investment to modernise the science base; vigorous promotion of the commercialisation of university research; promoting awareness of the potential of information and communication technologies to SMEs; and the Foresight Programme. The fostering of a culture of entrepreneurship is also a government priority. Under the heading of collaboration, the Government's priority is to act as a catalyst to "promote creative collaboration between business and within regions". Specific measures here concentrate on the issue of "clusters" of related industrial and research activity. Regionality forms another major pillar of Government policy, particularly in the context of developing regional innovation action plans and policies.

2–12 A clear Government target is to improve and enhance the way in which the results of research conducted in the higher education and independent research sectors are translated into commercial products, processes and services. This objective is founded on a strong, well-resourced research base with a strong culture of entrepreneurship, which exhibits close links in terms of the transfer of expertise, knowledge and trained personnel, with the business and enterprise sectors. A major guide in this policy of "partnership" between the links in the innovation chain is furnished by the continuing Foresight Programme. The biotechnology sector forms a particular target for measures aimed at the promotion of clustering and intra–sectoral

---

[5] And the Department's responsibilities in the area of nuclear and energy technology which were taken on after it subsumed the Department of Energy.

co-operation. In recognition of the past failure of U.K. companies to translate the successes of the science base into internationally competitive products, processes and services, U.K. innovation policy has increasingly focused on ways in which this situation may be remedied and a number of measures designed to foster links between the science base and industry have been introduced and strengthened. In addition the Government is sensitive to the particular needs of SMEs and the role they play in the economy.

More specific government support to meet the aims of *Our Competitive Future* was provided in a number of new or expanded schemes:

- an extra £1.4 billion[6] investment into science funding over three years[7];

- the creation of eight university centres of enterprise under a £25 million Science Enterprise Challenge;

- a two-fold increase of the Government's contribution to the Teaching Company Scheme;

- the announcement of a second round of the Foresight Programme;

- £10 million for a second round of Foresight LINK awards;

- the provision of specialist advice services through new Business Link Centres of Expertise.

An overview of the U.K.'s science and technology structures is set out in Figure 1.

FUNDING OF R&D

2–13 While there is no overarching series of priorities for the allocation of Government reseach funding according to socio-economic objectives (regardless of political rhetoric), over the last ten years or so there have been some relatively large changes in the patterns of this funding (see Table 1). For example, the proportion of total public funding for R&D has increased in areas such as environmental protection, health, social development and services, and the advancement of knowledge. At the same time public funding has declined for defence, industrial development (industrial production and technology) and energy research. Other R&D areas, including agriculture, forestry and fishing, earth and atmosphere, and civil space, have remained relatively stable.

The allocation of funding is achieved through a process introduced by the Labour Government in 1997 and known as the Comprehensive Spending Review (CSR). This replaced the previous annual Public Expenditure Survey (PES) and is conducted by the National Audit Office on a three year basis.

2–14 According to OECD (1994) definitions, the U.K.'s Gross domestic Expenditure on Research and Development (GERD) — that is, funding from all sectors, including government and business enterprise sources — exceeds £14.5 billion annually. In terms of Gross Domestic Product (GDP), this represents a proportion of about 1.8 per cent,[8] a ratio which has fallen consistently since 1993. In international comparisons, this figure

---

[6] Government sources tend to use the U.S. definition of a billion here, *i.e.* one thousand million.
[7] This was in partnership with the Wellcome Trust, the U.K.'s largest medical research charity, which contributed £400 million. OST/DTI provided £1.1 billion. The Department for Education and Employment (DfEE), in announcing its own CSR results, proposed an allocation of £300 million for research infrastructure, to be channelled through the higher education funding councils.
[8] Figures for 1997.

2-15 **Table 1.** Trends in Government funding of R&D by socio-economic categories (1986–1998): per cent.

	1986–87	1987–88	1988–89	1989–90	1990–91	1991–92	1992–93	1993–94	1994–95	1995–96	1996–97	1997–98
Agriculture, forestry and fishing	4.7	4.5	4.7	4.1	4.0	4.3	5.1	5.3	5.1	5.0	4.5	4.6
Industrial development	10.5	9.4	8.9	9.5	9.7	7.9	7.8	8.5	3.5	2.9	2.5	1.7
Energy	4.4	3.9	3.9	3.3	2.9	2.6	2.4	1.8	1.1	0.9	0.7	0.7
Infrastructure	1.5	1.5	1.5	1.5	1.5	1.3	1.7	1.8	1.9	1.7	1.7	1.7
Environmental protection	1.1	1.3	1.3	1.1	1.4	1.4	1.4	2.0	2.3	2.3	2.2	2.3
Health	4.5	4.6	4.9	5.5	5.9	5.9	6.7	7.1	7.6	13.4	14.5	14.4
Social development and services	1.5	1.6	2.1	2.1	2.2	2.3	2.8	2.8	2.7	2.4	2.1	1.9
Earth and atmosphere	1.9	1.9	2.2	2.6	2.9	2.9	2.4	1.8	2.1	1.9	1.7	1.4
Advancement of knowledge	21.5	22.8	23.3	22.1	22.7	24.0	25.6	22.9	31.4	29.4	29.7	28.9
Civil space	2.9	3.0	3.3	3.1	3.1	2.7	2.9	3.5	3.1	2.7	2.8	2.8
Defence	45.1	45.2	43.5	44.7	43.5	43.9	40.8	42.0	38.9	37.0	37.2	39.2
Not elsewhere classified	0.3	0.3	0.3	0.3	0.2	0.7	0.5	0.6	0.4	0.4	0.4	0.4
%	100.0	100.0	100.0	100.0	100.0	100.0	100.0	100.0	100.0	100.0	100.0	100.0
Total (£ million)	4254.8	4407.8	4496.5	4771.8	4955.1	5027.4	5077.6	5402.3	5200.4	5642.0	5759.3	5892.0

Source: Office of Science and Technology, SET Statistics 1999.

falls well short of expenditure by Japan and the United States, is below that of Germany and France, but exceeds that of the remaining G7 nations of Italy and Canada.

As can be seen in Table 2, almost two-thirds of the U.K.'s expenditure on R&D originates from non-government sources, including charities, overseas sources and private industry, the latter contributing by far the largest share (49 per cent). The major part (around 85 per cent) of the U.K.'s research effort is categorised as civil R&D. In 1997, total civil GERD amounted to £12.54 billion, equivalent to 1.54 per cent of GDP. About 65 per cent of this was contributed by the private sector. Figure 2 and Table 2 provide a broad overview of the flow of R&D funds between the major players in the U.K.'s "system of innovation".

In terms of Government funded R&D[9] (see Table 2), in very broad terms, spending on the Science and Engineering Base (comprising the Science Budget and expenditure by the higher education funding bodies) predominates, accounting for around 38 per cent of expenditure, Civil departments account for about 20 per cent of total expenditure, and the majority of the remainder — 36 per cent — is spent through the Ministry of Defence. About 5.7 per cent goes to European Union science and technology commitments.

Using the OECD Frascati definitions it is possible to categorise government R&D funding according to the type of research supported (Table 3). In terms of performers, the Ministry of Defence (MoD) is responsible for the greatest share (71 per cent) of experimental development, through its programme of defence equipment procurement, whilst the research councils and Higher Education Funding Councils (HEFCs) fund most of the basic research, largely through the higher education (university) sector.

2–16 Table 3. Government funded R&D (1997/98) by Frascati categories.

Frascati type	Research Councils	Civil Departments	Defence
Pure Basic[1]	20.9%	2.9%	-
Orientated Basic[2]	38.9%	2.4%	-
Total Basic	**57.8%**	**5.3%**	-
Strategic Applied[3]	31.2%	41.2%	6.7%
Specific Applied[4]	8.2%	46.4%	21.6%
Total Applied	**39.4%**	**87.6%**	**28.3%**
Experimental Development[5]	0.8%	7.1%	71.6%
Total Funding (£ million)	*1264.8*	*1282.4*	*2311.4*

Source: Adapted from: Office of Science and Technology, SET Statistics 1999.

Notes:
1. Experimental or theoretical work primarily to acquire new knowledge of the underlying foundations of phenomena and observable facts, without any particular application or use in view.
2. As above but where a potential area of application is foreseen.
3. Original investigation to acquire new knowledge but directed primarily towards a general practical aim or objective. "Strategic applied" and "oriented basic" together comprise "strategic research" which is work that has evolved from basic research and where practical applications are likely and feasible but cannot yet be specified, or where the accumulation of the underlying technical know-how will serve several diverse purposes.
4. As above, but work has more specific applicability.
5. Systematic work, drawing on existing knowledge gained from research and practical experience directed to producing new materials, products and devices; to installing new processes, systems and services; or to substantially improving those already produced or installed.

---

[9] In its statistics the Government makes a distinction between funding for R&D and funding for SET (Science, Engineering and Technology). The latter actually covers expenditure on research and development; technology transfer activities; and scientific and technical education and training. It excludes all other "scientific, technical, commercial and financial steps. . . necessary for the successful development and marketing of new or improved products, processes or services" (DTI/OST 1997).

2–17   Table 2.   Gross expenditure on civil and defence R&D performed in the U.K. in 1997[1].

	Sectors carrying out the work						
Sectors providing the funding[2, 3]	Government departments	Research Councils	Higher education	Business enterprise	Private non-profit	Totals	Abroad
Government departments	1166	78	161	919	18	2343	236
Research Councils	21	405	691	8	10	1135	119
Higher Education Funding Councils	-	-	1033	-	-	1033	
Higher education institutions	0	3	118	-	1	123	
Business enterprise	203	37	207	6770	35	7252	
Private non-profit	6	28	438	1	107	579	
Abroad	30	38	248	1856	19	2191	
TOTAL	**1427**	**590**	**2896**	**9553**	**191**	**14656**	**n/a**
Civil							
Government departments	547	70	123	273	18	1032	160
Research Councils	20	405	691	8	10	1134	119
Higher Education Funding Councils	-	-	1033	-	-	1033	
Higher education institutions	0	3	118	-	1	123	
Business enterprise	139	37	183	6385	35	6781	
Private non-profit	6	28	438	1	107	579	
Abroad	9	38	248	1543	19	1857	
TOTAL	**722**	**582**	**2834**	**8209**	**190**	**12538**	**n/a**
Defence							
Government departments	619	8	38	646	0	1311	76
Research Councils	0	-	-	-	-	0	-
Higher Education Funding Councils	-	-	-	-	-	-	
Higher education institutions	0	-	-	-	-	0	
Business enterprise	64	-	24	384	-	472	
Private non-profit	-	-	-	0	-	0	
Abroad	22	-	-	313	-	335	
TOTAL	**705**	**8**	**62**	**1343**	**0**	**2118**	**n/a**

Source: Office of Science and Technology, SET Statistics 1999.
Notes:
1 – includes research in the social sciences and humanities
2 – Uses OECD terminology for the breakdown of GERD by sector
3 – Some numbers estimates

GOVERNMENT DEPARTMENTS

2–18 If measured by civil R&D expenditure on science, the Department of Trade and Industry (DTI) is by far the most important of the civil departments (see Table 4). The remit of the DTI is extremely wide-ranging but its overall objective is to stimulate innovation,[10] promote the development of a climate favourable for innovation and to enhance the competitiveness of U.K. business both at home and at the international level. The DTI is also responsible, through the OST, for the science budget, the research councils and cross-departmental co-ordination and development of government policy on SET.

DTI programmes and schemes

2–19 Whilst the DTI once formed an important source of funds for the support of technology development, direct sponsorship of R&D has declined in importance over recent years and has been largely replaced by measures to promote the growth of SMEs, the transfer of technology and the spread of management best practice. This shift was in part prompted by the recognition that, with U.K. industry spending some £9 billion a year on R&D, the Government could only fund a fraction, at best, of this and that its role would need to change to support innovation by a more considered application of its resources. Much of this activity is funded under the Department's Innovation Budget, which will be increased by about 20 per cent to almost £230 million in 2001/02.

Technology transfer and access have been identified as crucial factors to competitiveness. Specific DTI-sponsored schemes to promote these factors include **Business Links** — a series of over 200 centres, or "one-stop shops", providing local advice and support, particularly to SMEs, on business and innovation needs. Individual Business Links are operated in partnership by the local Chambers of Commerce, Training and Enterprise Councils (TECs), local authorities and other relevant organisations.

Departmental support for the promotion of collaboration between industry and the Science and Engineering Base is focused through a variety of schemes. Several of these are described below. As already noted, a strong level of interaction between the science base on the one hand, and industry, business and commerce on the other, forms a major focus of U.K. innovation policy. To monitor this interaction, the DTI has commissioned a number of surveys of industry/university co-operation and the main results of the 1996 survey are given in Box 1.

2–20   Box 1. Key findings of the 1996 DTI Industry:University co-operation survey.

- External research income for 111 U.K. universities was £1,349 million

- £145 million of this came directly from industry

- Of 89 reporting universities, 63 reported having at least one wholly or partly owned subsidiary for exploiting research

- Research income for these companies was £121 million for the period 1994-95

- In the same period the 89 universities filed 546 patent applications, of which 397 were new

- Total income for licenses, patents and other IP was £14.9 million

---

[10] Defined as "the successful exploitation of new ideas" *Forward Look*: HMSO, Cm. 3257–I, May 1996 (p. 82).

- During the same period again, 46 new spin-out companies had been formed. A previous (1995) survey found that 277 spin-out companies were in existence in the 80 responding universities

- U.K. higher education funding councils reported that 2,315 of the 101,685 full and part-time staff and 79,103 of the 1.3 million under- and postgraduate students in 111 universities had industry as their main source of support

LINK

2–21 Since 1986 the LINK programme, a government-wide initiative, has been the Government's principal support mechanism for collaborative research projects between the industry/commercial sector on one hand and the science base (HEIs, research council institutes, Government research establishments, hospitals or independent research organisations). Programme goals are defined by the sponsors, in consultation with industry and the research base. Account is also taken of priorities identified by the Foresight Programme. LINK programmes cover pre-competitive research in a wide range of technology and generic product areas, within defined technology or market sectors. These broad categories and the programmes open within them (as of May 1999), are shown below:

Electronics, Communications & IT
> Information for Fraud Control, Security and Privacy
> Broadcast Technology
> Sensors and Sensor Systems for Industrial Applications
> Advanced Sensors for Ocean Applications

Food and Agriculture
> Sustainable Arable Production
> Competitive Industrial Materials from Non-Food Crops
> Sustainable Livestock Production
> Advanced and Hygienic Food Manufacture
> Earth Observation
> Horticulture
> Aquaculture
> Eating, Food and Health
> Food Quality and Safety

Biosciences and Medical
> Medical Technologies (MedLINK)
> Genetic and Environmental Interactions in Health
> Integrated Approaches to Healthy Ageing
> Analytical Biotechnology

Materials and Chemicals
> Applied Biocatalysis
> Applied Catalysis and Catalytic Processes

Energy and Engineering
> Ocean Margins
> Future Integrated Transport
> Integration in Design and Construction
> Standardisation in the Construction Industry
> Foresight Vehicle
> Oils and Gas Extraction

2–22   **Table 4.**   Expenditure on SET by Government Departments (£ million - cash terms).

	1997–98	Estimated Outturn 1998–1999	Plan 1999–2000
OST- DTI [1]	27.5	43.4	147.3
Research Councils [2]	1303.7	1336.0	1349.2
Total Science Budget SET	**1331.1**	**1379.4**	**1496.5**
Higher Education Funding Councils	1032.7	1072.3	1153.2
Total Science & Engineering Base SET	**2363.8**	**2451.7**	**2649.7**
Ministry of Agriculture, Fisheries and Food	142.2	141.4	142.2
Department for Education and Employment	75.5	104.4	132.3
Department of Environment, Transport and the Regions	154.0	149.8	168.6
Department of Health [5]	59.1	61.7	70.1
National Health Service	401.3	400.5	400.6
Department of Social Security	3.9	4.6	4.0
Health and Safety Commission	26.4	24.6	24.9
Home Office	17.8	19.3	19.5
Department for Culture, Media and Sport [6]	9.8	9.7	10.5
Department for International Development [7]	112.1	115.8	119.3
Department of Trade and Industry (excluding OST and Launch Aid)	343.3	322.0	339.9
Net launch Aid	−119.2	−180.2	−170.5
Northern Ireland departments	23.6	31.5	34.0
Scottish Office	72.5	73.7	73.8
Welsh Office	17.6	17.9	17.9
Other departments (including Forestry Commission)	22.0	23.2	22.7
Total Civil Departments SET	**1361.8**	**1319.8**	**1409.6**
TOTAL CIVIL SET	**3725.6**	**3771.5**	**4059.3**
Ministry of Defence [8]	2311.4	2346.4	2618.7
of which: Research	656.0	609.7	600.9
Development	1655.4	1736.7	2017.8
Total SET	**6037.0**	**6117.9**	**6678.0**
Indicative U.K. contribution to E.U. R&D budget	335.4	371.9	408.2
GRAND TOTAL	**6372.4**	**6489.8**	**7086.2**

Source: Adapted from OST The Forward Look 1999: Government-funded science, engineering & technology.
Notes:
1. Excludes OST running costs, included in DTI.
2. Includes Pensions.
3. Excludes R&D programmes of the NHS Executive which appear under NHS.
4. Formerly the Department of National Heritage.
5. Formerly the Overseas Development Administration.
6. Figures reflect changes to estimates arising from Long Term Costing exercise.

Each programme receives sponsorship from relevant Government departments or research councils (see Table 4) and consists of a number of collaborative research projects involving industrial and academic/science base partners. Within each two to three year project, up to 50 per cent funding is received from the sponsoring department or research council with the industrial partner providing the balance of funds. Although participation by SMEs is particularly encouraged, multinational companies may also take part provided they have a significant manufacturing and research operation in the U.K., and the benefits of the research are exploited in the U.K. or European Economic Area. Government spend in 1997/98 was about £30 million (matched by industry), whilst estimated outturn for 1998/99 was over £37 million. Statistics to October 1996 show that over 1,100 companies (including around 350 SMEs) were active in LINK projects, together with 130 Science and Engineering Base institutions. Over 800 projects were either underway or completed.

In order to broaden the scope of the LINK scheme, the LINK marque has been franchised to schemes run by the Biotechnology and Biological Sciences Research Council and the Medical Research Council. This has enabled these research councils to fund LINK projects across their own selected research areas. In addition, the Innovative Manufacturing Initiative run by the Engineering and Physical Sciences Research Council is now run under a similar franchising arrangement.

Teaching Company Scheme (now TCS)

2–23 Also aimed at promoting the flow of knowledge between academia and industry, the TCS supports supervised technology transfer projects for young graduates within companies. Each student (TCS Associate) has an industrial and an academic supervisor and receives a balance of formal academic tuition together with practical commercial experience. The scheme receives support and finance from six Government departments and the research councils, and is delivered, promoted and managed on their behalf by the Teaching Company Directorate (TCD). Government funding is made to the knowledge base partner (usually a higher education institution) and is complemented by funds from the participating company. The budget for a TCS programme is determined largely by the number of TCS Associates to be employed, the size of the partner company and other details. However, for a first partnership, the annual company contribution per TCS Associate will generally be around £14,000 for companies with fewer than 250 employees and £21,000 for companies with 250 or more employees. A network of TCS Centres for small firms has also been set up. By the end of March 1997 over 600 TCS programmes were operating, involving more than 1,000 TCS Associates. Over three-quarters of these programmes involved SMEs. A £9m expansion of the scheme was announced in 1999.

College-Business Partnerships and Postgraduate Training Partnerships (PTP)

2–24 The TCD also manages, on behalf of the DTI, two other initiatives with broadly similar objectives. The College-Business Partnerships (CBP) scheme, is based on the same principles and concepts as the TCS, but primarily involves further education colleges in partnerships with SMEs. The PTP scheme, jointly supported by the DTI and the Engineering and Physical Sciences Research Council, provides support for industrially relevant research towards a higher degree. Projects involve RTOs[11] and universities who conduct co-ordinated research to address the research requirements of an industrial sector (with particular attention to the needs of SMEs). Students undertake their research under the joint supervision of academics and industrialists. PTPs have now been incorporated into the Faraday Partnerships initiative.

---

[11] RTOs (Research and Technology Organisations) include industrial research associations (*e.g.* Sira), independent contract research organisations (*e.g.* Huntingdon) and privatised Public Sector Research Establishments (*e.g.* NPL).

Faraday Partnerships

2–24a   The Faraday Partnership initiative is run by the DTI in conjunction with the Research Councils and other Government Departments. Faraday Partnerships are intended to foster the interaction between the U.K. science base and industry through the involvement of intermediate organisations (*e.g.* RTOs). The aims of the Faraday Partnerships, together with a list of ongoing projects, may be found at: http://www.dti.gov.uk/mbp/access/faraday.html. Collaborations are intended to become self-supporting, by attracting financial support from business and from existing U.K. Government, research council and E.U. schemes.

Shell Technology Enterprise Programme

2–25   The Shell Technology Enterprise programme (STEP) is a Shell U.K. Ltd National Placement scheme for undergraduates. It is part-funded by the DTI and aims to promote graduate employment by SMEs and to provide undergraduates with experience of work in smaller companies. Projects generally involve technology acquisition, implementation and usage. By 1997, the eleventh year of the scheme, over 5,000 undergraduates had been placed in companies.

SMART

2–26   While the primary focus of DTI support is centred on the fostering of an environment conducive to innovation, a number of technology development schemes are in operation. These concentrate largely on assisting SMEs facing difficulties in raising finance to develop products and processes, encouraging industry/academia collaboration and supporting sector specific programmes in energy, space and civil aeronautics. Support to SMEs is provided under SMART (Small Firms Merit Award for Research & Technology). This is a package of support which provides grants to help individuals and SMEs review their use of technology, access technology and research and develop technologically innovative products and processes. A range of support measures are available, including:

- Technology Reviews — grants of up to £2,500 towards the costs of expert reviews against best practice. Companies of 250 or fewer employees are eligible;

- Technology Studies — grants of up to £5,000 to help identify technological opportunities leading to innovative products and processes. Eligibility is as for Technology Reviews;

- Micro Projects — grants of up to £10,000 to assist with the development of low-cost innovative prototypes of products and processes. Applicants must have fewer than 10 employees;

- Feasibility Studies — competitive grants of up to £45,000 to cover the costs of feasibility studies into innovative technologies. Applicants must have fewer than 50 employees;

- Development Projects — competitive grants of up to approximately £122,000 for SMEs (with under 250 employees) undertaking development projects;

- Exceptional Development Projects — awarded, as their name suggests, in exceptional circumstances to help SMEs cover the cost of development projects up to approximately £366,000.

An extra £26 million of funding, over three years, was announced for this scheme in 1999.

Information Society Initiative

2–27 In 1996 the DTI launched the Information Society Initiative, a focused scheme which aims to assist businesses in responding to the challenges involved in moving toward the so-called "information society" — that is, being able to access, use and develop the dramatic advances in information and communication technologies which have occurred in recent years.

Funding for specific industry sectors

2–28 Energy In the field of energy, the Department supports programmes both in the field of nuclear and non-nuclear R&D. Key areas in the latter include offshore oil and gas, coal, fuel cell technologies and renewable energy, whilst those in the former are fusion and materials and radioactive waste. In support of the Government's domestic and international aims for the reduction of CO_2 and other greenhouse gas emissions, the department funds programmes such as the £43.5 million New and Renewable Energy Programme and the Cleaner Coal Technology Programme.

2–29 Aeronautics and space The DTI provides support for the aircraft industry through the Civil Aircraft Research and Demonstration Programme (CARAD). More specifically, the programme supports the development of civil aviation and the design and efficient use of civil aircraft. British national space R&D programmes are supported by the DTI through the British National Space Centre (an executive Agency of the DTI). In the area of space research, DTI support prioritises the development of commercial applications for space science, largely in the fields of telecommunications, meteorology and Earth observation.

2–30 Biotechnology The biotechnology sector has been identified as an area in which the United Kingdom has an established expertise and a number of internationally competitive companies. In order to reinforce this position and to assist in the translation of U.K. biotechnology research into competitive products, processes and services, and to promote the start-up of new biotechnology companies, the DTI has launched two initiatives.

The *Biotechnology Exploitation Platform Challenge* aims to capitalise on U.K. research strengths through encouraging syndicates of universities, academic institutions and intermediaries with complementary bioscience research to work together and to build portfolios of intellectual property. Eligible syndicates must include more than one research group and are also likely to include at least one of the following: Industrial Liaison Offices, University Technology Transfer Companies, Research Technology Organisations, Charities, Trade Associations, and intermediaries that offer the necessary skills in the management of bioscience intellectual property. The award covers up to 50 per cent of eligible costs for projects up to four years, after which time the successful syndicate is expected to become self-supporting. A £2.34 million pilot programme, which funded eight new consortia, was followed in August 1999 with the launch of a £6.45 million programme. The maximum award is £250,000.

The second initiative, the *Biotechnology Mentoring and Incubator (BMI) Challenge* offers support to organisations with relevant skills and/or premises to provide mentoring services and develop specialist bioincubators for biotechnology start-up companies. Its overall aim is to stimulate the creation and growth of high quality biotechnology companies, through overcoming the barriers to growth to such companies. These barriers include limited access to affordable specialist business and commercial advice, difficulties in building management teams and attracting adequate finance, and in locating suitable, affordable laboratory premises. Awards of up to £500,000 to a maximum of 50 per cent of project costs are available to successful proposals which should become self-supporting from the fourth year. These are awarded on a competitive basis. Proposals may be made by individual organisations or by consortia, which may include: companies, universities, other academic institutions, research and technology organisations, Training and Enterprise Councils, charities, trade associations, Business Links, Business Connect, Local Enterprise Councils, and other intermeidaries. The fourth call for proposals was announced in November 1999.

2–31   European and international activities   The DTI is active in developing U.K. participation in the European and international arenas, particularly in the European Union's Framework Programmes and the EUREKA initiative. U.K. firms are assisted in developing their awareness of technology developments and opportunities in other countries by the Department's Overseas Technology services.

A breakdown of DTI expenditure, by subject area, is provided in Table 5,below.

2–32   Table 5.   Department of Trade and Industry: R&D and SET expenditure by subject area (£ million).

Subject area	Estimated outturn 1998–99	Plan 1999–2000
Innovation and Technology Support:	**180.4**	**194.1**
Innovation Promotion and Support	62.4	58.9
Knowledge Transfer and Collaboration	43.0	63.7
Technology transfer and access	35.7	48.8
Standards, Statutory & Regulatory Support	61.6	62.0
Sector Challenge	13.5	11.6
Aeronautics:	**20.3**	**19.7**
Space:	**88.0**	**89.3**
National Space Programme	7.4	14.2
European Space Agency	80.6	75.1
Non-nuclear energy:	**13.9**	**14.7**
Offshore oil and gas; Industrial technology support	1.0	1.0
Offshore oil and gas; Enhanced oil recovery	1.2	1.1
Renewables	9.2	8.8
Clean coal technology	2.4	3.9
Nuclear energy:	**13.9**	**15.8**
Fusion	12.6	14.6
Safety and acceptability	1.3	1.2
OST administration:	**5.4**	**6.2**
TOTAL NET SET EXPENDITURE	**322**	**339.9**
Primary purpose breakdown of SET:		
Policy support	67.4	68.5
Technology support	211.6	209.7
Total R&D	**279.0**	**278.2**
Technology transfer	43.0	61.7
Total SET	**322.0**	**339.9**

Source: OST, *The Forward Look 1999: Government-funded science, engineering & technology.*

OFFICE OF SCIENCE AND TECHNOLOGY (OST)

2–33   The OST has responsibility for the Science Budget, the research councils, and the development and trans-departmental co-ordination of Government policy on science, engineering and technology. Headed by the Chief Scientific Adviser, and containing the Office of the Director General of the Research Councils (DGRC), OST provides the central focus for the development of U.K. Government policy on science and technology, both nationally and at the international level. The DGRC oversees the allocation of the Science Budget (see Table 6) which, since the move of the OST into the DTI, has been "ring-fenced".[12] The 1997/98 Science Budget totalled some £1.3 billion (49 per cent of civil expenditure on SET).

In addition to the tasks outlined above, OST is responsible for implementing the policies set out in the 1993 SET White Paper, for producing the annual *Forward Look* and for carrying forward and promoting the implementation of the outcomes of the Foresight Programme. Under this initiative 16 panels comprising representatives of government, business and commerce and academia have been created to examine market and technology sectors of the economy. The ongoing work of these panels is overseen by a steering group led by the Chief Scientific Adviser. Departmental co-ordination of activities in support of Foresight is undertaken by the Whitehall Foresight Group which consists of representatives from 16 government departments. Top level co-ordination of the programme across Whitehall is provided by a Ministerial Foresight Group.

The annual *Forward Look*, mentioned above, sets out government strategy for science, engineering and technology, together with the programmes of the research councils and each government department. It has been produced since 1994, and incorporates the statistics on government-funded science, engineering and technology. OST also provides the secretariat for the LINK programme.

2–34   Table 6.   The Science Budget, 1994/95–1997/98 (£ millions).

	1997/98	Estimated 1998/99	CSR Plan 1999/2000	% of grand total (1997/98)
OST - DTI[1]	27.5	43	147	2%
BBSRC	187.8	189	193	14.1%
ESRC	65.2	67	70	4.9%
EPSRC	386.6	389	396	29%
MRC	282.3	289	302	21.2%
NERC	159.9	165	169	12%
PPARC	195.9	205	185	14.7%
CCLRC	-	0	2	-
Pensions/Other	26.0	30	33	1.9%
Total	1331.1	1379	1497	100%

Source: Adapted from: Office of Science and Technology, SET Statistics 1999.
Notes:
1. Excludes OST running costs. Includes support for the Royal Society and the Royal Academy of Engineering, funds for the Joint Infrastructure Fund and initiatives not yet allocated to the Research Councils.

---

[12] *i.e.* separated from the spending priorities of the remainder of the DTI's budget envelope.

OST initiatives

ROPA

2–35 OST runs several initiatives in support of the objectives of the 1993 SET White Paper. The first of these, the Realising Our Potential Award (ROPA) scheme which derives its name from the White Paper's title, is intended to function both as a reward and an incentive scheme by making funds available to those researchers who have collaborated with U.K. industry and commerce. It operates across all the research councils (effectively as another arm of "responsive mode" funding) and since its introduction in 1994, over 1,200 awards have been made at a cost of in excess of £109 million. ROPAs are intended to fund curiosity driven, speculative research and are aimed at the development of long term strategic partnerships between academics and industry. A particular objective is to "pump-prime" new or highly speculative areas of research — a feature which particularly distinguishes them from standard research council grants. The awards are not, however, to be used to subsidise existing research which is being supported by industry or commerce. The research councils operate slightly different timetables for ROPA applications — some have fixed closing dates whilst others do not. At present the future of the scheme is under consideration by OST.

Foresight LINK

2–36 The second initiative was launched in 1995 as the Foresight Challenge, a competition intended to promote partnerships between the science base and industry on Foresight priority issues. Offering funds to attract collaborative bids with matching contributions from industry, the initiative was deemed to have been highly successful, with 24 winning projects worth a total of £92 million, of which £62 million was contributed by industry. Following a consultation exercise on the LINK programme, it was recognised that a greater degree of user-friendliness and effectiveness would result from drawing LINK and the Foresight Challenge closer together. Thus, the Government decided to consolidate and embed the Foresight Challenge principles in its current format, the Foresight LINK awards. These are based on LINK scheme guidelines, operate to published deadlines and are managed by the LINK Directorate. The maximum public sector contribution to a Foresight LINK project is 50 per cent of the total eligible cost. Two rounds of Foresight LINK have been funded to date — in 1998 and 1999. The DTI contributed £10 million to each round with additional funding from other government departments (see Table 7 at paragraph 2–38). First round projects covered research into micro-scale laboratory instruments and processes, fuel-control sensors to reduce aircraft pollution, and environmentally-friendly alternatives to cement and existing wood preservatives.

University Challenge Scheme

2–37 In March, 1998, the Government announced a further initiative aimed at commercialising the results of university research. The University Challenge Scheme is a £50 million venture capital fund open to competitive bids from those universities which can demonstrate a range of viable projects and entrepreneurially-oriented staff. Successful universities receive an award of £3 million, but must also contribute to the seed funding either from their own resources or from externally generated income. Of the total available funding, £20 million is government money whilst two major charities, the Wellcome Foundation and the Gatsby Trust provide matching funds. The OST forms the lead partner in the scheme's operation.

2–38   Table 7.   LINK Programmes: Government Department and research council expenditure (£000s).

Department	Outturn 1997-98	Estimated outturn 1998-99	Total spend to 1998/99
Biotechnology and Biological Sciences Research Council	2483	2840	18857
Economic and Social Research Council	59	215	274
Engineering and Physical Sciences Research Council	8394	14300	74861
Medical Research Council	861	1070	4457
Natural Environment Research Council	1154	1300	4509
Ministry of Agriculture, Fisheries and Food	3062	5170	23644
Department of the Environment, Transport & the Regions	963	854	6265
Department of Health	1029	1452	3685
Department of Trade and Industry	11716	9909	111053
Ministry of Defence	-	-	1645
Scottish Office	139	183	1119
Northern Ireland Office	-	120	120
Total	**29860**	**37413**	**2500488**

Source: Adapted from: Office of Science and Technology, SET Statistics 1999.

Infrastructure

2–39   A 1997 review of the Higher Education sector identified serious deficiencies in the research infrastructure (research equipment, facilities, libraries, etc.). In response, the Government announced two competitive schemes aimed at reversing this infrastructural decline. The *Joint Research Equipment Initiative* (JREI) provided £75 million for equipment in higher education institutions in 1998. Additional funds are required from external sources, such as industry or charity, to support the bid; these sponsors provided £40 million in 1998. OST administers the scheme.

In 1998 the Wellcome Trust and the Treasury agreed to create a joint fund to provide a one–off programme towards addressing the infrastructure problems of the universities — the *Joint Infrastructure Fund* (JIF). The JIF, with current funding of £750 million, is targeted at the biological, physical, engineering and social sciences, and provides for buildings, major equipment and other elements of infrastructure of the universities.

HERO-BC and Science Enterprise Challenge

2–40   Two recently introduced schemes also target the development of linkages between the higher education and business sectors. In partnership with the Higher Education Funding Council for England and the Department of Education and Employment, the DTI has launched a new "reach-out" fund. The Higher Education Reach-Out fund to Business and the Community (*HERO-BC*) aims to enhance university interaction with local businesses and the community. Initial funding for the scheme stands at £20 million per year.

Under the *Science Enterprise Challenge*, eight English universities have been jointly awarded £25 million to establish new enterprise centres which promote the commercial development of cutting edge science.

THE RESEARCH COUNCILS

2–41 Overall responsibility for the six research councils is exercised by the DTI through the OST. Each of the councils is an autonomous, non-departmental, public body principally funded by the Science Budget, with additional commissions from government departments and agencies, industry and international organisations. The research councils support research in the higher education sector directly through the provision of research grants, fellowships and postgraduate student support. Depending on the research council in question, these grants can support research projects, programmes or designated centres of research activity. Indirect support is provided by some research councils through the provision of large-scale facilities. A number of the research councils support research in their own institutes or units.

As may be seen from Table 8 below, despite some variation in most cases the majority of research council expenditure is on "basic – orientated"[13] and "applied – strategic" research. The notable exception to this is the Particle Physics and Astronomy Research Council (PPARC), where expenditure is almost entirely devoted to "basic – pure" research.

2–42 Table 8. Research council R&D expenditure by Frascati type of research activity and department (1997–98).

	Basic %	Applied %	Pure %	Experimental Development Orientated %	Strategic %	Specific %	Total £ million
OST- DTI	0	81.1	12.3	6.6	0	100.0	27.5
BBSRC	0	51.0	39.0	10.0	0	100.0	186.1
ESRC	16.6	24.8	44.2	14.4	0	100.0	59.1
MRC	0	61.8	26.5	11.7	0	100.0	281.3
NERC	3.0	48.4	31.5	10.0	7.0	100.0	152.3
EPSRC	18.0	31.0	44.0	7.0	0	100.0	362.7
PPARC	94.5	0	5.2	0.3	0	100.0	195.9
Total	20.9	38.9	31.2	8.2	0.8	100.0	1264.8

Source: Adapted from: Office of Science and Technology, SET Statistics 1999.

Biotechnology and Biological Sciences Research Council (BBSRC)

2–43 The aims of the BBSRC are to promote and support high-quality basic, strategic and applied research relating to the understanding of biological systems, and to advance knowledge and technology in biotechnology-related industries, including agriculture and other bioprocessing industries, such as chemicals, food, healthcare and pharmaceuticals.

A large proportion of the council's research is funded through project grants to researchers in universities or public sector research institutes. The council participates

---

[13] Frascati definition.

in initiatives such as LINK and ROPA which foster the links between academia and industry, runs its own Collaboration with Industry Scheme and conducts directed research programmes, sometimes in conjunction with other research councils, in priority areas. In 1997–98, from a total expenditure of £261.5 million, BBSRC funded £95 million of research in HEIs. Intramural research expenditure for the same year amounted to £158.2 million, whilst £7.9 million was spent in other Government departments or research councils.

The BBSRC provides support for research and training in four Interdisciplinary Research Centres, eight research institutes and several units and groups based within universities. The BBSRC's Interdisciplinary Research Centres and research institutes are listed in Box 2.

2–44 Box 2. Interdisciplinary Research Centres and Research Institutes of the BBSRC.

Institute	Location(s)
Advanced Centre for Biochemical Engineering (ACBE)	University College London
Centre for Genome Research (CGR)	University of Edinburgh
Oxford Centre for Molecular Sciences (OCMS)	University of Oxford
Sussex Centre for Neuroscience (SCN)	University of Sussex
Institute for Animal Health (IAH)	Compton and Pirbright, and the joint BBSRC/ MRC Neuropathogenesis Unit in Edinburgh
Institute of Arable Crops Research (IACR)	Rothamsted (houses the International Agricultural Development Unit), Broom's Barn and Long Ashton
Institute of Food Research (IFR)	Reading and Norwich
Institute of Grassland and Environmental Research (IGER)	Aberystwyth (Plas Gogerddan and Trawsgoed), North Wyke and Bronydd Mawr (Trecastle)
Babraham Institute	Cambridge
John Innes Centre	Norwich, Norfolk
Roslin Institute	Roslin, Midlothian
Silsoe Research Institute	Silsoe, Bedfordshire

Economic and Social Research Council (ESRC)

2–45 The ESRC promotes and supports basic, strategic and applied social science research and related postgraduate training, to gain a better understanding of social and economic change. Its research and training agenda is based on nine broad Thematic Priorities, namely:

(i)   Economic performance and development

(ii) Environment and sustainability

(iii) Globalisation, regions and emerging markets

(iv) Governance, regulation and accountability

(v)   Technology and people

(vi)   Innovation

(vii) Knowledge, communication and learning

(viii) Lifespan, lifestyles and health

(ix)   Social inclusion and exclusion

ESRC research programmes and research centres are targeted at one or more of these themes and postgraduate training will also be undertaken in theme areas. Although the council allocates over 65 per cent of its expenditure to theme areas, ESRC maintains its responsive mode funding (largely through the provision of research grants) which contributes to theme development and renewal.

Whilst ESRC has no intramural research facilities, it supports over twenty research centres based at universities. These centres form national focal points for long-term social science research projects. They are selected by an annual competition and are funded for an initial period of ten years, subject to a successful mid-term review, hence the number of centres and their research subjects may vary over time. The council also supports several resource centres which underpin the activities of its research community. In addition to its centres, the ESRC funds around twenty-three major research programmes together with numerous smaller scale research awards. Intramural expenditure in 1997/98 amounted to £3.9 million, whilst £53.2 million was spent through HEIs and £2.7 million through other Government departments and research councils. Total net expenditure in this period was £59.1 million.

Engineering and Physical Sciences Research Council (EPSRC)

2–46   The EPSRC is the largest of the research councils and supports basic, strategic and applied research, and postgraduate training in engineering and the physical sciences (*i.e.* Chemistry, Physics and Mathematics). It funds eight Interdisciplinary Research Centres (see Box 3) and manages the joint research council programme in High Performance Computing. EPSRC maintains a large responsibility for the provision of high-cost instrumentation. In 1997/98, of a total budget of £369.4 million, EPSRC funded £282.7 million of research in HEIs, £19.5 million intramurally and £49.6 million in other government departments and research councils.

The EPSRC programme of research is divided into Programme Areas based on the core disciplines of science and engineering or on generic technologies, thus:

Engineering programmes
    General Engineering
    Engineering for Infrastructure,
        the Environment and Healthcare
    Engineering for Manufacturing

Technology programmes
    Information Technology and
        Computer Science
    Materials

Science programmes
    Chemistry
    Physics
    Mathematics

Programmes interface
    Life Sciences

Responsive mode funding,[14] operated through research grants comprises around 60 per cent of EPSRC's total grant funding. The Council is also particularly active in

---

[14] That is, grants made to sucessful unsolicited applications from members of the U.K. research community.

funding schemes for the support of academic/industrial collaboration, such as LINK, CASE studentships, Faraday Partnerships and the Integrated Graduate Development Scheme (which provides modular, part-time training at Masters level to young scientists and engineers already in employment), together with ROPAs and the Teaching Company Scheme.

2–47   Box 3.   EPSRC Interdisciplinary Research Centres (IRCs).

Name	Location
IRC for Biomedical Materials	Queen Mary and Westfield College
IRC for High Temperature Superconductivity	Cambridge
IRC in Materials for High Performance Applications	University of Birmingham
IRC for Optoelectronics	University of Southampton
IRC for Polymer Science & Technology	University of Durham
IRC for Process Systems Engineering	Imperial College of Science Technology & Medicine /University College London
IRC for Semi-conductor Materials	Imperial College of Science Technology & Medicine
IRC for Surface Science	University of Liverpool

Medical Research Council (MRC)

2–48   The Medical Research Council is responsible for supporting research and related postgraduate training in the biomedical and related sciences. Its aims are to maintain and improve human health; to advance knowledge and technology and to provide trained researchers which meet the needs of users and beneficiaries including the providers of health care, and the biotechnology, food, health-care, medical instrumentation, pharmaceutical and other biomedical-related industries.

As a consequence of its numerous establishments (see Box 4) intramural expenditure is comparatively high — £161.1 million in 1997/98. R&D funded in HEIs amounted to £131.7 million, whilst that funded in private industry and public corporations cost £12.7 million.

2–49   Box 4. Institutes and Interdisciplinary Research Centres (IRCs) of the MRC.

Institutes:	
National Institute for Medical Research (NIMR) (London).	
MRC Laboratory of Molecular Biology (Cambridge)	
MRC Clinical Sciences Centre (Hammersmith Hospital, London)	
Interdisciplinary Research Centres:	Location:
MRC Interdisciplinary Research Centre for Cognitive Neuroscience	University of Oxford
Cambridge Centre for Brain Repair	Cambridge
IRC Cell Biology	University College London
Centre for Mechanisms of Human Toxicity	University of Leicester
Cambridge Centre for Protein Engineering	MRC Centre, Cambridge
Institute of Molecular Medicine	John Radcliffe Hospital, Oxford

Natural Environment Research Council (NERC)

2–50 The Natural Environment Research Council supports research, surveying, long-term environmental monitoring and related postgraduate training in: terrestrial, marine and freshwater biology; earth, atmospheric, hydrological, oceanographic and polar sciences; and earth observation. Its mission is also to advance knowledge and technology and to provide services and trained scientists and engineers to meet the needs of users and beneficiaries (which include the agricultural, construction, fishing, forestry, hydrocarbons, minerals, process, remote sensing, water and other industries). Intramural expenditure accounted for £118.4 million of a total of £152.3 million in 1997/98, whilst the Council supported R&D worth £70.4 million in HEIs. It supported £7.0 million worth of R&D in other research councils.

NERC supports research in universities through four funding modes — core strategic, thematic, non-thematic, and infrastructure. A number of schemes are also operated, aimed at developing collaborative links between academia and industry, for example, LINK, ROPA, CASE, and the Teaching Company Scheme. NERC also runs the CONNECT scheme, the objective of which is to initiate new interactions between industry, business, commerce or public sector bodies, on the one hand and the science base. There are two types of CONNECT award: CONNECT A (which provides up to £5,000) covers "proof of concept" proposals for research which has potential application but which may entail a high degree of technical risk and lack a clearly defined end user. CONNECT B, a variant of NERC's standard grants requires an agreement and commitment from the collaborating body to fund at least 50 per cent of the total projects costs. Minimum NERC funding is £25,000 over three years.

Particle Physics and Astronomy Research Council (PPARC)

2–51 PPARC has responsibility for the promotion and support of fundamental research and related postgraduate training in particle physics, astronomy and planetary science, thereby contributing to the greater understanding of the nature of matter and the origin and evolution of the Universe. It funds research assistants, technical support and students, together with the construction of instrument systems required for this research. Its role is also to ensure that the U.K. research community has access to the necessary major facilities, whether located in this country or abroad. The Council is also responsible for U.K. policy on space science through its membership of the British National Space Centre.

By the nature of the facilities it uses the Council's plans frequently involve international collaborations, the most significant of which are its membership of the Laboratory for Particle Physics (CERN) and the European Space Agency (ESA). International co-operation is pursued either in the context of intergovernmental agreements (such as participation in CERN and ESA) or in bilateral partnerships (for example, sharing of telescopes or space missions).

Although the Council's programmes are long-term and fundamental in nature, efforts have been made to improve links between itself, the research community it serves and industry. These include the PPARC Industrial Programme Support Scheme (PIPSS) and longitudinal studies of the careers of PPARC research students.

By virtue of its international commitments, the Council's overseas expenditure accounted for £93.9 million of a total of £195.9 million in 1997/98. R&D supported in HEIs cost £49.8 million and that undertaken in other research councils cost a further £23 million.

Council for the Central Laboratory of the Research Councils (CCLRC)

2–52 The CCLRC was set up in 1995 with the mandate to serve the other research councils and their client communities, particularly universities, Government departments and executive agencies, industry and the U.K.'s international partners. It comprises the Daresbury Laboratory in Cheshire and the Rutherford Appleton Laboratory in Oxfordshire, both of which were formerly administered by the EPSRC, and the Chilbolton Observatory in Hampshire. The Central Laboratory does not maintain its own research programme, rather in compliance with the contractor-customer principle, all its facilities, services and activities are provided for paying customers.

RESEARCH COUNCIL SCHEMES

2–53 In addition to the schemes already covered under the DTI and the OST above (such as LINK, TCS, ROPA, and the Faraday Partnerships), the research councils run several schemes to promote research partnerships between companies and university research groups.

Innovative Manufacturing Initiative (IMI)

2–54 The IMI is an industry-led partnership with universities, research councils (EPSRC, BBSRC and ESRC) and Government departments aimed at the development of multidisciplinary research (strategic and applied). It is primarily intended to encourage the spread of best practice in business processes and to fund an associated programme of research and postgraduate training. Industry partners are expected to contribute at least 50 per cent of the funding in cash or in kind. The IMI targets four sectors: Integrated Aerospace manufacture; Process Industries; Construction as a Manufacturing Process; and Road Vehicles.

Industry fellowships

2–55 The EPSRC, BBSRC and the Royal Society jointly fund an Industry Fellowships scheme which provides industrial scientists and engineers with the opportunity to undertake research or course development in universities or other HEIs, with reciprocal arrangements for academics to pursue projects within industry. Similar schemes are run by the Royal Academy of Engineering, for example, its Industrial Secondment Scheme.

Interdisciplinary Research Centres (IRCs)

2–56 The IRCs are university-based centres of excellence. Each is of sufficient size to make a significant contribution to their relevant research areas. One of the most important objectives of the IRCs has been the establishment of collaborative arrangements with industry. Each IRC is expected to have a ten year life span after which research council core funding will be withdrawn and researchers will be required to rely on responsive mode funding and alternative sources of research income.

Co-operative Awards in Science and Engineering (CASE)

2–57 The research councils also support a number of industrially relevant training schemes. Of these, the longest running scheme (in place for over twenty years) is the Co-operative Awards in Science and Engineering (CASE). Run by all six research councils, CASE students undertake a one to three year programme of research towards a higher degree (usually a PhD), with a collaborating organisation (industrial and commercial organisations in the private and public sectors, local authorities and research council institutions and laboratories). At least three months is spent at the collaborating entity's premises and supervision is shared between the academic and non-academic supervisors. Formerly, CASE awards were only allocated to the host university. However, in 1994 the research councils launched Industrial CASE awards in which the industrial partner is allocated the studentship. These enable the company to take the lead in defining projects for students (provided they are within the remit of the research council funding the studentship) and in selecting an academic partner of their choice. Students are then chosen in consultation with the HEI concerned. The initiative for CASE projects may come from the university department or the co-operating body. However the identification of the CASE project and company are not required until the nomination of the student. Students receive at least £3,000 from the co-operating body and any travelling or out of pocket expenses incurred by the student, and tuition fees are paid by the research council. Co-operating HEI departments receive the standard research training grant, plus an additional £1,400 from the public/private sector collaborator.

OTHER GOVERNMENT DEPARTMENTS

Department of the Environment, Transport and the Regions (DETR)

2–58 DETR[15] has a total annual expenditure on SET of some £154 million (1997/98). As might be expected, the department has a wide ranging policy remit covering: planning, roads and local transport; housing, construction, regeneration, wildlife and countryside; local and regional government; railways, aviation and shipping; and environmental protection.

Ministry of Agriculture, Fisheries and Food (MAFF)

2–59 MAFF is a significant contributor to the funding of United Kingdom SET; about £125 million will be allocated to commissioning research in 1999/2000. In terms of its responsibilities, MAFF has an extremely diverse remit but it is predominantly concerned with the farming, food and drink, and fishing industries. It also deals with issues ranging from public safety and animal welfare to the protection of the rural and marine environments.

The Ministry has four policy aims and a fifth aim which relates to its operational efficiency. The aims are broad and each encompasses a number of objectives, thus, MAFF aims to:

- protect the public by promoting food safety;

- protect and enhance the rural and marine environment;

- improve the economic performance of the agriculture, fishing and food industries;

- protect farm animals;

- ensure the best use of internal resources in support of the Ministry's business.

---

[15] DETR was formed by a merger of the Department of the Environment (DoE) and the Department of Transport (DoT) in June 1997.

MAFF funds science — via external contracts, intra-departmentally or through its Executive Agencies — in order to identify and investigate problems, develop policy options, implement solutions and assess their effectiveness. This encompasses research, surveillance and monitoring, analysis, dissemination and technology transfer. The Ministry has an extensive research programme in support of its departmental aims and research expenditure for 1997/98, including grant-in-aid to the Royal Botanic Gardens, Kew, totalled some £140 million.

Department of Health (DoH)

2–60   The Department of Health is responsible for the improvement of health and well-being of the population and for ensuring the provision of high quality health and social care through the National Health Service (NHS). Given the Department's remit, it is unsurprising to find that research and development play a pivotal role in its activities, for underpinning the science and technology upon which medical services and health care are based, for providing a sound basis for the formulation of Departmental and NHS policy, and for the provision of ministerial advice.

The Research and Development Directorate co-ordinates its research with the activities of other relevant bodies, such as the research councils, the medical charities and industry, other Government departments and the European Union. The combined R&D budgets of the DoH and the NHS totalled in excess of £460 million in 1998/99. DoH research strategy is based largely around two complementary R&D programmes: the National Health Service R&D Programme and the Department's Policy Research Programme (PRP). With a strong focus on the needs of the health service, the NHS R&D Programme is aimed at developing sound information on research findings and scientific developments which can be used to inform clinical, managerial and policy decisions.

Department for Education and Employment (DfEE)

2–61   The two broad aims of the DfEE are to raise the levels of educational achievement and skill of the population and to promote an efficient and flexible labour market. The Department has overall responsibility for education and training matters at all levels, from schools to higher education and vocational training. The Department has a central research budget of some £7 million and an annual R&D budget of around £104 million (1998/99) — this excludes the funds disbursed through the HEFCE and FEFC.

Ministry of Defence (MoD)

2–62   The Ministry of Defence is the highest spender of R&D funds and the largest single customer for U.K. industry (see Table 4, above). As such, its decisions on defence science and technology policy and on the procurement of defence equipment have major implications for the U.K.'s science and technology base.[16] Despite Government intentions to substantially reduce the level of defence-related R&D expenditure by around one-third by the turn of the century (as stated in the 1993 SET White Paper), expenditure on defence related SET still accounted for almost 37 per cent of the total in 1998/99. Government defence R&D is dominated by expenditure on development which is directed at the MoD's programme of defence equipment procurement. In 1998/99 this expenditure accounted for almost three–quarters of all defence SET

---

[16] James, A.D. and Gummett, P. "Defence", in Cunningham, P. N. (ed.) *Science and Technology in the U.K.* 'Cartermill Guides to Science and Technology, Cartermill Publishing, November 1998'.

spending. As noted, the defence industries are highly dependent on the MoD as a source of funds for R&D — of a total defence SET spend of £2,118 million in 1997, 22 per cent was met by U.K. industry and a further 16 per cent from overseas sources. The remainder was funded by Government (*i.e.* the MoD).

In the face of research budget limitations, and as a result of the findings of the Government's Strategic Defence Review, the MoD focuses on improving the two-way transfer of technology between the civil and defence sectors, between industry and Government and with the U.K.'s allies. Priority is also accorded to maximising the spin-off from defence research into civil applications and to "spinning-in" innovations from the civil sector into military applications.

2–63 The MoD has several executive agencies of which the largest is the Defence Evaluation and Research Agency (DERA). This brings together the core of the MoD's non-nuclear scientific and technical assets. It is organised into 15 business sectors which offer a range of services, from the highest level of operational studies and analysis, through the various categories of basic and applied research, to consultancy-type advice on the procurement process and the testing and evaluation of specific equipment in both the development phase and during actual operations. With around 12,000 staff and a turnover of approximately £1 billion, DERA is one of Europe's largest research organisations. Discussions are currently under way concerning the Agency's partial privatisation.

Alongside DERA, the Defence Diversification Agency (DDA) has been set up to provide U.K. SMEs with appropriate access to the technology, expertise, knowledge and facilities available within DERA. The DDA encompasses a range of developments already under way at DERA sites including dual-use technology centres, science parks and innovation centres.

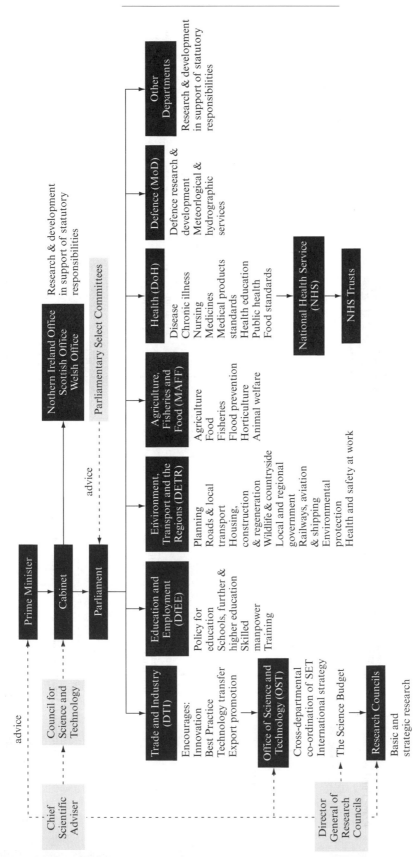

2–64 Figure 1. Structure of U.K. science, engineering and technology, and primary sources of policy advice. *Source:* British Council, *Guide to Organisation of Science and Technology in the United Kingdom,* 1997.

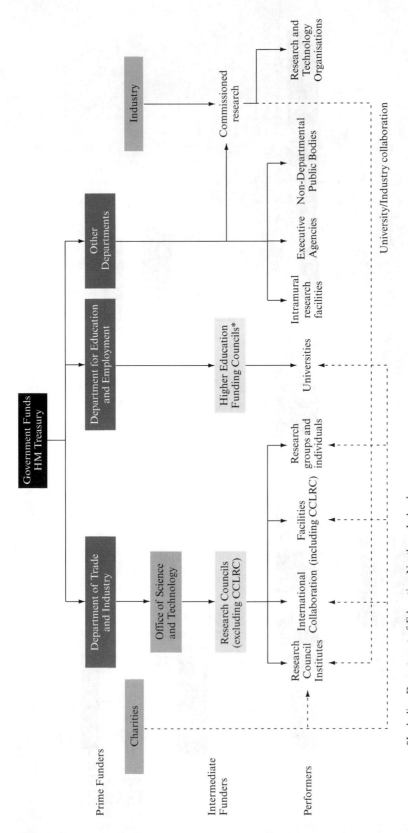

2–65   Figure 2.   Flows of R&D funding in the U.K. public sector
Source: British Council, *Guide to Organisation of Science and Technology in the United Kingdom,* 1997.

*Including Department of Education Northern Ireland

CHAPTER 3

VENTURE CAPITAL

Dr Paul Castle, Founder and Chief Executive, MTI Partners Ltd

THE NEED FOR VENTURE CAPITAL

3–01 During the formative years of a new business, cash is absorbed by purchases of fixed assets; by the growth of working capital in the form of stocks and work-in-progress, debtors and creditors; and by trading losses. By way of illustration, assume sales turnover in the first three years amounts to £100,000, £300,000 and £800,000 respectively, then the aggregate demand for working capital will amount typically to approximately £150,000. If, in addition, trading losses in the first two years amounted to £300,000 with a profit of say £50,000 in the third year and total capital expenditure during the three years approaching £100,000, then the total demand for cash during the three year period would be around £600,000.

In such a hypothetical but typical situation of a new business which required £600,000 of funds in its first three years, it is normally possible to secure support from bankers, provided that some sort of security is available in the form, for example, of directors' and founders' personal guarantees, the Government's Small Firms Loan Guarantee Scheme, or (once trading has commenced) debtor cover. Government grants might be available, depending on the region in which the business is located, the sector in which it is operating, and the status of the development of the business. In addition it may be possible to lease certain items of capital equipment. The founders themselves or their friends or relations may also be able to provide some of the funding from their own resources.

3–02 In any event, even under the most favourable circumstances, the sum total of all such sources of finance is unlikely to contribute an amount that comes anywhere near to meeting the cash demands of the business. Furthermore it is unwise, particularly during the start-up and early stages of a business's development, to burden the company with significant amounts of debt finance. Debt finance appears as a liability on the balance sheet and incurs interest charges, usually payable as they arise, which young emergent companies can ill-afford to service. Thus there is a need to provide such companies with a source of long-term funding in the form of equity share capital, the term for which is venture capital.

SOURCES OF VENTURE CAPITAL IN THE UNITED KINGDOM

3–03 The existence and growth of sources of venture capital within a first-world economy is a direct result of the funding gap which exists between the aggregate of non-equity related sources of finance and the total cash requirement of the new business, as described above. The world's leading venture capital nation is the United States, whose venture capital industry dwarfs those of all other countries put together. Nevertheless the United Kingdom's venture capital industry is second only to the United States, and by some measures (for example aggregate venture capital investment per unit of GDP) the United Kingdom beats even the United States.

But such statistics can be misleading. Many of the United Kingdom's sources of funding for "private equity" (a term used to describe investment in any and all companies other than those quoted on a recognised stock exchange) are far removed from "venture capital" in the classical American sense of the term. Statistics prepared by The British Venture Capital Association (BVCA), whose member firms represent essentially all the providers of private equity capital in the United Kingdom, show that at the end of the 1990s less than eight per cent of the total funds invested are in early stage situations. Furthermore only five per cent is invested in early stage technology companies. These proportions are vanishingly small compared to the corresponding amounts invested in the United States.

3–04   Reference to "early stage" in the previous paragraph alludes to the practice in the venture capital community of identifying investments under a number of broad categories: Early Stage; Development or Expansion; Buy-Out. Early Stage covers a variety of stages in a business's formative period of development, ranging from Seed through Start-Up and up to and beyond the establishment of trading. Once sales revenues are regular and growing on a month by month basis, losses are beginning to reduce, and the business is on track for sustained profitability, the company would then be classified as being at the Development or Expansion phase. A significant majority of venture capitalists in the United Kingdom will not consider investment in a company until it has reached at least this phase of its growth and development.

Early stage investments

3–05   In the Early Stage category, Seed Capital refers to investment in a situation where the business, its products or technology exist as no more than an idea or a concept or a project. A significant amount of research and development, of a technical and/or commercial nature, will need to be undertaken before the product or service around which the business is to be built is in a tangible or demonstrable form. When such a stage is achieved, the investment opportunity would be classified as a Start-Up. In product terms, using the example of a physical item of hardware or a suite of software, there would then exist a version to at least a pre-production prototype stage. That is to say that it must be possible to use the product in a realistic test environment and be able to demonstrate its functional performance to a defined technical specification. At this stage, it should also be possible to show that the eventual production version of the product will be capable of being produced to a certain commercial specification.

There are only a small number of venture capital firms in the United Kingdom which are willing to consider Early Stage investments. Furthermore only a minority of such firms will consider Start-Ups and even fewer will contemplate Seed funding, although with the advent of the e-business revolution a new breed of "incubator fund" has emerged, specifically targeted at financing embryonic dot.com and e-commerce businesses. The shortage of Early Stage venture capital firms is exacerbated by two further factors. Firstly, several venture capital fund managers will not invest in high technology businesses, perceiving them as adding an excessive and unacceptable degree of risk to an already high risk stage of a company's development. Secondly, almost all venture capital firms have a minimum level of investment, which in the majority of cases is measured in units, rather than fractions, of millions or tens of millions of pounds. Thus the climate within which early stage technology businesses in the United Kingdom seek venture capital funding is very inclement, and gives rise to the expression "equity gap". The phenomenon is not new, having first been identified as such and coined by Harold Macmillan many decades ago.

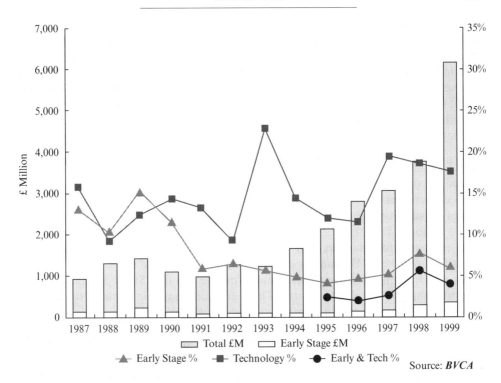

Figure 1. Private Equity Investment in the United Kingdom

3–06 The appetite of the United Kingdom venture capital community for early stage and technology deals is illustrated in the chart at Figure 1 above. Investment in the technology sector relative to the total funds invested in United Kingdom private equity shows an increasing if erratic trend in the second half of the 1990s. Investments in early stage situations, however, show little sign of growth and it has taken until the end of the 1990s for the amounts invested to reach and exceed those achieved a decade earlier, although the trend of early stage technology deals shows an encouraging and consistent pattern of growth over the five years leading up to the new millennium.

INVESTORS

3–07 The BVCA maintains a directory of all its members showing for each one its preferred size and type of investment. These range from managers of very large Buy-Out funds (CINVen, CVC, Candover, Doughty Hanson), through large generalist fund managers (Apax, Schroder Ventures, Advent International, 3i), to smaller firms managing specialist funds dedicated to specific well defined areas of investment (MTI, Advent, Amadeus, Prelude). It is the constituents of this last category of specialist funds that are most likely to consider investments in early stage technology opportunities.

Such specialist fund managers will normally require to invest on a basis which enables and entitles them to participate to a greater or lesser extent in the management of the company. This style of operation is known in the venture capital industry by a variety of expressions including "hands on", "pro-active" and "adding value", and follows the American model. The investment executives of firms that operate on this basis will typically have a career which encompasses senior management experience in industry prior to them becoming venture capitalists. Thus firms that invest in this way can be thought of as classical venture capitalists in the American sense and might define venture capital as "Investing equity finance and management resources in manufacturing, high technology and related businesses at an early stage of their development or recovery".

3–08 Fund managers of the type mentioned above tend to be independent firms, owned by their investment teams, which raise their venture capital funds directly from conventional institutional sources such as pension funds and insurance companies. In addition there are a number of "captive" managers of venture capital that are an in-house department or subsidiary of an investment or clearing bank or insurance company, for example, and which have access to their parent organisation's funds for investment. Yet another type of venture capital fund is the Venture Capital Trust which is an investment trust quoted on the London Stock Exchange whose investors enjoy favourable tax status. There are also quoted investment trusts which specialise in investments in venture capital situations (of which 3i is the most notable).

The BVCA also compiles and publishes a list of so called "Business Angels". A business angel is a private individual who is prepared to invest personally in and to become directly involved in the management of unquoted companies, most typically at a very early stage of their development. Business angels who invest in this way benefit from generous tax reliefs on the amounts they invest and realise in capital profits. The business angel is therefore a very important source of venture capital in the United Kingdom. In fact, it has been suggested that the aggregate amount of investment made annually by business angels is of similar proportions to the corresponding total amount invested by all venture capital funds.

TYPES OF VENTURE CAPITAL

3–09 The venture capitalist, whether a professional fund manager or a business angel, provides high risk capital in the form of equity that is invested in businesses in exchange for a proportion of the share capital. Thus the investee company sells and the venture capital investor buys a proportion of the company's equity in exchange for the investment of cash. Founder shareholders and executives of investee companies often refer mistakenly to "giving away" equity to the venture capital investor, but nothing could be further from the truth. The investment transaction is a bargain struck between the existing shareholders and the new shareholders whereby the latter purchase (usually) new shares in the business in exchange for the injection of new equity capital.

Depending on the size and nature of the investment opportunity, and the style and preference of the venture capital house in question, the new cash to be invested in the business may take forms other than straightforward ordinary share capital. Even in the case of an investment that does consist only of ordinary share capital, the venture capitalist will almost certainly require that its ordinary shares be of a separate "Preferred" class. Such shares are different to and rank ahead of the class of share held by the founder and management shareholders in the event of a sale or liquidation of the company.

3–10 Preferred Ordinary shares are not and should not be confused with Preference shares, which are a class of share whose holders benefit from greater protection than ordinary shareholders in the event of the company becoming insolvent. Like all shares, both may be redeemable (that is to say repayable at their face or an enhanced value), but such redemption can only occur out of the company's distributable reserves. Alternatively Preference shares may be convertible into Ordinary shares at a pre-determined conversion rate, and with or without a conversion premium. Preference shares would normally carry a fixed dividend or coupon, and may enjoy "participating" dividend rights entitling the holders to a specified share of the distributable profits of the company. Share capital, whether or not in the form of equity or preference shares, strengthens the balance sheet of the company and appears there as the source of the company's permanent funding.

Valuing the company

3–11 In situations where the company receiving the venture capital is an early stage opportunity with no trading history and little more than plans and forecasts of a rosy future, the venture capitalist will have difficulty ascribing material value to the company at the date of investment. Accordingly investment by the venture capitalist of several hundreds of thousands or millions of pounds in the business will most probably correspond to a majority of the equity. For founder investors and management of a young company who believe that their business has a potential future value extending well beyond the dreams of avarice, to cede a majority of their equity to an incoming institutional investor is often problematic. Under such circumstances however such problems are usually emotional rather than rational in nature, but they need solutions nevertheless.

One approach to solving founder investors problems is to build into the investment structure the facility for the venture capitalist's equity to dilute in favour of the founder and management shareholders at some future date on a basis dependent on the success of the business. Such a mechanism is known in the venture capital trade as a "ratchet". A ratchet may also be constructed to operate in reverse such that the proportion of the equity owned by the founder and management shareholders starts at a relatively high level thus anticipating future success, but reduces in favour of the institutional shareholder if the business turns out to be less successful than originally thought. A second approach is to structure the funding package in the form of a small amount of equity (corresponding to only a small proportion of the issued share capital) combined with a large loan or redeemable preference shares, as described below. Yet another approach is to phase the total amount of funding required such that the initial tranche is made at a share price corresponding to a relatively low valuation with later tranches being made at progressively higher share prices, thus corresponding to smaller proportions of the equity. Such arrangements normally require that the business performs to certain pre-agreed milestones in order to trigger the later tranches at the commensurate higher prices. Subsequent tranches or rounds of funding might also bring in new venture capital investors who are more risk averse than the original investor, and who may therefore be prepared to accept a somewhat higher price and thus lower return.

Debt finance

3–12 In addition to share capital, the venture capitalist may also provide other types of finance, in the form, for example, of loans or debt finance. Debt finance is usually structured as a term loan (*i.e.* for a specified period of time), as opposed to the repayable-on-demand lending usually characterising bank overdrafts. Repayment schedules would normally extend over several years and might provide for a repayment "holiday" for an initial period. Depending on the duration and terms of the instrument used to establish the loan, such debt finance goes by many, often rather exotic, names such as Mezzanine Finance, Senior Debt, Loan Stock and Junior Debt.

Whatever its name however, the amount of the loan will normally be secured (by means of a debenture or mortgage) against the assets of the company. By this means, if the company fails then the proceeds realised from the sale of its assets will accrue first to such lenders in preference to unsecured creditors and ordinary shareholders (of whatever class). Furthermore the providers of that part of debt finance which would normally carry a relatively higher risk, such as mezzanine and senior debt, will normally require that their return is not restricted to an income stream in the form of an interest coupon. In addition they will also require some upside if the business performs to or better than expectations, which would be achieved for example by them being issued with warrants (*i.e.* options) to purchase a small proportion of equity at some future date at a significantly discounted share price.

Disadvantages of debt finance

3–13   All loan finance will show as a liability on the company's balance sheet. As a result the balance sheet will be substantially weaker than if the same amount of funding had been provided entirely as share capital, and the ratio of all sources of debt finance to the aggregate share capital of the company is described as "gearing" or "leverage". A company which is highly geared as a consequence of its initial capital structure will find it much more difficult to raise conventional bank lending if it is needed in due course, to fund increasing working capital requirements for example. A further disadvantage of loan capital is that interest normally has to be paid on the principal. Such interest charges are an operating expense of the company for the purposes of profit and loss account, and give rise to cash outflow, which during the early stages of the development of a company, it can ill-afford to pay.

There are however advantages to be gained from debt finance. As noted above, if a transaction is structured as a mixture of equity and debt, then the proportion of equity surrendered to the institutional investor is less than it otherwise would be if the funding was provided solely in the form of equity. Furthermore if the business is expected to be strongly cash generative in the short to medium term, then debt instruments put in place at the time of the initial financing can be redeemed, and maximum efficiency of, and return on, initial equity will be achieved.

Approaching the Venture Capitalist: the Business Plan

3–14   The initial approach to the venture capitalist can be either direct or via a professional advisor such as an accountant or a solicitor, the bank manager or a venture capital "marriage broker". In America it is almost unheard of for an investee company to approach a venture capital investor directly ("off the street") where its chances of even a hearing would be remote, although in the United Kingdom the practice is not uncommon and is acceptable. Whichever route is chosen it is preferable that the investee company has undertaken at least a small amount of research to determine that the investment criteria of the selected venture capitalists appear to be met by the business seeking investment.

A typical venture capitalist sees at least one business plan per day and only two to three per cent are successful in raising venture capital. It is essential therefore that the business plan is readable, well structured, well (but not lavishly) presented, concise and to the point. The quality of a business plan is gauged on its content and merits, not on its weight or thickness. A plan containing reams and reams of finely detailed spreadsheet workings is likely to be, at best, unread and, at worst, an instant "turn-off". The business plan is the single most important document influencing the venture capitalist's decision to proceed with an appraisal of the prospective investee company. It creates the first, and perhaps the only lasting, impression of the initial encounter between the venture capitalist and the investee company.

The business plan

3–15   The plan should describe the overall concept, strategy and objectives of the business and the reasons why additional equity capital is required. A summary of the history, the background and the *raison d'être* should be followed by a description of the technology, products, processes or the services to be offered by the business and the features and unique selling points that differentiate them from those of competitors. The market sectors and geographical markets which are to be addressed should be described and characterised in terms of market size, growth rates, competitors, and market shares, including where possible a summary of existing customers. The calibre and experience of the management team should be described, together with any weaknesses in the profile of the team and the steps to be taken to compensate for any deficiencies. Comprehensive CVs should be provided as an appendix to the plan.

Financial plans and historic performance should be presented in summary form, for three years forward and if appropriate three years back, showing profit and loss account, balance sheet and cash flow. Projections of sales revenues, gross profit, headcount, overheads, the trading result, capital expenditure, cash and/or borrowings should be cross-referenced to other parts of the plan by way of verification. To test the effects of the business failing to match expectations to a greater or lesser extent, particularly at the sales revenue level, the results of sensitivity analyses should be presented showing the effects of various alternative scenarios.

It is preferable for the founding management team to write the business plan in their own words taking advice, as required, from professional advisors. It is usually obvious to the experienced venture capitalist when a business plan has been written in its entirety by professional advisors and the contents of the stereo-typed and mass-produced document tend to be devalued as a result.

Dealing with the venture capitalist

3–16   Like the members of the investee company's management team, the venture capitalist is human. He is in business to invest the funds at his disposal and is therefore actively seeking, and will be very receptive to, good quality prospective investee companies. The venture capitalist should be made aware that he is in competition with one or perhaps two other sources of funds. The venture capitalist should also be aware that selection of the venture capitalist and the investee company is a two-way process and that he has to convince the investee that his skills and experience are relevant to the future of the business.

THE INVESTMENT PROCESS

3–17   As noted above, receipt and study of the business plan is the starting point for stimulating the venture capitalist's interest in the prospective investee company. An exploratory meeting then follows, preferably on site at the company's premises, and aimed principally at enhancing the venture capitalist's understanding of the technology, the products, the market opportunity and the management team. If the venture capitalist's interest is sustained following such an exploratory meeting then he will probably wish to initiate some preliminary commercial referencing with the company's customers, if such exist. At this stage the venture capitalist should then be prepared to indicate the terms on which he would consider concluding a transaction in principle, to establish with the prospective investee company whether such terms could be the basis for proceeding.

Due diligence

3–18   If agreement in principle can be reached, a due diligence process then starts in earnest, with detailed appraisals of the technology and products, the market and commercial issues, the management team and the financial history and projections. This process would normally entail the commissioning of specialist consultants in one or more of the areas of the appraisal. By its nature and thoroughness it is necessarily time-consuming, a major distraction to the company's management team, and expensive. Before commencing his appraisal, the venture capitalist will normally agree with the prospective investee on the manner in which, and the extent to which, the company will meet the cost of the due diligence in the case of the investment proceeding to completion or not as the case may be. Depending on the findings of the due diligence appraisal, the venture capitalist may decline to proceed or may propose to proceed either on the basis of modified terms or on the terms of the in-principle offer. If the prospective investee company agrees to the final deal terms the next stage is the negotiation of the legal agreements.

Legal negotiations

3–19   Legal negotiations are again a time-consuming and expensive process, with legal representation being required by the investor, by the management and by the company and its existing shareholders. Two principal documents will form the majority of the legal work, the subscription agreement which defines the commercial contract between the various parties and the company's articles of association which define and describe the company's constitution and its administration in legal terms. An important and usually rather contentious component of the subscription agreement is the section of warranties to be given by the investee company's existing management and shareholders (the "warrantors") to the in-coming investor. The warranties provide the investor with a guarantee that the investee company is all that it appears to be and that there are no skeletons in the cupboard of which the investor has not been made aware prior to investment. If it subsequently transpires that as at the date of the transaction a warranty has been materially breached, and had not been disclosed by the warrantors as having been so breached, then the investor will be entitled to bring a claim against the warrantors to recover their loss.

COMMON OBJECTIVES: CAPITAL GAIN AND EXIT

3–20   Whilst a number of private equity firms require a return on their investment on an on-going basis, for example, in the form of loan interest or dividends, the principal objective of the venture capitalist is to make a capital gain on funds invested in the equity of the company constituting the investee's business. The larger the capital gain the more the fundamental objective of venture capitalists is realised. Also the capital gain should be sustainable and indeed should progressively increase from the short through the medium to the long term.

The venture capitalist expects also that the principal return to the management and original or founder shareholders will be by means of a substantial capital gain in the same form as the return to the venture capitalist. In the short term the members of the management team must therefore be prepared to make significant personal sacrifices of their time, their security, their remuneration, and their benefits as well as facing the consequence of possible failure. The venture capitalist is particularly averse to investments in so-called "life-style" companies, in which the management and founder shareholders are content to run the business so as to provide themselves with a comfortable living, as opposed to the objective of maximising the value of shareholders' investment.

3–21   The capital gain is only of use to the venture capitalist if it is in a realisable form. That is to say it should be liquid (*i.e.* in a form which is readily convertible into cash) or readily tradable (*i.e.* be convertible into cash according to a market rate or valuation). Each of these aims can be achieved variously by a flotation and quotation on a recognised share market such as the Stock Exchange, the Alternative Investment Market or EASDAQ; by a trade sale in the form of a take-over by another company; by the acquisition of another company; by a merger with another company; or by the sale back to the founder shareholders most probably by means of a leveraged management buyout. Whatever the means of realising the investment in order to crystallise the capital gain, the generic term for the process is the "Exit". It is said that the Exit is the primary consideration a venture capitalist gives to a prospective investee company on first acquaintance, because unless there is a clearly defined exit route there is no point in the venture capitalist even considering an investment.

It is fundamentally important that the investor and the founding or original shareholders of the investee company have a common aim to create substantial wealth. The more millionaires the venture capitalist creates, the more his own objective of substantial capital gain is satisfied. What the venture capitalist most definitely does not want is a portfolio of small moribund investee companies whose sole purpose is to provide a comfortable living for their founders.

CHARACTERISTICS OF THE ATTRACTIVE INVESTEE COMPANY

3–22   The ideal characteristics of the perfect investee company are one which requires a minimum involvement by the venture capitalist, one which is self-funding after the injection of the first round of venture capital finance, and one which can achieve the maximum appreciation of its equity value by means of rapid organic growth. The venture capitalist typically identifies a compound annual growth rate target of the value of his investment of 60 per cent to 70 per cent against a budget of 50 per cent and in the light of a historical average achievement of 20 per cent to 40 per cent. The ideal investment should be readily realisable at any time and the business should offer guaranteed viability. In practice of course the businesses of investee companies are far removed from these ideals but it is useful to bear in mind the characteristics which the venture capitalist would identify as perfection.

Businesses which address niche markets are attractive in that there will be fewer competitors and there will be many opportunities for creating new business in other areas. In terms of market development, it is preferable that the business should be addressing a developing growth market which has already demonstrated its potential rather than either an emergent market which is unproven and unquantifiable or a long established market which is already well served by suppliers and which is approaching saturation or decline. However, a major technical innovation in an established market can clearly represent a potentially exciting business prospect.

3–23 Barriers to entry by competitors in the market sector being addressed by the investee company's products are an important feature in the venture capitalist's assessment of the business. Such barriers to entry might include the need for very large investment in fixed assets, very large investment (in both time and money terms) in establishing channels to market, or intellectual property comprising rights such as patents, copyright, design rights or proprietary know-how.

The ability to address international markets with the investee company's products is important. In terms of most industrial, scientific and consumer products in the markets of the free world the United Kingdom represents only four per cent of the total market size with mainland Europe offering 19 per cent, the United States 50 per cent, and the remainder 27 per cent. Businesses whose products rely heavily on the English language, are dependent on the existence of legislation or are covered by established trading standards or are regulated by technical standards, are bound to experience significant difficulty in developing export markets successfully so they are less likely to be attractive to the venture capitalist investor.

In terms of reducing commercial risk a prospective investee company which can point to a successful trading record is more likely to raise venture capital investment than a business which is either little more than a start-up or has a history of trading losses. In any event, whatever the status of the business when venture capital investment takes place, there must exist the prospect of realisation in a three to seven year timescale.

The management team

3–24 The management team is probably the single most important characteristic of the business to the venture capitalist. In assessing the management of a company the venture capitalist is essentially evaluating the managers as people and individuals. He is looking for compatible chemistry between himself and the management team and between the individual members of the management team. The management should be balanced in terms of their expertise and experience, their maturity, their temperament and stability, and their character and personality.

Leadership is sought from one or more of the managers. Resolve on the one hand and flexibility on the other is a combination which is likely to indicate to the venture capitalist that the management has the drive, the ambition, the stamina and the staying power which contribute substantially to eventual success, coupled with the essential ingredients of pragmatism and fleet of foot that are a vital part of effective day-to-day management. A declaration of any weakness in the management team is often perceived by the venture capitalist as a strength in that it demonstrates confidence, firstly to admit the weakness, and secondly to take steps to put it right. Total honesty and realism are important. The experienced venture capitalist will quickly see through any veneers or smoke screens and when any such attempts to cover-up weaknesses are discovered their effect is to present to the venture capitalist a strong case not to proceed.

The management team together should obviously be able to demonstrate detailed knowledge of the market sector and the technology which the investee company's products are addressing.

High technology products

3–25   In seeking to reduce the commercial risk associated with investing in unquoted companies the typical United Kingdom venture capitalist believes that the risk increases in proportion to the technological content of the investee company's products or processes. This view was particularly prevalent throughout the 1990s when prospective businesses based on state-of-the-art knowledge at the "bleeding edge" of a particular technology would have found it relatively difficult to raise venture capital in the United Kingdom. The opposite has probably always been true in America, where prodigious returns have been made in the high technology sector by investors operating in Silicon Valley in California and Route 128 around Boston. Whilst it remains the case that the average United Kingdom venture capitalist will be more receptive to low technology and no technology than to high technology and technology, the climate has changed markedly in the new millennium. Several new and existing venture capital fund managers, together with many e-commerce and dot.com incubator funds, enthusiastically embrace the TMT (Technology, Media, Telecoms) sector. In addition the United Kingdom Government has introduced a range of initiatives and fiscal incentives to encourage the development of and investment in emerging high technology businesses. The popular view which seemed valid towards the end of the 1980s that the higher the technology the higher the returns has a new resonance at the start of the new millennium, following its rejection throughout much of the 1990s. Investment by the venture capital community in technology situations should increase markedly throughout 2000 and beyond, continuing the trend established in the second half of the 1990s, compared to the first half, as the chart in Figure 1 shows.

The identification of which technologies or sectors are most attractive to venture capitalists can be determined as much by fashion as an analysis of the underlying fundamental viability and potential for long term success, as the dot.com and e-commerce euphoria demonstrated at the turn of the new millennium. Typically businesses which are fashionable command a higher price and find investment funds more readily available. This state of affairs is clearly advantageous for the investee company but in general can turn out to be disadvantageous for the unwary and uninitiated venture capitalist in the medium to long term. The nature of the business is not normally a determinant in so far as the likelihood of successfully raising venture capital is concerned. Marketing organisations, companies in the service sector and consultancies are just as likely to receive venture capital as manufacturing companies which are product or process orientated; although the prospects for traditional "smoke-stack" industries are not high.

TIMING, IF NOT EVERYTHING, IS IMPORTANT

3–26   Venture capital is the fuel and lubricant that eases the rapid growth of young emergent companies and as such is a vitally important and indispensable source of high risk investment funding. Raising venture capital is a time consuming activity and invariably takes longer than was originally anticipated. The process inevitably and ironically constitutes a distraction to the senior management of an emerging business from their principal role of managing the business and consumes a significant amount of nervous energy.

It is preferable, both from the point of view of the potential investee company and the venture capital investor, that the company should attempt to source venture capital well in advance of it actually being required. This permits the investee company to adopt a much stronger bargaining position during the fund raising activity and demonstrates to the venture capitalist shrewd management and the ability to think well in advance. On the other hand attempting to source funds as a staging post immediately preceding bankruptcy is almost guaranteed to deter even the most adventurous or imprudent venture capitalist.

The prospective investee company should always bear in mind that the objective of the fund raising exercise is to present the company as being irresistibly appealing to the venture capitalist, and not, as is so often the case, to finish up on bended knee appealing to the venture capitalist in desperation.

CHAPTER 4

EUROPEAN COMMISSION FUNDING

Graham Stroud, Research DG, European Commission

HISTORICAL CONTEXT

4–01 Support for research and development[1] has a relatively long history in European Community terms. Research on atomic energy matters was specifically provided for in the European Atomic Energy Community Treaty (the Euratom Treaty) and various other research programmes were brought into being over the years according to the political priorities prevailing at the time. By the time the Single European Act of 1986, modifying the Treaty of Rome, gave a standardised legal base for research in areas other than nuclear energy, research and development was firmly embedded as an area of Community activity. The subsequent Treaty on European Union (the Maastricht Treaty) and the Amsterdam Treaty of 1997 have refined the Chapter on research and development and the general legal framework is now reasonably stable and adapted to the particular nature of Community R&D.

OBJECTIVES

4–02 The objectives of Community R&D policy and the procedures and rules which govern the definition of that policy are set out in Articles 163 to 173 inclusive of the Amsterdam Treaty.[2] Article 163 sets out clearly that Community R&D policy has two main objectives:

- to strengthen the scientific and technological bases of Community industry and encourage it to become more competitive at international level;

- to promote research in support of other European Union policies.

In this respect, Community R&D policy is distinct from research support policies carried out at the national level or by Member States since it specifically excludes curiosity-driven, or "blue skies", research. All Community R&D activity is objective-driven with a clearly defined agenda. This is an important point that applicants to Community supported R&D programmes have to bear in mind.

The programmes

4–03 As indicated in the Treaty, there are two basic levels to Community R&D policy — the framework programme and the specific programmes that implement each framework programme.

---

[1] Often referred to in E.U. parlance as RTD — Research and Technological Development.
[2] At the time of writing, many people still persist in using the older Maastricht Treaty numbering — Articles 130f to 130p.

Framework programmes

4-04   The framework programme is, as its name suggests, a policy framework for a period of, usually, four or five years. It sets out the overall structure of all the R&D activities supported by the Community, giving the scientific and technological objectives to be achieved, fixing priorities and indicating the broad lines of the activities to be supported. In addition, it fixes the maximum overall budget for the research activities during the period in question.

Specific programmes

4-05   The framework programme is implemented through the specific programmes. The specific programmes set out the scientific and technological objectives which are in the framework programme and give more detail on the particular areas of R&D that are to be funded. The adoption of the specific programmes by the Council of Ministers allows the Commission to spend money on the research areas set out in those programmes.

Rules for participation

4-06   In addition to these two basic types of decision that are required before the Commission can spend money on its research programmes, one other Council decision setting out the ground rules for who can participate in the programmes, the means of doing so and the rules for dissemination and exploitation of the results has to be adopted. This is usually known as the "rules for participation".[3]

DEFINING THE CONTENTS OF FRAMEWORK PROGRAMMES AND
SPECIFIC PROGRAMMES

4-07   The process of defining framework programmes and the specific programmes implementing them is a highly democratic one, involving a large proportion of the research and industrial communities in the European Union and neighbouring countries. The framework programme is renewed every four or five years and the timetable for renewal is well known to policy makers and to scientific and industrial circles in the Member States. The current framework programme is the fifth since 1984 and will run from the beginning of 1999 to the end of 2002.

Proposals for framework programmes start their life well before the end of the previous framework programme and the actual beginning of the process is sometimes difficult to discern as it is intimately associated with the normal processes of evaluating and monitoring the running programmes and drawing the lessons from these evaluations.[4]

4-08   As an example, for the fifth framework programme, the Commission issued a short "political" document outlining its views some two and a half years before the

---

[3] In Community parlance this decision is still often known as the "Article 130j" decision, since it is required by Article 130j of the Maastricht Treaty (now Article 167 of the Amsterdam Treaty).

[4] The Commission has evaluations of existing programmes and framework programmes carried out independently by outside experts at regular intervals to learn lessons and to make adjustments and it is the results of these evaluations which form the basic building blocks for defining policy for the next four or five years. In addition, national and European scientific bodies, industrial associations, members of the European Parliament, Member State bodies and individual scientists all prepare contributions and suggestions and take part in the many discussions which lead to the preparation of framework programme proposals.

programme was finally adopted. This was followed by debates on the document with its regular advisory groups, the European Parliament and Member State authorities. Various academic and industrial bodies as well as individuals submitted written contributions in response to this political document. As a result, a second working document setting out more details was prepared, followed by a third, longer document detailing the main scientific and technical objectives. This last one was issued to coincide with a large conference, involving around 300 representatives of the industrial, scientific and political communities in Europe.

After listening to all the debates and sifting through the comments presented, the Commission prepared its formal proposal for the fifth framework programme and this then started on its adoption process in the European Parliament and Council.[5]

ADOPTING EUROPEAN COMMUNITY R&D PROGRAMMES

4-09   As with all Community policies, proposals are prepared by the executive body, the European Commission, and require examination by the European Parliament and the Council of Ministers before they can be adopted. In the case of R&D policy, there are three basic decision making procedures set out in the treaties for the three types of decisions that need to be taken to put an R&D policy into place (although nuclear energy research still has slightly different procedures arising from the original Euratom Treaty).

Following the adoption of the Treaty on European Union (the "Maastricht Treaty") in 1992, the framework programme was subject to a co-decision, with the European Parliament and Council acting jointly in its adoption.[6]

4-10   The discussions which led to the Amsterdam Treaty to amend the Maastricht Treaty recognised that the previous process for adopting framework programmes was too cumbersome. As a result, future RTD framework programmes will be adopted by co-decision but with the Council acting by qualified majority. This should mean that small differences of opinion with the Council will be unlikely to hold up the adoption of the framework programme in future as they have done in the past.

In the case of the specific programmes, the Council of Ministers adopts these by qualified majority after receiving the opinion of the European Parliament.[7]

The decision on the rules for participation ("the Article 130j rules") follows yet another procedure, with a Council common position after the Parliament's opinion

---

[5] While the Commission has clear ideas from the moment that the framework programme proposal is prepared as to what the contents of the specific programmes will be, the formal proposals are not presented until the debate on the framework programme is well advanced. In this way, the specific programmes can be proposed with some degree of certainty that they will not be radically altered in the subsequent legal process for adopting them.

[6] In the co-decision procedure, the Parliament's opinion of the Commission's proposal is sent to the Council, which then adopts a "common position", taking into account the Parliament's views. If the Parliament agrees with the Council's common position, this is then adopted as the final decision. If, however, the Parliament disagrees with the Council's common position, a "conciliation committee" is convened to work out the differences between the Parliament and the Council. Of the two framework programmes adopted using this procedure so far, it has been necessary to convene the conciliation committee in both instances, since the Maastricht Treaty required that the Council act unanimously at every step of the adoption process.

[7] As noted above, discussions on the specific programmes tend to take place in parallel with discussions on the framework programme and the specific programmes are usually adopted very soon after the adoption of the framework programme. In the case of framework programmes and specific programmes deriving from the Euratom Treaty, the Council can adopt these without any reference to the European Parliament, however, out of courtesy, the Parliament is always asked for its opinion before such programmes are adopted and the E.C. and Euratom framework programmes are generally adopted in one batch.

followed by a second reading in the Parliament before final adoption by the Council.[8]

CURRENT E.U. R&D POLICY

The old régime

4–11 Although the current framework programme is the fifth in the series which started in the 1980s and draws on experience from specific programmes dating back to the 1950s, it represents a clear difference with respect to previous framework programmes. In the past, the European Community RTD programmes have often been a continuation of the R&D policies carried out by Member States, where the European Community has acted, in effect, as a sixteenth Member State. It has always observed the principal of subsidiarity, which requires an activity to be undertaken at the level of the European Union only if that proves more effective than the national, regional or local level. Apart from the political imperative of only acting on European level when appropriate, there are sound financial reasons for observing the subsidiarity principal as the sums of money involved, although substantial overall, represent only a few per cent (approximately five per cent) of what can be mobilised at the level of all European countries.

At the beginning of 1997, a panel of high level personalities from the research and business world presented an evaluation of the impact of previous framework programmes. In this report, they accepted that the E.U. had done a good job in creating a culture of fostering European collaboration on science and technology through the framework programmes and the specific programmes implementing them. However, they criticised the end results of previous programmes for lacking visibility both at a political level and with the public at large and for spreading scarce resources too thinly by trying to cover too many areas of science and technology and cater for too many interests which did not necessarily all have the same priority at European level. To a certain extent, this was also the product of the decision-making process put in place by the Maastricht Treaty; the required unanimity in the Council deliberations was often achieved at the expense of the setting of clear priorities.

The message in the evaluation report coincided with the point of view of research Commissioner Edith Cresson, who subsequently piloted through a fundamental rethink of E.U. policy away from an approach driven by the expression of scientific and technological needs towards an approach driven by societal and policy problems.

The new approach

4–12 In essence, previous framework programmes had been defined from the point of view of which areas of science and technology it was necessary for Europe to have a capability in, whereas the fifth framework programme started from a statement of the most pressing societal problems which science and technology could be called on to help solve. For particular areas of science and technology to be included in the framework programme, they had to demonstrate their relevance to solving real soci-

---

[8] The upshot of the complexity of the different legislative procedures used to adopt framework programmes and specific programmes is that it can take a considerable amount of time between the Commission's formal proposal and the final adoption. In the case of the fifth framework programme, this process took more than one year from beginning to end and the programmes underwent a number of changes in the process of the discussions.

etal problems, not simply because they were addressing interesting areas of science and technology. With this fundamental shift in the premises behind the framework programme, it is now clearly differentiated from the research and development programmes sponsored by the individual countries of the E.U.

THE CURRENT R&D PROGRAMMES

4-13 The new approach can be seen in the structure of the fifth framework programme (FP5), which is broken down into seven specific programmes (as opposed to 18 specific programmes under the previous framework programme).

The fifth framework programme comprises four thematic programmes which cover a series of well identified problems and three "horizontal" programmes which respond to common needs across all research areas. The total budget allocated to the programme is 14,960 million Euros. This money funds three main types of activity:

- **The key actions**. These are one of the major innovations of the fifth framework programme and account for most of the spending in thematic programmes. Although largely concentrated in the thematic programmes, there is also one key action in the horizontal programme "Improving human research potential". Their aim is to concentrate resources and skills of all relevant disciplines, technologies and people on a series of well-defined, high priority, socio-economic problems. Each key action integrates research, technology demonstration, training and other activities and allows a better co-ordination of research between the Member States, countries outside the E.U. and other international initiatives. A total of 23 key actions have been identified dealing with concrete problems through multi-disciplinary approaches involving all the interested parties.

- **Generic research activities**. These activities support research work complementary to that undertaken within the key actions. The aim is to maintain flexibility in the fifth framework programme and to support research and development on generic technologies in up-and-coming sectors. Support for generic research activities is a continuation of the kind of research supported under previous framework programmes, although this now accounts for a much lower percentage of the total budget of the framework programme.

- **Support for research infrastructure**. Although the construction and operation of research infrastructure is beyond the means of the E.U.'s framework programmes and falls within the competence of national authorities, Community support is used on two levels: to ensure the optimum use of existing infrastructure and to allow rational and economically effective development of additional research infrastructure through transnational co-operation. Support can be given for activities such as networking of research facilities, facilitating access to facilities for other Community researchers, pooling collections of genetic resources, networking of high speed computers, access to databases, etc.

In brief, the areas covered by the specific programmes of the fifth framework programme are as follows:

Thematic programme 1 — Quality of life and management of living resources

4–14 This programme supports research aimed at improving health, reconciling economic progress with environmental requirements and improving the response to consumer needs in the life sciences. There is also a large potential for economic growth and job creation in this field. In the fields covered by this programme, European research adds value either because the subject is cross-border by nature,[9] because it is worth studying at the European level,[10] or because it has a direct link with Community policies.[11] To achieve the objectives of the programme, six key actions have been identified:

- Food, nutrition and health
- Control of infectious diseases
- The cell factory
- Environment and health
- Sustainable agriculture, fisheries and forestry
- The ageing population and disabilities

Longer term generic research activity focuses on chronic and degenerative illnesses, genomes, neurosciences, public health and health services, ethics and socio-economic aspects of life sciences, among others.

Thematic programme 2 — User-friendly information society

4–15 The aim of this thematic programme is to help reap the benefits of the new information society now emerging in Europe and at the same time to facilitate its advent by ensuring that the needs of both private individuals and companies are satisfied. Four complementary objectives have been set concerning the satisfaction of the needs and expectations of private individuals, innovation and improving productivity of companies and workers, multimedia content and the acceleration of the development and application of appropriate technologies. To achieve these objectives, four key actions have been identified:

- Systems and services for the citizen
- New methods of work and electronic commerce
- Multimedia content and tools
- Essential technologies and infrastructures

The programme also funds long term research on emerging and future technologies such as quantum computing and communications, personal bioinformation systems and nanotechnology information devices.

---

[9] *e.g.* environmental protection and management of living resources.
[10] *e.g.* epidemiology.
[11] *e.g.* bioethics and biosafety.

Thematic programme 3 — Competitive and sustainable growth

4–16 The objective of this programme is to support research on a range of goods and services which are both competitive and environmentally-friendly. The accent is on the production of high quality, clean and "intelligent" industrial goods and services. This can only be achieved, however, with effort also in the development of safe, economic and environmentally-friendly transport, as well as high quality materials and reliable methods of measurement and testing. To achieve these goals, four key actions have been identified:

- Innovative products, processes and organisation
- Sustainable mobility and intermodality
- Land transport and marine technologies
- New perspectives for aeronautics

In addition to these key actions, research activities in the following generic fields are supported: new and improved materials (their production and transformation), new and improved materials and production technologies in the field of steel, and measurements and testing.[12]

Thematic programme 4 — Energy, environment and sustainable development

4–17 This programme tackles two challenges which, in the context of sustainable development, are closely connected: to ensure satisfactory long-term energy supplies and reduce the impact of human activity on the environment. The problems addressed are best tackled at the European level since the majority of challenges are common to all European countries and the aim is not only to pool knowledge and know-how but also to take account of the intrinsically transnational character of certain problems. To achieve the objectives, six key actions have been identified for tackling under the European Community part of the programme, supplemented by two key actions which apply under the Euratom programme:

- Sustainable management and quality of water
- Global change, climate and biodiversity
- Sustainable marine ecosystems
- The city of tomorrow and cultural heritage
- Cleaner energy systems, including renewables
- Economic and efficient energy for a competitive Europe
- Controlled thermonuclear fusion
- Nuclear fission

In addition to the key actions, research of a generic nature is supported on major natural and technological hazards, earth observation technologies,

---

[12] This area is strongly directed towards support for Community policies and the standards-making process and towards the fight against fraud.

socio-economic aspects of environmental change, radiation protection, behaviour of radioactive materials in the environment and industrial and medical uses of radiation, among others.

Horizontal programme 1 — Confirming the international role of Community research

4–18 The principal aim of this horizontal programme is to make known the quality of the Community's research by opening it up to the world. More precisely, it helps European research centres and companies gain access to scientific and technological knowledge based outside the E.U., it encourages the implementation of strategically important activities with non-European countries,[13] prepares for accession of new Member States, increases the opportunities for researcher training[14] and improves co-ordination with other Community programmes and other European organisations and initiatives for co-operation, such as COST and EUREKA. Whilst the other programmes of the fifth framework programme are open to the participation of third country scientists, this programme provides overall co-ordination and a number of specific activities aimed at particular sets of countries. It also operates a system of fellowships for young doctoral level researchers from developing countries, including Mediterranean countries and emerging economies. The programme's activities cover the following:

- Countries preparing for accession

- Newly independent states and central and eastern European countries not preparing for accession

- Mediterranean partner countries

- Developing countries

- Emerging economies and industrialised countries

- Co-ordination with Community activities and other European initiatives.

For each of the above categories, a number of specifically designed initiatives have been put in place.

Horizontal programme 2 — Promotion of innovation and participation of SMEs

4–19 The fifth framework programme acknowledges clearly the importance of small and medium sized enterprises (SMEs) in the innovation process and as creators of wealth and employment and devotes considerable resources to encouraging the participation of SMEs in all the programmes. Each specific programme issues calls for proposals for special measures in favour of SMEs. In addition, this horizontal programme provides overall co-ordination of the implementation of these specific measures. It also aims to ensure the dissemination and exploitation of the results generated by framework programme projects and to stimulate technology transfer by co-ordinating activities promoted under thematic programmes. The following activities are contained within this programme:

---

[13] *e.g.* the problems of food and health in developing countries.
[14] *e.g.* through fellowships and other researcher mobility measures.

- Mechanisms to make better use of the results of Community research

- New approaches to technology transfer (*e.g.* methods for managing the introduction of new technologies, use of internet advanced brokerage services, fostering of "innovation projects" combining research and demonstration, etc.)

- Studies and good practices in innovation (*e.g.* surveys, workshops, studies, benchmarking, etc.)

- A single entry point for SMEs and joint support and assistance measures

- Economic and technological intelligence for SMEs

- A European support network for the promotion of research, technology transfer and innovation and electronic information services

- An information system on intellectual property matters and on access to private innovation financing (*e.g.* an intellectual property rights helpdesk)

- Mechanisms to facilitate the creation and development of innovative companies

Horizontal programme 3 — Improving human research potential and the socio-economic knowledge base

4–20 The framework programme recognises that, ever increasingly, it is knowledge that drives developed economies and shapes society, bringing with it new production methods and new types of organisation. As a result, Europe's best assets are, more than ever, its human resources and the quality of its researchers, its engineers and its technicians. In addition, there is a need to undertake work to enable us to understand the crucial problems which confront a changing European society, particularly the socio-economic impact of new technologies. To do this, one key action has been created:

- Improvement of the socio-economic knowledge base — improving understanding of societal trends and structural changes, particularly managing such structural change in technology, society and employment, governments and citizenship and new development models fostering growth and employment

and a number of activities are directed at improving Europe's research potential:

- Training and mobility of researchers through research training networks and "Marie Curie Fellowships"

- Access to research infrastructure

- Promoting scientific and technological excellence, through high level scientific meetings, awards and prizes for exceptional research work and raising public awareness of science and technology

In addition, work is supported on the development of scientific and technological policies in Europe.[15]

---

[15] Common scientific, technological and innovation indicators, analysis of the impact of and implications for scientific and technological policy, etc.

Other E.U. research activities

4–21 The brief descriptions of the various programmes above illustrate those areas of the framework programme which are open for competitive bidding for funds. However, a small proportion of the framework programme budget is also set aside for the European Community's in-house research laboratories, the Joint Research Centre. The various laboratories of the Joint Research Centre have primarily a public service and policy role in that they provide neutral advice and expertise on questions relating to, for example, health and environmental standards, hazard and radiation protection, etc. Potential participants in E.U. R&D programmes should note, however, that the laboratories of the Joint Research Centre are eligible to bid for funds in collaboration with any other researchers or research teams in Community countries or associated states and are a potential source of collaborators on research projects; many of the laboratories have equipment that is unique in Europe.

Administering E.U. R&D programmes

4–22 Unlike much of E.U. policy, its spending on R&D programmes is administered directly by the Commission from Brussels or Luxembourg. In budget terms, the E.U.'s R&D programmes come in a very distant third place after spending on agriculture and regional development, accounting for just a few per cent of the Communities' budget. R&D policy, though, is the largest block of Community spending to be administered directly by the Commission and not by the Member State administrations on behalf of the Community. This simple statistic also points up one highly important difference between E.U. R&D policy and other E.U. policies in that money is not passed back directly to the Member States to administer on behalf of the Community; in E.U. R&D programmes, there are no national quotas or calculations of *"juste retour"*. The bidding process for obtaining funding from the R&D programmes is highly competitive and only those proposals of the highest quality receive funding. In the past, an average of only one in five proposals submitted has been funded.

Administering the E.U.'s R&D programmes is a large business. During the fourth framework programme, around 25,000 proposals were received each year involving more than 100,000 participants. The number of research contracts in progress at any one time is in excess of 11,000.

Types of contract

4–23 The R&D projects that are part-funded by the E.U. (normally up to a maximum of 50 per cent of the total costs) fall into a number of categories that often have features that differentiate them from national R&D projects:

(i)   The major proportion are *shared-cost* R&D projects. These are almost always collaborative, *i.e.* they must normally involve at least two independent partners from different Member States.[16]

---

[16] Or from one Member State and an associated state.

(ii)   *Coordination projects* cover the costs of bringing together researchers and research teams working on existing projects (usually nationally funded) to create "European added value".[17]

(iii) In the case of *training and mobility of researchers*, the contracts are either with individual researchers, to enable them to spend a period of study outside their own country, or with the institutions that host visiting researchers.

(iv)   Because of the importance of *small and medium-sized enterprises* to the European economy, some special measures have been devised to attract them to the E.U.'s R&D programmes. There are "exploratory awards" which fund up to 75 per cent of the cost of developing a full-blown research proposal. These allow small groups of SMEs (a minimum of two) to carry out patent searches and feasibility studies, find partners, etc. In addition, there is a special variety of shared-cost projects — co-operative research or "CRAFT" projects — designed for consortia of SMEs which do not necessarily have any in-house R&D capabilities. In this case, they can subcontract the research itself to another organisation. Both of these special measures for SMEs have proved very popular and have given rise to some important innovations.

MANAGEMENT OF PROGRAMMES

4–24   The very particular nature of E.U. programmes and projects affects the way they are managed. The wide geographical spread of the participants and the diversity of languages and cultures involved are important factors in the management of the programmes. The Commission has to maintain a neutral position, and ensure and be seen to ensure equal access and opportunities for all potential participants and equal treatment of proposals. These have always been fundamental tenets of Commission research policy management.

The management of the programmes can basically be broken down into a number of steps, during the implementation of which, the Commission is assisted by committees of Member State representatives who act as the "eyes and ears" of the Member States to ensure that the administration is carried out correctly. These "programme committees" have a formal role to play in certain management decisions. The basic steps are as follows:

(i)   **Adoption of the work programme.** Work programmes are prepared on the basis of the content of each specific programme, in order to provide researchers with a detailed description of the objectives of the programme, the approaches to the research work which the Commission intends to favour and a provisional timetable for calls for proposals, including an indication of their content. The work programme is made available to researchers on the Commission's websites and in the programme information packages.

(ii)   **Preparation of information packages and informing potential proposers.** Each programme prepares one or several information packages, designed to help

---

[17]   For example, in the area of medical research, enlarging the scope of national epidemiological studies has clear benefits both at the European and the national level.

potential proposers prepare their submissions. Information packages include general details of the programme, indications of how to submit a proposal, the timing of calls, general information about the proposal evaluation process that will be followed, including the criteria that will be applied to judging the proposals, information on the Commission's research contracts and a list of contact points in each of the Member States or Associated States.

(iii)   **National Contact Points.** The job of the "National Contact Points" (NCPs) is to be the first "port of call" for potential proposers and to answer queries or redirect proposers to the appropriate contacts within the Commission. NCPs also provide "marriage broking" services enabling potential proposers to get in touch with possible partners from other countries. The National Contact Points are the privileged interlocutors of the Commission and receive first hand information to pass on to proposers throughout the framework programme, but particularly in its launch phase. The NCPs are the prime source of information and assistance for proposers.

(iv)   **Calls for proposals.** Calls for proposals are prepared on the basis of the work programmes and are published in the *Official Journal of the European Communities*. Information is also posted on the Commission's websites, particularly *CORDIS* which is devoted to the R&D programmes of the E.U. Currently, calls are mostly published on the 15th of each month (with the exception of August) or on the next publication date for the *Official Journal* after the 15th. Calls for proposals are open for a minimum of three months and specify cut-off dates by which proposals must be received by the Commission to be eligible for consideration.[18]

(v)   **Receipt and registering of proposals.** As outlined above, calls for proposals specify a cut-off date by which proposals must be received to be eligible for consideration. In the case of proposals submitted on paper, the rule is that they will not be considered unless they arrive within the deadlines for submission set out in the calls. The Commission also provides a software tool which allows proposers to prepare and submit their submissions electronically. In the latter case, proposals must be received within 48 hours of the deadline for submission.

Following receipt of the proposals, details are entered into internal databases and acknowledgements of receipt are forwarded to the proposers. Commission staff carry out checks on whether the proposals fulfil the eligibility requirements set out in the call for proposals before the proposals are evaluated on the quality of their content.

(vi)   **Evaluation of proposals.** Evaluation of proposals is the responsibility of the Commission, but in carrying out this task, it uses the services of many thousands of external experts. Although this system is essentially the same as the peer review process which is used by almost every research funding agency in the world, there are a number of particular features that are specific to the E.U. research programmes.

---

[18] Exceptionally, calls may need to be published on other dates; in this case an advance notice is published on one of the fixed dates beforehand. In some instances (*e.g.* special SME measures and fellowships), calls remain open for most of the framework programme and proposals are evaluated in batches at regular intervals announced in the calls.

First, the experts are chosen from lists of potential evaluators drawn up following a published call for experts. This is an open invitation published in the *Official Journal of the European Communities* to all suitably qualified individuals to put their names forward to be proposal evaluators. Generally, a minimum of three experts examine each proposal.

Second, the evaluation experts are usually brought to a central location in Brussels or Luxembourg for the evaluation sessions, rather than working at home or their place of work, as is often the case in national peer review systems. The reason for this is that, as the framework programme is predominantly targeted at questions of industrial competitiveness and exploitable technologies, commercial confidentiality must be given a high priority when examining the proposals. Where questions of commercial confidentiality are not an issue,[19] proposals are usually sent to the evaluators for examination.

(vii) **Evaluation of proposals — the criteria.** The criteria used for evaluating proposals are set out in the framework programme and in the Council decision on the rules for participation. These are sometimes supplemented by specific selection criteria for each programme, which are set out in the Council decisions on the specific programmes. These selection criteria are described in the information packages and guides to proposers prepared by each programme and, in addition, a general manual of proposal evaluation procedures for all the programmes of the framework programme is prepared and made available on the Commission's webservers. This manual describes the evaluation process in detail and, in particular, the marking systems to be used by the evaluators and any specific criteria or interpretations of the general criteria which will be used by the evaluators.

(viii) **Evaluation of proposals — the process.** Proposal evaluation is often split into two phases. First, the experts examine the scientific and technical content of the proposals, usually without a knowledge of the identity of the proposers. This is then followed by an examination of the strategic and socio-economic relevance of the proposals and the resources requested. The second phase is often carried out by a second set of expert evaluators with a broader-based knowledge. After examining the proposals individually, the evaluators are brought together to agree on their final marks and they provide the Commission with written comments of their assessment of each proposal. During this process the role of the Commission staff is to act as "moderators" for the panels of experts by providing information where necessary and helping them to reach a consensus. The Commission does not express an opinion on the proposals or attempt to guide the evaluators to any particular viewpoint.

(ix) **Drawing up of priority lists.** After the examination of the proposals by the external expert evaluators, the Commission staff draw on their advice to prepare ranked priority lists of the project proposals. In drawing up these priority lists, the primary consideration is the advice of the evaluators, but the Commission also has to take into account questions of programme coverage[20] and any other criteria fixed for the particular call. The funding available for the particular call is used to arrive at the cut-off point between those proposals which can be taken further and those which are ranked too low to

---

[19] For example, in the evaluation of requests for fellowships.
[20] For example, by not concentrating all funding in a particular programme area.

be considered for funding. The priority list will usually contain a small reserve of proposals which can be drawn on in the event that any projects on the main priority list are withdrawn during subsequent contract negotiations or if additional funding becomes available to a programme, through, for example, savings negotiated with project partners.

(ix)   **Rejection.** The Commission's intentions with respect to the priority list and those proposals to be rejected are submitted to the programme committee for information, following which, a formal Commission decision is taken to reject all those proposals not on the priority list. In their information packages and the evaluation manual, the individual specific programmes indicate what weighting they will give to the individual selection criteria. However, they also have the possibility of setting "threshold" marks for certain criteria. If a proposal fails to achieve this minimum threshold mark for any particular criterion, then it will not be selected for funding no matter how good the rest of the proposal. All proposals failing these thresholds are rejected at this stage. Following the Commission's rejection decision, proposers are informed and are given feedback on the reasons for rejection, taking into account the comments of the expert evaluators.

(xi)   **Contract negotiation and finalisation.** For the proposals remaining on the priority list for which funding is available, contacts are made with the proposal co-ordinators with a view to preparing a contract. Contract negotiations can vary in length, depending on the nature and extent of the changes that are required to the proposals[21] and on the size and complexity of the proposals. Sometimes proposal co-ordinators are invited to meetings to discuss the changes, or if there are only minimal adjustments, these can be dealt with by telephone or fax.

(xii)   **Financial considerations.** During this process, the Commission staff check on the financial viability of the project partners and that they have the necessary resources to start the project. A close check is made of the costs of the resources requested to carry out the work and adjustments may be made to the work programme to take into account the comments of the expert evaluators.

(xiii) **Conclusion of negotiations.** A major source of delays in this step is often the time taken by the members of the project consortium to agree between themselves when changes are required to the initial proposal. The Commission staff try to speed up the process of contract negotiation by setting time limits for replies to requests for further information. However, substantial changes to a large or complex initial proposal will inevitably take time to agree between all the contract partners. At the end of this stage, a formal decision is taken by the Commission to select those projects for which contract negotiations have successfully been finalised. For projects above a certain value, the programme committee must first give a favourable opinion before the Commission can take this decision.

(xiv)   **Contract signature and advance payment.** After the selection decision, contracts are sent for signature by the partners and then signed by the

---

[21] Suggested by the evaluators or by the Commission staff.

Commission, at the level of Director or Director General. Work on the contracts can then start, usually on the first of the month following the date of final signature or on a date agreed in the contract. An advance payment of up to 40 per cent of the Commission's contribution can then be made to the proposal coordinator. Contractually, the Commission is obliged to pay this within two months of contract signature but, in reality, advance payments are often made much sooner than this.

(xv) **Project monitoring and reporting.** Commission staff monitor the progress of projects through site visits and through the regular reports that the contract partners submit through the project co-ordinator. In addition, contractors can be invited to make presentations on project progress at specially convened meetings and the Commission sometimes uses the assistance of external "project technical auditors" to carry out checks on the progress of projects.

Most projects, unless they are very short in duration, provide for a mid-term review to check whether the original objectives and timetable are being adhered to. If there are serious problems, projects can be terminated or refocused at any stage. At the end of a project, contractors must submit a "Technology Implementation Plan" indicating how they intend to use the results generated by the project over the following period. These Technology Implementation Plans comprise two parts — a general part for the whole project and individual confidential plans for each of the partners. These latter are not divulged to the other partners, in order to protect the commercial interests of each project participant. In addition to the Technology Implementation Plans, contractors must submit a final report detailing all that has been achieved on the project.

(xvi) **Payments.** Regular intermediate payments are made throughout the life of the contract to the contractors following the submission and acceptance by the Commission of progress reports and cost statements for the various project partners. Payments are made to the project co-ordinator, who is contractually obliged to pass on the sums received to the different project partners immediately. A final payment of 15 per cent is withheld until the final report, cost statements and Technology Implementation Plans have been accepted by the Commission.

CONTRACTS

4–25 All projects under the R&D programmes of the E.U. are covered by standardised contracts, known as "model contracts". These set out standard contractual conditions, rights and obligations for the contract partners and are derived from the rules for participation. For the fifth framework programme, there are five main categories of model contract, each of which has the same basic structure and shares many common clauses:

- Shared-cost action contracts

- Specific contracts for small and medium-sized enterprises (SMEs)

- Fellowship contracts

- Co-ordination contracts

- Contracts for accompanying measures.

For each of these categories, there is a main model contract and a number of contracts which are derived from the main one but which have special provisions or clauses for particular cases.

4–26 For **shared-cost action** contracts, the main contract covers most collaborative R&D projects. The various derived contracts cover variations in payment procedures,[22] demonstration projects, combined R&D and demonstration projects, and support for access to large-scale facilities.

Under the **SME-specific contracts** come two contracts which cover activities reserved only for SME participants — exploratory awards and co-operative research projects. SME participants in ordinary shared-cost R&D projects are covered by the standard contract for that category.

The category of **fellowship contracts** covers the main "Marie Curie" individual fellowship contract, where a researcher goes to work at a host institution other than his/her home laboratory. The derived contracts in this category concern industry host fellowships, development host fellowships, experienced researcher fellowships, stays at Marie Curie training sites, International Cooperation fellowships and Joint Research Centre fellowships.

4–27 The main contract in the category of **co-ordination contracts** covers thematic networks and concerted actions, in which the Commission contributes to the cost of co-ordinating research which is already paid for from other sources. Derived contracts cover research training networks.

The category of **accompanying measures** covers activities which are not specifically research projects but which are undertaken in support of research activities and which are selected following calls for proposals. Derived contracts include those for Euroconferences and for advanced study courses.

The various model contracts under the fifth framework programme all have the same basic structure which is set out in Table 1, below.

Contract participants

4–28 Participants in any E.U. R&D contract fall into a number of different categories according to the type of contract and the nature of the participant's role in it:

- For R&D projects, demonstration projects and combined projects, a participant who has a wide-ranging role in the project throughout its lifetime is normally a **principal contractor**. A participant whose role is largely in support of one or several of these principal contractors is termed an **assistant contractor**.

- For activities dealing with access to research infrastructure, the facility being accessed is the principal contractor.

- In measures for SMEs, the SMEs benefiting from the projects are the principal contractors. Organisations performing the research for SMEs are classed as

---

[22] *e.g.* by fixed instalments or fixed contribution.

RTD performers and are not considered to be participants as they are fully paid for their work.

- Co-ordination activities such as concerted actions, research training networks and thematic networks distinguish between the principal contractors who lead the activity and the **members** who are associated with them.

- In general, participants in accompanying measures are all principal contractors; exceptionally some can be members.

4–29 In all "normal" R&D contracts the participants designate amongst themselves a single co-ordinating principal contractor ("**the co-ordinator**") who takes the leading role in representing the project consortium and is the principal point of contact for the Commission in the execution of the contract. The co-ordinator's tasks include responsibility for collecting, integrating and submitting deliverables, such as reports and cost statements, to the Commission and for distributing the funds received from the Commission. In certain circumstances, the costs incurred by the co-ordinator in carrying out these tasks are recognised as eligible project costs.

Subcontractors may be appointed to provide a service to a principal contractor, an assistant contractor or a member, who fully funds their activity. Subcontracting costs are eligible for reimbursement according to the contract provisions. Since subcontractors make no financial investment in the project, they do not benefit from any intellectual property rights arising from the work carried out.

RIGHTS AND OBLIGATIONS OF PARTICIPANTS

4–30 In general, principal contractors are jointly and severally responsible for the execution of the contract and receive royalty-free access to any new knowledge (including intellectual property) generated under the contract either for the purposes of performing the contract or for exploiting that knowledge. In addition, they may also have access on favourable conditions to the pre-existing know-how (including intellectual property) of the other principal contractors for the purposes of carrying out the research or for exploiting the results. Assistant contractors only benefit from these rights with respect to their principal contractors.

In all instances, contractors are expected to use the results arising from E.U. R&D contracts, either on further research, by exploiting them commercially, or, if this is not possible, by disseminating the results. As a last resort, the Commission may itself disseminate any results that the contractors have not made use of or disseminated.

The provisions of the model contracts with respect to intellectual property rights (IPR) vary depending on the type of project and the type of contractor. These are summarised in Table 2, below.

4–31 Table 1.   Structure of model contracts under FP5.

	ANNEX 1	ANNEX II	ANNEX II	ANNEX II
BODY OF THE CONTRACT[1]	**PROJECT PROGRAMME**[1]	**MODULE A: PERFORMANCE OBLIGATIONS**[1]	**MODULE B: INTELLECTUAL PROPERTY RULES**[2]	***MODULE C: ALLOWABLE COSTS**[3]
Object	Work Description (possibly also project deliverables)	*Definitions other than relating to IPR	Property of project results (knowledge)	Allowable costs:
Project Duration (including the entry into force of the contract)		IPR Definitions	Protection of project results (knowledge)	– direct costs – indirect costs
Estimated costs and maximum contribution (including amount of the advance, % of balance)		*Management of project (particularly role of co-ordinator)	Use/Dissemination Access Rights	**MODULE D:** Technical and financial control
Periods covered by the reports and by the corresponding cost statements or payment requests		Reports and cost statements or payment requests (content, time limits for delivery and approval)	(general principles, access rights for purpose of (i) carrying out of project (ii) use, exclusivity)	Contractual sanctions Technological control
Applicable law and jurisdiction		Payment of contribution (including time limits, late payment interest)	Technology implementation plan (principles, content)	**MODULE E:** Forms for cost statements
*Special Conditions for the project and the programme		*Participation of third parties (sub-contractors)	Publicity and information on results	
Amendments		*Liability (including *force majeure*)	Confidentiality Communication of data on knowledge (standardisation purposes)	
Content of the contract (including Precedence in case of conflict between contract and (i) special conditions, (ii) annexes)		Termination of contract	Incompatible or restrictive undertakings	
Signature and language				

[1]   Common to all model contracts: shared-cost actions, SME-specific actions, fellowships, co-ordination (networks, concerted actions), accompanying measures following calls for proposals.
[2]   According to the type of action.
[3]   According to the type of action and the cost calculation method.
*   Variable provisions depending on the nature of the indirect RTD action or on the situation under consideration.

4-32 **Table 2.** Summary of Intellectual property rights provisions in E.U. model contracts.

This table lists the access rights (licenses under patents, and other user rights) to knowledge (intellectual property created during the project) and know-how (pre-existing intellectual property) in relation to the different types of actions and participants

TYPE OF ACTION	PARTICIPANT	Knowledge (created during the project)		Pre-existing know-how necessary for the execution of the project or the use of the knowledge	
		Access rights for the execution of the project	Use*	Access rights for the execution of the project	Use
Research and technological development project	**Principal Contractor**	Royalty-free	Royalty-free[1] to all knowledge	Favourable conditions	Favourable conditions
	Assistant Contractor[2]	Royalty-free/Favourable conditions	Favourable conditions/Market conditions[1]	Favourable Conditions/Market conditions	—
	Principal Contractor of the same specific programme	Favourable conditions	Market conditions		
Demonstration Project	**Principal Contractor**	Royalty-free	To all knowledge under favourable Conditions for Exploitation only	Favourable Conditions	Favourable conditions for Exploitation only
	Assistant Contractor[2]	Royalty-free/Favourable Conditions	Favourable Conditions/Market conditions for Exploitation only	Favourable Conditions/Market conditions	—

Table 2. Continued.

This table lists the access rights (licenses under patents, and other user rights) to knowledge (intellectual property created during the project) and know-how (pre-existing intellectual property) in relation to the different types of actions and participants

TYPE OF ACTION	PARTICIPANT	Knowledge (created during the project)		Pre-existing know-how necessary for the execution of the project or the use of the knowledge	
		Access rights for the execution of the project	Use*	Access rights for the execution of the project	Use
Combined R&D/ Demonstration Project	**Principal contractor**	In general, IPR rules for R&D projects shall be applied to R&D workpackages, for Demonstration projects to Demonstration workpackages. If the identification of the various workpackages is impossible, IPR rules for R&D projects shall apply if the total E.C. contribution to the project as a whole is superior to 42.5 per cent, of its total cost. If the figure is equal to or below 42.5 per cent, IPR rules for Demonstration projects shall then be applied.			
	Assistant Contractor[2]				
Co-operative Research Project	**Principal Contractor (SME)**	Co-ownership[3]	Co-ownership[3] for Exploitation only	Royalty-free	Favourable Conditions Exploitation only
	RTD performer (non-participant)	Royalty-free		Royalty-free	
Concerted Action	**Principal Contractor**	The knowledge which is suitable for dissemination will be disseminated			
	Member				

Networks	**Principal Contractor**	The knowledge which is suitable for dissemination will be disseminated
	Member	
Fellowships	**Host Institution**	The ownership of knowledge will be determined by the Host Institution according to the applicable law. The knowledge which is suitable for dissemination will be disseminated.
	Grant Holder	
Accompanying Measures	**Principal Contractor & in particular cases Members**	The ownership of knowledge will be determined according to the Community financing level. Use or dissemination will prevail, as the case may be.

* Access rights to knowledge for the purpose of use are limited to knowledge generated under the project concerned.
[1] Contractors and Assistant Contractors unable to exploit their own knowledge may grant access rights at reasonable financial or similar conditions, instead of royalty-free.
[2] More favourable conditions when beneficiary requests access from its principal contractor or the other assistant contractors of the latter.
[3] SME contractors are the owners of all knowledge resulting from the research work carried out by the RTD performers.

CHAPTER 5

THE UNIVERSITY PERSPECTIVE

Dr Roger Holdom, University College London and Mr James Evans, University of Manchester

INTRODUCTION

5–01 Access to expertise and facilities in Higher Education Institutions (HEIs) by organisations outside the HEI sector occurs on a very large scale and features many different types of contractual relationship. Some institutions have a traditionally high emphasis on research and development work, whilst others concentrate their resources on teaching and vocational training. The majority of HEIs offer consultancy links and professional development courses to organisations and sometimes these lead to more substantial research relationships. All HEIs are involved in producing graduates and postgraduates who form an important part of "knowledge transfer" from HEIs.

There are many long-standing funding support schemes which promote external research collaboration between HEIs and outside organisations. By contrast, there are still only a few schemes supporting technology transfer and, since the more significant ones were introduced as recently as 1998/9, practical experiences are rather limited. All support schemes contain a mixture of positive and negative aspects, depending on whether they are being viewed by the supplier (HEI) or the customer (sponsor).

This Chapter covers the HEI viewpoint and experience of funding schemes over recent years. HEIs have become more dependant on the income from external sponsorship of research as a proportion of total income from all sources than they were about ten years ago and therefore have a strong interest in ensuring that funding schemes work well and provide what the customer wants. The Chapter also looks at what lies ahead in the near future. There are many new factors influencing changes both within the sector and external to it, which will affect not only how much work of different types can be undertaken by staff in HEIs, but also the financial and other terms governing the relationships with sponsors.

First, it is important to describe briefly the reasons why organisations are attracted to HEIs for contract research and other services.

KEY STATISTICS OF THE HEI SECTOR

5–02 HEIs are a major asset which is used to achieve the United Kingdom's overall research and development goals. Their basic and strategic research capability currently accounts for about 20 per cent of the total of £15 billion gross expenditure on research and development by all Government agencies and industries.

The unique "Dual Support Funding" of HEIs[1] with its annual investment of about

---

[1] The Dual Support system for HEIs consists of two funding streams. Stream 1 — Block Grants from the Funding Councils — covers recurrent and capital expenditure (permanent academic staff, infrastructure, buildings, facilities, equipment and support staff and services), the research component of which is intended to cover the costs of in-house research by academic staff. Stream 2 — Research Council Grants — are awarded to academic staff making bids, which are subject to peer review. Grants cover additional (marginal) costs of approved projects including non-permanent staff, equipment, running costs and a contribution to indirect costs (at 46 per cent of staff costs).

£2.5 billion from Funding Councils and Research Councils ensures the continuous replenishment of manpower, know-how and research infrastructure at the frontiers of research in all the major areas of science, engineering, economics and social science. For many external organisations it is this "well–founded facility" which most attracts them to collaborate with HEIs when they need research based solutions to problems or access to consultancy expertise.[2]

Size of the sector

5–03 There are 172 institutions within the sector, 156 of which are Universities in receipt of some research funding from public and private sources. Ten of the Universities are very research intensive. More than 50 per cent of their total budget derives from external research income. Historically these have received about 40 per cent of the research funding available to the sector as a whole. This tends to be a self-sustaining system because such institutions are able to build up levels of research staff and infrastructure which help to maintain both their attractiveness to external organisations for contract research and their competitive edge over other HEIs. However, the Universities which obtain typically less than 10 per cent of their total income from research are often very active in other areas, such as consultancy services and the provision of tailored training courses to industry, commerce and the Health Authorities in their region.

Quality of outputs

5–04 United Kingdom University research has an outstanding record. It is second only to the United States in terms of production of highly cited scientific publications across a broad range of fields, and is especially strong in the life sciences. From a financial viewpoint, whilst United Kingdom spending per capita on the science base is amongst the lowest of the developed countries, the United Kingdom has more highly cited papers per pound spent on research than any other country. The number of published papers involving collaboration with industry has increased threefold in the last ten years. Over the same period HEIs have increased the number and quality of support staff in "industrial liaison" offices which help external organisations unfamiliar with the HEI system to gain access to the right expertise.[3] The statistics from a research and development viewpoint are summarised in Table 1 which shows that HEIs are a large human resource base for research purposes with a unique high-investment infrastructure.

---

[2] HEIs: Whilst the most significant activities in terms of turnover are Research and Technology Transfer, HEIs also provide access to consultancy expertise, specialised testing rigs, pilot plant facilities and associated technicians, and a wide range of training programmes to satisfy the continuing education needs of industry and commerce. These services are all part of "knowledge transfer" and often complement or support existing research and development links with companies, or may lead to substantial research and development relationships with clients. Also, HEIs with Medical or Clinical departments usually provide a range of essential medical services to patients in the NHS, sometimes leading to complementary clinical research projects or clinical trials which could not otherwise have been done. These service activities, whilst relatively small in overall turnover terms compared with sponsored research, are thus an integral part of research and development activity in HEIs.

[3] These offices also help academic staff by guiding them through the maze of research sponsorship schemes, preparing viable bids to sponsors and negotiating fair contract terms. They also help in formulating and implementing central HEI policies dealing with issues such as terms and conditions on contracts, protection and exploitation of intellectual property, and the marketing of HEI research and consultancy expertise.

5–05   Table 1.   Human Resources and Infrastructure for Research in HEI sector.

RESEARCH STAFF IN HEI SECTOR	NUMBER	RESEARCH COUNCIL FUNDED	COMMENTS
Academic Staff active in research and teaching	45,000	–	No change 10 years
Active in teaching and services (non-research)	33,000	–	
Research Fellows and Research Assistants	33,000	10,000 (30.3%)	Up 44% over 10 years Experienced manpower available after primary contract ends
U.K. origin PhD output per annum	6,000	3,000 (50%)	1.8% into Academic posts 13% into Postdoctoral research (contracts with various sponsors)
Masters degree, Diploma, other post-graduate qualification, per annum	15,000	3,000 (20%)	
INFRASTRUCTURE			
Example HEI[1]			
Insurance value of property and contents (used for Teaching and Research)	£700m	–	Equivalent to £10,000 per staff member
Proportion of income (all sources) deployed into infrastructure and running costs per annum	48%	–	
JIF funds[2]	£30m	–	For new generation Engineering Science building

[1] UCL — a typical large, research-led HEI
[2] JIF — Joint Infrastructure Fund – a £700m research support scheme funded in 1999 by Government and Wellcome Foundation, aiming to help reduce the scale of the "funding gap". JIF bids are made competitively and will fund capital projects (buildings and/or equipment) relevant to research activities.

FOCUS ON RESEARCH

Types of research activity

5–06   Research in HEIs is predominantly fundamental, basic research (sometimes called pre-competitive) or strategic in nature, the largest proportion of which is supported by the Research Councils and United Kingdom Charities. However, the amount of contract research has been increasing in recent years and this may be basic, applied, or, in some disciplines, developmental research.

Sponsored research

5–07   Sources of income   There are several thousand different sources of research support, some quite small in award value (*e.g.* a few thousand pounds to support a special study in a non-laboratory based subject), some very large (*e.g.* a million pounds or more to support a multi-institute initiative in a new area of science). Table 2 below indicates the most significant groupings by type of sponsor and level of awards made to the HEI sector. More detail on some of these sources is provided in other Chapters. Comments on specific schemes are given at paragraph 5–16, below. Although the proportion of income deriving from Charities, Business and the European Union sources is rising compared with government sources, Table 2 shows that all sources make an inadequate contribution to indirect costs of research, especially Charities, which pay almost no indirect costs across a very large volume of projects. The European Union contributes a maximum of 20 per cent of marginal costs as overhead, with some schemes contributing much less than this, the average overall being 16.4 per cent. The effect of this problem and the recent initiatives to redress it are discussed in more detail at paragraphs 5–47 to 5–54.

5–08   Table 2.   Sources of research sponsorship in HEIs.

Source	Income £M	%	Contribution to Indirect Costs £M	%[1]
OST Research Councils	439	32	90	25.7
U.K. based charities	310	23	6	2.0
U.K. central government, hospitals	238	17	37	18.5
U.K. industry, commerce	155	11	31	25.2
E.U. government bodies	110	8	16	16.4
E.U. other	19	1.4	2	11.6
Other overseas	60	4.4	9	18.6
Other sources	34	2.5	2	6.7
Total 1,365			**193**	**Average 15.6**

[1] As % of total revenue. This contribution covers infrastructure, running costs and administrative support costs and should be approximately 51% of the total.

Note the traditionally low contribution from Charities, an area where the volume of contracts is increasing and therefore contributing most to the "funding gap".

There are some facts which may not be well known outside HEI circles, an understanding of which will particularly benefit organisations entering into collaboration with HEIs for the first time.

5–09 Staffing In sponsored research contracts the work is typically carried out by Research Fellows (postgraduate) and Research Assistants (graduates) specifically recruited for the purpose on short term (two to three year) annually renewable appointments supervised by academic staff acting as project leaders. Experience shows that appointments offered for less than one year do not attract quality staff. This impacts on certain terms and conditions sought by HEIs in research contracts, such as the sponsor agreeing to pay any unavoidable employment costs in the event of early termination of a contract.

5–10 Cost considerations Most research support schemes depend on the concept of HEIs providing the "well–founded laboratory", including core research staff, supporting staff (technicians and administrative staff) and infrastructure, with the sponsor providing funds for the additional or marginal costs of the collaborative work. The price to the sponsor covers at least the direct marginal costs of Research Fellows/Assistants, the associated non-staff project costs (equipment, consumables, travel, etc.) and some or all of the indirect costs (usually recovered as a percentage of the staff costs). The HEI may also charge the costs of academic staff time (direct and indirect costs), especially if those staff are conducting part of the programme rather than supervising only. Where the contract has restrictions on publication of results or in other ways is lacking in benefits to academic work, it is usually priced at full eligible costs, including academic staff time and full indirect costs, with an added premium or profit margin which is reinvested in the department doing the work.

5–11 Matchmaking Matchmaking a researchers' funding needs with the right potential sponsor(s), and the corollary whereby a sponsor's research need is matched with the right academic expertise, has become something of a profession in its own right. Most research-led HEIs have specialised administrators to assist staff and outside organisations in the matchmaking process and many produce regular newsletters listing current opportunities, new schemes, reminders of deadlines for bids, tips on preparing submissions, and sources of further information and advice.[4]

5–12 Nature of sponsored research The type of sponsored research most favoured by HEIs may take the form of a "push for new knowledge" to solve current problems or to underpin areas of perceived importance in the next decade or so (as in the Foresight Programmes of the Research Councils). It may also take the form of collaborative work with one or several partners each contributing their particular expertise towards a focussed applied outcome, as in European Union research contracts and industrial consortium contracts.

5–13 Industry supported work When industry or commerce commissions research the costs may be funded entirely from an organisation's own resources, or may be covered in part by government schemes, for example when the research is too speculative or long range for industry to fund it alone.

---

[4] A sample of UCL's "Research Funding Opportunities" monthly newsletter, which is sent to all staff, is available by request to the author.

A minor proportion of contract work is of a type where the HEI performs tasks largely set by the sponsor, normally tied to specific deliverables and/or milestone dates and requiring application of the academic expertise and/or facilities without necessarily any research component to the work. Normally such work would be on a full cost reimbursement basis at least. One example is clinical trials which may be conducted in a University Hospital Medical School for a pharmaceutical company; publication of the results is not usually possible and there may be little or no other spin-off value to the HEI, but it is essential that the work be done in a clinical environment.

Speculative research and scholarly work

5–14   Research in HEIs also includes a small but vital proportion of own-funded research, that is speculative or scholarly work carried out by established staff and not funded by any specific sponsor.[5] The internal resources for this derive mainly from the annual grant provided by the relevant Funding Council (England, Scotland or Wales). Such research represents an important aspect of "academic freedom" to pursue ideas to see where they lead, which in turn underpins the ever-changing knowledge base in the HEI sector. Such work often leads to innovative new lines of research which then attract additional project funding from Research Councils, industry or other external sponsors.

Postgraduate research

5–15   An integral part of academic staff research activity is the design and supervision of Postgraduate (PhD) projects and this is significant in three respects.

(i)   PhD projects are usually an extension of supervisors' own areas of basic research and often open up new opportunities for subsequent applied work of interest to industry. These projects are supported from the Research Councils' budgets (*i.e.* part of the dual support system).

(ii)   Of the fresh doctorates who gain employment outside the HEI sector, over 70 per cent move into industry and commerce and are therefore a key component of the technology transfer process. They also help to ensure that companies have the "recipient skills" they need to capitalise on new developments in technology.

(iii) With approximately 11,000 projects each year being concluded and published (in thesis form), postgraduate research represents a significant contributor to the national research effort.

HEI PERSPECTIVE ON SELECTED RESEARCH SUPPORT SCHEMES

Industrially oriented PhDs

5–16   A long established scheme for supporting industrially relevant PhD projects is the Co–operative Awards in Science and Engineering (CASE) scheme[6] and in 1998

---

[5] The available time for established staff to conduct own-funded research has been eroded over recent years, to a degree which has caused concern to be expressed by industries which are dependent on the basic research output from HEIs, most notably the pharmaceutical and biotechnology industries. It has been calculated that the level of annual block grant left over after salaries have been paid has now fallen to between £2,500 and £6,000 per academic staff member, which has to cover infrastructure and other support costs.

[6] In standard CASE awards the Department or University concerned applies to "convert" one or more of its annual quota of Research Council postgraduate studentships into CASE awards once it has identified

about 400 United Kingdom companies made use of it. In return for relatively minor financial contributions from the company towards project running costs and a top-up to the postgraduate student's maintenance grant, research of relevance to the company is pursued. In some cases this leads to the postgraduate later being employed by the company, either to further develop the original project or to work on new areas. CASE awards are sometimes used as a low cost "entry level" scheme to promote first-time links between an HEI and a company. The relationship often grows into longer term contract research, consultancy and/or the provision of training.

LINK projects

5–17 Another well established scheme, LINK, brings together HEIs and one or more companies in a collaborative programme where the DTI will fund a substantial proportion of the HEI's costs if the companies contribute resources in the form of cash and kind and/or parallel in-house development work relevant to the project goals. The average size of a LINK award is £430k. Special help is available to first-time industrial users of the scheme.[7] HEI research staff benefit from exposure to the commercial focus of a LINK project. In some cases, areas requiring new fundamental (pure) research are discovered during such applied work and the HEI may take these up under different sponsorship (*e.g.* Research Council grant).

Teaching Company Scheme (TCS)

5–18 TCS is the longest running scheme (25 years old in 1999) promoting interaction between industry and HEIs (and certain other technology providers) and has become highly respected by both academic staff and industrial users. There are currently about 700 projects operating across the United Kingdom and in 1999 the DTI increased the level of funding by a further £10 million to raise the number of projects to 1000 by 2001. TCS is especially designed to assist companies to prepare for and implement technological change, for example in the development of new products by harnessing the latest advances in science and engineering, or in the introduction of improvements to the manufacture or marketing of existing products.

From the HEI viewpoint, TCS features many good practices which could be applied or adapted to other schemes with great advantage. These are outlined below.

Consultant support

5–19 TCS has its own Directorate (TCD)[8] with about 20 part–time consultants who operate regionally and assist the collaborating partners at every stage of the process, from preparing quality bids to running projects to maximum benefit. Decisions on proposals are made quickly; the consultants rate the projects and decide which to recommend for funding. The consultants participate in review meetings and are often the first to identify valuable intellectual property arising from projects.

---

a collaborating company and agreed a mutually acceptable project. In a useful variant of the scheme, certain companies can also apply in advance for a quota of "Industrial CASE awards" where the company decides internally which research areas it needs to pursue collaboratively and then finds the most relevant HEI partner for each project.

[7] LINK is now organised into a Directorate divided into subject groups with published research priorities, which helps a company to determine how the scheme fits in with its business strategy. The Directorate also provides excellent guidance and support to prospective partners in the pre-bid stage, which helps to reduce the chances of rejection, thereby avoiding the waste of time and other resources when bids are aborted.

[8] TCD also manages two related schemes — PTP (Postgraduate Training Partnerships) and CBP (College-Business Partnerships) though these are on a smaller scale than TCS.

Training at all levels

5–20 TCS projects encompass not only research and development work but also the training of the company's existing personnel, including top management, and the grooming of the Research Associates appointed under the scheme for possible employment in the company at the end of the project, thereby ensuring continuity of technical contributions. There is a separate budget for the training of Associates. The training is provided by TCD in important complementary areas not directly related to the science (*e.g.* presentation skills, business administration, etc.) and by the HEI partner (*e.g.* customised course work in relevant technical fields). Some of the latter training may be purchased from and undertaken at a different HEI if the partner HEI is unable to offer it.

Dual supervision of research associates

5–21 Each Associate recruited to the project works primarily on site in the collaborating company but is an employee of the HEI with full access to library, collegiate support from other departments, and other services (*e.g.* central computing). The Associate has two formal supervisors, one academic and one industrial. The regular review meetings during the life of the project are attended by senior company managers and the Chairman/MD is also obliged to attend.

European Union support schemes

5–22 A comprehensive description of the many European Union schemes which support collaborative research and development projects across Member States and certain non-Member States, is provided in Chapter 4. The most important schemes for HEIs are the Framework Programmes for research and technology development. A new Framework Programme is launched every four to five years, each having its own budget and research priorities. It usually contains some improvements to the administrative and/or contract terms following extensive consultation over procedures, success rates, and problem areas identified by contractors in the previous Framework Programme. For example, in Framework Programme 5 a number of new features were introduced which are of particular benefit to HEIs:

(i)   The cost of protecting new intellectual property arising from European Union projects can now be recovered, albeit from a different budget and by separate application for the funds. This enables HEIs to protect new discoveries before attempting to interest industrial licensees. In return for this provision, contractors have to include outline exploitation plans in their original bids, which is a useful exercise in itself and helps to alert all the research partners to be constantly on the look out for exploitable results and inventions.

(ii) A proportion of the additional costs incurred by a contractor who takes on the co-ordinating role in multi-partner projects can now be recovered. Unfortunately, where HEIs play this role, this benefit cannot be realised if they use experienced academic staff and/or support staff in the department to carry out co-ordination tasks since only marginal costs can be recovered; the academic and support staff costs are not eligible. However, if the HEI uses part of the time of a person recruited for the duration of the project to carry out these tasks, the costs can be recovered.

All projects are multi-partner and inter-state and they often forge long lasting relationships between research groups who would probably not otherwise have had the opportunity to collaborate. Also, the European Union supports research in areas which cross national boundaries, such as solving environmental problems, and which no one Member State is likely to support adequately on its own.

Cost recovery on E.U. projects

5–23 Generally, HEIs are not eligible for European Union funding on the so-called "full costs" financial model whereby up to 50 per cent of the full costs of carrying out the work are reimbursed. Full costs include the costs of all staff, including (in HEIs) the time of permanent academic staff, and the true indirect costs (overhead) calculated as a percentage of the staff costs at a rate which must be agreed with the European Commission and is subject to audit. The main reason why HEIs are not eligible for this model is that staff do not keep conventional records of how they spend their time on different activities, or if they do keep limited time sheets the data are not integrated with the internal costing system, as happens in industry and certain other organisations. This means that HEIs are currently unable to calculate to the satisfaction of the European Commission the proportion of their costs which are attributable to research activity only, which in turn means that the labour rate and overhead for research cannot be substantiated. The same problem exists in universities in most other European Union countries and so the European Commission introduced the "marginal costs" financial model (for HEIs only) whereby 100 per cent of the marginal costs of the work are reimbursed, including salary costs of non–permanent staff but excluding academic or other establishment staff, and 20 per cent of total costs is paid as a contribution to indirect costs. However, as will be seen from Table 2, the actual indirect costs rate realised from European Union contracts is only 16.4 per cent of total revenue and is a major factor in the appearance of the "funding gap" in HEIs (see paragraph 5–50).

Non-financial benefits of participating in European Union contracts

5–24 Despite the relatively poor financial return from European Union contracts, United Kingdom HEIs continue to bid for contracts and manage to find the financial subsidy from other resources because of the scientific and other benefits which can result from participating in multi-partner projects. After contracts have ended the scientific links between partners often continue and may grow. Also, important opportunities arise for participating scientists to join the various committees and panels responsible for shaping the content of the next Framework Programme, a role which is valued highly by academic scientists.

E.U. CRAFT program

5–25 This program, although not aimed at HEIs, deserves special mention. CRAFT provides up to 50 per cent of project costs where companies from two or more E.U. Member States are involved. Where the companies are not able to carry out research themselves and use a collaborating HEI for this purpose, the European Commission will sometimes pay 100 per cent (instead of the normal 50 per cent) of the HEI's charges. This is especially helpful where the companies are SMEs (Small and Medium Enterprises) which would otherwise not be able to afford the research subcontract with the HEI.

British Telecom funding

5–26 BT's Short Term Research Fellowship scheme is a good example of how larger companies and HEIs can stimulate links between research staff in both organisations, to mutual long term benefit. The scheme is open to academic staff and research assistants working in any field relevant to BT's interests and allows a short project (usually six weeks' duration) of mutual interest to be carried out in one of BT's internationally renowned research centres. In 1999, 40 Fellowships were supported covering various technical fields, management problems, economics, art and design, and linguistics. BT has several other support schemes running in parallel, including longer term collaborative research projects and also grants of up to £100,000 for general university development.

PRACTICAL ARRANGEMENTS

Basic terms and conditions in research agreements

5–27 The majority of research funding schemes have been established long enough for standard terms and conditions to have evolved and both HEIs and sponsors now operate under terms where generally only the resources, time scale and costs are negotiated. With most of the schemes, if the published terms and conditions are not acceptable to an HEI and experience shows that the sponsor is not responsive to negotiation of changes, no bid is made to that sponsor. Some Government Departments, for example, pay unacceptably low indirect costs and, unless there are strategic reasons for the HEI to subsidise the work or other benefits accrue out of the collaboration, HEIs may not submit proposals.

IPR arising from externally sponsored research

5–28 Where intellectual property arises from research with an external sponsor, the terms of the research contract usually provide a pre-determined route and method for transfer of the intellectual property to the sponsor. The terms finally negotiated with sponsors vary from case to case, largely depending on whether the sponsor has paid full costs or only partial costs (as would be the case in most collaborative research).

The starting position for most HEIs is to attempt to retain ownership of arising intellectual property, at least in the first instance, and to agree privileged access by the sponsor in the form of either assignment of ownership of intellectual property which the sponsor identifies as commercially important or a licence to use the intellectual property, on agreed terms.

In the case of a licence, the HEI recognises that it must initially bear the responsibility and expense of protecting the intellectual property (*e.g.* by patent applications). However, the HEI will normally expect the licensee to take on the defence of patents and action against infringers, based on their commercial judgement, and also to cover the costs involved on the basis that the HEI does not have a source of income to cover such costs since it has no manufacturing activity. The HEI may have to compromise by accepting a lower level of royalty in return for the licensee bearing the patent and related costs.

First time contractors

5–29 Companies or other organisations that have not funded contract research in an HEI before, often present the HEI with their "standard agreement", which may contain a number of features which the HEI finds unacceptable in the circumstances. For example they may contain severe restrictions on publication, or intellectual property terms which do not distinguish between ownership of "foreground" property and "background"[9] property, or termination clauses which do not reflect that HEIs

---

[9] See Chapter 20 for an explanation of these terms.

normally have to recruit Research Fellows/Assistants for a minimum appointment of one year, renewable annually thereafter, with the consequent financial commitment to pay the salary for this period. Negotiation of terms can become rather protracted in such cases.

Company schemes

5–30 Many companies have organised their procedures for setting up collaborative arrangements with HEIs in a way that ensures the best possible match between HEI research ideas and the companies' needs, which clearly suits both parties. Some companies define the most relevant areas of research each year by internal consultation with their directors of research and then proactively search out the most appropriate HEI scientists in the field and negotiate programmes of work. Usually such companies do not encourage spontaneous approaches from academic staff to fund work. On the other hand, other companies take a different view and encourage academic staff to submit proposals in broad areas of interest to the company, believing this leads to more innovative approaches to research. In both approaches the advantage is that the company will have set aside a budget for external collaborations.

Problem areas

Competition for funds and the cost of aborted proposals

5–31 Virtually all research income is now won by fierce competition on the technical quality of proposals and value-for-money. HEIs compete with one another in peer-reviewed applications to Research Councils, Government Departments and some collaborative research programmes arranged by industrial companies. In European Union research programmes the competitors are drawn more widely still from organisations across Europe. The success rate in terms of the number of "bids" converted into contracts can be as low as one in 20–30 "bids", but nevertheless a high proportion of the "failed bids" are of high quality.[10] In several of the Research Councils many of the rejected proposals are nevertheless rated very highly (alpha or alpha plus) by peer review and in years gone by these would have received automatic funding, the remainder of the budget going to good but lower rated projects. There are now insufficient funds to support all alpha-rated proposals, which eventually discourages many staff from applying at all. The time and resources wasted when proposals are not accepted has become a problem for HEIs and sponsors alike.

FOCUS ON TECHNOLOGY TRANSFER

5–32 Many HEIs have invested very substantially in active world-wide marketing of the intellectual property rights (IPR) arising from their research. This proactive stance has recently been encouraged and expanded by three new financial support schemes from Government: Challenge scheme 1999; Science Enterprise Challenge 1999; Reach-out Programme 2000 (see paragraphs 5–39 to 5–42, below).

---

[10] European Union procedures help in that submission of short outline proposals is encouraged and feedback is given on the degree of relevance to programme objectives and on any obvious weaknesses. Assistance is also given with finding partners in other European Union countries and in forging inter-laboratory links where similar or complementary work is ongoing. Without this support the failure rate would be even higher than it is at present.

Types of technology transfer from HEIs

5–33 Technology transfer, from an HEI's perspective, is a multifaceted process comprising the transfer of trained people and their skills and know-how, conclusion of licences granted to companies for use of specific intellectual property rights (IPR) and, increasingly, the creation of start-up companies in which the HEI and staff inventors may have a shareholding and/or directorships on the board of the company.

The different facets of this process may occur on quite different time scales, largely depending on whether the HEI has an existing relationship with a relevant company and a fast-track route to senior management, or is marketing the intellectual property to selected companies or companies more widely, in the hope of attracting potential licensees.

The act of searching for licensees and obtaining feedback from companies which show interest and undertake an evaluation of the commercial potential of the intellectual property, are often the primary sources of information available to an HEI about the real value of the property they hold. This helps when prioritising the use of staff time and central expenditure on different cases, for example when deciding on the level of investment in patenting or the costs of legal advice on draft agreements.

Start-up companies

Intellectual property assets

5–34 In start-up companies set up by HEIs the main (or only) asset owned by the company is usually a package of intellectual property, typically comprising patents and specific know-how or rights to future know-how in the HEI. Importantly for the HEI, the company is normally expected to bear the cost of patent prosecution and other patent-related expenses. It will also take responsibility for patent management and litigation. If the company has insufficient resources to do so, which is often the case, the HEI may agree to bear the costs for a while but in such cases will normally retain ownership of the intellectual property and give the company a licence. The licence would normally be exclusive for an agreed period with provision for termination of exclusivity (or of the licence) if the licensee fails to meet performance criteria such as development "milestones". There may also be provision for the licensed intellectual property to be assigned to the company later, although much will depend on the financial state of the company at that time and its perceived capability to exploit the technology successfully.

Access to HEI resources

5–35 In addition, start-up companies usually enjoy preferential access to HEI staff and laboratories under agreements of various kinds, such as contracts for staff appointments to the board, further research and development, consultancy, or testing work. The terms of shareholder agreements and/or board appointments may include "ring-fencing" of ongoing HEI research output in specified fields, especially where the academic staff concerned hold shares in the company. The company may enjoy at least first refusal rights to any new intellectual property arising within the "ring-fence". Where the HEI itself has a substantial shareholding interest in the company, the company is sometimes given automatic ownership of new intellectual property in the field, either for no additional consideration or in exchange for an increased shareholding or other consideration, such as a royalty on product sales.

Licences back

5–36 Normally there is provision for the HEI to be given a licence of the intellectual property assigned to the company in order to pursue any specific field of application in which the company has no commercial interest, sometimes including revenue–sharing with the company if the HEI successfully sub-licenses in that field. Royalty–free licences to the HEI for use of such intellectual property for research purposes[11] is a normal condition, subject to possible restrictions on its use in research where sponsorship by another party is contemplated.

Return on investment

5–37 It can be several years before the HEI starts to obtain any return on its investment in start-up companies. In many cases the commercial potential behind the technology is never realised and the company has to be wound up. Nevertheless, it is important that attempts to commercialise new discoveries continue for the sake of the major technological breakthroughs which do emerge from time to time.

Support scheme

5–38 A particularly useful support scheme for start-up companies is the DTI SMART Award, which provides up to 50 per cent of the costs of early development work, "proof of concept" demonstration projects, and the costs of patenting or other protection measures.

Recent funding schemes in technology transfer

5–39 During 1999–2000 three new support schemes directed at various aspects of technology transfer from HEIs were launched.

University challenge

5–40 This scheme was set up as a competitive bid scheme and features inter-institutional collaboration on technology transfer. The funding, which derives from the DTI and the Wellcome Foundation, aims to provide researchers in HEIs with access to seed capital to fund the early stages of commercialising exploitable research results and fills a long-standing gap in the innovation process. In March 1999 £45 million was awarded between 15 winners which included specific groupings of 27 HEIs and eight Research Council institutes.

From an HEI viewpoint the main concerns evident in the first year of operation of the scheme are that: (i) the HEI has an obligation to more than match the external funds from its own resources and, unless it has private funds which it can use for this purpose (*e.g.* endowments), it may be endangering its charitable status if it uses any of its public income for what may be classed as speculative commercial investment in technology transfer; (ii) restrictions apply to the Wellcome money which can only be spent on new research associated with the field, which means that only technology transfer projects linked with further research are eligible for investment. There are also serious concerns that there will not be enough United Kingdom venture capital providers to cater for the larger investments required at the next stage of development of the more successful projects.

---

[11] To the extent that this is deemed necessary.

Science enterprise challenge

5–41   This scheme was also set up as a competitive scheme and awarded £25 million in September 1999 between eight winners involving groupings of 18 institutions. The funding supports new Centres in HEIs which provide a focus both for teaching scientific entrepreneurship in science and engineering curricula and commercialisation of research or new ideas. Whereas the Challenge fund addresses the "equity gap" in technology transfer, the Enterprise scheme tackles more the "management gap".

Both these schemes were threefold oversubscribed with quality bids, which demonstrates the potential still existing within the system for more technology transfer to happen, given the right seedcorn funding.

The reach-out fund

5–42   The first awards in this scheme were made in November 1999 to 85 HEIs which shared £60 million over a four year period. The intention is that all HEIs will eventually draw some support from this new *permanent* third stream of funding which complements the existing "Dual Support" system for teaching and research. The aims are to enhance the capability and infrastructure in HEIs for responding to business and community needs, creating a culture and climate within HEIs where working with businesses and the wider community becomes a respected third core activity alongside teaching and research. This covers a range of needs including student placements, professional training, consultancy advice, secondments of staff to/from industry, and technology transfer of the kind supported by University Challenge and Science Enterprise Challenge schemes.

Difficulties with technology transfer

"Not-invented-here" syndrome

5–43   Whilst some companies actively pursue technology transfer opportunities in HEIs, others do not think of approaching HEIs when they need to innovate, or they rarely look outside their own resources for innovative ideas. Many companies prove to be unresponsive when HEIs target them with specific licence opportunities in their fields of interest, especially companies in the United Kingdom. Companies in Japan and the United States tend to be much more responsive. Many actively search for new technologies world-wide with a view to licensing-in and give this activity a high priority, often headed up by a board member or a person with full authority to negotiate deals on behalf of the company.

Better practice – more proactivity

5–44   The Pharmaceutical/Biotechnology sector is generally more proactive towards HEIs in their search for technology transfer opportunities than is the case in other sectors, such as Engineering. For example, Bio-Pharm companies often maintain permanent links with selected HEIs world wide. They may have formal research support schemes to fund projects which, although not central to their commercial programme, are of general interest and may lead to useful know-how or patents. Many of the larger companies openly state their dependence on basic research in HEIs for generating the breakthrough discoveries which, history has shown, can lead to and underpin major commercial developments.

New incentives – new hope

5–45   The relatively poor performance in exploiting inventions in the United Kingdom
has been recognised and addressed in a recent Government policy document which
aims to stimulate a culture of innovation and wealth creation.[12] One of the schemes
now operating under this initiative[13] features steady long term funding for the majority of HEIs, whereas previous schemes have featured one-off injections of funding
which were bid for competitively and supported a few institutions only. This new
scheme will help to ensure that knowledge transfer in its various forms has a higher
profile and status within all HEIs and is better resourced and managed than hitherto.

Costs of early stage protection of IPR

5–46   A major problem for HEIs is that there are very few schemes which assist with
the initial costs of protecting new discoveries. According to Funding Council rules,
funds derived from the block grant should not be used for such "speculative investments". This is a classic "pump-priming" problem. Once the HEI has created a net
profit from its technology transfer activities, a portion of the proceeds can be reinvested in a fund to cover future patenting and related costs, but this may take several years to achieve. Most HEIs are therefore severely limited in their outlay on
patenting which restricts the size of portfolio they can support. There is less of a problem if the value of the new technology is already proven or self evident. In such cases
companies will often reimburse some or all of the costs as part of a right to an exclusive evaluation period under a licensing option agreement. In other cases where the
commercial value of the technology is unproven or attracting licensees proves to be
unduly slow, the patent costs cannot be recouped and they increase substantially after
the provisional year of cover. The costs escalate when cover is applied for in relevant
territories world-wide. On a reasonably sized portfolio (*e.g.* 30–40 patented technologies) the "steady-state" outlay on patents alone, where in a given year some existing
projects will fail and some new ones will be supported, can easily reach £0.5 million.

THE FUTURE ENVIRONMENT FOR FUNDING SCHEMES

Factors influencing change

5–47   The research and development system, including the wide variety of schemes
for supporting HEI work described in other Chapters of this book, is undergoing a
period of fundamental reform, largely driven by external pressures from government
or its agencies. In particular, the conventions for costing, pricing and valuing of grant
and contract research are having to change in the wake of an enhanced awareness of
the full costs of research and a demand for greater transparency in the uses of the public funding of HEIs. Inevitably there will be many tensions associated with the reforms,
but both HEIs and sponsors should benefit in the long term if they work together to
facilitate change as rapidly as possible.

   These matters are already beginning to affect the perspective of HEIs and their staff
to research support schemes including the well established ones. By the same token,
these issues will cause all research sponsors, publicly funded and non-publicly–funded,
to review their current policies for collaborative schemes with HEIs, although at the
time of writing only a few organisations have begun such reviews.

---

[12]   White Paper 1999: "Our Competitive Future: Building the Knowledge-driven Economy".
[13]   Reach-out scheme – see para. 5–42.

Recent trends in research funding

5–48   The figures for a typical research-led University show that research grants and contracts activity increased by 9.4 per cent between 1994–5, 14 per cent 1995–6 and 26 per cent 1996–7. Over the same period core income for teaching and research from the Funding Council increased by 18 per cent, 17 per cent and then by only 1.8 per cent indicating that activity levels and consumption of resources have increased whilst core income for staff and infrastructure has not kept pace — a rundown situation for the HEI concerned if the trend continues. Figure 1 illustrates the steady decline in funding from Funding Councils to the HEI sector as a whole.

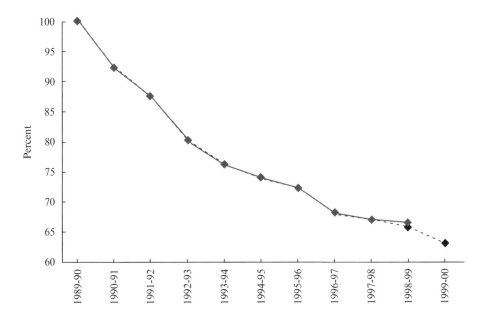

5–49 Figure 1.   Reduction in Unit of Funding to HEIs.

The size and nature of the "funding gap"

5–50   The gap between actual costs to the HEI of sponsored work and the income received from those sponsors has been widening over the last five years or so, caused primarily by unwillingness of sponsors to contribute sufficiently to actual indirect costs (overheads) in a period when Dual Support income for the "well-founded laboratory" has been declining.[14] HEIs have largely covered the deficits by not investing in modern equipment or infrastructure. The extent of the funding gap, variously estimated between £130 million and £720 million *per annum*, poses a serious threat to the national research base. The Funding Councils' own estimate of the gap for 1996–97 (approximately £480 million) could be bridged if, on average, sponsors paid 25 per cent more than at present towards the true indirect costs of research. In the present financial realities, the best performing HEIs in terms of total income from sponsors may simply be nearest to an operating deficit through under-recovery of costs.

The gap was eased by an emergency payment of £1.43 billion from the government's 1998 Comprehensive Spending Review but this was contingent on the sector meeting

---

[14] See also Table 2.

two new requirements: (a) greater transparency and accountability of research[15] and (b) more appropriate pricing of external contracts with all sponsors, including Research Councils and Charities, in order to recover the true costs of the work and to ensure that the funding gap remains permanently closed.

Effect on pricing of HEI research

Determination of true costs

5–51   One of the benefits from the Transparency Review will be a common approach throughout the HEI sector to the determination of the true full costs of a research project, before bids are made to external sponsors. In particular the labour and overhead rates applicable to different grades of staff employed on the project, including academic staff, will be known accurately and will have been reported to and checked each year by Government. It should be possible to obtain prior acceptance of these rates with a number of regular sponsors (especially Government Departments). Industry has tended to agree readily the direct marginal costs of projects but has negotiated hard on the level of indirect costs they regard as appropriate for them to pay, a level often as low as 40 per cent of staff costs. Unfortunately, HEIs have erroneously allowed the indirect costs rate to be a negotiated financial variable in a contract, whereas it should be a non-negotiable fixed cost. The argument typically put up by industry for its stance is that a proportion of a HEI's true indirect costs of carrying out the contract work has been supplied by the Higher Education Funding Council for England through the block grant. Whilst this was probably true two decades ago when different rules applied, it has long since ceased to be the case. Nowadays the conditions of award of the block grant stipulate that it is to be used in support of HEI–funded research (*i.e.* where there is no external sponsor) and research which is aided by supplementary grants from Research Councils. It further stipulates that full indirect costs should be recovered from all sponsors, unless there are strategic or other considerations for pricing the work below cost. No research contractor, other than an HEI, permits its indirect costs to be the subject of negotiation with an external client, although it may of course have to justify which of its costs are eligible for inclusion in the price, as happens under Government Accounting Conventions applied to non–competitive contracts with government departments. The contractor may negotiate price and may price the work at above or below costs, depending on economic or strategic considerations, but it would normally not negotiate on the basis of its costs.

The future of pricing

5–52   Negotiations by HEIs with sponsors will in future concern three areas:

(i)   the human and other resources required to achieve agreed research goals in the required time scale;

(ii)   price;

(iii)   value (to both parties).

---

[15] This Review aims to develop a uniform and transparent approach to the costing of research and other activities within HEIs and to achieve transparency in the use of public funds given to HEIs under the "Dual Support" system (see para. 5–02). The full implementation of the Review requirements in all HEIs will occupy at least three years to July 2002. Work began in August 1999 in eight research-intensive pilot institutions who are expected to achieve full cost transparency by July 2001. This initiative will cause the most significant change in pricing of external work for several decades.

Pricing strategy will be informed by costs; if the price first quoted is more than the sponsor can afford, negotiations will centre on negotiating a lower acceptable level of input resources and/or reduced time scale for the contract, *but on the same cost basis.*

Cost-sharing

5–53 If the sponsor still cannot afford the costs of the lowest acceptable resource input, negotiations will shift to the extent to which the HEI can share both the costs (subsidy) and the benefits, such as profits from exploitation of intellectual property. In some cases, such as collaboration with industrial partners, cost-sharing should take into account the value of any indirect contributions from the sponsor, such as the loan or donation of equipment to the HEI, management time spent with research fellows, training of project staff on new techniques. Also, academic departments will have to weigh up the strategic value of accepting a contract at less than cost. For example, they may acquire a significant foothold in a new area of technology. The department will also need to monitor the aggregate level of subsidy across all its contracts since any shortfalls between price and cost will be a charge on the departmental budget, not the HEI centrally. Some contracts should be priced at more than full costs, as is normal in industry and commerce, and the "profit" used towards the subsidy of other work.

Value

5–54 In contracts with industry there will need to be more emphasis on assessing the value of the project outputs, as perceived by the two parties, as an aid to determining a fair price. Price may be agreed as "cost plus" (a term common in industry where the contract price is based on actual costs plus a premium or profit margin), break even (equal to actual costs) or cost-shared. Where companies value highly the confidentiality of results and require the academic partner not to publish the work, this may harm the department's position in the next Research Assessment Exercise causing a direct loss of income from the Funding Council over a number of years. The HEI's price may be "cost plus" in such a case. Where the academic department values highly the access given by a company during the project to, for example, special software or other proprietary information and the right to use it for general research purposes, the HEI may be prepared to subsidise the work by agreeing a price which is below full costs.

The changes to the way projects are costed and negotiated predicted in the above scenario will require new ground rules and procedures to be developed, both in HEIs and sponsor organisations.

An implicit caveat in the continued public funding of the "well-founded system" will be that all external users of the system must pay a price which ensures that the true costs are recovered. Only in this way can the quality of research outputs from the system mentioned at the start of this Chapter be retained, and the alarming "funding gap" which threatened research infrastructure in the late 1990's, prevented from recurring.

CHAPTER 6

THE INDUSTRY PERSPECTIVE

Dr Campbell Wilson, Director, Discovery Alliances, AstraZeneca

6–01 This Chapter gives an industry view on the value of, and issues related to, collaborations. Specifically, the perspective provided is from the pharmaceutical industry, a business sector which has a long history of research collaborations, in particular with academia.

WHY DOES INDUSTRY COLLABORATE?

Aims of industry

6–02 Most major pharmaceutical companies aspire to annual growth of at least 10 per cent. To achieve this, it is likely that the number of successful product launches per year will have to rise significantly. These product targets demand an expansion of the development pipeline which is likely to be accomplished by both an increase in the companies' own research output and, significantly, an increase in in-licensing of compounds in clinical development. Furthermore, the trend to increased external interaction is not restricted to the development phase: research collaborations, as an alternative to, or in conjunction with, in-house research are viewed increasingly as essential for the continuing growth of the industry. Across the pharmaceutical industry there is probably not a single company that does not have academic or inter-company collaborations. From start-up biotechnology companies with a handful of staff to multi-national pharmaceutical companies employing many thousands, collaborations are critical to the business.

Variation within industry

6–03 While the degree of collaborative activity will vary between companies depending on the nature of their business, there is undoubtedly, and not surprisingly, a trend to fewer collaborations relative to company size. For example, small start-up companies spun-out of academic research will often be highly dependent upon the collaboration with their parent academic institution. Other collaborations will be formed to sustain the operation of the company during its infancy. But as companies grow they will achieve a critical mass at which point it may be considered that collaboration in a particular area is not necessary as the activity can now be supported in-house.

Different philosophies

6–04 However, we should recognise the distinction between needing to collaborate and choosing to collaborate. There are marked differences between companies in their philosophy and attitude towards collaboration. At one end of the spectrum there are those with an "if we can do it ourselves, we will do it ourselves" mentality. Occupying the middle ground are many companies who, on the basis of careful cost-benefit analysis, out-source certain activities. At the other extreme are the "virtual companies"; companies whose research and development operations are conducted entirely by external partners.

Virtual companies

6–05   The concept of the virtual company is relatively new among United Kingdom pharmaceutical companies, although Vernalis (formerly Vanguard Medica) has carried out drug development successfully for a number years using this business model. The approach has also been adopted more recently by companies such as Alizyme, initially for research phase activities. Many other companies, particularly in the United States, operate on this basis. The perceived advantage of this model is one of financial efficiency; the ability to conduct a significant business operation on minimal fixed costs. The potential disadvantage is, however, lack of control. The companies are highly dependent upon the project management skills of their few employees and, critically, upon the quality of their collaborative partners.

Benefits of collaboration

6–06   Collaborations are generally instigated with the expectation that they will produce valuable data. The data may either be completely novel or of higher quality, higher volume or produced more quickly, or any combination of these, than by conducting the work in isolation. It is difficult to envisage a collaboration where the anticipated benefit is not related to the delivery of results. Other reasons to collaborate tend to be additional, rather than alternatives.

Enhanced credibility

6–07   One such reason may be the kudos associated with research carried out within a highly respected university department and having company employees as authors on publications alongside academic leaders in the field. Publications resulting from collaborative work with quality academics are often held in higher regard than publications emanating solely from a company and can serve as a valuable endorsement of a company product or the company's research in general.

TYPES OF COLLABORATION USED BY THE INDUSTRY

6–08   Collaboration can best be defined as "an interaction between two or more parties involving shared input with the aim of providing mutually beneficial output". There are a number of business models that could be categorised under the general term of collaboration, but which vary considerably in their proximity to the definition above.

Research collaborations

6–09   Research collaboration between a company and a university, or between companies, most closely matches the definition of true collaboration. Generally, both parties will contribute on both a practical and intellectual level to the project. In a collaboration between a company and a university, the company obtains the opportunity to commercialise the output of the collaboration whereas the university may have the ability to publish the results, receive research funding and also further financial return when the company develops a product. Although much less common, this model can also apply to a collaboration between a large pharmaceutical company and smaller biotechnology company, although the structure of the division of benefits may differ.

The subject matter of a research collaboration will usually be basic research in a defined, usually narrow, area. For example, Company X may collaborate with University Y on "The role of endothelin in blood pressure control".

Strategic alliances

6–10 A strategic alliance is a semi-formal arrangement between parties who may bring different skill sets to the association. In a research context, a strategic alliance can be simply a research collaboration on a larger scale. Higher levels of funding are involved and usually more personnel from both parties are engaged in the collaborative work. The subject matter of the alliance is usually broader than for a research collaboration. For instance, Company X may form a strategic alliance with University Y on "The physiological role of endothelin".

Contract research

6–11 Contract research may again involve the interaction between a company and a university or a commercial enterprise, such as a specialist contract research organisation (CRO), but the nature of the interaction will be rather different. The project will generally be of shorter duration than a basic research collaboration and the subject matter and project specification will be tightly defined by the sponsor. The contribution of the contractor will be more at the technical than intellectual level. The sponsor will also want to have ownership of the output, including any intellectual property and publication rights, but will be prepared to fund the work at a premium in exchange for these privileges. A typical contract research study commissioned by Company X may be "The effect of endothelin antagonist X12345 on blood pressure in the rat".

Consortia

6–12 Consortia are multi-party collaborations in which the subject matter is often too expansive and/or expensive for any of the partners to exploit adequately on their own. Participants in the consortium therefore trade off exclusivity against cost. Although a consortium may operate on the basis of co-operation between a series of equal partners, the consortium model seen more frequently is that of the consortium members as satellites interacting primarily with a central hub company or institution, rather than with each other. An example of this model is the Dundee Signal Transduction Consortium in which the pharmaceutical companies, SmithKline Beecham, Pfizer, Novo Nordisk, and AstraZeneca (formerly Astra and Zeneca) fund research into cell signalling mechanisms at the University of Dundee's Division of Signal Transduction and share output of the research on a semi-exclusive basis.

Joint ventures

6–13 Joint ventures are essentially strategic alliances which are more formal in legal terms. Joint ventures tend to be more common for development and commercialisation activities than in research.

Other collaborations

6–14 Three other types of business interaction should also be mentioned although they are not collaboration in its true sense: out-sourcing, licensing and mergers/acquisitions.

Out-sourcing

6–15 While contract research can be considered as one type of out-sourcing, in general the term out-sourcing refers to an entire function rather than specific studies or projects. It also covers areas other than research. For example, Company X may have no, or insufficient, internal toxicology capability and may out-source all toxicology on its compound X12345 to a contract research organisation.

Licensing

6–16 Licensing is a vast and complex topic which is addressed in detail elsewhere in this publication. In the context of a discussion on collaborations, it is sufficient to say that the concept of licensing is somewhat remote from that of a true collaboration. Licensing is the granting of rights to a particular subject matter by Party A to Party B. Although the opportunity for licensing may have arisen from a collaboration between A and B, the term licensing does not invoke or necessitate any future collaboration (although this may be possible or even advisable). As an example, Company X may have insufficient internal resource to develop X12345 and therefore licenses the rights to the future sales of the compound to Company Z in exchange for an initial cash payment and a share of the prospective sales income.

Mergers and acquisitions

6–17 The merger of the businesses of two parties, or the take over of one party's business by another may, under certain circumstances, be the preferred route to gaining access to a particular research capability.

WHICH TYPE OF COLLABORATION DOES INDUSTRY PREFER?

6–18 Not surprisingly, there is no single type of collaboration which is most appropriate to the wide variety of needs of the pharmaceutical industry. Usually circumstances will dictate whether a research collaboration, an out-sourcing arrangement or a licensing deal is most appropriate. However, there are many options within these categories. Let us consider the many models available, but with a primary focus on research collaborations, reviewing the advantages and disadvantages of each in the eyes of the pharmaceutical industry.

Research collaborations

Postgraduate studentships

6–19 The main attraction of either a fully-funded or jointly-funded (see below) postgraduate studentship is the cost, often little more than £15,000 per annum fully–funded. Studentships are also a useful recruitment vehicle for industry, providing employees with industrial experience and a known track record. The drawback of the studentship as a means of collaboration is that, by its very nature, it is a training period in research for the postgraduate and useful output may only be forthcoming in the later stages of the project. The fixed three year timescale of a studentship may also be seen as a disadvantage. In a fast moving field such as pharmaceutical research, three years is now considered to be a long time to perform basic research on a single topic. The subject matter of the student's research may diminish in priority within the company before the project is complete. Studentships are, and are likely to remain, a popular vehicle for collaboration between companies and universities but are best used to fund more speculative research rather than research which applies directly and is fundamental to a company's own discovery programmes.

Research fellowships

6–20   The post-doctoral research fellowship is a frequently used form of collaboration which is highly favoured by the pharmaceutical industry. Post-doctoral collaborations offer an acceptable balance between cost and return. In the United Kingdom a typical post–doctoral fellowship will cost the company £50,000–£80,000 per annum depending on the experience of the researcher, the overheads charged by the university and the nature of the project. For this money the company receives the research output of a high calibre scientist and may find added value from interaction with the researcher's professor or other department members. Flexibility in the duration of the collaboration is another important factor. A period of two years now appears to be most appropriate to optimise output with topicality. Post-doctoral collaborations may be used for basic research to generate ideas which the company may exploit or for research linked directly to the company's own projects.

Joint funding of research collaborations

6–21   An alternative to the full funding of academic research by industry is the sharing of funding with a research council such as the Medical Research Council (MRC), the Bioscience and Biotechnology Research Council (BBSRC) or Engineering and Physical Sciences Research Council (EPSRC) or a charity such as the British Heart Foundation or Cancer Research Campaign.

The attractiveness of joint funding to industry varies depending on the particular situation but is generally determined by the balance of level of funding versus control of the research and ability to exploit the output. Probably the most popular joint funding schemes are the MRC, BBSRC and EPSRC Combined Awards in Science and Engineering (CASE). The main attraction is financial, allowing a company to sponsor academic research at around half of the fully-funded rate.

Another source of joint funding is the E.U. with its Framework programmes. In the past, pharmaceutical companies have not used these programmes to any great extent because of a rather complex application process combined with a concern over a lack of any exclusivity in the exploitation of results generated. The latest programme, Framework 5, appears to be more "industry friendly". It will be interesting to compare industry participation with previous Framework programmes.

Strategic alliances with academic centres of excellence

6–22   A strategic alliance between a company and a recognised academic centre of excellence can be an extremely cost-effective means of building or strengthening the company's research capability in a particular area. Typically, the company will fund three or more post-doctoral researchers within the same university department on a specific scientific topic. There will also usually be a major contribution from the department head, a leader in the field. One clear advantage of this type of collaboration is the access to a significant volume of high quality science. The disadvantage may be a lower degree of control of the work than the company would exercise in respect of its own in-house projects.

Strategic alliance models may differ and another variation on the theme is that of funding several fixed term projects, generated in response to a "call for proposals", within a longer term alliance. This format has the potential to generate many new ideas although these are less likely to be developed as fully as in the previous example where the larger team works on a specific subject. As with any collaboration, value for money of a strategic alliance can only be judged after the event. However, with such alliances costing upwards of £100,000 per annum, the expectation from the industry of real deliverables is considerably higher than from a single post-doctoral collaboration.

Collaborations between companies

6–23   The examples above refer to collaborations between companies and universities. However inter-company collaborations offer an alternative. The ever increasing number of start-up companies in the biotechnology sector, each determined to court a well-established partner with a suitably large bank balance, has led to a wealth of research and development alliances within the industry. The pages of publications such as BioWorld are packed with news of collaborations between major pharmaceutical companies and small biotechnology companies, and increasingly, between small biotechnology companies.

In the past, inter-company collaborations may have focussed on downstream research and development activities leaving early stage target discovery, the identification of novel molecular mechanisms that may be modulated by drugs, as the major preserve of academia, but no longer. The genomics revolution has transformed drug target discovery into a highly mechanised, highly commercialised activity. When a biotechnology company such as Incyte can form alliances, each running into millions of dollars, with around twenty major pharmaceutical companies, the academic community is likely to be concerned about the impact on funds available for university collaborations.

However, the delivery of new targets by genomics is still at a relatively early stage and most pharmaceutical companies will continue, at least in the near term, to hedge their bets and continue to use the more traditional approaches in parallel. Thus, collaboration between industry and academia on target discovery is not under threat, but quite the reverse. The genomics revolution is likely to lead to greater collaboration with academia as the industry seeks to characterise and assign function to the proliferation of new molecular targets that arise from genomics. Target validation, possibly moreso than target discovery, offers academia the opportunity for collaboration with the pharmaceutical industry for many years to come.

Academia or industry as collaboration partner?

6–24   Genomics has been singled out because of of the enormous impact it has made, or is likely to make, upon the pharmaceutical industry. However, most inter-company collaborations deal with other aspects of science and technology. The subject matter of many inter-company research phase collaborations could, in theory, be equally well exploited as an academic collaboration. The converse is also true and many companies offer a research capability that academia has provided historically. So when both a small company and a university appear capable of providing the same collaboration opportunity, who does the large company choose and why?

Cost

6–25   Of the many factors that will affect choice, cost is the most obvious. However, a choice made on the basis of cost alone is rare, since rare is the occasion when precisely the same package is offered by different parties. With cost as the overriding factor, the choice of collaborator in the past would undoubtedly have been academia. But the situation is changing. Coincident with the trend for universities to seek to recoup shortfalls in funding from other sources and charge ever-increasing overheads for collaborative work, is the realisation by biotechnology companies that they need to reduce their funding demands for collaborative research in order to remain competitive. The choice on the basis of cost alone is now less clear cut.

Other factors

6–26 Ease of implementation of a collaboration is an important consideration. Although universities, particularly those with their own technology transfer company, are becoming increasingly professional in their dealings with industry in negotiating collaboration contracts, small biotechnology companies are generally more proficient in this respect. The small company will usually have an experienced business development manager with a direct line of communication to the chief executive. Thus, decision making tends to be fast. In the academic sector, there are often several stakeholders and, in this situation, contractual negotiations will inevitably be slower.

Once a collaboration has commenced, appropriate management of the alliance is essential to ensure that the maximum benefit is derived. Many of the factors that influence the successful running of a collaboration are different from those that influence the choice between a university and a small company. Geographical proximity, historical relationships and the personalities involved will all contribute to the success or failure of a collaboration, be it with a university or with a small company.

Quality

6–27 Finally, and most importantly, there is the matter of results. It is of little consequence that a collaboration is cheap, easy to implement and easy to run if it fails to deliver the anticipated benefits. Unfortunately, however, there are no statistics to indicate whether the small company or academia has the better track record in achieving the aims of collaborations. This is not altogether surprising since measuring successful outcome of a collaboration is often subjective and will therefore remain a matter for the individual to judge based on his/her own experience.

In summary, the large company is likely to find different advantages and disadvantages in collaborating with a university or with a small company and will continue to collaborate with both. The choice of partner is likely to be motivated by the particular circumstances surrounding the research.

HOW DOES THE INDUSTRY ASSIGN ITS COLLABORATIONS BUDGET?

6–28 A major research-based pharmaceutical company will typically re-invest around 15–17 per cent of its sales income on research and development, a remarkably high percentage in comparison to other industry sectors. As an example, the former Zeneca Pharmaceuticals (now merged to form AstraZeneca) spent £498 million on research and development from its 1998 sales income of £2.8 billion. Of this £498 million, £114 million, *i.e.* 23 per cent was spent on research which again is representative of the industry whose expenditure in the development phase runs at around three to four times that of the research phase.

In 1998, Zeneca Pharmaceuticals spent approximately £15 million on research collaborations worldwide which was apportioned in a ratio of almost 3:1 in favour of company collaborations as opposed to academic collaborations. This ratio contrasts sharply with the actual number of collaborations in the two sectors which favoured academia by greater than 10:1. The most frequent type of academic collaboration was the post-graduate studentship, particularly the CASE studentship (see Figure 1, below).

The assignment of research budget to collaborations and the subsequent allocation of that sum externally will vary considerably even among peer group companies. Zeneca Pharmaceuticals' external expenditure of around 13 per cent of research budget is not atypical, although other major drug companies claim to spend up to around 25 per cent of budget on sponsored research. Whatever the current figures, the budget for external research within the industry is likely to rise both in absolute terms and as a percentage of research budget as the pharmaceutical companies strive to fill their research and development pipelines with potential new products to sustain growth.

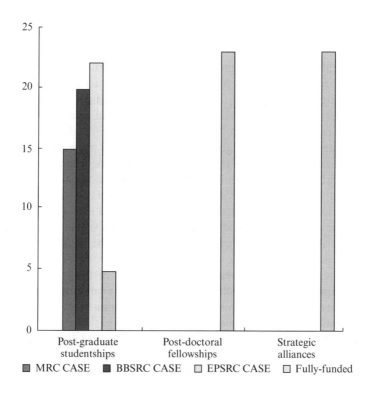

6–29 **Figure 1.** Zeneca Pharmaceuticals' academic collaborations.

WHAT ARE THE MAJOR CONSIDERATIONS OF INDUSTRY AT THE VARIOUS STAGES OF THE COLLABORATION PROCESS?

6–30 Collaborations can be considered in five distinct stages: identification, implementation, operation, termination and post-termination. Each stage raises different issues which must be addressed to gain maximum benefit from the collaboration. The issues will also be very different for the collaboration partners. The major considerations from the industry perspective are reviewed below in the context of, as an example, the collaboration with a university where the company is funding a post-doctoral research fellow.

Identification

6–31 At the identification stage, the overriding concern is that of matching the strategic needs of the industrial partner with the scientific excellence of an appropriate prospective collaborating partner. However, it would be an over-simplification to assume that the choice of partner would be on the basis of scientific reputation alone. Even at this preliminary stage it is important to give due consideration to the practicalities of running the collaboration. Geographical considerations will have some bearing. It will always be easier to operate a successful collaboration if physical proximity allows frequent face-to-face interactions. Another important consideration at this stage is that of a previous working relationship. A collaborative partner who has provided successful collaborations previously will clearly be a less speculative proposition than a new partner. Human relationships should not be neglected.

Implementation

The contract

6–32 Moving ahead to the implementation phase, the focus moves from the scientists to the lawyers, patent attorneys, accountants and contract negotiators. The implementation stage covers the period from the time when suitable collaborating partners have been identified to the time when the collaboration can actually begin. The trigger for the commencement of the collaboration is the signing of the agreement. The importance of the terms of the agreement can not be over-emphasised as it is these which will determine the value of the output of the collaboration to the company.

It should also be stressed that the collaboration should not start until the agreement has been signed by both parties. Often the scientists are keen to proceed and can be frustrated by extended delays caused by prolonged negotiations. However, any temptation to allow the collaboration to start in the absence of a signed agreement should be resisted. Even with the most amicable of negotiations, the possibility of an eleventh hour "deal-breaker" remains. It is not unheard of for a collaboration to have started without agreement having been reached on the terms of the contract and, in some instances, collaborations have finished without contract signature. In this situation, the funding company is unlikely to have all the necessary rights to the intellectual property that has been generated. Thus, at the implementation stage, the major consideration for the company will be to have a suitable agreement in place as quickly as possible, and the suitability of the agreement will, of course, be dictated by the terms negotiated.

Intellectual property rights, costs and control of publications are the most important aspects of the agreement. Their relative importance will depend on the nature of the collaboration. The three are also interlinked: often a higher level of funding will command more favourable terms relating to intellectual property and a greater degree of control over publication.

Intellectual property

6–33 Ownership of intellectual property is not generally an absolute necessity for a major pharmaceutical company when funding basic or "blue sky" research. It is possible for an academic partner to retain ownership of the intellectual property, although the industrial sponsor may well ask for a first option to take a licence to any novel inventions which may arise. When the subject matter of the collaboration is directly related to the company's own areas of research, ownership of intellectual property is of primary importance. Theoretically, it should not be necessary for the industrial partner to own the intellectual property, even if the subject matter is of strategic importance to the company. The collaborating partner could own the intellectual property and grant the company an exclusive licence. In reality, the pharmaceutical industry is generally uncomfortable with such an arrangement where the invention could be an important part of a company's research portfolio. Only the company will truly understand the value of any particular intellectual property to its own business and hence is likely to be more aggressive in patent prosecution or defence than would a third party licensor. Therefore, in areas of key research, the company will negotiate very hard to ensure ownership of intellectual property.

Cost

6–34 Cost, although an important factor, is generally not the major barrier to collaborations between industry and academia, although it remains a significant issue for collaborations with small companies. Universities rarely ask for unrealistic levels of funding. Since there are a vast number of such collaborations, there are many sources of reference pricing. Any exorbitant funding demands are very visible.

However, one potentially difficult aspect of funding academic collaborations is the overhead rate charged by the universities. The debate on overheads is likely to continue for quite some time and many different views will be expressed. While it would be neither correct, nor possible, to provide a single view from the industry, it is unlikely that many companies would be comfortable with overhead rates of 100 per cent or greater that may be asked for collaborative research. As a general premise, the pharmaceutical industry appreciates that it must fund the true cost of a collaboration and not just salaries and material costs. However, overheads on collaborations paid by industry should not be an alternative to government funding in providing more general financial support for a university.

The easiest way to ensure that negotiations around funding do not slow down the implementation of the collaboration is for the partner to provide comprehensive, detailed costings. Negotiations are more likely to run smoothly if the sponsor company knows exactly what it is being asked to pay for.

Publication

6–35 In the past, agreement on publication was probably the single biggest hurdle to overcome in establishing academic collaborations. To many academic research scientists, publication is their *raison d'etre*, and certainly of more personal importance than the possibility of some long term financial reward from prospective intellectual property rights. Happily, things are changing. Many university technology transfer offices now work closely with their academic staff and help balance the academics' need for rapid publication with the university's or funding company's need to delay publication in order to allow patent filing.

The industrial sponsor will usually ask for a delay to publication of 90, or possibly 60, days from notification of a new "invention" in order to file a patent application. A further delay may be requested if it is probable that the research will lead to an improvement of the invention and to a further patent application. In this case, any publication of the subject matter made after the first filing but before the second could be cited as prior art material in the second (and possibly more important) application. In exceptional circumstances, a delay of up to two years may be requested, although it is in the interests of all parties to minimise delay.

Thus, intellectual property, funding and control of publications are important aspects of the contract which must be negotiated before implementation of the collaboration. As mentioned previously, these factors are inter-related. Partial, low level funding of a collaboration may not secure ownership of intellectual property. This may not be a problem for the sponsor company if the subject matter is basic research. Full funding, as in the case of more applied research, should secure ownership of intellectual property and a degree of control over publications. A higher funding rate may be agreed for "contract research" where the subject matter is of critical importance to the company. This level of funding should secure ownership of all intellectual property and complete control over publications.

Operation

6–36 For the duration of the research project, the main concern of the sponsor company is, not surprisingly, that the partner produces the anticipated results. While the success of the collaboration will be highly dependent upon scientific progress, non-scientific matters can also be influential. Fundamental to the success of the collaboration is frequent communication between the two parties. In the case of larger collaborations, a management team with representation from both parties is highly advisable. Collaborations can not be left to run themselves. Just like internal projects, they need to be managed. An arrangement whereby the collaborator is simply provided with funding and there is no subsequent interaction is highly unlikely to achieve the desired benefits. Goals should be set and progress towards those goals reviewed by both parties at regular intervals.

The goodwill of the collaborating partners will be important for the resolution of problems as and when they arise. However, a sound contract, which is fully understood by both parties, may avert many of these problems. A detailed research programme attached to the contract as a schedule, with which both partners are completely comfortable, should go some way to avoid disputes borne out of a misunderstanding of the aims of the project.

Another potential problem area is confidentiality. If the subject matter of the collaboration is of a sensitive nature, the company will clearly want to ensure that information related to the project is not disclosed more widely than is necessary. Again, a close working relationship between the partners will help. But a little time taken at the outset spent discussing and, if necessary, explaining the confidentiality clauses of the contract is time well spent.

Termination

6–37 A collaboration may terminate either as planned, at the end of the funding period, or rarely, it may terminate prematurely. In either case, provision should be made for activities (*e.g.* patenting) initiated during the period of the collaboration that may need to continue after the project has ceased.

Normal completion

6–38 Where the collaboration carries through to the end of the agreed funding period, it is important that there is a clearly defined end to the work which has been funded under the collaboration. Thereafter, a final report should be produced and due consideration given to whether any new intellectual property may have been generated. In a well–managed collaboration where regular progress reports have been written, there are unlikely to be problems at the conclusion. However, where the collaboration may have been less interactive than is desirable, an effort should be made to at least obtain a report upon completion, otherwise any benefits, tangible or otherwise, may be lost to the sponsor.

Premature termination

6–39   Some collaborations may terminate prematurely since the project has failed to reach milestones which were agreed between the parties and defined in the research plan of the contract. In this situation, the parties should be able to part amicably and share any benefits that may have accrued.

Unfortunately, some collaborations end in rather more acrimonious circumstances. Although this is a very infrequent event, it is in the interests of both parties to protect themselves against such an outcome by including contingency terms in the contract. For example, it should be made clear in the agreement that confidentiality obligations and intellectual property ownership arrangements (which should be clearly articulated in the contract) must survive termination. There must also be a common understanding between the parties of what payments are due at the different stages of the collaboration and again this should be explicit in the contract.

Anticipation of the possible undesirable outcomes and making provision for these within the contract should limit the damage, and possibly rescue some benefit from a collaboration which terminates prematurely.

Post-termination

6–40   Although the funding period of a project may have ended, the collaboration between the parties may continue for many years thereafter. For example, it will be in the interests of the collaborator to follow events within the sponsor company should the collaboration have generated intellectual property. Development of products based upon this intellectual property may trigger milestone and/or royalty payments to the collaborator many years into the future. Intellectual property matters may also be grounds for the sponsor company to follow the continuing research activities of the collaborator. The company may have secured rights to any improvements to an original invention and will need to ascertain when such improvements have been made. The company will also wish to maintain the relationship with the collaborating partner where a licence to a patent has been taken and the patent, which by then may underpin an important product, has been challenged by a third party.

Thus, with reference back to the identification stage, the choice of a suitable collaborator can not be overstated: choose your partner carefully as they may be your partner much longer than you think.

CONCLUSIONS

6–41   Historically, collaborations with academia have proved invaluable to the pharmaceutical industry. In recent years the explosion of the biotechnology industry has provided large pharmaceutical companies with another source of collaborations. As advances in science and technology continue to accelerate within both of these sectors, pharmaceutical companies will become increasingly dependent upon external alliances in order to access these new developments.

Collaboration is no longer an option: it is a necessity.

CHAPTER 7

INTRODUCTION TO INTELLECTUAL PROPERTY

Iain Purvis
Barrister, 11, South Square, Gray's Inn

INTRODUCTION

7–01 "Intellectual property" is the collective name for the miscellaneous collection of rights which are the tools provided by the legal system to protect innovation, creative work, goodwill and other intangible assets of a business. Any business engaged in technological research and development should have a strategy for establishing and managing intellectual property rights, since without those rights the fruits of such research and development may become entirely worthless.

The individual rights themselves are explained and discussed in detail in the following Chapters. The Chapters on the individual rights do not deal with ownership or assignment of those rights. These matters are dealt with globally in separate Chapters, because many of the issues raised apply to all intellectual property rights, and we wished to avoid repetition.

For newcomers to the subject, a short overview and summary of the intellectual property rights covered in this book is provided below.

THE RIGHTS

7–02 The term "intellectual property rights" covers the following major rights available in the United Kingdom: patents, copyrights, registered designs, unregistered design rights, registered trade marks, unregistered trade marks (or passing off). It also covers a series of more minor rights covering particular fields: for instance protection under the Plant Varieties Act, the Design Right (Semiconductor Topographies) Regulations, and the Copyright Rights in Databases Regulations. Some of these rights (copyright for instance) cover a huge range of human endeavour, not all of which is within the remit of this book. Others only concern matters outside that remit. It should thus be noted that the ensuing Chapters deal in detail only with those aspects of these rights which are likely to be valuable in the protection of the results of technological research and development.

The status of confidential information

7–03 Classically defined, the term "intellectual property rights" probably does not cover the legal protection of trade secrets or confidential information, since information is not truly a "property right" at all. However, the term is often used loosely, and given the obvious importance of confidential information in the field of technology, we make no apologies for including it as an intellectual property right for the purposes of this book.

Foreign intellectual property rights

7–04 It should be remembered that outside the United Kingdom, other species of intellectual property right exist. Most significantly, the smaller brother of the patent known in many countries as the "utility model", which gives a shorter term monopoly, typically of five or 10 years.

Registered/Unregistered rights

7–05 Some intellectual property rights arise naturally. Others require an act of registration with a public body. Those which require registration (patents and registered designs for instance) are true monopoly rights — in other words, they can be infringed even though the infringing article is produced independently and with no intention to imitate. Those which do not require registration (copyright and unregistered design right for instance) are not monopoly rights, and generally require some intent to imitate.[1]

Distinguishing the rights

7–06 Even amongst the informed public, there is a great deal of confusion as to the distinctions between different intellectual property rights. It is still common to read in the press references (for instance) to "registering the copyright in a name", or "copying a patent". This confusion is even sometimes to be found in businesses for whom intellectual property rights are their life-blood. As part of this brief introduction, it may thus be useful to give a brief synopsis of the defining features of the various rights which will be considered in more detail in the succeeding chapters.

Patents

7–07 Patents are registered rights, that is to say they derive from a document lodged with the Patent Office of the country where they apply. A patent is available in respect of any technological development, capable of industrial application, which is new and inventive. "New" means that no-one has previously made the invention public, including the patentee himself. The patent contains definitions of the invention called "claims". Where the invention is a product, the patent gives the patentee a monopoly over the manufacture, use, keeping or disposal of any product falling within the definition in the claims. Where it is a process, the monopoly is in respect of the use of the defined process or the use, keeping or disposal of any product made directly from that process.

A patent may be granted to the inventor, his employer or their successors in title. It is subjected to an examination process by the Patent Office, but even after grant may be revoked by the Office or the Court if it turns out that it should not have been granted. It is limited in duration. In the United Kingdom and most other countries, the period of protection is 20 years from the date of filing of the application. At the expiry of that period, anyone is free to use the invention.

Registered designs

7–08 Registered designs are also (as the name suggests) registered rights. They are obtained by filing an application bearing a representation of a design (a drawing or photograph), and an indication of the article to which it is applied, at the Patent Office. Again, registration can only be granted for a design which is "new". This means that it must differ substantially from any other design which was published in the United Kingdom before the date of application. It also means that the applicant must make his application before he publishes the design by marketing it in the United Kingdom. Registration is only permitted where the design appeals to and is judged by the eye. It is not permitted where aesthetic considerations played no part in the design process (a piece of machinery, for example). Registration gives a monopoly which covers the application of the design to the article for which it is registered. It also extends

---

[1] Or at least a "causal connection" with the work which is protected by the right.

to variants which are "not substantially different" from what is registered. Being a true monopoly, it can be infringed regardless of whether there has been copying.

A design may be registered by the designer, his employer, the person who commissioned the design, or their successors in title. It is subjected to a cursory examination process, and may be revoked after grant. It lasts for 25 years from date of filing. At the expiry of that period, anyone is free to use the design.

Design right

7–09 Design right is an unregistered right, which gives protection against the copying of aspects of shape and configuration of an article (other than articles which are "artistic works"). It exists whether or not the design could have been registered. There is no design right in "surface decoration", or aspects of shape determined by the shape of other articles with which the article in question is designed to fit or match. The right is infringed by substantial copying of the design.

The right only belongs to "qualified persons" — essentially United Kingdom or E.U. residents, including companies registered in the E.U., who employ or commission the designer to produce the work, or who first market the work in the E.U. It lasts for between 10 and 15 years, depending on when it was first marketed. In the last five years of protection, anyone is entitled to a licence under the design, at a royalty to be fixed by agreement or imposed by the Patent Office.

Copyright

7–10 Copyright is another unregistered right. It protects "artistic" "literary" and "musical" works, though these terms do not connote any aesthetic or other merit. It covers highly technical works such as computer programs and databases. To attract copyright protection the work must be original in the sense that some skill and labour was expended in producing it and it was not slavishly copied from another work. In the industrial field, its relevance (protecting the copying of drawings, for instance) has been reduced by the introduction of unregistered design right. Design right and copyright are mutually exclusive. Copyright prohibits the copying (direct or indirect) of the whole or a substantial part of the work in question.

The right belongs to qualified persons, but the qualification rules extend to residents or nationals of almost every country in the world and their employers. It lasts for the life of the author plus 70 years, save in the case of designs which could have been registered, which are protected for 25 years only.

Registered trade marks

7–11 These are registered rights giving the exclusive use of "signs". "Signs" are names, devices, get-up, or even the shapes or colours of articles where these are distinctive of a particular enterprise. Registration may be obtained by the person who is known by the mark or (in the case of marks which are inherently distinctive) by anyone who intends to use the mark in the future. The right is infringed by use of the mark for the goods for which it is registered, or the use of a similar mark for similar goods where this is likely to cause confusion. It may last indefinitely, provided that the owner continues to pay the fees due to the Registry, and the mark continues to be distinctive of his business.

Unregistered marks/passing off

7–12 Any mark which has become distinctive of a particular business with goodwill in the United Kingdom may be protected by passing off, whether or not registered.

The business owning the goodwill associated with the indicia in question may take action to prevent the use of similar indicia likely to cause confusion amongst a substantial proportion of customers. Again, the right of action will persist indefinitely so long as the indicia remain distinctive.

Confidential information

7–13 The common law protects any relationship of confidence between two individuals. Valuable information passed pursuant to such a relationship may not be used save for the purposes for which it was passed, nor disclosed to a third party. Recipients of information disclosed to them in breach of confidence will also be prohibited from using it. The obligation of confidence will be protected so long as the information remains confidential and does not enter the public domain.

THE INTERNATIONAL POSITION

7–14 At present, intellectual property protection is almost entirely a national affair. Each country maintains and enforces its own bundle of rights, which are valid only within its territory. A United Kingdom intellectual property right is only valid within the United Kingdom and can only be infringed within the United Kingdom. If a business wishes to protect its technology throughout Europe, it must obtain separate rights in each country within Europe.

There are however, a large number of international treaties and documents related to intellectual property rights. There are three categories:

(a)   those which attempt to unify or harmonise the laws of different countries;

(b)   those which provide administrative routes to simplify the task of applying for separate registered rights in every industrialised country;

(c)   those which lay down general principles of law to be recognised by signatory countries, and require those countries to give reciprocal protection to nationals of other signatories.

In category (a) are the Directives and Regulations harmonising the laws of countries in the European Union, particularly on database protection, semiconductor topography protection, and trade marks. In category (b), the most important are in the field of patents: particularly the worldwide Patent Co-operation Treaty and the European Patent Convention. In category (c), the best known is the Berne Convention concerning copyright.

7–15 Unfortunately none of this has improved the position of parties seeking to sue for infringement of rights in different states. The basic position is that they must sue in the state where infringement took place. Some limited inroads have been made into this, so far as unregistered rights are concerned, by the Brussels Convention, which permits a defendant to be sued in the state of its domicil for infringement in different states within the E.U. But this does not apply in the case of registered rights.[2]

We have chosen in this book to concentrate on intellectual property rights in the United Kingdom. Exceptions have been made to this rule in the copyright Chapter (which deals with the different rules for subsistence of copyright in continental

---

[2] At least where validity is in issue — which is nearly every case.

Europe) and the patent procedure chapter (which refers to the principal differences between United States and United Kingdom patent law). In the case of the copyright chapter, this is because harmonisation of copyright law between Member States of the E.U. is taking place only in a piecemeal way,[3] and it is important for anyone dealing with the creation or transfer of copyright in the United Kingdom to have some understanding of the differences that still exist. In the case of the patent procedure Chapter, the United States (as the major industrial power in the world) is of enormous importance, and its patent laws are unique.

THE FUTURE

7–16   The intellectual property rights dealt with in this book are in a state of constant flux. The rapid pace of technological development experienced in the modern age has put a great deal of pressure on a set of rights which date back (in the most part) several hundred years. In particular, biotechnology and computer technology have not fitted well into the existing framework of laws. Furthermore, the European Union is constantly attempting further to harmonise intellectual property rights across its Member States.

   As a result of all this, hardly a year goes by without some significant change in intellectual property law. The law set out in the succeeding chapters is thus correct at the time of going to press, but anyone reading a book on intellectual property should be careful to verify, before relying on it, that no significant new laws had been introduced since its publication.

---

[3] And is lagging behind the progress made in patents and trade marks.

CHAPTER 8

THE LAW OF PATENTS

Iain Purvis
Barrister, 11, South Square, Gray's Inn

INTRODUCTION

8–01 This Chapter seeks to explain fairly concisely the law relating to the validity and enforcement of patents. This includes (following the pattern of the Chapter):

(a) the general philosophy underlying the patent system;

(b) the structure of a patent as a document, and the purpose of each part of that document;

(c) the legal approach to construing a patent — in particular the meaning to be given to the "claims" which set out the patent monopoly;

(d) the requirements of a valid patent;

(e) amendment of patents;

(f) infringement;

(g) crown use;

(h) rewards for employees under the Patents Act;

(i) supplementary protection certificates.

The legal, administrative and practical aspects of applying for and obtaining patent protection are dealt with in the succeeding Chapter.

PATENT THEORY

8–02 A patent is a monopoly granted by the state to the maker of an invention. It permits him the sole and exclusive rights to exploit the invention for a limited period of time.

The concept of patent protection is often characterised as a bargain between the state and the inventor. The economic interests of the state are served by encouraging innovation, and still more by encouraging the free interchange of ideas. The inventor wishes to make money from his ideas. The patent system serves the interests of both sides by providing an incentive for the inventor to make inventions and to publish them.

The rationale can be best understood by considering what would happen in the absence of a patent system. Most inventive work arises out of consistent investment in research and technology. This investment costs money. Without the protection of a monopoly right, much of this investment would be wasted, since as soon as a new product was publicised, competitors could take advantage of the new ideas embodied in it. Not having the extra costs of research and development to recoup, such competitors would be able to undercut the inventor and thus gain all the benefit of his work for themselves.

This would amount to a disincentive to making inventions. Furthermore, where inventions were made, inventors would seek to keep them secret as far as possible so that their competitors could not take advantage of them. Such secrecy is generally a bad thing, since it denies others the opportunity to learn from new ideas and improve upon them.

8–03   The patent system is the state's attempt to combat this problem. By the grant of a patent, the state awards the inventor a monopoly for a limited period (20 years in the United Kingdom and Europe) over the exploitation of his invention. In return, the state demands that the invention be made public in the form of a description and drawings enabling others working in the field to understand the invention and to put it into effect. Thus, after the monopoly has expired, the rest of the world is in a position to exploit the invention itself.

This philosophy underlies many of the basic concepts of patent law discussed below. In particular:

(a)   The restriction on availability of monopoly protection to inventions which represent something new and clever. There is no need for an incentive or a reward for doing what is old or obvious.

(b)   The requirement that the patent gives a full and clear description of the invention. This is the consideration given to the state by the inventor in return for the grant of the patent.

(c)   The limits on the scope of the monopoly available under a patent, such limits being defined by reference to the nature and quality of the invention which has been made.

NATIONAL AND INTERNATIONAL DIMENSIONS TO PATENTS

8–04   In the United Kingdom, the law of patents is governed by a United Kingdom statute — the Patents Act 1977. However, there is an increasing international element to patent protection. The Patents Act itself was enacted in the light of two international treaties: the European Patent Convention (1973) and the Community Patent Convention (1975 and 1989).

The European Patent Convention is the more important of these, and the only one which is actually in force. It did two things. It established a trans-national authority for the grant of patents across practically the whole of Europe (the European Patent Office). It also established a common set of rules governing patentability by which the whole of Europe[1] is supposed to abide.

The Community Patent Convention is more ambitious. The European Patent Office merely grants a "package" of national patents. In the event of infringement in a number of different states, the patentee must sue on the relevant national patent for each state in which infringement has occurred. Assuming validity is challenged, he must litigate separately in each country, and each legal system must come to its own conclusion (these may differ, for instance, as to whether infringement has occurred). The Community Patent Convention on the other hand seeks to establish a "community patent" — a single right of action which does not recognise national boundaries within the Union. It may be enforced in a single place, and a judgment obtained which

---

[1] The member states of the EPC are the states of the E.U. plus Finland, Liechtenstein, Monaco and Switzerland.

would be enforceable across the whole E.U. Perhaps unsurprisingly, the CPC remains to be ratified by all Member States, and is presently bogged down in wrangles over translation costs and the establishment and staffing of a central court of appeal. However, the CPC still has some relevance in United Kingdom law, since the infringement sections of our Act were drafted in accordance with its provisions.

The history of these two treaties illustrates the truism that it is easier to generate international co-operation and understanding at the administrative end of patent protection than the legal end. Indeed, the administrative stage of filing applications, searching for prior art and even making preliminary decisions on patentability is becoming increasingly harmonised internationally (in particular through the World Intellectual Property Organisation — WIPO — see Chapter 9 on patent procedure). The legal stage of enforcing patents through actions for infringement, or challenging patents on the ground of invalidity, has so far remained resolutely the province of national laws and national courts.

THE STRUCTURE OF A PATENT

8–05 The precise content to be expected of a United Kingdom patent may vary depending on whether it originated in the European Patent Office or the United Kingdom Patent Office. However, in general terms, it may be explained as follows.

A patent is a document in four sections: the title page, the description, the drawings and the claims.

Title page

8–06 The title page of a patent contains important information. It first of all bears the number of the patent. This will be followed by a letter. The function of the letter is to identify the status of the patent. When the application for the patent is first published, it will have the letter A. Upon grant this is replaced by the letter B. Upon any further amendment, it will change again.

Underneath the number of the patent is the "title of invention". This will indicate very briefly the nature of the invention and the field in which it is made. Further down, the "priority data" are given: this comprises a date (or a series of dates). The date or dates are what the patentee claims to be the priority date or dates of his invention — *i.e.* the relevant dates so far as assessing the validity of the patent is concerned (see the section on "priority" below at 8–30 for more detail). It also gives the date of the publication of the patent application (which tells us the date at which it became part of the "state of the art"), and of the patent itself (the date of grant), together with the filing date of the application.

The rest of the title page is given over to further useful information as to the documents cited by the patent office examiner during prosecution of the patent, and the fields in which searches were made (using internationally recognised codes), the name of the inventor, the name of the proprietor and the address for service of his patent attorney.

Description

8–07 The description is the most substantial part of the document. Its role is to explain what the invention is and how it works in sufficient detail for the average skilled man to put it into practice. It tends to be written to a fairly rigid format. It commences by describing what the patentee claims to be the "background" to his invention. This may explain in general terms the products or processes which were in use at the date of the invention and upon which the invention attempts to improve. It will then usually explain the problem which the invention is claimed to solve or alleviate, and at the same time or later discuss the closest "prior art": *i.e.* the nearest (publicly available) attempt to solve the same problem of which the patentee is aware. The patentee will commonly seek to distinguish his invention from the prior art by emphasising the important features of his invention which are not present in the prior art.

The description proceeds with a "summary of the invention". This may consist of one or a number of "objects of the invention", setting out what the invention is meant to achieve by way of improvements over what is known, followed by what is sometimes called the "consistory clause". This is a lengthy sentence starting with something like "accordingly . . ." or "the present invention provides . . .". This is no more than a pre-view of claim 1 — explaining the monopoly claimed by the patentee. If there is more than one independent set of claims (see explanation of "claims" given below at 8–10) there will be an equivalent number of consistory clauses, sometimes headed "in one aspect . . ." and "in another aspect . . .".

8–08 There is then a "brief description of drawings" which explains what each of the drawings accompanying the description represents. Finally the patentee gives a detailed description of the "preferred embodiment" or embodiments of his invention. He is only required to give a description of a single embodiment, but (particularly if his claims are wide in scope) it is common to find descriptions of several different embodiments of the same invention.

The description of the preferred embodiment(s) is given by reference to the drawings. Each feature of the drawings which is referred to in the description (and often many which are not) is given a key number which is used in the text. The aim of the description is to enable the skilled man to understand the nature of the invention and to perform it without undue experimentation. There is no longer any obligation on the patentee to disclose the best method of performance known to him, but he is not entitled to leave obvious holes or problems which the skilled man could not solve save with undue effort. As we shall see, the description may also be important in "construing" the true meaning of the claims of the patent.

The drawings

8–09 The drawings are an integral part of the description, and should be clear and unambiguous. They are presented as a number of "figures". These are generally figurative plans, at different levels of detail, of the device being described, or parts of it. However, the patentee may sometimes also draw the prior art, to illustrate that part of the description where it is being distinguished from the invention. Where the invention is for a process, it is common to use flow-diagrams.

The claims

8–10 These are the numbered paragraphs at the end of the document. They are crucial to all questions of infringement and validity because they set out the scope of the monopoly which is given by the patent. Claims may be to a product or article of manufacture ("product claims"), a method of making a product or carrying out a technical process ("process claims") or a composite of the two, defining a product by reference to the process by which it was made ("product by process claims").

Claim 1 defines the monopoly in the broadest way which the patentee believes he can justify. The structure of the claim is rather formal, and nowadays tends to be governed by EPO practice. It will commonly start with a phrase defining the field in which the invention has been made. It will then set out what the patentee acknowledges to be the "state of the art" at the date of the invention. This is called the "precharacterising" portion of the claim. It concludes with the "characterising" portion, which is supposed to define that which is new and inventive. To take an example, the inventor of the pedal-driven bicycle might have framed his claim as follows:

A transportation device for a human being comprising . . . [*field of the invention*]

two or more wheels, at least one wheel arranged in front of at least one other wheel, a frame to which the wheels are rotatably connected, and a manually operable handle bar arranged to change the position of the front wheel or wheels with respect to the back wheel or wheels whereby the device may be steered [*precharacterising portion*]

characterised in that at least one of the wheels is rotatable by foot-operable means [*characterising portion*].

8–11   The importance of the precise wording will be apparent. At the beginning of the pre-characterising portion, the phrase "two or more wheels, at least one wheel arranged in front of at least one other wheel" is used. Had the claim simply said "two wheels", this would have excluded a tricycle. Because the claim is meant to include a tricycle (or quad cycle) it simply refers to "at least one" wheel at front and back. Similarly, the wording of the characterising portion is not limited to a rotating pedal (*cf.* "*at least one of the wheels is driven by rotation of the feet*"), but would include driving the axle by an up-and-down movement of the feet.

Claim 1 is usually followed by a succession of claims known as "subsidiary claims" each one of which places an additional limit on the invention. Thus, in our example, claim 2 might read:

a device according to claim 1 [*i.e. with all the features of claim 1*] in which there are only two wheels, arranged as front and rear wheels in the same plane.

This excludes tricycles. Of course, anything which fell within claim 2 would necessarily also fall within claim 1. So why bother with claim 2 at all?

The reason is to give the patentee a fall back position in the event that the broadest claim be found invalid. If for instance a pedal powered tricycle were shown to have been disclosed before the date of the patent, then claim 1 would be invalid, but the patentee could defend claim 2. The consequences of this are explained in more detail below.

8–12   Each new claim in the "cascade" of subsidiary claims places a further limit on the scope of the monopoly. Thus claim 3 might read:

a device according to claim 2 in which the front and rear wheels are of equal diameter

and claim 4:

a device according to claim 3 in which the foot operable means are connected to one of the wheels by means of a chain

It will be noted that the extra limits on the scope of the claims are achieved by adding new features, or by selecting features out of a number of possibilities envisaged by the wide claim.

In more complex arrangements of claims, there may be two or three separate "cascades". This is achieved by "starting again" with a completely independent claim (say claim 9). This will define the basic invention in a slightly different way, and then have its own cascade subsidiary to it.

8–13 Sometimes the series of claims ends with a "substantially as described" claim:

a bicycle substantially as described with reference to and as illustrated in the accompanying drawings.

The point of this is to give the most narrow possible protection as the ultimate fall-back position — limiting the scope of the monopoly to the particular embodiment depicted in the drawings.

CONSTRUCTION OF CLAIMS

8–14 The majority of disputes about patents turn crucially on the meaning to be given to the claims. In particular, whether the words used in the claims are wide enough to catch the alleged infringement, or whether they are so wide that the invention claimed covers the prior art and thus lacks novelty. It is therefore useful to consider the construction of claims as a separate topic on its own.

General principles

8–15 In general, construing a claim does not differ a great deal from construing a contract or a statute.[2] A patent is to be read through the eyes of the typical skilled man to whom it is addressed. For this purpose the claims should not be read in isolation but rather as part of the document as a whole. They should be given a meaning which is appropriate to their context. The Patents Act and the EPC require that the body of the specification and the drawings should be taken into account.[3]

However, as a matter of reality, the wording of the claim itself is of the most crucial importance. The claim is where the patentee has expressed the form of his monopoly. The words used in the claim are words "of the patentee's own choosing",[4] and he should normally be held to what they say. Clear wording in a claim should not be overridden by different or even contradictory wording in the description.[5]

The overriding principle

8–16 The overriding principle to be applied where there are difficulties in determining the scope or meaning of a claim is that it should be construed so as to give effect to what the skilled man would conclude were the intentions of the patentee, given what he has said about his invention in the document as a whole. As Lord Diplock put it[6]:

---

[2] See Balcombe L.J. in *Daily v. Berchet* [1993] R.P.C. 357.
[3] See Article 69 of the EPC: "*The extent of the protection conferred by a European Patent or a European Patent Application shall be determined by the terms of the claims. Nevertheless, the description and drawings shall be used to interpret the claims*".
[4] *Catnic v. Hill & Smith* [1982] R.P.C. 183.
[5] *Brugger v. Medic-Aid* [1996] R.P.C. 635.
[6] *Catnic v. Hill & Smith* [1982] R.P.C. 183.

"A patent specification should be given a purposive construction rather than a purely literal one derived from applying to it the kind of meticulous verbal analysis in which lawyers are too often tempted by their training to indulge."

Common issues in construction

Acknowledgements of prior art

8–17   As we have seen, a patentee will commonly give an account of the prior art in his specification. It is a reasonable approach to construction to assume that he did not intend to include such prior art within his claims, assuming that a construction which excludes such art can be achieved without doing undue violence to the language of the claim.[7]

On the other hand, a patentee is not bound in legal proceedings by apparent admissions as to the state of the art if they turn out to be untrue and he wishes to disavow them.[8]

"For"

8–18   It is common for claims to specify a particular integer "for" a particular purpose. It has been the universal practice in the United Kingdom courts to treat "for" as meaning "suitable for". This enables infringement and validity to be judged by purely objective criteria.[9]

Assertions during prosecution

8–19   The patent prosecution files of the United Kingdom Patent Office and the EPO are open for inspection by the public. Commonly, a patentee will have made representations in correspondence or in recorded telephone conversations as to the scope of his claims, or his intention in making amendments. In the United States, from issuance, such representations are treated as public statements from which the patentee has obtained an obvious benefit. By a doctrine known as "file wrapper estoppel" he may be precluded in legal proceedings from making submissions contradictory to or inconsistent with what he has told the United States Patent Office.

There is no such doctrine in the United Kingdom. However, occasionally the courts have been prepared to have recourse to the prosecution file in order to attack a patentee's assertion of a broad construction. The circumstances in which this may be done are unclear and the whole question lacks coherent analysis. However, the best that can be said is that it may be permissible to examine the file in some cases in order to resolve "puzzles in the specification" and to cut down the apparent scope of a claim.[10]

Dependent claims

8–20   The courts will assume that a later dependent claim is narrower than the claims from which it claims dependency. Thus claim 1 will be assumed to cover more than claim 2 which is dependent on it. This may occasionally assist a patentee who wishes to contend for a broad meaning of a particular integer ("it must have been intended to cover more than x because x is specifically claimed in claim 2").

---

[7] *Beloit v. Valmet* [1995] R.P.C. 705.
[8] *Gerber v. Lectra* [1995] F.S.R. 492.
[9] United Kingdom courts have always followed *Adhesive Dry Mounting v. Trapp* [1910] R.P.C. 341.
[10] See *Furr v. Truline* [1985] F.S.R. 553 (generally regarded as wrong on this point); *Bristol Myers Squibb v. Baker Norton* [1999] R.P.C. 253.

Expert evidence

8–21   Construction of claims is a matter of law to be decided by the judge. Experts are not permitted to give evidence as to the meaning of claims. The only exception to this is where the claim includes "terms of art" — words which have a particular and special meaning in the relevant trade. In such a case experts may educate the court as to the meaning of the word in their trade.

This does not mean that experts have no other role when it comes to construction. A claim cannot be construed in a vacuum. It has to be given a meaning within the "factual matrix" in the context of which it was written: *i.e.* the technical background to the invention.[11] For example, a skilled man would not naturally assume that the claim had been drafted so as to include matters which were within the "common general knowledge" of everybody in the trade. Evidence from experts as to the scope of common general knowledge may therefore be an important factor bearing on construction.

Subjective intention of the patentee

8–22   What the patentee actually meant is strictly irrelevant. This should be contrasted with an objective assessment of intention (what a skilled man would have concluded — from the patent itself and from his knowledge of the art generally — that the patentee meant).

Variants, mechanical equivalents and "pith and marrow"

8–23   The most difficult questions on construction concern the extent to which the claim may be taken to cover products or processes which do not fall within the actual wording of the claim, but which have taken the essence of the invention as described in the specification as a whole. On the one hand, it is important that there be certainty in patent law, and a competitor is entitled to assume that the monopoly granted by the patent is limited by the words of the claims. On the other hand, it would be wrong to permit competitors to take the whole benefit of the invention by simply replacing some unimportant element of the claim with an exact mechanical equivalent. The United Kingdom courts have struggled with this dichotomy for many years. In former times, doctrines of "mechanical equivalence", "essential integers" and "pith and marrow" were developed to deal with the problem. The old authorities on such doctrines can now be taken as having been superceded.

The Protocol

8–24   The modern approach to the question of variants is that set out in the Protocol to Article 69 of the EPC.[12] This is worth setting out in full:

> "Article 69 should not be interpreted in the sense that the extent of the protection conferred by a European Patent is to be understood as that defined by the strict literal meaning of the wording used in the claims, the description and the drawings being employed only for the purpose of resolving an ambiguity found in the claims. Neither should it be interpreted in the sense that the claims serve only as a guideline and that the actual protection conferred may extend to what, from a

---

[11] *Glaverbel v. British Coal* [1995] F.S.R. 254 at 264.
[12] Part of United Kingdom law by reason of section 125(3) of the Patents Act 1977.

consideration of the description and the drawings by a person skilled in the art, the patentee has contemplated. On the contrary it is to be taken as defining a position between these extremes which combines a fair protection for the patentee with a reasonable degree of certainty for third parties."

The perceptive reader may have detected in the Protocol a political compromise between two apparently different points of view. The points of view as expressed are in fact little better than parodies of the approaches to claim construction thought to be taken at the date of the Convention in Germany (the broad "guideline" approach) and the United Kingdom (the "strict literal meaning" approach).

United Kingdom approach to the Protocol: "Catnic"

8–25 Perhaps unsurprisingly, the United Kingdom courts have never accepted that their pre–Protocol approach was anything approaching the doctrine of "strict literal construction". In *Catnic v. Hill & Smith* [1982] R.P.C. 183, the House of Lords considered the law of construction under the old law of the 1949 Patents Act (not governed by the Protocol). The issue in that case concerned the angle of part of a lintel. The claim required it to be "extending vertically". In the defendant's product, the relevant component was not quite 90° to the horizontal components. However, on the evidence, the difference was not material to its effectiveness. Lord Diplock reviewed the authorities on how to construe claims, and in particular those authorities which considered the position where the defendant had taken the "pith and marrow" of an invention but not precisely what was claimed.

He laid down a test for establishing whether a defendant who had not "strictly complied" with a word or descriptive phrase in a claim, but had used some variant, ought nonetheless to be found to infringe. The test was happily simplified by the later decision of *Improver v. Remington* [1990] F.S.R. 181 as a threefold approach:

(i) Does the variant have a material effect on the working of the invention? If yes, then the variant is outside the claim. If no:

(ii) Would this (*i.e.* that the variant had no material effect) have been obvious at the date of publication of the patent to a person skilled in the art? If no, the variant is outside the claim. If yes:

(iii) Would the reader skilled in the art nevertheless have understood from the language of the claim that the patentee intended that strict compliance with the primary meaning was an essential requirement of the invention. If yes, then the variant is outside the claim. If no, then there is infringement.

8–26 This test is often misunderstood. The first two tests are not intended to establish a kind of *prima facie* finding of infringement wherever the defendant's "variant" is a true mechanical equivalent — *i.e.* it has no material effect on the way the invention works. They merely establish a pre-requisite which needs to be established before the court can make a finding of infringement by something outside the strict wording of the claim. Even when they are established, it will commonly be the case that the wording of the claim is so clear that the variant must be treated as excluded from its scope.

The *Improver* case itself is a classic example of this — the claim required a "helical spring", and the defendant's product produced precisely the same effect using a rubber rod with slits in it. This was held to satisfy the first two tests for "variant" infringement, because it made no difference to the way the invention worked. However, the judge held that a skilled man would assume that the patentee had limited his claim to a "helical spring" for good reason (even though the actual reason may not have been apparent to him). There was thus no infringement.

How to interpret the Catnic guidelines

8–27 How do you tell whether the patentee intended that strict compliance with the primary meaning of the claim was an essential requirement of the invention? Often this will be apparent from the specification itself. As we have seen, the patentee will generally (in accordance with EPO practice) discuss in the description the objects of the invention, the closest prior art, the distinction between the closest prior art and the invention and the advantages which the invention provides. These will generally give a clue as to what are the "essential requirements" (Lord Diplock's words) of the invention, and whether certain elements of the claim might be replaced by mechanical equivalents without doing violence to the inventive concept. The position of the feature within the claim is also of some importance. Features which appear in the "characterising" portion of the claim may be taken to be regarded by the patentee as the essence of his invention — indeed they are likely to be stressed as such in the body of the description. Features in the "pre-characterising" portion on the other hand, being by definition part of the background art, are less likely to be an essential element of the invention.

Recent developments

8–28 The United Kingdom courts have, since 1982, affirmed that the approach to claim construction laid down in *Catnic* is in accordance with the Protocol to Article 69. Since it provides a more "user-friendly" code than the somewhat woolly wording of the Protocol, they tend to apply the *Catnic* test (as set out in the *Improver* case) by way of substitute.[13]

The experience of the last few years has not in general suggested that the courts are taking a liberal approach to construction following the Protocol. With some exceptions,[14] the courts have preferred to hold the patentee to the words of his claims, on the basis that any other approach would promote uncertainty and chaos. Two citations may be taken as exemplifying the typical approach to these issues.

In *STEP v. Emson* [1993] R.P.C. 513 at 522, Hoffmann L.J. had to consider whether the feature of a claim requiring a "conduit means extending upwards within [a] cylinder to define therewith a pump chamber" could be taken to include a construction in which there was no conduit means extending upwards within the cylinder, but merely a bore through the solid block which formed the bottom of the pump chamber. He held that it could not, and the fact that it made no difference to the operation of the device was irrelevant:

> "The well known principle that patent claims are given a purposive construction does not mean that an integer can be treated as struck out if it does not appear to make any difference to the inventive concept. It may have some other purpose buried in the prior art and, even if this is not discernible, the patentee may have had some reason of his own for introducing it. In my judgment therefore, the claim

---

[13] See in particular the Court of Appeal in *Kastner v. Risla* [1995] R.P.C. 585, refusing to follow the spirited denunciation of the *Catnic* approach by a differently constituted Court of Appeal in *PLG v. Ardon* [1995] F.S.R. 116. *Kastner* has subsequently been followed without exception.
[14] Notably *Kastner v. Risla, ibid.,* and *Johnson v. Mabuchi* [1996] F.S.R. 93.

requires that something which can fairly be described as a conduit must extend upwards within the cylinder and its outer surface must, together with the walls of the cylinder, define the pump chamber."

8–29 In *Brugger v. Medic-Aid*[15] the issue was whether a claim whose characterising portion required a "cylindrical insert [disposed at the lower end of an air inlet flue] with a baffle screen extending outwardly and downwardly" was satisfied by a baffle screen in the same physical location but not extending from a cylindrical insert at the end of the flue. The effect of the baffle was identical. Laddie J. said as follows:

"it is the baffle screen which is the functional item so far as aerosol production is concerned and it is improvement in the aerosol production *by use of the baffle* which is said to be the heart of the invention. However the fact that the insert may be viewed as having no, or the most trifling, technical value does not entitle the plaintiffs, in effect, to ignore its presence . . . If a patentee has chosen to define the characterising part of his claim in narrow terms it is not for the court to rewrite it in broader language simply because it thinks a wider form of wording would have been easy to formulate. No doubt it could be said that a limitation to an insert which it cylindrical is very narrow. It could be avoided by having the inlet flue and insert with square cross-section. But that is how the patentee has chosen to define his monopoly."

PRIORITY

What is meant by "priority"

8–30 Dates are very important in patent law, particularly when considering validity. The most important date for a patent is what is called the "priority date". All issues of validity, in particular novelty and inventiveness, are judged as of the priority date of the patent in question. This means that it is of immediate practical importance for the patentee to know when the priority date is: from that date he can make his invention known to the public without the risk of invalidating his patent.

Prima facie the priority date of a patent is its date of application.[16] On this date, the patentee will have filed with the patent office a specification describing his invention and draft claims setting out his proposed monopoly. However, it is possible to backdate priority by up to a year in certain circumstances, to claim the date of an earlier application.

How to claim priority

8–31 An earlier priority date can be claimed by the patentee by making and filing a declaration along with his patent application. This declaration must identify one or more "earlier relevant applications" within the previous 12 months. "Earlier relevant applications" may either be applications for a patent in the United Kingdom or the EPO, or applications for a patent or the equivalent in any "convention country" — *i.e.* a member of the Industrial Property Convention, or other state specified by order. Nearly all industrialised countries are "convention countries" within the meaning of the Act.

---

[15] see n. 5 above.
[16] s. 5(1).

Which applications can support a claim to priority?

8–32 The test in the Patents Act is that the invention must be "supported" by the earlier application.[17] This means that the earlier application disclosed the invention in such a way that it could be performed (it was an "enabling disclosure").[18] However, it goes further than that. It also means that the earlier application was sufficient to justify the breadth of the claims which claim priority from it. In other words, a single limited disclosure will not give priority to a very broad claim which includes products or methods which cannot be deduced from that disclosure.[19] The test for "support" in the law of priority is the same as the test for sufficiency of disclosure (see below at 8–75).

Multiple claims and priority

8–33 It is quite common that some claims of a patent are able to claim priority from an earlier document but others are not. For instance, in the case of the single limited disclosure discussed in the preceding paragraph, the earlier document would give priority to a claim to the specific product or method which it disclosed, even though it did not give priority to wider claims.

The practical effect of claims to priority

8–34 Claims to priority are "claims" and no more than that. The burden of establishing entitlement to the priority date in any litigation is on the patentee.[20] If he fails to establish entitlement to priority, then by default the priority date becomes the date of application for the patent. The priority date often becomes important because someone (usually the patentee himself) has disclosed the invention by exploiting it between the claimed priority date and the actual date of application. If it is held by the court that he is not entitled to the claimed priority date, the patent will be invalid for lack of novelty.[21]

It is therefore very important for patentees considering taking action for infringement, and for those who have been sued, to pay close attention to the question of entitlement to priority and, if there is any question mark over such entitlement, to investigate whether the patentee has used or otherwise published his invention in the period between the claimed priority date and the date of application.

VALIDITY OF PATENTS

8–35 One of the most important things to understand about the patent system in the United Kingdom is that the validity of a patent is constantly reviewable throughout its life, both in the Patent Office and in the courts. The mere grant of a patent does not mean that it will necessarily be enforceable against defendants with the means to challenge it. An action for infringement will usually induce a counterclaim for revocation on the grounds that the patent is invalid. Such counterclaims and other applications to revoke often succeed.

---

[17] s. 5(3)(a).
[18] *Asahi's application* [1991] R.P.C. 485.
[19] *Biogen v. Medeva* [1997] R.P.C. 1.
[20] Although the evidential burden in establishing specific facts may be on the defendant.
[21] *Biogen* (n. 19 above.) is a classic example of this.

It is therefore crucial to understand the criteria of a valid patent. The best place to start is probably to identify the circumstances in which validity may be challenged.

CHALLENGES TO VALIDITY

8–36 The validity of a patent may be challenged at different stages in its life in different ways.

Pre-grant

8–37 A patent is subjected to a process of investigation by the Patent Office (or European Patent Office) before grant. The purpose of this is to ensure that the patent complies with the requirements for a valid patent laid down in sections 1–4 of the Patents Act, and the requirements for a valid application laid down in section 14 (or the equivalent articles of the EPC where the application is made before the EPO). The procedural details of this are explained in more detail in Chapter 9 on practice and procedure.

It should be noted that at this stage validity is essentially a matter between the Office and the applicant. Interested persons such as competitors have no formal right to intervene. The most that can be done to challenge the grant of a patent pre-grant is after publication, when "observations" in writing may be sent to the Office under section 21 of the Patent Act, or, in the EPO, under Article 115 of the EPC. The Office is obliged to consider such observations but there is no right on the part of the observer to be heard or even to be informed of the Office's response.

Opposition proceedings

8–38 In the EPO, there is a specific procedure for formally challenging the validity of a granted patent, provided that the challenge is made within nine months of publication of the grant of the Patent in the Official Bulletin. This is called "opposition".

Applications to revoke in the Patent Office

8–39 At any time after grant of a United Kingdom Patent, or a European Patent (U.K.) an application may be made to the Patent Office to revoke it on the grounds that it is invalid.

Applications to revoke in the United Kingdom Courts

8–40 Alternatively, anyone may apply to revoke a United Kingdom Patent or a European Patent (U.K.) in the Patents Court or in the Patents County Court. This may be done directly, by way of a claim to revoke, or (more commonly) by way of counterclaim when an action for infringement is brought.

The matters which may be raised in any challenge to validity brought post–grant are laid down in section 72 of the Patents Act. They are essentially as follows:

(a) That the invention was not a patentable invention (*i.e.* it failed to comply with the requirements of sections 1–4 of the Patents Act in that it was not new, obvious, incapable of industrial application, or fell within one of the exclusions, and should not have been granted).

(b) That the patent was granted to a person not entitled to be granted it.

(c) That the patent did not disclose the invention clearly enough and completely enough for it to be performed by a person skilled in the art — usually called "insufficiency".

(d) That the matter disclosed in the specification extends beyond that disclosed in the application for the patent as filed — usually called "added matter".

(e) That the protection conferred by the patent has been extended by an amendment which should not have been allowed.

Save for (b) — the substance of which can be found in the Chapter on ownership of intellectual property rights[22] and is an objection which can only be brought by the true owner — these matters are dealt with below.

PATENTABILITY

8–41 The Patents Act addresses the necessary requirements for a patentable invention in section 1. This is done by reference to certain positive requirements: novelty, inventive step and capability of industrial application,[23] and certain exclusions.[24]

It is convenient to deal with the exclusions first.

The excluded matter

Discovery, scientific theory or mathematical method "as such" (section 1(2)(a))

8–42 Insofar as the invention consists of a discovery, scientific theory or mathematical method "as such", it is not patentable. The phrase "as such" is crucial here. An invention is often founded in a discovery, but this does not in itself render the invention non-patentable, provided that the claims are to something other than the discovery "as such". This is the case even where the "inventiveness" justifying the patent is in the making of the discovery.

For example the discovery of the efficacy of a particular plant extract in treating a disease may give rise to a perfectly valid patent for a manufactured drug. The fact that the plant extract is therapeutically valuable is a discovery and not patentable as such. It would also be impermissible to claim all possible practical applications of the discovery. However, the manufactured drug is an invention, and may be patented even though making it required no ingenuity after the discovery was made.

The distinction between "discovery and invention" is thus a somewhat artificial one rather more relevant to the scope of claims than to the analysis of what comprises an invention.

The exclusion for scientific theories is similar. It would have prevented Faraday from taking out a patent which claimed a monopoly in electromagnetism. It would not have prevented him from patenting particular products whose development derived from and whose operation depended on electromagnetism.

The exclusion of mathematical methods prevents (for example) claims to monopolies in ways of manipulating statistical information.

[22] Chapter 14.
[23] s. 1(1).
[24] s. 1(2) and s. 1(3).

Literary dramatic musical or artistic works or any other aesthetic creation "as such"
(section 1(2)(b))

8–43   Patents relate purely to technical matters. The main purpose of this exclusion is to ensure that artistic and aesthetic matters are governed solely by the law of copyright.

Scheme, rule or method for performing a mental act, playing a game, or doing business or a program for a computer "as such"

8–44   Again, the mere fact that the "clever bit" lies in a new idea for a game (for example), does not prevent a patent being granted for the game itself, comprising pieces, a board and rules.

This distinction between the nature of the innovative step and the actual product or process which is sought to be patented has given rise to great conceptual difficulties in the field of computer programs. The Act prevents a patent being granted for a "computer program as such". In order to get around this, prospective patentees used to try to claim "a computer system programmed" in such and such a way. Whilst this would appear to get around the objection on its face, to allow all such claims would clearly be the equivalent of allowing all computer programs to be patented, which cannot have been the intention of the Act.[25]

The way in which the Patent Offices of the United Kingdom and Europe now deal with the question of patenting computer programs is not to look at the form of the claim but rather at the substance of what has been invented. Provided that the patent discloses a "technical contribution" to a particular field (which is novel and non-obvious) then the claims to "a computer system programmed" in a particular way will be allowed. And, more recently, both the EPO and the United Kingdom Patent Office have recognised that, where a technical contribution has been made, there is no point in drawing an artificial distinction between "a computer system programmed . . ." and the program itself.[26] So, claims to a computer program "as such" now seem to be allowed, always assuming that it comprises a novel and non-obvious technical contribution to the art.

Presentation of information "as such"

8–45   No claim will be allowed to the presentation of information as such. As with the other exclusions, the phrase "as such" is crucial. There is no difficulty with claiming a machine whose purpose is to present information (a novel overhead projector, for example). However, the mere provision of novel information about an old product (for instance instructions on the packaging for a new kind of use) will not constitute a patentable invention.[27]

Offensive, immoral or anti-social behaviour

8–46   No patent will be granted for an invention the publication or exploitation of which would be "generally expected" to encourage offensive, immoral or anti-social behaviour.[28]

---

[25] See *Gale's Application* [1991] R.P.C. 305.
[26] See Patent Office Practice Note April 19, 1999: "the Patent Office's practice will in future be to accept claims to computer programs, either themselves or on a carrier, provided that the program is such that when run on a computer it produces a technical effect which is more than would necessarily follow merely from the running of any program on a computer and which is such that claims to the computer when programmed would not be rejected under section 1(2)(c) under the existing practice."
[27] *Ciba-Geigy (Dürr)* [1977] R.P.C. 83.
[28] s. 1(3)(a).

It is not sufficient for refusal under this ground that the product or process claimed is prohibited by law in the United Kingdom or any part of it.[29] The exclusion has been given a very narrow scope by the United Kingdom courts and by the EPO.[30] Despite having been raised a number of times, particularly in recent years by those opposed to the patenting of human genes or genetically engineered animals, the objection has never been upheld in any reported case.[31]

Animals, plants, etc.

8-47 Section 1(3)(b) excludes from patentability "any variety of animal or plant or any essentially biological process for the production of animals or plants, not being a microbiological process or the product of such a process".

The varieties provision is reasonably clear. Plant or animal varieties, whether created by selective breeding or by genetic engineering, may not be claimed.[32] The exclusion for biological processes is in practice quite limited. Provided that at least one step in the process requires human intervention, then it will be held not to be biological.[33] The exception for microbiological processes permits patents for any novel and inventive process, capable of industrial application, involving microscopic organisms.

Capable of industrial application

8-48 Under section 1(1)(c) this is a positive requirement for patentability. Section 4 defines what is meant by "capable of industrial application" in the following way:

An invention shall be taken to be capable of industrial application if it can be made or used in any kind of industry including agriculture.

The scope of modern industry being so wide, it is quite hard to imagine product claims which would fail this test. The usual context in which the question arises concerns method claims. For instance, a new method of tying a neck-tie would clearly be excluded under this section.

Medical processes are specifically dealt with under section 4(2) and (3). Essentially, these exclude methods of treatment, diagnosis and therapy practised on the human body or on animals, with the proviso that this does not exclude substances or compositions (such as drugs) which are intended to be used for such purposes.

NOVELTY

8-49 To be valid, a patent must be for a "new" invention.[34] Since the invention is defined by the claims, each claim must satisfy this requirement in order to be valid. That is to say that the claim (when properly construed — see the analysis at 8-14 to 8-22 above) must not include within its scope anything which was not new at the priority date.

---

[29] s. 1(4).
[30] EPO Guidelines suggest that the appropriate test is whether the public would regard the invention as so abhorrent that it would be inconceivable that a patent should be granted for it.
[31] See *Howard Florey/Relaxin* [1995] E.P.O.R. 541; *Harvard/Onco-mouse* [1991] E.P.O.R. 525.
[32] Though plant varieties are protectable under the Plant Varieties and Seeds Act 1964.
[33] T356/93, [1995] E.P.O.R. 357.
[34] s. 1(1)(a).

Definition of novelty

8–50   Novelty is defined by section 2 of the Act as being that which does not form part of the "state of the art".

"State of the art"

8–51   The state of the art is all matter which has at any time before the priority date of the invention[35] been "made available to the public" anywhere in the world. Matter can be made available to the public by written or oral description, by use or in any other way.[36]

"Made available to the public"

8–52   Matter which forms part of the state of the art is often called "prior art". The most common form of prior art which is used to challenge validity is a disclosure in writing. This is often in the form of a patent specification, but may be a technical abstract or an article in a journal, an extract from a text-book or a commercial brochure. However, prior art may also consist of oral disclosures or of the commercial sale or demonstration of a product or a method of production.

In each of these examples, "matter" or "information" was disclosed to the public before the priority date. The question of novelty is resolved by asking whether this information included the subject matter of the claims of the patent in question. Put another way — would following the teaching of the prior art result in infringement?

Various problems with establishing lack of novelty commonly arise, and are discussed below.

Enabling disclosure

8–53   There can be no effective challenge to novelty unless the prior art is "enabling" — that is to say that it provides enough information and detail to enable the skilled man to make something or do something falling within the claim in question. This is a question of degree. There is no requirement for the prior art to spell out in precise detail what would immediately be apparent to any skilled man. On the other hand, the prior art may be so vague that the skilled man might have interpreted its instruction in a number of different ways. If some of these interpretations would have resulted in something which was not within the claim, then the prior art will not destroy the novelty of the patent (although it may well give rise to a successful obviousness attack). This case should be distinguished from that in which the prior art specifically sets out various alternatives. There, each of those alternatives can be taken as a separate disclosure, any one of which may be novelty-destroying.

The relevant test is often said to be whether the prior art gives "clear and unmistakable directions" to do or to make something within the claims.[37]

Each and every element of the claim must be disclosed for a successful attack on novelty. If any element is missing, then the attack becomes one of obviousness only. However, the disclosure does not have to spell out explicitly what the result of following its instructions would be. Provided that the "inevitable result"[38] of following the instructions would be to produce something with all the features of the claim, it does

---

[35] Save for the s. 2(3) exception — see below.
[36] s. 2(3).
[37] *General Tire v. Firestone* [1972] R.P.C. 457 at 485.
[38] *ibid.*

not matter that the author of the prior disclosure did not say so (or never realised himself).[39]

"Black box" disclosures

8–54 Sometimes, prior art takes the form of a disclosure where the details of how it works are not apparent. Thus, for example, a patentee may test a novel engine in a motor car by driving it down a public road. The details of the engine are hidden by the bonnet and cannot be deduced from an observation of the car itself. In this case, only the external details of the car are made available to the public. This will not destroy the novelty of any claim filed for the engine.[40]

A more difficult case arises where a product is sold or made available for hire or publicly tested in such a way that an interested skilled man could (if he had been sufficiently interested) have carried out an inspection from which he would have deduced the inventive matter. To take our example in the preceding paragraph, assume the maker of the engine had hired out cars containing it to the public as part of a "blind" performance test. As a matter of probability, it is unlikely that the car would have fallen into the hands of a skilled man with sufficient interest or ability to work out the invention. Nevertheless, it is technically possible, and the law regards this as enough to amount to making the invention available to the public.[41]

Foreign disclosures

8–55 Any disclosure, anywhere in the world, may invalidate a patent. This marked the most dramatic departure of the 1977 Act from the old law, under which only disclosures in the United Kingdom were relevant.

Extent of disclosure

8–56 Although the test is whether the invention was made available "to the public", disclosure to a single person may suffice. The important thing is whether the information was put into the public domain. Disclosure to one person can achieve this, provided that such person was free to publish the information himself.

Circumstances of disclosure

8–57 An invention is not made available to the public if it is disclosed only to persons who are not free as a matter of law to use the information for their own benefit or to disclose it to others. Thus disclosure to employees of the company where an invention is made will not invalidate a patent, assuming that they are bound by the usual implied terms as to confidentiality. Similarly, with disclosure to persons who have signed a binding confidentiality agreement.

The law will on occasion be prepared to imply terms as to confidentiality between parties in a commercial relationship. For instance, a new prototype provided to a testing laboratory for carrying out specific tests is so obviously a sensitive and confidential matter, that there would be no question of it being found to invalidate a subsequent patent application. However, many cases are far from clear-cut. Disclosures to potential customers may be deemed to amount to promotional activity, which is unlikely to be regarded as confidential.

---

[39] Though cf. the *Mobil* case referred to in more detail at 8–60 below.
[40] *Merrell Dow v. Norton* [1994] R.P.C. 1.
[41] See *Availability to the Public* [1993] E.P.O.R. 241, EPO Enlarged Board of Appeal.

The difficulties of the law in this area mean that companies who wish to disclose something which is potentially the subject matter of a patent application to someone other than an employee should always ensure that this is done under explicit terms as to confidentiality.

Unlawful disclosure

8–58   The law makes a specific exception for matter disclosed within six months of the date of filing of an application for a patent in various circumstances.[42] Essentially, these exceptions protect the inventor from disclosure by someone acting in breach of confidence, or disclosure in consequence of the matter having been obtained unlawfully (such as by theft) either from the inventor himself or from someone who had himself obtained it from the inventor.

Selection patents

8–59   In the case of chemical patents, there is a particular problem in that prior art disclosures commonly identify a range of compounds within a class: for instance the disclosure of a new compound will tend to include the disclosure of a vast number of derivative and analogous compounds. The discloser will not have actually made all these compounds, and will not have detailed knowledge of their properties, but their availability logically follows from the discovery of the main compound by applying the rules of chemistry.

It might appear at first sight that the disclosure of such a class or range of compounds would amount to a novelty-destroying disclosure of all compounds within the range. This is not necessarily so. The law has developed so as to permit claims to chemical compounds within a previously disclosed range in certain circumstances. The justification for such claims is that the mere disclosure of a range of compounds as a class does not in itself give any specific technical teaching (or incentive) to the skilled man to make individual compounds within that class. Thus an inventor who discovers a new and useful property or advantage for a particular compound or set of compounds within the previously identified class will be allowed to claim those compounds. However, it is not permissible to claim products within the class which are not proved to have the new property, nor to rely on a property which the prior art disclosure has already identified as characteristic of the class.[43]

New use for old product

8–60   Logically, the mere discovery of a new use for an old product cannot give rise to a valid claim to the product itself. If the product is already known, then any claim to the product itself lacks novelty. The new use may of course give rise to a good "process" claim to the use itself.

This rule has always been applied in the United Kingdom courts. Patentees who tried to get around the rule by claiming the old product "for" the new purpose failed because the courts refused to give any meaning to the word "for" other than "suitable for" (thus effectively depriving it of any meaning at all).[44]

The EPO has in recent years taken a slightly different view. In *Mobil/Friction*

---

[42] s. 2(4).
[43] There are a large number of EPO decisions touching on this point. But the principles can perhaps best be derived from *Hoechst/Tricholoformates* [1979–85] E.P.O.R. 501 and *Akzo/Bleaching activators* [1996] E.P.O.R. 558.
[44] See *Adhesive Dry Mounting v. Trapp* [1910] R.P.C. 341.

reducing additive [1990] 2 E.P.O.R. 73, the Enlarged Board of Appeal held that a product previously known as an anti-rusting additive in lubricant compositions could be
patented again in the form of a claim for "use of [the product] as friction reducing
additive in a lubricant composition". This was even though the use of the product in
a lubricant as taught by the prior art would have had the inevitable result of reducing
friction. The difficulties caused by such claims so far as assessing infringement is concerned were recognised by the EPO but regarded as beyond their remit.

The difficult legal issues raised by *Mobil* are beyond the scope of this Chapter, which
aims only to be a brief introduction to patents. However, it should be noted that no
United Kingdom court has ever directly approved a *Mobil*-type claim (despite the citation of some parts of the *Mobil* decision with approval by the House of Lords in
Merrell Dow[45]). It may be that some compromise along the lines suggested by Jacob J.
in *Bristol-Myers Squibb v. Baker-Norton*[46] — drawing the distinction between a claim
for a new use (permissible) and a claim for more information about an old use
(impermissible) will be arrived at, but this would be hard to reconcile with the facts of
the *Mobil* case itself.

Section 2(3)

8–61   There is one exception to the rule that all prior art relied on to attack novelty
must be published before the priority date of the patent in suit. This is the case of prior
patent applications — specifically dealt with in section 2(3) of the Act.

For the purpose of questions of novelty (though *not* obviousness) the state of art is
deemed to include matter contained in a patent application published on or after the
priority date. This is subject to two provisos: the prior application must have an earlier priority date than the patent in suit, and the matter in question must also have
been contained both in the original application and in the application as published.

Patent applications for this purpose include United Kingdom applications,
European patent applications (U.K.) and international patent applications (U.K.).
Only those applications which proceed to the stage of publication count.

OBVIOUSNESS

8–62   For a claim to be valid, it must involve an inventive step. This means that it was
not obvious to a person skilled in the art having regard to any matter which forms part
of the state of the art.[47] The test for what forms part of the state of the art is the same
as that applied for the purposes of novelty (see above) — *i.e.* it must have been made
available to the public.

The modern approach to assessing obviousness is that set out by Oliver L.J. in
Windsurfing International v. Tabur Marine.[48] It comprises four steps:

(a)   identify the "inventive concept" embodied in the patent in suit;

(b)   assuming the mantle of the normally skilled but unimaginative addressee of
the patent at the relevant date (the priority date), impute to him what was the
common general knowledge in the art at that date;

---

[45] [1996] R.P.C. 76.
[46] [1999] R.P.C. 253.
[47] s. 3.
[48] [1985] R.P.C. 59 at 73.

(c) identify what differences exist between the prior art cited and the invention;

(d) decide whether, viewed without any knowledge of the invention, those differences constitute steps which would have been obvious to the skilled man, or whether they required any degree of invention.

Inventive concept

8–63 This can be a confusing idea. On the face of it, the phrase suggests a distillation of the "essence" of the invention. This is not the case. The patentee is obliged to state his invention in his claims. If one product within the claim, even right at the margin of the claim, is obvious, then the whole claim is invalid. The inventive concept must therefore encompass every embodiment falling with the claim, whether or not these have the particular technical advantage which the patentee regards as important.[49] Normally, then, the "inventive concept" will be defined by the words of the claim itself.[50]

The skilled man

8–64 The relevant skilled man for the purposes of obviousness is the person to whom the patent is addressed. That is to say, the person in an interested company who would be given the job of developing and working out the technology disclosed in the patent. Exactly what qualifications he will have will depend a great deal on the precise nature of the patent. In the case of inventions made in the context of modern technical industries such as microcomputing and biotechnology, the "skilled man" will in fact tend to be a whole "team" of skilled men, no one person having the necessary skills to perform the invention on his own.

The most important characteristic of the skilled man is, ironically, a negative one. It is his lack of imagination. He is deemed to be devoid of insight, abjuring lateral thinking in favour of a drone-like ability to follow instructions and to do what is well-established in the art.

Nonetheless, he must be taken as being interested in the prior art which he is shown, and willing to improve it. He is technically capable of improving it, but only by using standard techniques which anyone familiar with the art would immediately recognise as giving a potential improvement.

Common general knowledge

8–65 The skilled man is not expected to approach the prior art with an empty mind. He is taken to have in his mind when considering the prior art all those matters sometimes said to be the "stock in trade" or "mental equipment" of those working in the field at the priority date. In patent jargon, this is called "common general knowledge".

The establishment of what was common general knowledge is crucial to most questions of obviousness. This is because the application of matters of common general knowledge to the prior art will generally be accepted by the courts as something which is fair to expect the skilled man to do without the exercise of undue imagination. Thus for example, in *Brugger v. Medic-aid*,[51] it was part of the common

---

[49] *Brugger v. Medic-Aid* (n. 5 above).
[50] *ibid.*; *Bourns v. Raychem* [1998] R.P.C. 31.
[51] See n. 5 above.

general knowledge of the skilled man in the field of respiratory nebulisers that the introduction of a baffle to the air stream tended to increase the proportion of respirable particles reaching the patient. The patent in suit differed from the prior art only by the addition of a baffle. It was held that even an unimaginative skilled man presented with the prior art would have realised that it could be improved by the addition of a baffle. Therefore the invention was obvious. On the other hand, a patent which teaches the opposite of what the common general knowledge would suggest as the solution to a problem is likely to be found to be inventive.

The mere fact that something is published in an article in a reputable journal or is otherwise available generally does not make it common general knowledge. Indeed it may be suggested that the fact that it was chosen for publication means that it was not generally known to all skilled men. Even publication in a text book does not necessarily confer the necessary status. On the other hand, the information does not need to be something which all skilled men have actually memorised. It is enough if they know the information to exist, and know where to find it: for example the boiling points of metals would have the status of common general knowledge even though most scientists would not carry the specific information around in their heads.[52]

Identifying the differences between the prior art and the invention

8–66   As we have seen, a claim will be bad for obviousness even if only one embodiment falling within it was obvious. Thus, when making the necessary comparison, one may take any product or process within the claim for comparison with the prior art.

The differences between the prior art and the invention are all those features of the claim not actually disclosed, either explicitly or implicitly, in the prior art.

Inventive step

8–67   Whether the differences between the prior art and the patent required any degree of inventiveness must be judged as of the priority date of the patent in suit. It is of course a value judgment on which reasonable men can (and often do) differ. When it comes to litigation, it has been held by the Court of Appeal that the primary evidence to be used by the judge in reaching this decision is the opinions of experts. However, various "secondary" indications may be relied on as well.[53] The following guidance may be useful.

"Degree of invention"

8–68   On many occasions, the courts have tried to paraphrase the term "inventive step" in an attempt to explain the degree of inventiveness actually required for a patent. These attempts are not always consistent and can be confusing. However, it is worth stressing that the grant of a patent does not require any great leap forward or spectacular exercise of inventiveness. Any step which required a "spark of imagination",[54] or went beyond a mere "workshop alteration"[55] is enough.

---

[52] The best modern discussion of this topic is in *Bourns v. Raychem*, n. 50 above.
[53] *Molnlycke v. Procter & Gamble* [1994] R.P.C. 49.
[54] *Genentech* [1989] R.P.C. 613.
[55] *Samuel Parkes v. Cocker Brothers* [1929] R.P.C. 241.

Hindsight

8–69 The courts have often warned of the dangers of "hindsight" analysis in this respect. In other words, it is often all too easy to dismiss an invention as obvious once one has been shown the solution. Obviousness must be judged at the priority date with only the benefit of the prior art and the common general knowledge.[56]

Equally, the courts have criticised the "familiar step by step analysis" whereby the prior art is transformed into the claim of the patent by a series of steps each of which in itself seems to involve no real inventive contribution. An invention may well consist of the recognition that each of those steps in combination produced a useful result.[57]

"Would" or "could"?

8–70 There is a danger in asking "could the skilled man have made the invention from the prior art"? It is often easy to propose hypothetical circumstances in which the invention would be likely to be made. Put another way, many patents may be criticised as "an invention waiting to happen". This is not a legitimate criticism of a claim, any more than the "monkeys tapping at typewriters" analogy is a valid criticism of the genius of Shakespeare. The better question is whether the skilled man "would" have made the invention, given the prior art and his common general knowledge.[58]

Obscure prior art

8–71 The mere fact that the prior art was obscure and would have been unlikely to have come to the attention of the skilled man at the priority date is irrelevant to the question of obviousness.[59]

However, the skilled man is not to be taken as an automaton, considering each piece of prior art without reference to its context. For instance, it is certainly legitimate to say that he would be likely to pay far less attention to a piece of prior art from a completely different field, which on the face of it would appear irrelevant to his work. He is not expected to search through every document, however apparently peripheral, and to act with equal interest and enthusiasm on all suggestions, even those which appear absurd.[60]

Age of prior art

8–72 It is no answer to a case of obviousness to say that the prior art was very old. One should beware of "uncritical ageism."[61] On the other hand, the age of the art may provide some "secondary evidence" that the invention was not obvious. For instance, if the step from the prior art to the invention was technically feasible before the priority date, and a considerable time elapsed during which it was not taken, it is legitimate to ask "why not before?"[62] However, it should be remembered that there may be many answers to this question which do not favour the patentee. For instance, the prior art may have been obscure and little known in the trade. Or there may have been commercial reasons why the improvement was not worth pursuing at the time.[63]

---

[56] *Vickers v. Siddell* [1890] R.P.C. 292.
[57] *British Westinghouse v. Braulik* [1910] R.P.C. 209.
[58] See *Technograph v. Mills & Rockley* [1969] R.P.C. 395.
[59] *Windsurfing*, n. 48 above.
[60] *Technograph*, n. 58 above.
[61] *Brugger v. Medic-aid*, n. 5 above.
[62] *Buhler v. Satake* [1997] R.P.C. 232.
[63] Obviousness means "technically obvious", not "commercially obvious" — so that one cannot get a patent simply for recognising the commercial value of a particular obvious step.

Commercial success and long felt want

8–73   The best secondary evidence of lack of obviousness is a commercially success-ful invention in circumstances where the improvement over the prior art had been needed for a long time. It may be presumed from those facts that skilled men, armed with the relevant common general knowledge, were in fact turning their minds to the problem solved by the patent, and had failed to come up with the invention.[64]

However, a word of warning is necessary before jumping to too many conclusions based on the supposed commercial success of a patented product or process. There have been relatively few cases in the courts in recent years where a plea of "commer-cial success" has succeeded. This is because the reasons for the success of a product may be unrelated to the actual invention: for instance a successful advertising cam-paign, or a new production process rendering the product for the first time cheap enough for the public to afford. Furthermore, even where the commercial success is due to the product itself, the patentee has to show that what made the product suc-cessful was what is actually claimed as the invention.

"Problem and solution" approach

8–74   In the EPO especially, a short cut to the question of obviousness is often taken, known as "problem and solution". It assumes that the patent is an attempt to solve a technical problem existing with the prior art. On this assumption, the prior art is examined, and the problem identified. It is then asked whether the solution proposed by the patent was an obvious answer to the problem. This approach is rarely adopted by the United Kingdom courts, and even in the EPO it has been criticised in recent years as not being a universal formula.[65]

SUFFICIENCY OF DISCLOSURE

8–75   It is a ground of revocation of a patent that the specification does not "disclose the invention clearly enough and completely enough for it to be performed by a per-son skilled in the art".[66] This is usually referred to as "insufficiency".

This objection takes two forms. One is a pure attack on the description of the inven-tion given in the specification — that it is insufficiently clear to enable the skilled man to carry out the invention at all. The other is an attack on the scope of the claim itself: that the breadth of the monopoly is not "supported" by the matter actually disclosed in the description. It is convenient to look at these separately.

Inadequate description

Lack of detail or clarity

8–76   The skilled man must be given sufficient information in the description of the patent to perform the invention claimed. It will immediately be apparent that it is important before addressing this question to identify the qualities of the skilled man in question. The skilled man for the purposes of sufficiency is the same as the skilled man for the purposes of obviousness (see above).[67] Thus he may well be a team of experts in different fields of activity.

---

[64] *See Samuel Parkes v. Cocker Brothers*, n. 55 above.

[65] *Alcan/Aluminium Alloys* [1995] E.P.O.R. 501.

[66] s. 72(1)(c) and EPC, Articles 83 and 84.

[67] This sometimes permits a "squeeze" argument: if the patentee has left his description vague, he implies that the skilled man has little difficulty in getting round problems which may arise in the performance of

The degree of detail which must be given will clearly depend on the nature of the invention and the nature of the field of activity. The skilled man does not have to be told what anyone skilled in the art would already know. Nor does he have to be told the details of materials or processes which he would simply buy in from others who are experts in that area. For instance, it is common for specifications to involve the processing of information to reach a particular result. Such processing will be within the normal capacity of anyone skilled in computing. It is not necessary for the patentee to spell out the details of the programming involved, even though the primary addressee of the specification may have no idea how to do it.

Research and experiment

8–77 It must be expected that the skilled man will have to carry out simple and straightforward trials in order to exploit the invention. He is assumed to have the "will" to carry these out. However, he is not expected to have to exercise any invention himself, nor to have to engage in a prolonged programme of research or experiments.

Errors in the specification

8–78 Sometimes, a specification contains mistakes such that literally following the instructions would result in inevitable failure. This does not necessarily invalidate the patent. In many cases (for instance, writing MV (megavolts) instead of mV (millivolts)) the mistake will be immediately corrected by the skilled reader.

The test is whether both the existence of the error and the way to correct it would both be quickly discovered by the skilled man. If so then the error is not enough to render the patent insufficient.[68]

Breadth of claims

8–79 This is an extremely important objection to validity, because it places a restriction on the scope of the monopoly which the patentee may claim on the basis of his invention. In other words, he has to provide sufficient "consideration" by way of disclosure in the description for the monopoly of the claims.

Under the pre-1977 law, this objection was explicitly set out in the statutes and known as "lack of fair basis" — *i.e.* the description did not provide any fair justification for the monopoly of the claims. It did not appear explicitly in the 1977 Act. However, it was revived as an objection in the United Kingdom under the heading of "insufficiency" by the House of Lords in the landmark case of *Biogen v. Medeva* [1997] R.P.C. 1.

There are two ways in which a claim may be found to be too broad. The first is that where a claim clearly covers more than one product or process, the description must enable the production or performance of each one. The second is where the description is of specific products or processes, but the claim is generic. If the claim is so wide that it covers products or processes which may be developed in the future but would "owe nothing" to the teaching of the patent, then it will be invalid.

The classic case of this is where the patentee is the first to reach a particular goal, for example the synthetic production of a naturally ocurring substance, but does not limit himself in his claims to his own method of reaching the goal. Instead he claims all methods of reaching the goal (or even the goal itself).[69] Even if at the date of the

the invention. This means that he cannot rely on the existence of such problems as a barrier to making the invention in the first place.
[68] *Valensi v. British Radio Corporation* [1973] R.P.C. 337.
[69] Such claims are sometimes called "free beer claims".

patent the method achieved by the patentee was the only method of reaching the goal known to science, it is conceivable that other methods might be developed in the future, many of which might be quite different from that of the patentee, and which might be better or more efficient. It would be an unwarranted barrier to technical progress to award a monopoly to the first inventor which would prevent any alternative methods being developed, and such claims will be held invalid.

ADDED MATTER

8–80 The prosecution process permits considerable amendment to specifications. However, section 72(1)(d) of the Act places one very serious restriction on what may be done. A patentee is not permitted to introduce additional matter into the specification, going beyond what was disclosed in the original application. The rationale for this is that the date of the application for the patent is the basis for all claims of inventiveness and novelty. Almost invariably, it is in the patentee's interests to file as soon as possible, to gain the benefit of as early a date as possible (thus restricting the amount of prior art which may be cited and the level of common general knowledge of the skilled man). If he is permitted to introduce new matter into his specification after filing, this would give rise to an obvious potential for abuse, in that the patentee could claim the benefit of an application date which actually pre-dated the making of his invention.

The test under section 72(1)(d)

8–81 The disclosures of the application and the patent as granted must be compared, and any new matter not found in the original application (either explicitly or implicitly) means that the patent is invalid.[70] Stated so simply, the question would seem to be an easy one. However, as usual, there are certain complexities that need to be addressed.

Trivial or inconsequential matter

8–82 Trivial differences which amount to little more than a tidying up of language or clarification will not be treated as additional matter within the meaning of the section. It should be remembered that one of the jobs of the Patent Office is to ensure that patents are granted in a reasonably clear form. Such clarification should not have the result of rendering the patent invalid.

Discursive material

8–83 It is common for the Patent Office (and more particularly the EPO) to require the patentee to cite the most relevant prior art in his specification and to explain how his invention differs. Very often, such prior art only turns up during the prosecution process. Thus the patentee is obliged to introduce a new paragraph or so dealing with it. Again, this will not usually be treated by the courts as added matter within the meaning of the section. The only case where a problem might arise is where the new material gives a potentially new slant on the nature of the invention itself: for instance it is something which could affect the construction of the claims.

Amendments to the claims

8–84 Modern applications for patents contain claims. It is very common to amend these in the course of prosecution. The mere amendment of a claim, although it

---

[70] *Bonzel v. Intervention* [1991] R.P.C. 553.

necessarily changes the nature of the monopoly sought by the patentee, does not in itself amount to objectionable added matter. This is the case even where the claim is widened so that it covers products not previously said to be within the scope of the monopoly.[71]

However, in order to avoid added matter objections, the amendments must be derived from matter already to be found in the body of the specification. The problem comes where the claims are amended to introduce a new inventive concept nowhere to be found in the original specification. Take the case where the original specification discloses an invention only in the context of a vehicle with four wheels, and the original claim was limited to the invention in the context of such a vehicle. An amended claim to the invention in a vehicle with "no more than six wheels" would be objectionable. It would introduce a new inventive concept, namely that the invention was not limited to a four wheeled vehicle, but extended to one with any number of wheels up to six.

A more subtle problem arises where the patentee has simply removed a feature from the original claims. In some cases, the existence or otherwise of this feature would have been recognised by the skilled man reading the original application as being a matter of little importance. Its presence in the original claims was out of an excessive abundance of caution, and it can be removed without adding matter. However, in other cases, the feature is taught in the description as being essential to the successful operation of the invention, and no alternatives to its use are described. There, it is very dangerous to remove the feature from the claims during prosecution, since it was an inherent part of the original inventive concept. Its removal will amount to the addition of a new inventive concept and will thus be objectionable under section 72(1)(d).

AMENDMENT

8–85 A patentee will commonly apply to amend the claims of his patent in order to deal with an actual or perceived challenge on the basis of particular prior art. The point of amending the claims is to introduce further limitations, which will serve to distinguish them from that art.

Amendment is generally made by application to the Patent Office. However, it must be made to the court where there is pending litigation in that court. Very commonly an application to amend is made in the face of prior art cited by a defendant in an infringement action, or by someone bringing proceedings to revoke the patent.

Whether or not amendment is granted is a discretionary matter, subject to two legal bars. No amendment may be made where it:

(a) will result in the specification disclosing added matter; or

(b) will extend the protection conferred by the patent.[72]

(a) is unsurprising, given the bar to disclosure of added matter between the original application and the patent as granted. (b) establishes the rule that claims may not be broadened post-grant. As we have seen, failure to comply with this rule gives rise to an objection to validity.

The extent of the Patent Office's or court's discretion to permit amendment has been

---

[71] *A.C. Edwards v. Acme Signs* [1990] R.P.C. 621.
[72] Patents Act 1977, s. 76.

the subject of a great deal of debate in recent years.[73] The present position can be summarised as follows.

Duty of full and frank disclosure

8–86   A patentee applying to amend is under a duty to put before the court or the Patent Office all matters which might be relevant to the question of discretion. However, he need not disclose privileged material (such as documents passing between himself and his patent agent relating to the invention). Refusal to waive privilege will not (contrary to the view expressed in some old cases) be held against him.[74]

Misconduct by patentee

8–87   The most relevant issue is the conduct of the patentee. In particular, the court or the Patent Office is interested to see whether he has acted covetously or recklessly in framing his claims prior to grant, or abusively in asserting his patent against competitors, knowing it to be invalid.

This commonly raises the following issues:

(a)   Whether the patentee knew of the prior art before grant, and if so why he allowed the claims to proceed to grant in the invalid form;

(b)   Whether the claims were in any event drawn recklessly and graspingly wide, when the patentee should have realised that they were highly unlikely to be valid;

(c)   At what point post-grant the patentee realised that the patent was invalid;

(d)   The period of delay between such realisation and the application to amend;

(e)   The patentee's conduct in that period: did he continue to assert the invalid patent against competitors, and/or continue to collect licence fees?

Deletion of claims

8–88   There is a very important distinction between an amendment which amounts simply to the deletion of one or more broad claims (falling back on later claims which are narrower in scope) and one which seeks to rewrite the claims. The former type of amendment is much easier to achieve. The point is that the discretion on amendment is there to protect potential infringers who had looked at the patent in its unamended form and concluded that it was invalid and could thus be ignored. Such infringers must be taken to have been aware of the narrower claims of the patent, and the risk that they would be found valid even if the wider claims were invalid. In those circumstances, there is no conceivable prejudice to them in deleting the broad claims at the request of the patentee. However, they could not have been aware of the scope of any new claims subsequently framed by the patentee. Such amendment could well cause serious prejudice.

The upshot of this is that the courts always permit amendment by deletion save in the most exceptional cases of "very grave" misconduct by the patentee.[75] Amendment by "rewriting" claims on the other hand has often been refused.

---

[73] See the recent authoritative decision of the Court of Appeal in *Kimberley-Clarke v. Procter & Gamble* [2000] I.P. & T. 27.
[74] *Oxford Gene Technology Ltd v. Affymetrix Inc.* CA, November, 2000 (unreported).
[75] *Van der Lely v. Bamfords* [1964] R.P.C. 54.

Amendment leaving patent invalid

8–89   The courts or the Office will refuse any amendment which would result in the patent as amended being invalid.

INFRINGEMENT

8–90   The term "infringement" tends to be used in practice to cover two quite separate issues: whether a particular product or process falls within the claims of the patent, and whether the particular act done in relation to the product or process is one which is reserved exclusively to the patentee. This section is only concerned with the latter issue — the former having being dealt with above under the heading "construction" (8–14 to 8–29).

The acts which constitute infringement of a patent are set out in section 60 of the Patents Act 1977. The section distinguishes between infringement of product claims and infringement of process claims.

Product claims

8–91   These are infringed by the making, importation, keeping, disposal, offer to dispose or use of a product falling within a claim in the United Kingdom. Liability is strict. There is no requirement of *"mens rea"* or knowledge (actual or constructive) of the fact that the product infringes the patent, or even of the existence of the patent itself. (Though there is a limited innocence defence to a claim for damages — see below).

These acts of infringement give rise in practice to few problems of interpretation. Two points may be worth some explanation.

The concept of "offering to dispose", like the rest of the section, comes from the Community Patent Convention. It is not meant to import the same meaning as the word "offer" in the English law of contract. Thus, advertising infringing products in brochures and the like will be an act of infringement.[76] Since all the other acts of infringement are required to take place in the United Kingdom, it would seem that, to amount to an infringement, the offer must be to dispose of the product in the United Kingdom.[77]

"Keeping" is not meant to catch innocents like carriers. It comports some idea of retaining in stock for future commercial use[78]

Process claims

8–92   These are directly infringed by the use of a process falling within a claim.[79]

There may also be infringement of a process claim where the process is offered for use in the United Kingdom.[80] Here, however, there is an element of *"mens rea"*. The offeror must either have knowledge himself, or it must be obvious to a reasonable person in the circumstances, that the use of the process in the United Kingdom without the consent of the proprietor would be an infringement. This seems to require knowledge of the existence of the patent and at least some idea of its scope.

Under section 60(1)(c), a process claim may also be infringed by doing any of the

---

[76] *Gerber v. Lectra* [1995] R.P.C. 383.
[77] *Kalman v. PCL* [1982] F.S.R. 406.
[78] *SKF v. Harbottle* [1980] R.P.C. 363.
[79] s. 60(1)(b).
[80] s. 60(1)(b).

acts prohibited in relation to product claims (making, keeping etc.) in relation to a product obtained directly by means of the process. The word "directly" is an important limitation here. It requires that the product dealt with is the immediate result of the application of the patented process. It is not enough that the patented process was used as an intermediate step in production.[81]

Secondary and contributory infringement

8–93 Under section 60(2), it is an infringement to supply or offer to supply in the United Kingdom means "relating to an essential element of the invention" for putting the invention into effect in the United Kingdom, knowing or it being obvious to a reasonable person in the circumstances that (a) the means are suitable for putting the invention into effect; and (b) the means are intended to put the invention into effect in the United Kingdom.

Thus someone supplying a printed circuit board designed to operate a machine in a particular way will infringe a patent which claims the machine operating in that way (assuming he fulfils the requirements of knowledge). There is no requirement here (unlike offering a process for use — see paragraph 8–92 above) of knowledge of the existence of the patent or the scope of its claims.

The phrase "essential element of the invention" seems to import little more than the requirement that the invention cannot be performed without it. It would on the face of it catch a whole range of possibilities — right down to nuts and bolts. However, section 60(3) excludes from liability the supply or offer to supply of "staple commercial products" unless such supply or offer is deliberately intended to induce the person to whom the supply or offer is made to infringe the patent.

DEFENCES TO INFRINGEMENT

Exclusions from infringement

8–94 Section 60(5) sets out a range of acts which cannot infringe a patent. The most important of these are:

(a) Acts for private and non-commercial purposes;

(b) Acts for experimental purposes relating to the subject matter of the invention.

The courts have placed severe limitations in the ability to rely on the "experimental use" exemption. The key distinction is between collecting evidence about properties of the product or process which are already known, and seeking to establish new information about the product or process. Thus, experiment for the purposes of section 60(5) does not include the carrying out of trials aimed at collecting data which is intended to be used to promote a product to customers, or to impress a regulatory body.[82]

Consent of the proprietor

8–95 Section 60 specifically excludes from infringement the case where the act is carried out with the consent of the proprietor. Such consent may of course be express, but

---

[81] *Pioneer v. Warner Music* [1997] R.P.C. 755.
[82] *Monsanto v. Stauffer* [1984] F.S.R. 559.

there are also circumstances in which consent will be implied. One such case is that commonly referred to as "exhaustion of rights" in which the patentee has himself put the patented product on the market in the United Kingdom or elsewhere in the E.C., or has consented to the product being put on the market. He is deemed to consent to any future transactions in the product, within the E.C. This subject is considered in more detail in Chapter 23 ("Licensing").

Repair

8–96   A species of consent is identified by the general term "right to repair". By selling a product, the patentee must be taken to have agreed to it being repaired. This extends to the purchaser having the product repaired by a third party. However, there is an obvious limit to this right: where the repair is so extensive as to amount to the creation of a replacement article.[83]

In the most recent formulation of the "repair" rule, the House of Lords has said that the only issue is whether, taking into account the nature of the invention and the activities of the defendant, it can be said that the defendant has "made" an article within the meaning of section 60(1).[84]

Right to continue use commenced before priority date

8–97   Under the pre-1977 law, even "secret" use of the invention by a third party before the priority date could invalidate a patent. This rule was abolished in the 1977 Act in favour of a straight publication rule. However, the interests of the individual secret prior user were protected by giving him a defence to infringement under section 64.

Where the defence applies

8–98   The defence applies in two circumstances:

(a)   where a person has done in good faith before the priority date an act which would have constituted an infringement had the patent then been in force;

(b)   where a person has made "serious and effective preparations" to do such an act.

In either case, the right is limited to the ability to do or continue to do the act after the coming into force of the patent. The person may continue to do the act himself, authorise the doing of the act by a partner in his business, or assign the right to do the act as part of his business. Furthermore, any purchaser of a product made in furtherance of the rights conferred by section 64 has a good defence in respect of any dealings he makes with the product. No licensing of the right to use the invention is permitted.

The nature of the "act" which may be continued

8–99   The scope of section 64 is very limited. It does not confer a right to do all acts falling within the claims. On the contrary, it only permits the repeated doing of the particular act which was done before the priority date, or for which serious and effective preparations were made. Although some minor changes in detail may be

---

[83] *Sirdar Rubber v. Wallington* [1906] R.P.C. 257.
[84] *United Wire v. Screen Repair Services* (unreported).

permitted, the product being produced or the process being used must remain "substantially the same".[85]

Serious and effective preparations

8–100   Where no act falling within section 60(1) has been done before the priority date, the preparatory steps must have been serious and effective preparations for the act which is now being done: which in the case of a product is likely to be the manufacture and sale of the final version. There is obviously a line to be drawn somewhere between mere "blue skies" research, with no particular aim in mind, and making a final production prototype. Precisely where this line falls in any individual case is often a difficult decision.

Rights of co-owners

8–101   Where a patent is jointly owned, each co-owner has a limited right to exploit the invention without the consent of the other. This is set out in section 36. The co–owner, by himself or by his agents, may do any act which would otherwise be an infringement. He may not, however, license any other person to do such an act.

The reference in section 36(2) to the co-owner's "agents" has been held to cover any sub-contractors employed by the co-owner to do the work for him.[86]

Compulsory licences

8–102   In certain circumstances, where the invention is not being exploited fully by the patentee, it is possible to apply to the Patent Office for a licence on terms to be fixed (including payment of royalties).[87] These are known as "compulsory licences", and are discussed in Chapter 23 ("Licensing").

THREATS

8–103   It has long been recognised that a patent can be a powerful weapon for abuse. In particular, it is easy for a patentee to cause a great deal of damage to a non-infringing competitor without ever having to justify himself in court, simply by threatening the competitor's customers with litigation.

To counter this potential abuse, statutory provision is made under section 70 of the Patents Act giving a person aggrieved by such threats various remedies, provided that the threats are unjustified.

A cause of action for threats may arise where any person (whether or not the proprietor of a patent or any right under a patent) threatens another person with proceedings for infringement of patent.[88] It is not a threat merely to "notify" the person in question that a patent exists.[89]

This is a very broad provision. It covers threats made to the infringing manufacturer, as well as threats made to his customers.[90] Any "person aggrieved" may bring proceedings for threats, even where he was not the person threatened. Thus a person

---

[85] *Lubrizol v. Esso Petroleum* [1997] R.P.C. 195.
[86] *Henry Brothers (Magherafelt) Limited v. Ministry of Defence* [1997] R.P.C. 693.
[87] s. 47 and s. 48.
[88] s. 70(1).
[89] s. 70(5).
[90] Subject to the exclusion for "making" and "importing" — see below.

may bring an action for threats made to sue himself, his supplier, or his customers, regardless of to whom those threats were made.

It should also be remembered that, where a threatening letter is written by a legal adviser or patent attorney, a threats action can be brought not only against the patentee on behalf of whom the letter was written but also against the adviser/attorney who wrote it. This may sometimes cause embarrassment and a conflict of interest between professional and client.

What is a "threat"?

8–104 The nature of a "threat" of proceedings is difficult to define with precision, particularly given the somewhat arbitrary exclusion for mere "notification". A threat can be express or implied. The court will look at the whole of any correspondence or other matter alleged to amount to a threat, from the point of view of a reasonable recipient.[91] As a practical matter, it should be assumed that any letter which goes beyond (a) a simple notification of the existence of a patent; or (b) a general warning about patent rights, unrelated to any specific product, is likely to be held to be a threat. A reasonable recipient of a letter which refers to the author's patent rights in relation to a product made by the recipient is likely to understand that there is an implicit threat of legal action, unless there is a clear statement that no action will be brought.

No action can be brought in respect of a threat made after proceedings have actually been brought. This is for two reasons: if proceedings have already been brought, they can no longer by definition be "threatened": no damage can be caused by the "threat" in such circumstances, since what has been threatened has already taken place, thus there is no person "aggrieved".

Making and importing

8–105 By section 70(4), no action may be brought where the threat is to bring proceedings for infringement by making or importing an infringing product for disposal or using a process. This permits threats to be made against a primary infringer with impunity, provided that those threats do not go beyond the specific acts of making, importing or using. For instance, if the letter also seeks undertakings or damages from the recipient in relation to the selling of the infringing item, then it would still be actionable as a threat.

Justification

8–106 Where a threat is made that would otherwise be actionable, the maker of the threat can avoid liability by showing that it was "justified": namely that the acts in respect of which proceedings were threatened did amount to patent infringement, or would have amounted to patent infringement if carried out.

It has been held that where the patent was not actually granted when the threat was made, no argument of justification can be put forward. It is no answer to say that the act complained of did infringe the patent as eventually granted.[92]

"Without prejudice" threats

8–107 Threats made in the course of without prejudice negotiations or in without prejudice correspondence are not actionable by the person aggrieved because the

---

[91] *Cavity Trays v. RMC Panel Products* [1996] R.P.C. 361.
[92] *Brain v. Ingledew Brown (No. 2)* [1997] F.S.R. 271.

contents of the negotiations or the correspondence are privileged and may not be referred to in court.[93]

Those planning to use the "without prejudice" label to write a letter before action without the risk of a threats action should beware, however. The letter must be a genuine "without prejudice" offer, which means that it must be a genuine attempt to negotiate a settlement. This requires some degree of compromise: for instance an offer to agree to waive damages if the recipient undertakes to stop infringing.

Available relief

8–108   Under section 70(3), the person aggrieved by a threat may seek a declaration that the threats were unjustifiable, an injunction against further threats, and damages in respect of any loss sustained by reason of the threats.

Damages may be difficult to prove. The burden is on the person aggrieved to show that the threats have resulted in lost sales or other damage to his business. Where the threat was made to him rather than to his customers, this will obviously be problematic. Where the threat was to customers, who subsequently withdrew their custom, he must prove that this was caused by the threat rather than by anything else. Not only is it notoriously difficult to persuade customers to assist in this kind of litigation, there is also the problem, where the threat was followed by an action for infringement, of proving that the withdrawal of custom was caused by the threat rather than by the actual proceedings themselves.

CROWN USE

Entitlement to use inventions

8–109   Under United Kingdom patent law, the government is in a privileged position. Section 55 of the Patents Act provides that any government department or any person authorised by a government department may for the services of the Crown do, without the consent of the proprietor, any of the acts which would otherwise infringe under section 60.

The term "for the services of the Crown" is a broad one which would appear to encompass most of the public services provided by government departments. For the avoidance of doubt, various matters are specified in section 56 as being within the meaning of the term. These are:

(a)   the supply of anything for foreign defence purposes;

(b)   the production or supply of drugs and medicines specified in regulations by the Secretary of State for the NHS;

(c)   matters relating to atomic energy.

Compensation

8–110   Where the government makes use of a patented invention, it is required to pay compensation to the patentee, the terms of which are to be settled by the court if they cannot be agreed.[94] In a case where the use of the invention by the Government or a person authorised by the Government has resulted in a contract to supply the

---

[93] *Unilever v. Procter & Gamble* [1999] F.S.R. 849 (upheld in CA — unreported).
[94] s. 55(4) and s. 58.

patented product or process being lost by the patentee, this can include payment in respect of the profits thereby lost.[95] The Government is obliged to inform the patentee of the use of his patent, save where it would not be in the public interest to do so.[96]

Exceptions to entitlement to compensation

8–111 There are two cases where compensation will not be paid. The first is where the Crown successfully challenges the validity of the patent. The second is where it establishes trial or recordal of the invention before the priority date by or on behalf of a government department (or by the United Kingdom Atomic Energy Authority), otherwise than in consequence of a relevant communication made in confidence. This is to deal with cases where the government or its servants made the invention themselves before the priority date but did not disclose it to the public such as to render the patent invalid.

REWARDS FOR EMPLOYEES

Circumstances in which rewards may be applied for

8–112 By sections 40–41 of the Patents Act, Parliament provided a new scheme for rewarding employees for making inventions in two circumstances:

(a) where the invention belonged from the start to the employer and the resulting patent was of "outstanding benefit" to the employer;

(b) where the invention belonged originally to the employee and subsequently passed to the employer by way of an assignment or exclusive licence, and the benefit obtained by the employee from that assignment, licence, or any ancillary contract was "inadequate" in relation to the benefit obtained by the employer from the patent.

The employee may apply to the Patent Office or to the court for his monetary reward (called "compensation" in the Act). The employee must show in either case that by reason of the benefit gained by the employer from the patent, and (in the latter case) the inadequacy of the remuneration already obtained, it is "just" that the employee should receive compensation.

"Outstanding benefit"

8–113 It is important to remember that the outstanding benefit must be caused by the existence of the patent. The employee cannot rely on any benefit due to the making of the invention *per se*. Thus, for instance, an invention may save an employer millions of pounds in previously wasted energy. However, the patent derived from the invention may have been of no benefit at all, since the particular process involved was not applicable to anyone else's business. In such a case, the employee would be entitled to no reward under section 41.

Exactly when a patent has proved of outstanding benefit to an employer is obviously a question of degree. The only guidance given by the Act is that the size and

---

[95] s. 57A.
[96] s. 55(7).

nature of the employer's business is relevant. In the reported cases, words and phrases such as "unexpected" and "out of the ordinary" have been used.

Scale of reward

8–114   Under section 41, the award to be made must be such as to secure for the employee a "fair share" (having regard to all the circumstances) of the benefit which the employer has derived, or may reasonably be expected to derive, from the patent.

A patent may benefit an employer by increasing his competitiveness (other competitors being prevented by the patent from obtaining the savings represented by the invention). Or it may bring in licence fees, or may be assigned to a third party for money or money's worth. If the patent is assigned by the employer at an undervalue to a person connected with him, it will be deemed to have been assigned for the amount which it would have been reasonable to expect it to fetch in an arm's length transaction.[97]

Fair share

8–115   The factors to be taken into account when assessing a fair share are laid down in section 41(4) and (5). In the usual case of an invention which has always belonged to the employer, these include: the nature of the employee's employment, his pay and other aspects of his remuneration — including remuneration already given for the invention (for example an award under an internal incentive scheme); the effort and skill he put into making the invention; the effort and skill put in by other employees; the contribution of the employer to the success of the invention.

In the case of an invention assigned to the employer, the following are to be taken into account: any conditions in licences granted under the patent; the extent to which the invention was made jointly with another person; the contribution of the employer.

Payment may be ordered to be made in the form of a lump sum or periodically.

Collective agreements

8–116   The only way in which an employee can be contracted out of the provisions of sections 40 and sections 41 is by a "collective agreement" within the meaning of the Trade Union and Labour Relations (Consolidation) Act 1992. This must be made between the employer (or an association to which it belongs) and a trade union to which the employee belongs.

SUPPLEMENTARY PROTECTION CERTIFICATES

Introduction

8–117   The scheme of the 1977 Act was to give patents for all inventions the same length of protection: 20 years from date of filing. Unlike earlier Acts there were no provisions for applying for extensions of the monopoly period in special cases.

This was perceived to be unfair to certain industries, where it could sometimes take a long time before a product could be put on the market, because of regulatory requirements. In effect, in the case of such products, the period during which the patent monopoly could be enjoyed was not 20 years, but 20 years minus the length of time needed to obtain regulatory approval.

---

[97] s. 41(2).

As a result, the E.C. have introduced by direct legislation[98] a scheme for extending the term of the protection granted by patents for "medicinal products" and "plant protection products". This extension is obtained by applying for a "supplementary protection certificate" (or "SPC") from the relevant Patent Office.

Products for which SPCs are granted

8–118 SPCs are only available for "medicinal" and "plant protection" products. A medicinal product is defined in Article 1 of Regulation 1768/92. Essentially it is any substance (or combination of substances) for treating or preventing disease, or making a medical diagnosis, or correcting or modifying physiological functions, in humans or animals. Plant protection products are defined in Article 1 of Regulation 1610/96 and essentially include products designed to protect plants from pests or diseases, including herbicides and pesticides, or to regulate growth.

How to apply

8–119 An SPC can only be applied for once a "basic patent" covering the medicinal or plant protection product in question is in force in the relevant country, and authorisation qualifying as an "administrative authorisation procedure" under E.C. Directives has been granted by the relevant authority to put the product on the market.[99]

An SPC must be applied for from the local patent office (in the United Kingdom the United Kingdom Patent Office) within six months of the date of the grant of authorisation. If the "basic patent" is still not granted at that time, it should be made within six months of the grant of the patent.[1]

Effect of SPC

8–120 The SPC gives a "hybrid" right. It is defined as the same right as granted by the basic patent (Article 5) but it is limited to the product actually covered by the authorisation (Article 4). So where the claims of the basic patent go wider than the scope of the authorisation, the SPC cuts down the scope of protection during the extra period after the expiry of the basic patent.

Length of protection

8–121 The length of protection granted by an SPC depends on the length of time it took to obtain authorisation. It takes effect after expiry of the basic patent and lasts for the period which elapsed between the date of application for the basic patent and the date of authorisation, minus five years. However, it cannot last for more than five years from the date on which it took effect.[2]

Thus if the patent was applied for in 1980, and authorisation was given in 1988, the SPC would take effect in the year 2000 and last for three years. If however, authorisation was not given until 1993, it would last for five years.

Lapse or revocation of SPC

8–122 An SPC will be deemed invalid if the basic patent lapses before its allotted term has expired, or is revoked. It can also be challenged on the grounds that it was

---

[98] Reg. 1768/92 and Reg. 1610/96.
[99] Article 2.
[1] Article 7.
[2] Article 13.

granted in breach of the Regulation, or on the grounds of the invalidity of the basic patent (even after that patent has expired).[3]

LITIGATION

8–123   It is not the purpose of this Chapter to consider the remedies available for patent infringement, nor the legal procedures for obtaining them. These matters are dealt with in detail in Chapter 24.

However, for the sake of completeness, it may be worth summarising here the court proceedings available in the case of patent disputes.

Infringement proceedings

8–124   These are commenced in the Patents Court (High Court) or the Patents County Court. The defendant may counterclaim to revoke the patent or some claims of the patent on any of the grounds of invalidity set out in paragraph 8–40 above.

Revocation proceedings

8–125   These may be commenced in the Patents Court or the Patents County Court in the same way as an infringement action (the CPR has swept away the old cumbersome procedure of petitioning to revoke). An application to revoke can, as set out above, be brought by way of counterclaim to an infringement action, or in a threats action where the defendant has sought to justify on the basis of infringement.[4]

They may also be brought in the Patent Office.

Declarations of non-infringement

8–126   This is a useful way for a party who believes he may be sued on a patent, or who wishes to demonstrate to his customers or others that he does not infringe, to obtain a judgment to that effect without having to wait for the patentee to issue proceedings. Under section 71 of the Patents Act, he may bring an application for a declaration that a particular product or process does not infringe. This may be done in the Patents Court (or the Patents County Court) or the Patent Office.

The pre-requisite for bringing such an action is to make an application in writing to the patentee, giving him full particulars in writing of the act in question, and requiring him to acknowledge that it does not infringe. This means that the applicant must provide sufficient details of his product or process for the patentee to be able to make a proper judgment as to whether or not it has the features of the claims of the patent.

Only if the patentee refuses or fails to give such an acknowledgement can proceedings be issued under section 71. However, it should be remembered that the Patents Court has a general jurisdiction to make declarations even where the provisions of section 71 have not been complied with.[5]

Applications for declarations of non-infringement can be combined with invalidity attacks. The most common form is to ask for a declaration that the product or process in question "does not infringe any valid claim" of the patent.

---

[3] Article 15.
[4] s. 74.
[5] This is expressly preserved by s. 71(1). The court will only use this power where it can be shown that there was an existing dispute or issue between the parties (*e.g.* the patentee had made an allegation of infringement). Section 71 is thus wider in that the prospective defendant can "force the issue" even without any allegation having been made.

CHAPTER 9

PATENTING PROCEDURE AND PRACTICE

Derek Chandler, Chartered Patent Agent

INTRODUCTION

9–01 To achieve the grant of a patent an application has to be filed at a Patent Office where the application is examined and, if acceptable, granted as a patent. This Chapter describes some national and international patent systems and summarises the steps in patent procedure by the applicant and a Patent Office. The fees required and the time allowed for the steps are outlined. Special terms used in the procedure are identified by italics. The Chapter is based on procedures and fees at the time of writing and changes in procedures and fees are inevitable so a check must be made for the current conditions. The procedural steps often involve more detailed actions than can be set out here so the use of a professionally qualified adviser, a *patent attorney*, is strongly recommended. All the stages of patent procedure must be carried out accurately and in time. It is important to remember that many time periods for action or payment of fees are inextensible and failure to meet the deadline will cause loss of rights.[1] Even where extension is possible there is generally a penalty fee. The best practice is to act in good time and as completely as possible.

Professional advice and costs

9–02 It is possible for an individual to file an application for a patent at the Patent Office and then deal with all the procedural steps to the grant of a patent. However without professional advice the resulting patent may not be the best that could have been granted and may even have a fatal flaw. Furthermore the individual generally has to be resident in the territory of the Patent Office. For example an individual resident in the United Kingdom can not file an application in the United States Patent Office without providing an address in the United States for correspondence, called an *address for service.* Many Patent Offices strongly advise applicants to have professional advice as early as possible.[2] Professional advice involves costs for the time of the adviser. These costs, and the fees required by a Patent Office, mean that patent action will involve significant expense. It is not easy to quantify the expense for the grant of a patent but at the time of writing the total expense to grant for a United Kingdom patent would be from about £1,800 for fees and professional costs for a straightforward application. The expense will be higher for complicated subject matter and if the application has procedural difficulties, such as a dispute as to the ownership of the invention.

Patents are usually national in effect, that is to say that they generally only relate to one country, and when granted generally require the payment of renewal fees to keep them in effect for whatever term is permitted. To secure patent protection in several

---

[1] If a fee payment fails, *e.g.* no funds to meet a cheque, rights are usually lost.
[2] For example the European Patent Office states in its "Guide for Applicants": "As in any patent procedure and particularly in the case of one involving prior examination, applicants must be thoroughly familiar with patent matters if they are to steer their way successfully through the European procedure. **Those not having the requisite experience are therefore advised to enlist the services of a professional representative before the EPO.**" (Such a representative is generally a European Patent Attorney.)

countries a patent application effective in every country in which protection is required must be made. In most cases it is not practical, as well as being very expensive, to file all these applications for one invention at the same time in each country but if they are not filed at the same time the protection is likely to be different in each country as the later applications will be examined against a later state of the art (including patent applications filed by others in the meantime). To achieve the same effective application filing date in a number of countries it is necessary to take advantage of the international systems which provide this feature.

INTERNATIONAL AGREEMENTS

9–03 Several international agreements now exist which can assist an applicant who wishes to secure patent protection for an invention in several countries. The effect of these is that an *application filing date* in one country can be claimed as a *priority date* for applications for the invention in many other countries or *regions* (a group of countries). *Priority*, based on a priority date, is a valuable feature of patent procedure. Priority enables an applicant having filed an application, the *priority application*, in one country to then file the application, usually within twelve months, in many other countries or regions as if, for many purposes, it had been filed on the priority date. Usually *priority must be claimed when filing* an application. (Priority can exist within a country.)

The International Convention

9–04 The oldest agreement is the *International Convention for the Protection of Industrial Property*, known as the *Paris Convention*. This provides that in the twelve months from the date of the first application in a country of the Convention (the priority date) a *similar application* can be filed in each of the other countries of the Convention and be given the priority date as the date of the application there, provided the claim is made *at filing*. There are now some one hundred and forty Convention countries and further signatories are added from time to time. The major omission is Taiwan, which now has bilateral agreements with the United Kingdom and some other countries.[3] Political issues such as recognition of the existence of one country by another occur from time to time so it is essential to check the current list.

Note that the term "similar application" is used above. The eventual application in the Convention country does not have to be identical to the one establishing the priority date. The Convention requirement is that the document(s) of the priority application(s) "as a whole specifically disclose"[4] the content of the eventual application in a Convention country. This is a valuable feature of the Convention as work on the invention can continue during the twelve months and developments can be included in the eventual Convention application. However professional advice is advisable as to what can safely be included and what should be the subject of a separate application. Several applications, even from more than one Convention country, can be combined into one eventual Convention application at the end of the twelve months from the earliest application but professional advice is essential in these situations to avoid loss of rights.

Applications must be in the language of the relevant country and conform to the

---

[3] Some other countries, including Bangladesh, Pakistan and Thailand, can be regarded as "convention" countries in the U.K. because of specific arrangements that are not part of the International Convention.
[4] Article 4H of the Convention.

rules of that country so there is substantial cost and effort at this early stage in the life of the invention, arising from the need to prepare numerous documents in different languages in a short time. For this reason it is convenient to describe the Convention route as the "multiple national application route".

The Patent Co-operation Treaty (PCT)

9–05   When applications are required in a large number of countries a more convenient route is to make an *international application* using the Patent Co-operation Treaty. Instead of a Convention application having to reach each required country within the Convention year, and often to be translated there before timely filing, an application through the PCT route is filed within the Convention year as one application in a single patent office, generally the applicant's national office, specifying (*designating*) the regions and countries (which are *contracting members* of the PCT) where the applications are to be made and claiming the priority date. Appropriate fees[5] for the number of designations are required but the costs and work needed at this time are greatly reduced when compared with the Convention route. The designations can include the country of the initial application.

Some one hundred countries and regions are members of the PCT and more join each year. Taiwan is not a member. The procedures within the PCT are followed by national and regional procedures which result in national patents in countries selected from among those initially designated. As the PCT procedures are carried out in a single office for all the designations, the investment of money and effort in the early stages of the application is reduced. The PCT procedure includes a search which can indicate the scope of an eventual patent. The final choice of countries from among the initial designations and the substantial expense of translations and national procedures in other languages can usually be delayed for up to 30 months from the earliest date claimed for the application, usually the priority date. Very often by this time the commercial possibilities for the invention will be reasonably clear, and much unnecessary expense may be avoided.

Regional patent systems

9–06   Another development in international patent arrangements is that of *regional patent systems*. A regional patent system is formed when a group of countries agrees to join as a *region* in which at least some of the stages to patent grant are carried out in common, often in one central patent office. Usually national patents are retained but often subject to restriction as to those outside the country who can apply. As with the PCT, designations and appropriate fees are needed on entry to a regional patent system. It is possible to enter a region directly by an application within the Convention year. There are four regional patent systems at present. These systems are the European Patent Organisation (EPO), the Organisation Africaine de la Propriété Industrielle (OAPI), the African Regional Industrial Property Organisation (ARIPO) and the Eurasian Patent Organisation (EAPO). The EPO region established by the European Patent Convention at present includes 20 countries with 6 more countries to which extension can be requested. Note that the EPO is *not* co-extensive with the European Union. The EPO procedure results in grant of a "bundle" of national patents which can be established in their respective countries. The OAPI region at present includes 16 Francophone African countries. The ARIPO, also called the Harare Protocol, region at present includes 12 Anglophone African

---

[5] A maximum designation fee equivalent to six designation fees is required.

countries. The EAPO region at present includes nine of the countries of the old Soviet Union.

A valuable feature of regional patents is that generally the region counts as one designation in the PCT, reducing designation fees on entry to the PCT. Also when the application is taken from the PCT into a region the work and costs are reduced as at least some of the subsequent stages are carried out in the central office of the region, rather than in each national office individually.

A person in the United Kingdom can file a PCT or EPC application as the first application, instead of filing a United Kingdom national application, **provided** it is filed in the United Kingdom Patent Office. See section below.

Community Patent Convention

9–07 A Community Patent Convention exists but has not yet come into force. This would produce a patent system to grant a single patent for the whole of the European Community (now the European Union). The patent would be enforced or revoked in a single court, whose judgment would be effective throughout the Community.

FUNDAMENTALS OF PATENT PROCEDURE

9–08 Patent law is national or regional in origin, so many parts of the procedure for a particular jurisdiction are specific to that jurisdiction. However the main steps of most patent procedures are similar and are now summarised. Details for some specific regions and countries then follow.

Main steps in prosecuting a patent application:

9–09 The main steps are five:

- preparation of an application including a description and usually claims, respectively describing and defining the invention;

- filing the application, with any fee required, at a Patent Office;

- a search by an examiner for documents relevant to the invention;

- an examination of the invention with respect to the search documents to determine whether it fulfils the necessary requirements for grant and a dialogue with the applicant to attempt to settle what can be protected;

- the grant of a patent, sometimes subject to opposition by third parties.

More fully these steps are, from application to grant, as follows:

The application:

9–10

- The applicant prepares a *patent application* which consists of a *request for the grant of a patent,* a *specification* of an *abstract,*[6] a *description,*[6a] *drawings* if

---

[6] A brief description of the invention.
[6a] See 8–07 *et seq.*

appropriate and *claims*,[7, 8] (with a priority claim if appropriate), all in the specified form, *e.g.* page size, type size, number of copies, etc. and usually an *application fee*. (The inventor may not have to be named at this stage.)

- The applicant *files* all of the above at the Patent Office.

Filing:

9–11 The Patent Office:

- checks the application for obvious defects and if satisfied that the application meets the *formal requirements*

- allocates an *application number* and *application filing date* of the date of receipt of the application, and

- sends a *filing receipt* to the applicant.

- Defective applications may be rejected.

- The filing date is the *priority date* of the application for subsequent applications elsewhere and for later relevant applications in the same country.

The Search:

9–12 The Patent Office:

- passes the application to a *search examiner* who searches for documents which relate to the invention set out in the claims. The extent of the search is defined by the particular Office but usually extends to everything published anywhere in any way before the date of the application. In practice the search often extends only to patent documents of the major Patent Offices and to the major relevant journals.[9]

Examination:

9–13
- An examiner examines the description and claims in detail with reference to the search result to determine whether a patent can be granted for the invention set out in the claims.

- *Examination fees* may be required.

- The examination generally uses the criteria for novelty and obviousness in Chapter 8.[10]

---

[7] Before preparing an application the applicant can have a pre-filing search made by a commercial searching organisation and some patent offices, *e.g.* EPO, Sweden and U.K. The EPO may refund part of the cost if the EP application search is reduced by the pre-filing search.

[8] Applications relating to nucleotide or amino acid sequences or to biological substances such as micro organisms involve special requirements, usually sequence listing in written and electronic form or the deposit of the micro organism in a recognised culture collection.

[9] Many Patent Offices publish the application at this stage, often with a search report. If this publication is in the language of the country a form of provisional protection may exist. If the application is granted with claims substantially as first published it may be possible to then sue for damages for infringement from the date of publication.

[10] *Unity of invention* is also considered. If claims relate to more than one invention a divisional application is possible, retaining priority of the first, parent, application. A further fee is due.

- If the examiner is not satisfied with the application the Patent Office sends an *official action* to the applicant advising of the *objections* to grant.

- The applicant can file a *response* including *amendments* within the scope of the application and/or *arguments* to meet the objections, but no *new matter* is permitted.

- The response has to be filed within a set period or the application can fail.

- (Further objections and responses are possible until the examiner is satisfied or *rejects* the application.)

- The applicant can usually *appeal* against rejection of an application.

Grant:

9–14 When the examiner is satisfied and approves the application the Patent Office:

- *grants* a patent with a *serial number*, which is the number of the patent, and a *date of grant*,

- *publishes* the patent specification in the form as granted with the serial number and date of grant and *advertises* the grant in an Official Journal.

- *Grant and/or printing fees* may be required.

- The patent is *enforceable* once granted.

- The patent will continue *in force* as long as all necessary *renewal fees* are paid until the *full term*, usually twenty years from the application date, when the patent *expires*.

Time taken to reach grant

9–15 The procedure from application to grant typically takes between two and four years. However it can be much longer if the objections are not easily resolved.

Withdrawal of an application

9–16 An application can be *withdrawn* during the procedure and, if withdrawn soon enough before preparation for publication, will generally not become available to the public. However an application used to support a priority claim in an application in another country may eventually become available to the public.

Opposition

9–17 An opposition is a procedure by which another person can object to the grant of a patent on specific grounds. The grounds could be that the opponent can produce evidence that the invention in the claims as granted (in some countries the claims when published before grant) is not new or is obvious. The opponent's case will be considered in the Patent Office and the applicant will also be able to advance arguments and amendments. The opposition can result in the loss of the patent or the claims of the patent being amended, if the opponent's objections are effective, or the opposition may fail and the patent survive unchanged. A failed or only partially successful opposition can help to increase the perceived strength of a patent. An opposition often delays the time when a patent may be effectively enforced.

Application after publication or disclosure

9–18   The basic requirement of novelty means that in most jurisdictions (the United States is an exception) the subject matter of the claims must not have been available to the public before the priority date of the application. However, there are some exceptions to this rule. If the disclosure is made in contravention of the applicant's rights, for example in breach of confidence, then, in most countries, an application made within six months will be unaffected by the disclosure. Furthermore, display at certain exhibitions recognised by international agreement and which took place up to six months before the application will not affect the application, provided the occurrence of the display is stated when the application is filed. These exhibitions are rare and the recognition has to be established by the time the exhibition opens.

UNITED KINGDOM PATENT PROCEDURE

9–19   Some documents must be filed in all circumstances. Other documents may be required as indicated.

The application:

9–20   The essential initial documents to be filed at the United Kingdom Patent Office for a United Kingdom application not claiming priority are:

- a description of the invention together with a *request* for a patent identifying the applicant; desirably the request should include the *name, address, United Kingdom address for service* and *signature* of the applicant and a *title* for the invention.

- *No fee* is required for the application itself.

- If the application is the first for the invention, claims need not be filed at this stage. However it is essential that the specification is as detailed as possible to provide adequate basis for the claims when they are filed.

If the application is to claim priority:

- this *must be claimed with the United Kingdom application* which must be filed within twelve months from the earliest priority application date claimed, giving the date and country of the priority application.

- *Supporting documents* are required within set periods, including the number and a *certified copy* of the priority application and a *translation* into English.

- An application claiming priority usually must include claims and an abstract and the fee for preliminary examination and search (see below).

Filing:

9–21   The United Kingdom Patent Office, on receipt of the application:

- checks the application for correct form and, if correct, issues in a few days a filing receipt with the date of filing and an application number, and

- *publishes* the number, date of filing, applicant and title in the *Official Journal* and the index in the Library.[11]

The applicant must, to maintain the application, by the end of twelve months from the earliest date for the application:

- file claims and an abstract, if not already filed[12];
- request the *preliminary examination and search*[13];
- pay the preliminary examination and search fee (£130).
- Also the applicant must, if not the inventor, within sixteen months from the earliest date for the application, supply the *name and address of the inventor* and the applicant's *derivation of title* to the Office.

The preliminary examination and search:

9–22

- A Patent Office examiner searches for documents relevant to the novelty and inventiveness of the features in the claims.[14]
- The relevant documents found by the examiner, called *citations,* form a *search report* sent to the applicant.
- The relevance of a citation is indicated by various letters; the most important are X, Y, A and P. X is for novelty or, when taken alone, inventive step, Y for inventive step, A for background art and P for an *intervening publication*.[15] (A copy of each citation is supplied by the Patent Office.)
- If formal defects are found in the application these will also be notified to the applicant.

The applicant on receiving the search report, which may include citations which *anticipate* (destroy novelty) or make *obvious* the invention, must consider its effect and:

- choose to continue or withdraw the application. (To prevent publication of the application withdrawal must be notified to the Office before preparations for publication are complete.)
- make amendments to the claims and to the description and drawings, (provided they are within the scope of the application as filed and do not add new matter), and/or prepare arguments to distinguish the invention claimed. These can be filed at once or held back until the substantive examination is due.

---

[11] The Office will require the prompt completion of omitted formal matters when giving a filing date.
[12] The claims must have basis in the description.
[13] The U.K. examination is in two stages, *search* and *substantive,* and a fee is required for each. If grant is required quickly the examination procedure can be started at once and the two stages combined by filing the relevant documents and fees with the application.
[14] The examiner may decide that more than one invention is claimed. The application can then be divided, with a further application fee, etc. and each proceeds. The divisional can usually have the priority date of the parent application. Costs double as there are now two applications.
[15] Intervening publications are those applications filed or effective in the U.K. with a date earlier than the application but published after its application date.

Publication of the application:

9–23 Meanwhile the Patent Office:

- prepares the application for publication, allocating a serial number (with suffix A), which is the number for the patent if granted;

- *publishes* the application eighteen months[16] from the earliest filing or priority date claimed for the application, listing citations and adding any amendments to the claims filed in time. The documents for the application now become public;

- publishes the serial number, date, applicant and title in the *Official Journal.*

After publication of the application, but before grant, any other person can submit a reasoned observation to the Patent Office as to whether the invention is a patentable invention. Such observations are sent to the applicant and will be considered as part of the examination if submitted in time, but the person making the observation has no right to a hearing.

Substantive examination:

9–24 The applicant must, within six months of publication of the application:

- request *substantive examination* and pay the fee (£70). Otherwise the application becomes abandoned.

- If examination is requested, a Patent Office examiner determines whether the invention claimed is novel, inventive, and whether the patent otherwise fulfils the requirements for patentability. The search report citations, and possibly further citations, are considered.

If the examiner is satisfied the application is allowed, and if not

- the examiner issues an *official action* to report the adverse result of the examination to the applicant.

The applicant must reply to the official action to continue the application:

- by filing a response with amendments, which must be within the scope of the application and not add new matter, and/or arguments against the objections to bring the application to the state for allowance.[17]

Grant:

9–25 When the examiner is satisfied, the Patent Office:

- grants the patent;

- publishes the patent as granted (under the same serial number as the application, but now with suffix "B", as the patent number); and

- publishes number, date, applicant and title in the *Official Journal.*

---

[16] Publication is delayed until the search report is ready.

[17] Report/response sequence is repeated if necessary until agreement is reached. (If agreement can not be reached the applicant can ask for a hearing. Agreement has to be reached within a set period from the application date, unless a hearing is requested or the applicant proceeds from an unsuccessful hearing to the Patents Court.)

Once granted the patent is enforceable. There is no opposition procedure in the United Kingdom.

Renewal fees have to be paid for the fifth and subsequent years from the filing date of the application to keep the patent in force. These are due before the end of the previous year. The maximum term is twenty years from the filing date of the application. Renewal fees are £50 for year five rising to £400 in year 20.

United Kingdom law requires that anyone resident in the United Kingdom must first file a patent application in the United Kingdom or have been given permission by the United Kingdom Patent Office to file outside the United Kingdom before so filing. However this does not prevent an applicant making use of the European Patent Convention (EPC) or the Patent Co-operation Treaty (PCT) for the first filing of the application as applications for these jurisdictions can be filed in the United Kingdom Patent Office.

Within twelve months of the date of filing an application in the United Kingdom Office it is possible to file a fresh similar application claiming the filing date of the earlier application. Provided that the subject matter of the new application was disclosed in the earlier application, the new application will be entitled to claim priority from the filing date of the earlier (see previous Chapter for more detail on ability to claim "priority"). The 20–year term will however run from the date of the fresh application, giving a potential expiry date 21 years after the priority date.

EUROPEAN PATENT CONVENTION PROCEDURE

9–26 Some documents must be filed initially in all circumstances. Other documents may be required as indicated.

The application:

9–27 The essential initial documentation for an EPC application not claiming priority is:

- a description, drawings if appropriate, an abstract and claims;

- a request for the grant of a patent using the European Patent Office (EPO) form, signed and including the title of the invention, the name and address of the applicant, the designation of the countries within the EPC region and the countries with which the EPO has *extension agreements*[18] to which the application is to apply and appointing a representative if appropriate[19];

- the appropriate fees.[20, 21]

---

[18] Extension agreements bring countries which are not members of the EPC within the scope of the EPC procedure.
[19] The application may be filed by any natural or legal person or any body equivalent to a legal person regardless of nationality or place of residence or business. The applicant can appoint a representative but this must be someone entitled to act before the EPO. The EPO advises applicants to appoint a representative unless experienced in EPO procedures. An applicant from outside the EPC region must appoint a representative. The EPO sends correspondence to the representative.
[20] Several fees are required with the application; these are a filing fee (£77), a search fee (£417), a designation fee (£46, maximum designation fee £322); designations in excess of seven are free; extension country fee £62) for each country designated and, for each claim above ten, an excess claims fee (£24).
[21] The fees should be paid with the application but all can be paid later within set terms. Failure to pay a fee in time will almost certainly result in the application being deemed to be withdrawn and any hope of a patent being lost. As a safeguard against faulty designation an "all designations" request can be made and resolved later by paying the appropriate designation fees six months after the publication of the search report.

- The rules for the content of the specification and claims are detailed and strict and failure to follow these will make the subsequent procedures more difficult and can prejudice the application. The regulations for European applications are also very detailed as many different combinations of applicants, inventors, nationality and residence have been allowed for. The regulations must be properly understood and observed to avoid the risk of the application failing.

- The official languages of the EPO are English, French and German. One of these languages must be used for the application. However, applicants from an EPC country where a language other than English, French or German is an official language may use that language but must file a translation into one of the three languages within a set term. The official language used initially is the *language of the proceedings.*

- If the applicant is not the inventor the inventor must be identified together with the applicant's right to the patent on a specific form. This must be filed within sixteen months from the earliest date of the application.

If the application is to claim priority, a claim to that effect, including at least the date and country of the priority application, must be made at the time of filing of the application.

- The number and a certified copy of a priority claim application must be filed within 16 months of the priority date. A translation can be required within a set period.

Filing:

9–28 United Kingdom residents may file a European patent application at the United Kingdom Patent Office or, with permission given as above, at the EPO in Munich or The Hague. The rules for other countries depend on their national law.

When the application is filed at the EPO or the appropriate recognised Office:

- the Office makes a preliminary check of the application for conformity with the regulations and to see if it is in proper form;

- issues an application number and filing date and filing receipt.

- In the case of an application filed in a recognised office, it is forwarded to the EPO.

The search and search report:

9–29 At the EPO:

- an examiner in the Search Division searches for documents relevant to the application[22];

- prepares a search report listing the citations in categories including X, Y, A and P (see United Kingdom procedure above).

---

[22] The search is among everything made available to the public by means of written or oral description, by use, or in any other way, before the date of filing of the European patent application (or priority date). For the purpose of determining novelty, European applications for a designated country of a date earlier than the application but published after the application date are also searched.

- The search report is sent with a copy of each citation to the representative, if appointed, or the applicant. The report is prepared and sent out as soon as possible, although this may take several months.

The applicant, on receiving the search report, must decide whether to proceed with the substantive examination or withdraw the application:

- As with United Kingdom applications, amendment of the description, drawings and claims is permitted at this point provided it is within the scope of the application and does not add new matter. Amended claims are included in the published application if received in time.
- Renewal fees are due while the application is pending.[23]

Publication by the EPO:

9–30

- The application is published, with the search report if available, under a serial number eighteen months after the earliest filing or priority date claimed for the application.[24] The serial number is that for the patent, if granted.
- The publication of the search report is *mentioned* in the European Patent Bulletin and the EPO *communicates* the date to the applicant.

After publication of the application, any person can submit a reasoned observation to the EPO as to whether the invention is a patentable invention. Such observations are dealt with in the same way as in the United Kingdom (see above, 9–23).

Substantive examination at the EPO:

9–31 If the applicant has decided to proceed:

- substantive examination must be requested within six months of the date of the Bulletin and the fee paid (£866). (Extension of one month with surcharge.);
- the application is sent to the *Examining Division* for substantive examination based on the search report and any comments and amendments filed by the applicant;
- the examiner considers novelty, inventiveness, industrial applicability and other criteria for suitability for patent protection;
- the examiner assesses whether the application is in condition for grant. If the application is in condition for grant the Examining Division decides to grant a European patent and invites the applicant to approve the text (see below).
- If not, the objections are reported to the applicant in a *first reasoned communication* with a time limit for reply (usually two or four months).

The applicant on receiving the communication

- must then file a *full reply* to the objections. This may include amendments to deal with those objections, and/or arguments against the objections. The time limit can usually be extended but this may delay the whole procedure.

---

[23] Year 3 £232, rising to £618 in Year 10 and thereafter.
[24] The EPO publishes the search report with the application if it is ready, under a serial number prefixed with A1. If the search report is not available the publication has the prefix A2 and the search report is published when available with the prefix A3.

9–32 EPO:

- The examiner considers the reply. If the application is in a condition for grant the Examining Division decides to grant a European patent and invites the applicant to approve the text of the specification including the claims.

- If the examiner is still not satisfied a *further communication* is issued.

The applicant continues the procedure:

- If a further communication has been sent, a written response or telephone discussion (and, usually, a face-to-face discussion[25] if required) are used to continue the procedure until agreement is reached and the applicant is invited to approve the text.

- If the applicant is not reaching agreement with the examiner the applicant can request *oral proceedings*[26] (no fee).

- Failing agreement with the examiner the Examining Division decides whether the application is to be refused.

- If grant is refused the applicant can file an *appeal*[27] (£618).

The applicant when invited to approve the text:

- If in agreement sends approval to EPO within four months. Amendments may be allowed. Failure to agree on a text for which approval has been invited may lead to refusal of the application.

- Following approval of the text by the applicant, the Examining Division sets a term (usually three months) for the applicant to complete the grant procedure.

Grant of European patent:

9–33 The applicant, if requiring grant, must within the term set:

- pay the fees due[28];

- file a translation of the claims of the approved text into the other two Official languages;

- file a translation of the priority application if not already filed.

---

[25] Face-to-face discussion is usually with the representative but can include the inventor and/or applicant.
[26] "Oral proceedings" is a formal procedure, usually before three examiners, and requires the preparation of arguments and evidence. Alternative claims need to be prepared as the procedure often operates by eliminating claims during the hearing. The examiners can raise a wide range of issues during the hearing so very thorough preparation, forseeing all likely issues, is needed. This requires a considerable expenditure of time and effort by the applicant and the representative as well as the cost of attending the hearing, usually in Munich but sometimes at the Hague Office of the EPO. As well as the representative it is often appropriate to instruct Counsel to present the case. Video conferencing for oral proceedings is being developed.
[27] Appeals are to the Legal or Technical Board of Appeal. Major legal issues may go to an Enlarged Board of Appeal. An appeal is a substantial action and requires time and expense in preparing the arguments and evidence.
[28] The fees are: grant and printing fee (£433 + £6.20 per page above 35), any excess claim fees now due and any renewal fees and additional fees due.

When the applicant has satisfied the grant requirements the EPO:

- mentions the grant in the European Patent Bulletin. The date of grant is the date of the Bulletin, which is also the date of the publication of the approved text of the specification with the above serial number, but now with the identifier B1, as the patent number;

- sends the European patent certificate with an attached specification to the *proprietor* (hitherto the applicant).

Transfer to the National patent systems:

9–34

- The proprietor decides in which of the designated countries and extension countries patent protection is required and requests the *validation* of the European patent in each of these. (The patent is already granted.)

Each country has specific requirements for fees and translations and the periods for completing these, to ensure validation of a patent. The proprietor is responsible for observing them, without notification. The patent term runs from the filing date of the application which is granted as a European patent, not the priority date. Once validated as a national patent, national law and procedure applies in respect of revocation, infringement and other matters.

Opposition

9–35   The European patent system provides for opposition by third parties to the grant of a European patent as part of the procedure in the European Patent Office. While it is possible to attack a European patent in any country in which it has been validated, this attack will only be effective in that country and would have to be repeated in each validated country. By successfully opposing the grant in the EPO it is possible to restrict the scope of the patent or to prevent the patent being effective for all of the designated countries using a relatively simple procedure. The opposition is thus a very significant part of the European procedure, both for an applicant and potential opponents.

Timing and procedure for oppositions:

9–36   An opposition is made by filing a notice of opposition and fee at the EPO within nine months of the mention of the grant in the Bulletin. The notice of opposition must include a reasoned statement[29] of the grounds of opposition, details of the opponent and any representative. Use of the EPO form is advisable.
  On receipt of the notice the EPO:

- checks the opposition for admissibility;

- communicates deficiencies to the opponent with time for correction;

---

[29] The reasoned statement must give at least one ground for opposition of those set out in Article 100 EPC, which, briefly, are that the invention is not novel, inventive and industrially applicable, that the description is not sufficient and that the subject matter extends beyond the application as filed. The preparation of the statement thus requires substantial time and effort during the 9 month opposition period. The fee is £371.

- communicates the opposition to the proprietor;

- invites the proprietor to file observations and amendments in a set term.

The EPO then:

- requests the Opposition Division to examine the opposition to determine whether it is significant.

- The Opposition Division continues the procedure with observations and amendments from the parties and, usually, oral proceedings before reaching a decision.

- An appeal against the decision is possible.

- If the opposition succeeds the European patent is revoked.

If the patent is not revoked the Proprietor can continue the procedure:

- If the patent requires amendment the proprietor has to pay a printing fee and file translations for the revised specification. Translations may be needed for designated countries as well.

EPO hearings:

9–37 All proceedings in the EPO require very thorough preparation. Material which the EPO considers relevant can be introduced at a late stage, even at the hearing, as the aim of the EPO is to ensure that all relevant issues are dealt with at the hearing. (This applies to oral proceedings, appeals and oppositions.) Such material need not be in the language of the proceedings. As several parties are usually involved, a hearing is rarely postponed or adjourned. The proceedings are very intensive. A professional representative is considered able to bind the client (the applicant) and will rarely be able to seek instructions during a hearing. The comments above about the preparations for oral proceedings during examination apply even more strongly to appeals and oppositions. Extensive searching and the collection of substantial amounts of evidence may be required. The presence of expert witnesses as well as the briefing of Counsel to present the case is possible. An opponent must also be prepared for this scale of involvement. It is usually essential that alternative claim strategies (known as auxiliary requests) acceptable to the applicant/proprietor are prepared in detail for presentation during the procedure in case the Office decides that the original claims arc too broad . Equally the opponent must be ready to counter alternatives proposed by the applicant/proprietor.

PATENT COOPERATION TREATY PATENT APPLICATION PROCEDURE

9–38 Procedure under the PCT is conveniently divided into two phases, the international followed by the national/regional. Both phases are needed to achieve grant of a patent, which will be national or regional. The Patent Cooperation Treaty text is in several chapters. In the international phase Chapter I deals with the international application and the international search, Chapter II deals with the international preliminary examination. The terms Chapter I and Chapter II are widely used to identify these parts of the procedure.

The international phase

9–39   The international phase is conveniently divided into four main steps. These are in Chapter I the filing and processing of the application and allocating an application number in a *Receiving Office*, the *establishing* of the *international search report* by an *International Searching Authority*, the publishing of the *international application* with the international search report under a publication number and, at the option of the applicant where permitted, in Chapter II the establishing of the *international preliminary examination report* by an *International Preliminary Examining Authority*.

Chapter I

The application:

9–40   The essential initial documentation for a PCT application not claiming priority is:

• a description, drawings if appropriate, a title, an abstract and claims, a request for the grant of a patent using the PCT form, signed and including the title of the invention, the name, address and nationality of the applicant,[30] the designation of the countries and regions within the PCT to which the application is to apply and appointing a representative if appropriate, together with the fees.[31, 32]

• The International Searching Authority must be selected if a choice is possible.

• The inventor should be identified, and must be if the USA is designated, as the inventor has to be the applicant in the USA.

• The specification must be in a language acceptable to the receiving office but a translation may be required for the search authority.

• An assignment from the inventor to the applicant need not be submitted at this stage but may be required at entry to the national phase.

• If the application is to claim priority this claim in the form of the date, serial number and country or regional office of the priority application must be made at the filing of the application.[33]

• A certified copy of the priority claim application is normally required within 16 months of the priority date.

---

[30] The application may be filed by any natural or legal person or any body equivalent to a legal person having nationality of or residence or place of business in a PCT contracting member. The applicant may have to appoint a representative and this must be someone entitled to act before the receiving office. The PCT authority advises applicants to appoint a representative unless experienced in PCT procedures. Correspondence is sent to the representative.

[31] For applications filed in the U.K., several fees are required with the application; these are a transmittal fee (£55), a basic fee (£264), a designation fee of £56, for each country designated with a maximum of six fees (£336) — designations in excess of six are free — a search fee (£605) and, for each specification sheet above 30, an excess sheet fee (£6).

[32] The fees should be paid with the application but all can be paid later within set terms. Failure to pay a fee in time will almost certainly result in the application being deemed to be withdrawn and any hope of a patent being lost. As a safeguard against faulty designation a "precautionary designation" request can be made and resolved later. The fees can be reduced by filing in a suitable electronic form on a computer diskette.

[33] As the PCT is based on the International Convention the basis for the priority claim should be in a Convention country, otherwise not all countries may recognise the priority claim.

The regulations for PCT applications including the content of the specification, claims, drawings and abstract and the request are detailed and strict as many different combinations of applicants, inventors, nationality and residence have been allowed for. The regulations must be properly understood and observed to avoid any risk of the application failing.

United Kingdom residents may file a PCT patent application at the United Kingdom Patent Office or, with permission given as above, at the EPO or the *International Bureau of WIPO (World Intellectual Property Organisation)* Geneva. The rules for other countries depend on their national law.

Filing:

9–41 The application must be filed at an appropriate Receiving Office.
The Receiving Office:

- makes a preliminary check of the application for conformity with the regulations and, if in proper form,

- issues an application number and filing date and filing receipt.

- The Office transmits appropriate documents to the International Searching Authority and the International Bureau. Limited bibliographic detail of the application can be published by a designated Office.

The International Searching Authority Search:

9–42 At the International Searching Authority:

- an examiner searches for relevant documents[34];

- checks for unity of invention and will require further search fees if more than one invention is present and the applicant wants the others to be searched;

- prepares a search report classifying citations as X, Y, A and P (see United Kingdom procedure above).

- The search report is prepared as soon as possible; the aim is to issue the report within about four months from filing the application.

- The search report is sent to the representative, if appointed, or the applicant and to the International Bureau.

The applicant, on receiving the search report, must decide whether to proceed or withdraw the application:

- amendments to the claims, but not the description, within the original scope, and a brief explanatory statement, can be filed after receipt of the search report. These will be included in the publication, if timely.

---

[34] The search is among everything made available to the public by means of written or oral description, by use, or in any other way, before the date of filing of the PCT patent application (or priority date) subject to the availability of the material. A "minimum documentation" is specified by the PCT. For the purpose of determining novelty, respective applications **for designated countries and regions** of a date earlier than the application but published after the application date are relevant.

The International Bureau:

9–43   Meanwhile the International Bureau:

- publishes the application, and the search report if available, with the *international publication serial number* 18 months after the earliest application date;

- sends copies of the publication to the applicant and the Patent Offices of designated countries and regions;

- publishes details of the application, and the abstract and drawing in the *PCT Gazette*.

Chapter II

9–44   Chapter I, the filing and processing of the application in a Receiving Office and allocating an application number, the establishing of the international search report by an International Searching Authority, the publishing of the international application with the international search report with a publication number, is now complete. The applicant can either enter the national/regional phase or choose to continue with the application in Chapter II.

Chapter II or national phase — time for action:

9–45   An applicant must normally proceed with the national phase (described below) within 20 months of the earlier of the application and priority dates. Some members may allow more time. An applicant should start the Chapter II procedure within 19 months from the earlier of the application and the priority dates as then the substantial costs and work required for entry to the relevant national/regional phases can be delayed until 30 months from the earlier of the application and priority dates. Some members may allow more time.

Chapter II preliminary examination:

9–46   The international preliminary examination of Chapter II is defined as "a preliminary and non-binding opinion on the questions whether the claimed invention appears to be novel, to involve an inventive step (to be non-obvious), and to be industrially applicable". The report of the examination will assist the applicant in deciding whether to continue to the national/regional phase. The costs of entering the national/regional phase are deferred until nearly 30 months from the earliest date for the application, giving a valuable extension of time for assessing the prospects for the invention and seeking possible licensees before a major expense is incurred. There are of course costs involved in "buying" this extension of time but these are much smaller than the costs for a large filing programme in the national phase or national applications *ab initio*. The search and examination can also lead to amendments which may reduce the costs in each subsequent national/regional application. Much duplication of work is avoided as amendments are in a single document rather than several parallel documents in different languages and jurisdictions.

The applicant can file the *demand* for international preliminary examination at any time in the international phase but to make best use of the procedure the demand should be filed as early as possible (in practice usually by 19 months from the filing date or priority date) as the report of the examination has to be established within 28 months of the earlier of the application and priority dates. Conveniently it is filed when the search report has been considered and any amendments to the specification and claims settled, although this must not be delayed as the time for the examination is reduced by such delay and the full benefit of the examination lost. It is of course possible that the applicant on considering the search report will conclude that the effect of the citations is to make the prospect of a patent so poor that the applicant withdraws the application. If the application is withdrawn before preparations for publication are complete (usually at about 17 months) it is not published.

Chapter II procedure:

The Demand:

9–47   The applicant prepares the Chapter II demand.

- The demand identifies the applicant and the application and includes a petition for the examination and

- the *election* from among the designations of a member or members of the PCT bound by Chapter II.

- Amendments to the description and claims can be made before the examination and must be mentioned in the demand. If earlier amendments to the claims were filed the relevance to these of the later amendments must be stated. Amendments must be submitted in proper form so that the content of the amended specification is clear.

- The demand and the fee[35] are filed before the end of the 19 month term with the International Preliminary Examining Authority appropriate to the receiving Office and the language of the application.

The International Preliminary Examination:

9–48 The International Preliminary Examining Authority:

- examines the application for lack of novelty, inventiveness or industrial applicability of the claimed invention and checks for other deficiencies and shortcomings such as lack of clarity in the description, drawings or claims, improper claim structures and amendment beyond the disclosure of the application as filed.

- A written opinion commenting on any such matters is sent to the applicant with a term for reply, usually two months.

The applicant considers the opinion:

- A response with amendments, arguments and requests for clarification, can be made if desired but is not compulsory. The applicant can decide to leave changes until later, particularly if he considers that the changes suggested by the opinion are suitable for the later phases of the application and accepts that this may mean a more severe International Preliminary Examination Report by the Authority.

If the Authority has no comments, no opinion is issued and the Authority proceeds with the Report.

---

[35] The fees are £927 for the examination and £94 handling charge when the EPO is the International Preliminary Examining Authority.

The International Preliminary Examination Report:

9–49   The International Preliminary Examining Authority:

- considers any response from the applicant if time permits before the 28 month deadline for the report and then prepares the report. (There may be time for more than one written opinion and response by the applicant.)

- The report includes a simple yes/no for each claim examined on the criteria of novelty, inventive step and industrial applicability. The relevant citations are identified and concise reasons given. New citations may be made. Unity of invention is also considered. The report indicates which amendments were taken into account.

- The report is sent to the applicant and to the International Bureau.

9–50   International Bureau:

- Copies of the report are sent to the Offices of the elected members.

- The applicant considers the report, and

- decides on entry to the national phase, the international phase being complete.

- Sometimes the applicant sends comments on the report to the Offices elected for the national phase.[36]

The National/Regional Phase

9–51   Entry to the national/regional phase before or after Chapter II:
The applicant has to start the entry:

- National/regional phase entry is the **exclusive responsibility** of the applicant.

- Failure to meet the time for action is usually fatal to the application.

- In the absence of preliminary examination the time for action is within 20 months of the earliest date for the application.

- With preliminary examination the time for action is 30 months.

- (Some members of the PCT allow one or two months longer.)

- Entry requires the filing in each Office in which the application is to continue of a translation, if required, and the payment of the national/regional fee.

- Other actions not subject to the PCT time limit, for example filing documents executed by the applicant and/or the inventor, may be required within set time limits by individual Offices and there is no obligation to advise the applicant of these.

- Annual or renewal fees may also be due in a designated or elected Office.

- The name and address of the inventor, if not already provided during the International Phase, may be required by certain Offices.

- From this point the procedure is that of the respective national/regional Office.

Although no special form is required many Offices have suitable forms available and their use is recommended where possible.

---

[36] This is comparatively rare.

UNITED STATES PATENT PROCEDURE

9–52 United States patent procedure has many aspects in common with the basic outline given earlier. However it is important to bear in mind that United States patent law differs from that of Europe in many important respects. Most significantly, the United States utilises a "first to invent" system, under which patents are awarded to those who first *conceive* and then *diligently reduce to practice* a claimed invention. Thus, while the actual filing date of a patent application may give some general indication of priority, it is by no means dispositive. Secondly, the United States has no procedure for a party to appear before the Patent Office to oppose the issuance of a pending application.[37] This means that most validity issues can only be tested judicially — often by a jury trial — after an infringement case has been filed by the patentee.[37a] In judicial proceedings, any issued patent has a presumption of validity, which can only be rebutted by clear and convincing evidence.

There are other significant differences in procedure, and these are outlined below.

The United States application:

9–53

- The initial applicant must be the *inventor*, not an assignee.

- The request is in the form of an oath and declaration, signed by the inventor.

- The assignment can be filed with the application or up to grant to come into effect on grant, which is then to the assignee.

- The specification must include the *best mode* of performing the invention.

- A United States patent specification usually includes a much more detailed description than one for the United Kingdom or the EPO and failure to use the preferred form can delay the progress of the application and/or lead to loss of rights.

- The claim structure for United States patents also is different, usually having many more independent claims.

- The *duty of candour* requires the applicant and anyone associated with the applicant to disclose to the Patent Office anything known that could be *material* to the consideration of the application.

- The applicant must provide an *information disclosure* statement of **all** the known information which could be material to the consideration of the application by the Patent Office.

- Failure of candour is *fraud on the Patent Office* and usually results in the loss of the application or patent. There can be other penalties including *multiplied damages* and action against attorneys.

- The information disclosure must be updated with any fresh information, usually within three months during prosecution, for example with citations on the application in other countries.

---

[37] "Re-examination" is a mechanism whereby an opposing party can put additional prior art references before the Patent Office post-issuance of a patent. The re-examination procedure does not permit the opposing party to appear before the Patent Office, however. All re-examination hearings are conducted by the Patent Office, on an *ex parte* basis with only the patentee participating. See 9–59 below.

[37a] In certain circumstances, a potential infringer may also commence "declaratory relief" proceedings.

- The application can usually be filed up to twelve months after the applicant has published details of the invention anywhere.

- The application fee is $690 and excess claim fees are levied.[38]

Patent Office procedure after filing:

9–54

- A combined search and substantive examination starts at once. The examiner's first communication to the applicant, or assignee if recorded, lists citations, gives details of any objections to the form of the claims or to grant on the basis of lack of novelty and inventiveness or other grounds and expresses these as a

- *rejection*[39] with a time limit for reply (usually three months); or

- if there are no objections the application is *allowed*.

The applicant considers the rejection and:

- files a response which must be *complete*, refuting or amending to deal with every objection, otherwise a second communication of a *final rejection* is likely.

- Arguments and admissions during prosecution can constitute an *estoppel,* preventing a later change of position during (for instance) infringement proceedings. This is commonly known as "file wrapper estoppel".

Patent Office:

9–55

- The examiner considers the response and if satisfied can allow the application. But the examiner can raise further objections relating to the response and issue a second communication which is likely to be a *final rejection*.

The applicant has one more opportunity to amend:

- after final rejection the applicant can file a further response which does not raise *new issues* and makes amendments to fully meet objections. This can result in allowance. A brief *interview* with the examiner can be requested and is usually permitted.

- If the applicant does not achieve allowance after the second communication has been considered, the application can be *continued* on the basis of the original application and any amendments already made by paying a further application fee. The examination restarts and again two further communications may be possible before final rejection if allowance is not achieved.

- Alternatively an *appeal* (notice fee $300) to the Board of Appeals, within the Patent Office, can be filed. This requires a written submission (brief fee $300).

Further appeal is possible, to a Federal District Court.

---

[38] "Small entities" and "not for profit" organisations may pay lower fees.
[39] U.S. applications often include several groups of claims each headed by an independent claim. If the examiner decides that an independent claim is too different in scope from the others a *restriction requirement* is made. The applicant then *elects* the claims to remain in the application and can file a divisional application for the unelected claims.

Grant of United States patent:

9–56   Allowance is followed by grant (issue fee $1210) and publication of the allowed patent.

   Renewal fees are due at 3 and a half, 7 and a half and 11 and a half years and are $830, $1,900 and $2,910. The term now runs from the earliest United States application date.

   Patent term restoration, similar to a Supplementary Protection Certificate, (see Chapter 8, paragraph 8–117) is available. Some extension may be given for interference delays.

Objections to applications and patents in United States procedure:

9–57   While there is no procedure called opposition, two other procedures can achieve similar results. These procedures are called *interference* and *re-examination*.

Interference

9–58   Because the United States is a "first to invent" jurisdiction, there needs to be a mechanism to determine which patentee did actually invent the claimed subject matter first. An interference can be declared when a claim in an application is for the same or substantially the same subject matter as a claim in an issued patent or an earlier-filed application. This generally arises in one of two ways. First, an interference can be declared by the Patent Office when a patent application is made which would interfere with a pending application or an issued but unexpired patent. Alternatively, an interference can be brought about by a pending applicant that wishes to challenge the priority of another's newly-issued patent (typically when the claims in the pending application could support or overlap claims in the granted application). The pending applicant can then copy claims from the granted application to provoke an interference.

   When an interference is declared, the parties submit evidence and arguments to support the priority date and patentability of their claims. Establishing the priority date involves *swearing back* to a date before the date of an application. Swearing back is based on evidence of *conception* and diligent *reduction to practice* of the invention at an earlier date. Evidence can be in the form of signed and dated laboratory notebooks, testimony from those involved and documents, for example orders for the manufacture of material relevant to the invention. Conception is the earliest date at which the inventive concept was formulated; reduction to practice is the work of implementing the concept to achieve or make the invention, and can be the date of filing the relevant patent application. Progress from conception to reduction to practice to filing an application must be *diligent* in order to maintain a priority. Put differently, this means that there must not have been inexcusable delay.

   The evidence and arguments are considered by the Patent Office and a decision given as to the claims, if any, allowable to each applicant or patentee. The Patent Office will also consider the patentability of the claimed interfering subject matter. The Patent Office's interference decision can be judicially reviewed on appeal. Parties can settle an interference but must file any settlement agreement with the Patent Office.

Re-examination:

9–59   Re-examination occurs when prior art that could affect the patentability of allowed claims is brought to the notice of the Patent Office, with a fee, and the Office asked to examine the allowed claims to see whether they are still patentable. Anyone, including the patentee, can request re-examination.

Reissue:

9–60 *Reissue* is another significant element of United States procedure. A patentee may apply for a reissue patent to correct errors made without deceptive intent that cause the patent to be wholly or partly inoperative. The most typical reason for seeking reissue is the patentee claiming more or less than he has a right to claim. A reissue application that seeks to broaden any claim must be filed within two years of the original patent's issue. Reissue claims must be for an invention disclosed in the original patent specification.

The original patent is surrendered upon reissue.[40] The reissued patent is effective as of its issue date and lasts for the original patent's unexpired term.

Dedicated to the public:

9–61 Invalid claims can be *disclaimed* leaving valid claims in the patent. *Dedication to the public* is a way for a patentee to abandon a patent by making the content free of patent cover.

Changes in United States procedure

9–62 Recent changes in United States law provide for provisional applications and the publication of some pending applications. Professional advice on these matters, as with all aspects of United States patent procedure, is strongly recommended.

JAPANESE PATENT PROCEDURE

9–63 Japanese patent procedure has much in common with the basic outline while having some distinct features.

Deferred examination permits the examination of an application to be deferred for up to seven years from the date of the application in Japan. From October 2001 the deferral is only three years. The application is published, with amendments if made, 18 months from the earliest date for the application. After the application is examined and granted it is again published and an opposition can be filed.

The term of a patent is 20 years from application. Annual fees accumulate until grant when they become payable.

GENERAL OBSERVATIONS ON ISSUES RELATED TO PATENTING PROCEDURE

Laboratory notebooks

9–64 Records of work leading to an invention can be very valuable evidence, particularly in post grant proceedings and in United States Patent Office proceedings. It is therefore essential that verifiable contemporary records are made and preserved. Records must be kept in hard-bound books with numbered pages. If possible, different work should be recorded in different books. Every page must be dated and signed by the inventor(s) and an independent witness who verifies that the page was read and understood and who preferably saw the work that is recorded. Full details of the work should be written up as soon as possible, preferably as carried out, and should be restricted to facts without opinions. A daily record of work is the longest safe interval. For some work intervals of only hours or minutes may be

---

[40] The reissued patent is printed with brackets and italics to indicate changes.

needed for a full record. Written records should be in black permanent ink with any attachments to a page permanently glued in place and signed over and witnessed. No blank spaces or pages are allowed. These should be crossed through, signed and witnessed. Errors should be crossed through to remain legible, an explanation added and signed and witnessed. Pages must not be removed or obscured. Record material that cannot be glued into the book must be preserved securely and a cross-reference made between the material and the book recording the work. All the record material must be kept together and preserved carefully.

Searching

9–65 Searches are an important part of patent procedure but their compass and limitations need to be understood. Searching has developed alongside the patent system because patentability is established by comparison with what is known. Accordingly existing knowledge had to be organised for easy retrieval. Classification systems developed and as time passed material was added, particularly patent abstracts or abridgements. Later on indexes and abstracts of technical journals were produced. Much of this material was on paper but when punched cards and then computers became available these were used to produce databases. The computer databases permit much more thorough searching including full-text searching and keywords logically linked into a search profile. Before requesting a search, several questions should be answered: why is the search being made, where is the search to be made, who is to make the search and who is to assess the material found?

Why are searches made?

9–66 The term search is usually used for a search to determine whether a proposal is both novel and inventive. This may be during the application procedure, by a patent office examiner, or before filing an application or in respect of a proposed commercial activity. Novelty exists if the proposal as such can not be found in any publication of any form anywhere. Inventiveness is the determination whether the "distance" of a proposal from what is known, as judged by the notional person skilled in the art of the proposal, is great enough to have required more than routine skill to produce the proposal.

A search before filing a patent application can avoid wasting effort on an invention which can not mature into a useful patent. Money and effort spent before filing frequently saves much more than would otherwise be spent prosecuting the application to no avail. Another widely used search is that for freedom to use. Typically a new product or process is being considered and those responsible want to assess whether patent or other infringement problems exist. A standing or watching search is often maintained by a manufacturer in a technical area or with reference to competitors of commercial importance to identify competitive activity. A less common search is one made to predict technology trends by quantifying inventive activity in various areas.

Where are searches made?

9–67 The PCT refers to a *minimum documentation* for the International Search Procedure. This includes, broadly, published post–1919 patents and applications of France, Germany, Japan, the Soviet Union, Switzerland, the United Kingdom and the United States of America together with some 135 technical periodicals from various dates but if other material is available it should be included. This body of material is still only a part of that available to searchers if they are provided with sufficient time and money. The aim of this minimum documentation is to have a search base large and diverse enough to give confidence that any search made in it will find relevant material if it exists. Each of the major patent offices has a very extensive and usually larger collection of material for searches by their own examiners and some make these collections open to the public to make searches. However, the body of material that could be searched is enormous. Some material, such as a university thesis, exists in only two or three copies of which only one is likely to be in a library. However if a thesis has been made available to the public even in only one copy in a university library it could well be properly included in material that should be searched. This extreme example is given to show the difficulty of carrying out a comprehensive search. Before on-line computer databases became available searching depended on the physical examination of indexed printed material in a library. Now on-line databases contain patents or abstracts for most countries covering the last twenty or so years and in some cases going back to the beginning of the twentieth century. Technical and scientific journals and conference papers are also available on databases. Historically, various patent offices devised their own classification systems for inventions but now there is an international system used by almost all offices, sometimes in parallel with their own one. Despite this organisation of the information an exhaustive search is clearly not feasible. The limiting factor is cost.

Who does the search?

9–68 Patent attorneys are particularly well-equipped to make searches, especially where they already have an intimate knowledge of the subject matter, and then assess the results. There are also numerous firms who make searches on a fee earning basis. These firms employ staff qualified in appropriate technologies (and nowadays in computer-aided search skills) who can efficiently find material relevant to a search brief. Generally they can advise on search strategies but also offer several fixed price searches in specified major databases. Usually the search result is a number of documents in which the searcher has identified the parts seen as relevant to the search brief. Patent office examiners are skilled searchers and competent to assess the results of a search but their work is mainly for their office. However some patent offices do offer a search service which is in addition to the searches carried out for patent applications. Sometimes the cost of such a search can be offset against the search fee for a subsequent relevant patent application. Some companies have their own searchers who specialise in the products of the company and both carry out watching searches and build up specialised databases.

Who assesses the results?

9–69 The result of a search is usually a bundle of photocopies of documents of various ages and origins, often in different languages. The assessment is the crucial part and should be done by someone with a thorough knowledge of the subject of the search as well as of the patent law in the relevant jurisdictions and who adapts the assessment to the purpose of the search. Novelty assessment can be straightforward. Assessment of inventiveness requires a knowledge of the state of the particular art and the skill level of those in it. Freedom to use searches involve the assessment of existing patent claims (in theory every existing claim) to determine the risk of infringement and the validity of the claims in view of the relevant prior art. It is important to remember that decisions involving millions of pounds, euros or dollars can depend on the assessment of the search result. For example the choice of the process for a synthetic chemical production plant can be decided by the freedom to use assessment of the different processes against the profitability of each process. A proper assessment should be made by appropiately qualified professionals, and they should involve the inventor or other technical staff where appropriate.

Registration Patents (Patents of Importation)

9–70 Granted patents can sometimes be registered in other countries without further examination. However this procedure is now rare and availability should be checked when filing an application if protection in smaller but industrially significant countries may be needed.

Accelerated Prosecution

9–71 An applicant can reduce the time to the grant of a patent in various ways. Clearly a prompt and full response to every communication will help but as applications are often dealt with in number order this will not move an application up the queue. The EPO offers accelerated prosecution if the guidance issued by the EPO is followed. In the United Kingdom a combined search and examination procedure is available on request. The requests and fees for the preliminary examination and search and the substantive examination are filed at the same time. (These could be filed with the initial application for best saving of time but then the application must include a claim.) A disadvantage is that the substantive examination is done before the applicant sees the search report and before all the possible prior art is available to the search examiner. Against this is the possibility, if the substantive examination allows claims appropriate to the applicant's needs, of a patent being granted in significantly less than two years. Other forms of accelerated prosecution in the United Kingdom are only available if a reasoned request, for example significant infringement, can be made.

Summary flow charts

9–72 It is not possible to show all of the available patent procedures on a summary flow chart of reasonable size. However there follow two flow charts to help in providing an overview. The first shows in parallel columns against elapsed months the main steps in four widely used routes to patent protection; the second shows the sequence of major choices which have to be made, without any time scale.

9–73 Figure 1.   Procedure Steps Outline

Applicant searches, prepares application, selects initial filing route(s)			
initial route(s)			
month / minimum or full U.K. national	EPC via U.K. Office	PCT via U.K. Office	overseas
0 (A) file request, description, drawings, (with claims, abstract for full) [no priority]	file request, description, drawings, claims, abstract & designations, search request [no priority]	file request, description, drawings, claims, abstract & designations, search request [no priority]	with permission file to local regulations using attorney [no priority]
6 receive search reports			
0–11 review invention development, make further searches, decide designations, settle description & claims, prepare revised application			consider further action
12 (A+12) file request, revised description, claims, abstract, priority claim (or proceed with earlier full)	file request, revised description, claims, abstract, designations & priority claim	file request, revised description, claims, abstract, designations & priority claim	possibly file revised application with priority claim or file convention application using attorney
16 on receive search reports, consider citations			
17 last chance for withdrawal before publication and for claim amendments for publication			
18 on application published when search report ready (UKP)	application published, with search report if ready, if not search report published later (P)		prosecute national applications
pre-examination amendments possible			
19 pre-examination amendments possible	pre-examination amendments possible	Chapter II due, advisable to file International Preliminary Exam demand and election	using local attorney
20		national & regional phases outside Chapter II due	
24 examination request due (UKP+6)	examination request due (P+6)	19–28mo IPEA written opinions and responses	

28	U.K. official actions and responses to grant,	EPO communications and responses to grant/rejection (no deadline), appeal, oral proceedings opposition then establishing national patents	International Preliminary Exam report due
30	rejection appeal, etc.		most national and regional phases due
31			Australia, EPC phases due
32			last national phases due
54	earliest deadline for U.K. in order		

A = application date; P = publication date of search report.

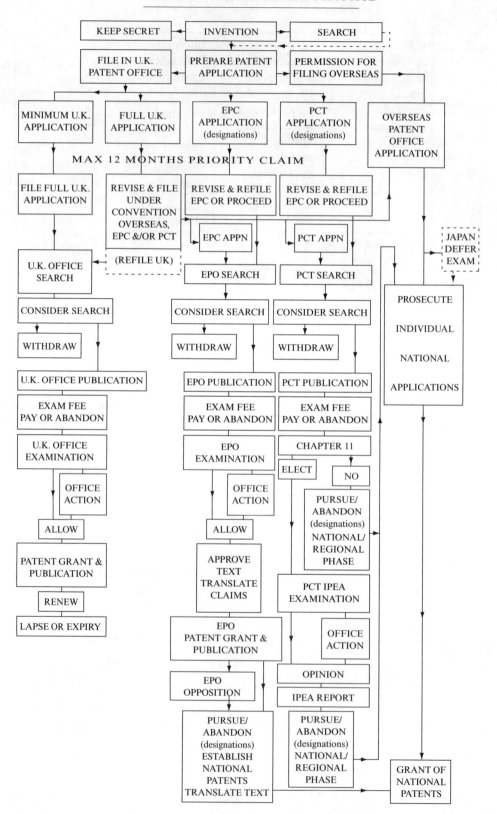

9–74   Figure 2.   Major choices flow chart.

CHAPTER 10

THE LAW OF CONFIDENCE

Iain Purvis
Barrister, 11, South Square, Gray's Inn

INTRODUCTION

10–01   The courts of the United Kingdom have from mediaeval times recognised the importance of enforcing obligations of secrecy. Such obligations arise in a variety of different contexts: commercial and business relationships; employment; solicitor and client; doctor and patient; even marriage. The common element to all of them is the existence of a relationship of trust between two parties. It is a fundamental component of any relationship of trust that where important information is imparted from one person to another within that relationship, it will not be disclosed to others or used in a way which was not contemplated by the discloser.

Against the requirement to protect the interests of those within relationships of trust, various other considerations have to be balanced by the courts. In particular, the economic and social health of the country demands that certain limitations are placed on the use of the law of confidence to restrict obviously desirable activity.

10–02   Thus, as we shall see, the law will not prevent anyone from making use of trivial information or information generally known to the public, even where it was disclosed to them in confidence. Furthermore, freedom of labour demands that employees may carry their accrued skills and knowledge with them from one employer to another, even where that involves the use of information which the employer would not wish to make public. Finally, the courts will not permit the enforcement of confidentiality where the public interest in disclosure of the information is greater than the public interest in protecting private confidences.

The common law recognises a number of different categories of obligation, enforced respectively by the laws of tort, contract and equity. In the past there have been a number of attempts to "pigeon-hole" the law of confidence into one or more of these categories. These arguments can be rather arid and unhelpful in practice. Against the background of the general principles outlined above, the courts will use whatever tools are most readily at their disposal to enforce confidences. If the dispute arises in the context of an existing contractual relationship (whether a commercial contract, employment, consultancy or some other), then the courts will use the language and tools of contract to enforce the obligation. If it arises in the context of some equitable relationship (solicitor and client, trustee and beneficiary, man and wife), they will tend to use the language and tools of equity.

The ensuing chapter seeks to explain how the policies set out above are enforced by the courts, in particular in the context of the industrial exploitation of technology.

LEGAL BASIS OF AN ACTION FOR BREACH OF CONFIDENCE

10–03   There are three basic elements of a breach of confidence action:

(a)   That the information in question was truly confidential in nature;

(b)   That it was imparted in circumstances which required the person receiving it to keep it confidential;

(c)   That the recipient has breached the obligation of confidence.

PROTECTABLE INFORMATION

10–04   The courts will not protect any information against disclosure or misuse unless it is truly confidential in nature.

The classic statement of the requirement is that of Lord Greene in *Saltman v. Campbell*[1]:

> The information, to be confidential, must . . . have the necessary quality of confidence about it, namely it must not be something which is public property and public knowledge. On the other hand, it is perfectly possible to have a confidential document, be it a formula, a plan, a sketch, or something of that kind, which is the result of work done by the maker upon materials which may be available for the use of anybody; but what makes it confidential is the fact that the maker of the document has used his brain and thus produced a result which can only be produced by somebody who goes through the same process.

Lord Greene's test of whether the information has the necessary "quality of confidence" can be subdivided into two conditions which must be satisfied in any confidential information dispute:

(a)   the information must not be publicly available; and

(b)   the information must be of sufficient importance to justify protection.

A further condition arises in the particular context of disputes between employers and ex-employees — can the information be readily separated from the general skill and knowledge which is part of the employee's personal stock-in-trade. These three conditions are dealt with separately below.

PUBLICLY AVAILABLE INFORMATION

10–05   As Lord Greene's exposition of the test makes clear, the first issue in any confidential information dispute is whether the information is truly secret or whether it is "public property and public knowledge".

On the face of it this might appear to be a very straightforward and obvious test. Public information is the antithesis of secret information. However, in practice, the dividing line is not always so clear as might appear.

The classic case on public disclosure is the House of Lords decision in *Mustad v.*

---

[1] [1948] R.P.C. 203 at 215.

Dosen [1963] R.P.C. 41. There, the plaintiff sought to prohibit by injunction an ex-employee of a company of which they were the successors in title from using information obtained in the course of his employment relating to a machine for making fish-hooks. After the issue of the Writ, the plaintiff was granted a patent for the very same machine. On grant, of course, the patent was published to the world by exposure for inspection on the shelves of the Patent Office. The plaintiff failed to prove that there was any particular information being used by the defendant which was not disclosed in the Patent. The House of Lords duly rejected the plaintiff's action.[2]

Mustad was of course a straightforward case of publication of the very same matters which were being sought to be protected under the law of confidence. However, difficult issues often arise as to whether material is "in the public domain".

The "public domain"

10–06 Exactly what constitutes the public domain? This is extremely difficult to define. Lord Goff, in his seminal Judgment in the *Spycatcher* litigation, was constrained to resort to the circular definition that "in the public domain" meant no more than that the information was so generally accessible that in all the circumstances it could not be regarded as confidential. This may not be as unhelpful a test as it seems. Much of the law of confidence turns on the courts' perception of what an honourable businessman would regard as reasonable. A rough general test of whether material is truly in the public domain is whether the honourable businessman, receiving the material in question in a commercial context, would regard himself as free to use it because it was already publicly known.

Disclosures to third parties

10–07 Not all disclosures to third parties will destroy the "quality of confidence" in information. Disclosures to persons themselves bound by obligations of confidence (such as employees or professional advisers) will not destroy the confidentiality of information. Similarly, a disclosure to a limited class of people, or in a brief or inconsequential manner, may not necessarily destroy the secrecy of a piece of information, so long as the information did not become generally known in the trade.

Constructive disclosures

10–08 Information may enter the public domain without being expressly published as such. Where the secret is a technical one, concerning the making of a particular product which is then sold on the open market, a question may arise as to whether the secret has implicitly been disclosed by the sale of the product. The ingredients of a particularly effective chemical, for example, might be able to be identified by spectroscopic analysis. Similarly, reverse engineering might identify the way in which a particular machine had been built. At what point does the sale of the ultimate product amount to constructive disclosure of information about its composition or how it was made?

---

[2] Publication of a patent specification will be treated by the courts as having made all the information in it available to the public. As such, it will not be protected by the law of confidence. Essentially, therefore, a business must elect between protecting its technology by a patent and protecting it by the law of trade secrets. Of course, the mere existence of a patent specification does not exclude the possibility of trade secret protection for information about the same subject matter which has not been included in the patent. However, the patentee must be careful: he is obliged under the law of patents to make sufficient disclosure of his invention that the skilled reader of the specification can make it work. If he keeps anything important back, he may risk invalidating his patent.

In the law of patents, a similar issue often arises as part of the question of whether an invention was "made available to the public" before the priority date. In this area, the courts (and the European Patent Office) have adopted the "without undue burden" test. If a product was put on the market before a particular date, its composition and/or internal structure will be taken to be known to the public even where they are not immediately apparent, provided that it was possible for a skilled man to have discovered and reproduced them "without undue burden". This is so even if some destructive analysis of the product would have been required.[3]

10–09 The ultimate question asked by the courts in the law of confidence in this context would appear to be the same as that applied in patent law — namely "is the information available to the public". However, the tests which have been applied are usually much more generous to the party claiming the right.[4] Essentially, provided that some degree of work has to be done to obtain the information from what is available on the market, then the information is not "publicly available".[5] The relative ease with which the information could be obtained may affect the length of time the obligation of confidence might last, or the remedies which are available,[6] but, provided that some degree of effort is needed to obtain the information, then the obligation of confidence is not wholly destroyed.

Of course, if the claimed secret is not in the composition or structure of the product, but is simply the idea of making it or some fact which is apparent simply from seeing the product, then any pre-existing obligation of confidence will be destroyed when the product is put on the market.

Selections from publicly available information

10–10 Similar questions arise where the "secret" sought to be protected is a selection or collation of pieces of information, all of which individually were publicly known or available. There is no reason why such a selection or collation should not have the necessary quality of confidence. The courts recognise the value of the work which went into the compilation of information, from whatever sources, and are prepared to protect it. Thus in *Roger Bullivant v. Ellis* [1987] F.S.R. 172, the Court of Appeal granted an injunction in relation to the use of a card index with the names and addresses of customers, the evidence of the defendant being that he only used the cards to provide the addresses of customers already known to him (such addresses being in the public domain). The Court of Appeal noted that:

> The value of the card index to Mr Ellis and the other Defendants was that it contained a ready and finite compilation of the names and addresses of those who had brought or might bring business to the Plaintiff and who might bring business to them.

Equally, much valuable technical information could, on microscopic analysis, be reduced to a set of individual pieces of information which could each be discovered

---

[3] *Availability to the Public* GO1/92 [1993] E.P.O.R. 241; *Lux Traffic Controls Ltd v. Pike Signals Ltd* [1993] R.P.C. 107.

[4] See in particular *Yates Circuit Foil Co. Ltd v. Electrofoils Ltd* [1976] F.S.R. 345 at 387.

[5] *Ackroyds (London) Ltd v. Islington Plastics Ltd* [1962] R.P.C. 97 at 104. But Morritt J. in *Alfa Laval v. Wincanton Engineering* [1990] F.S.R. 583 at 591 distinguished between the internal measurements of a machine (these could not be ascertained without dismantling the machine and thus were confidential) and external measurements (these could be ascertained without dismantling and thus were confidential).

[6] See below under "springboard".

from public sources. This sort of analysis does not find favour with the courts and is no bar to the protection of the compilation.

There is of course a risk of unfairness here, in that the defendant may be prevented from using information which the rest of the world would be free to use, having compiled it for itself. A common approach to deal with this is to provide "spring-board" relief, protecting the plaintiff by injunction or compensating him in damages for the use of the information simply during the time which it is fair to assume the defendant would have taken to compile the information himself. See paragraph 10–20 below.

TRIVIAL INFORMATION

10–11   The second limitation placed by the courts on protectable information is that it must have some inherent importance or interest. Essentially trivial information is not protected, even when it was acquired in circumstances which would otherwise have given rise to an obligation of confidence.

It is important here to distinguish between information which is actually trivial and information which is merely simple. The simplicity of an idea is no bar to its protection. Indeed it could be (and has been) said that the more simple the idea the more protection it needs.[7]

Nonetheless, there comes a point where simplicity degenerates into vagueness. In *De Maudsley v. Palumbo*,[8] Knox J. rejected a claim to confidentiality in a set of ideas concerning a new night club venture. Discussing Lord Greene's judgment in *Saltman v. Campbell*, he identified as essential before information would be protected in law that it:

> [had] at least some attractiveness to an end user and [was] capable of being realised . . . as a finished product . . .

This meant that the person claiming confidentiality must have gone *far beyond identifying a desirable goal*. Ideas such as "opening legally all night", "big with high tech industrial warehouse decor", "featuring top disc jockeys from the United Kingdom and around the world", were too vague to justify protection.

THE SPECIAL POSITION OF THE EMPLOYEE

10–12   Employment is by far the most common context in which disputes over confidential information arise. In particular, they arise between an employer and an ex-employee over information which the ex-employee is alleged to have taken with him to his new employer, or information which he is alleged to be using on his own account. Such disputes raise particular difficulties of their own since they squarely raise the two conflicting public interests of the desirability of protecting secrets and freedom of trade which are discussed in paragraph 10–02 above. The courts have developed particular rules concerning protectability of information disclosed to employees which are *sui generis*. These are considered below.

The legal basis of the protection of confidential information in the employment context is now well settled as a contractual one.[9] The courts will imply a term into the contract of employment which will prevent the use or disclosure of certain types of

---

[7] *Coco v. AN Clark* [1969] R.P.C. 41 at 49.
[8] [1996] F.S.R. 447 at 455–7.
[9] *Faccenda Chicken Ltd v. Fowler* [1986] 1 All E.R. 617.

information by an employee after he has left his employment. Insofar as any explicit terms in the contract of employment purport to extend the categories of information which the employee cannot use beyond those which the law will recognise as being within the implied term, such terms will not be enforceable since they are in restraint of trade.[10] There is thus a single test covering the types of information which will be protected by the law in this context (though explicit terms in contracts of employment dealing with post-employment obligations do nonetheless have some value — see below at paragraphs 10–53 to 10–59).

10–13 The rules regarding the protection of information disclosed to employees were set out most clearly in the modern context in *Faccenda Chicken v. Fowler.*[11] At first instance, Goulding J. divided such information into three categories.

The first of these was information excluded from protection on first principles because of its trivial character or its "easy accessibility from public sources of information". This is not confidential information at all, for the reasons given above, and will not be protected against use by an employee on his own account either during or after the period of his employment.

More importantly, he then divided up the remainder of the information obtained by an employee in the course of his employment into two further categories:

(a) Information which the employee must learn as an integral part of doing his job, but which, once learned, necessarily remains in his head and becomes part of his own skill and knowledge. Whilst employed he can only use this information on his employer's behalf: otherwise he is in breach of his obligation of fidelity. However, once he has left his employment, he cannot be expected to divide such matters from his general working skills, and he can use them for the benefit of his new employer or on his own behalf;

(b) Specific trade secrets so confidential that they cannot be used post-employment for the benefit of a new employer or on the employee's own behalf.

This categorisation was approved by the Court of Appeal, and has commonly been used since.

As will be appreciated, category (a) amounts to information which in other contexts would be regarded as capable of protection in law. However, in the peculiar context of the employment relationship it will not be protected after termination of that relationship. The distinction between category (a) and category (b) information is therefore extremely important to employers and employees alike.

10–14 There is nothing new about this distinction. The courts have consistently refused (since the decision of the House of Lords in *Herbert Morris v. Saxelby* [1916] 1 A.C. 688) to grant any injunction or other relief in relation to the knowledge, skill and experience which an employee has acquired for himself as a result of his employment. The justification is that such restrictions would make it impossible for the employee to seek and obtain work at a rival business (making him effectively a slave of his employer as well as placing a considerable fetter on legitimate competition).

This category of unprotectable information has been held in individual cases to cover the scheme of organisation and methods of business of the employer[12]; the

---

[10] *Faccenda Chicken Ltd v. Fowler* [1986] 1 All E.R. 626.
[11] [1985] F.S.R. 105; [1986] F.S.R. 291, CA.
[12] *Herbert Morris v. Saxelby* [1916] 1 AC 688.

identity of the suppliers of particular products of the employer[13]; the employee's memory of particular features of his employers' plant and equipment[14]. It is to be contrasted with information which does not seep into the employee's general skill and labour and which the employer has a legitimate interest in keeping for himself.

It is easier to state the distinction in generalities than to give a catch-all test to identify information as falling into one category rather than another. Sometimes the courts have approached the question on the rough and ready distinction between information which the employee inevitably carried away in his head and that which he had specifically written down or was contained on documents which he removed.[15] This is particularly the case with information as to customers or suppliers, and may be a useful test in some cases.

10–15 However, there are many cases where obviously secret information is in such common use at an employer that the employees cannot help but remember it. That does not necessarily mean that the information is not confidential. At this point, one has to resort to the more general question asked by Cross J. in *Printers & Finishers v. Holloway*, namely whether:

> the information in question can fairly be regarded as a separate part of the employee's stock of knowledge which a man of ordinary honesty and intelligence would recognise to be the property of his old employer, and not his own to do as he likes with.

There is a moral element to this (as there is with much of the law of confidence). If a reasonable man would think it to be wrong to make use of a particular piece of information obtained in the course of one's employment after that employment had finished, then the courts will not hesitate to protect it as a trade secret. The duty of confidence was said in *G.D. Searle & Co. Limited v. Celltech* to be a duty imposed by "the conscience of the employee".[16]

Reported decisions have identified a number of different factors as highly relevant to assessing the side of the line on which a particular piece of information falls.

Ease of separation of information from general mass

10–16 One of the most significant is the extent to which the information can be easily isolated from other information which the employee is free to use or disclose. In *Printers & Finishers,* Cross J. was prepared to accept the confidentiality of a particular formula or some particular thing said at a confidential meeting. However, in respect of features of the construction and arrangement of the employer's machines he said as follows:

> The employee might not realise that the feature or expedient in question was in fact peculiar to his late employer's process and factory; but even if he did such knowledge is not readily separable from his general knowledge of the flock printing process and his acquired skill in manipulating a flock printing plant, and I do not think that any man of average intelligence and honesty would think that there

---

[13] *Worsley v. Cooper* [1939] 1 All E.R. 290.
[14] *Printers & Finishers v. Holloway* [1965] R.P.C. 239 at 254–7.
[15] *e.g. Coral Index Ltd v. Regent Index Ltd* [1970] R.P.C. 147 at 149.
[16] [1982] F.S.R. 92 at 99, *per* Cumming-Bruce L.J.

was anything improper in his putting his memory of particular features of his late employer's plant at the disposal of his new employer.

The point there was that the employee had gained considerable experience in the optimal arrangement of flock printing equipment. He was clearly entitled to use that for the benefit of any employer. He could not separate out in his own mind particular solutions to particular problems, which his employer may have solved in an unusual way, from his general skill in the industry.

10–17 Another good example is *Balston Limited v. Headline Filters* [1987] F.S.R. 330. There, the defendant was a highly skilled maker of industrial filters. He had an immense amount of experience of process controls, formulations, dimensions and tolerances, most of which had been built up whilst he was employed by the plaintiff. Upon his setting up in competition, the plaintiff sued him for breach of confidence. At the interlocutory stage,[17] the defendant offered to give undertakings until trial that he would not make the specific formulations used by the plaintiff. The plaintiff sought a much wider injunction covering such formulations or any formulations "substantially the same". Scott J. rejected the plaintiff's claim, on the basis that such an order would be trespassing on the defendant's right to use his skill and knowledge of how to make filters. He put it like this:

> the defendant when manufacturing filter tubes to meet certain specifications is using his knowledge of the mixes likely to meet those specifications. This knowledge includes knowledge of the plaintiff's mixes. It is not, in my view, practicable to regard the second defendant's knowledge of the plaintiff's mixes as something that can, like cream from milk, be separated from the sum total of his knowledge of the manufacturing process.

The modern attitude of the courts to employers seeking to restrain the use of information outside the narrow scope of specific formulae or documentary lists of customers is illustrated by the concluding words of the judge in that case at 351:

> This case is, in my view, yet another example of an attempt by an employer to use the doctrine of confidential information to place fetters on the ability of ex-employees to compete. Technologically based industries abound. All have what they regard as secrets. Employees, particularly those employed on the scientific or technical side of the manufacturing business, necessarily acquire knowledge of the relevant technology. They become associated with technological advances and innovations. Their experience, built up during their years of employment, naturally equips them to be dangerous competitors if and when their employment ceases. The use of confidential information restrictions in order to fetter the ability of these employees to use their skills and experience after determination of their employment to compete with their ex-employer is, in my view, potentially harmful. It would be capable of imposing a new form of servitude or serfdom, to use Cumming-Bruce L.J.'s words, on technologically qualified employees. It would render them unable in practice to use their skills and experience after leaving.

He went on to say that the appropriate way for employers to seek a wider restriction on their employee's ability to use his skills was to impose a restrictive covenant, the

---

[17] When the plaintiff was seeking an injunction pending trial.

enforceability of which would be subject to strict examination by the courts. (As to this see below at paragraphs 10–53 to 10–59).

Extent to which employer has asserted secrecy

10–18 Another relevant factor which the courts have relied on in the past is the attitude taken by the employer to the confidential information. Although the employer cannot make something confidential which is not simply by asserting its secrecy, nonetheless his attitude will be regarded as powerful evidence. Thus in *United Indigo v. Robinson* 49 R.P.C. 178, an injunction was refused against an ex-employee in circumstances where he had never been warned that his employer regarded a particular process as confidential, and where it had become common knowledge amongst the employees of the company. In those circumstances, it was right to regard the information as having become part of the employee's own skill and knowledge which he was able to use having left his employment. This case illustrates the importance of both limitation of access to secret information and explicit instruction as to confidentiality.

Position of employee

10–19 Another factor is the position of the employee within the company. As we have seen, the obligation of confidence is related to the "conscience" of the individual employee. The courts are more prepared to recognise an employee's duty to maintain confidence if he is in the kind of senior or highly technical position within the company where he might be expected to come across confidential information on a regular basis, and to recognise the importance of secrecy. Presumably, also, the expected ability of someone in his position to distinguish the confidential from the non-confidential would also be taken into account.[18]

"SPRINGBOARD INFORMATION"

10–20 Confidential information is not a homogenous subject. There are different kinds of information, not all of which will be protected in the same way. Some, such as secret formulae, might be called "real trade secrets". The recipe for making Coca Cola is a good example. This is information the possession of which gives the user something which is unavailable to the rest of the world and which he could never have created for himself.

Other types of information can only be said to give the user a "head start" over the rest of the world. These include:

(a) a compilation of publicly known information, which any member of the public could put together given the necessary time and resources: for example a list of all potential purchasers of zinc oxide in the South East of England;

(b) information which will in time become public knowledge, but for the moment is secret: such as an unpublished patent specification for a new method of drying hops.

Naturally enough, these different types of information give rise to different rights on the part of their owners. Real trade secrets will be protected for ever, so long as

---

[18] *Faccenda Chicken*, n.10 above.

they remain secret. The other types of information attract a more limited degree of protection.

Both cases (a) and (b) above are clearly entitled to some form of protection, but it would be wrong to protect the information for ever. To do so would place the person who had illicitly obtained it in a worse position than the rest of the world. The rest of the world would in case (a) be able to create the information for themselves, and in case (b) would be given the information on publication.

In order to give no more protection to the compiler of the information than what is reasonable, the courts must calculate the advantage which has been given by possession of the information. This advantage over the rest of the world is often referred to as a "springboard".

10–21 The principal difference between "springboard" cases and cases of "classic trade secrets" is in relation to the scope of the relief which may be given by the courts. In particular, the courts will not restrain the misuser of the information from using it for ever, but will seek to frame an injunction which is sufficient to deprive him of any advantage which he has gained by its possession over the rest of the world. As Roxburgh J. put it in the case of *Terrapin Limited v. Builder Supply Co. (Hayes) Limited*[19]:

> the possessor of such information must be placed under a special disability in the field of competition in order to ensure that he does not get an unfair start

One common result of this doctrine is that an injunction may be granted against a misuser of confidential information which actually extends beyond the date at which the information has become public. See in particular *Speed Seal Limited v. Paddington* [1986] F.S.R. 309 and *Cranleigh Precision Engineering v. Bryant* [1966] R.P.C. 81. What the courts have attempted to do in these cases (and others in which the information has become public after the defendant has commenced using it) is a two-stage process. First they consider the extent to which the defendant had developed his business and his trade using the confidential information by the date it became public. Second, they estimate the length of time it would take a rival trader to get to that position if he had started on that date. An injunction will be granted for that length of time to place the defendant in the same position as the "innocent user".

This would at first sight appear to transgress the basic rule that publicly available information is not protectable[20], but should be regarded as an anomaly developed as the only practical way of protecting the value of certain categories of information.

Damages will be assessed on a similar basis: the loss which has been caused to the plaintiff through the "springboard" which the defendant has illicitly obtained.

CIRCUMSTANCES IN WHICH AN OBLIGATION OF CONFIDENCE MAY ARISE

10–22 The second limb of the breach of confidence action concerns the circumstances in which the information was disclosed, and in particular the relationship between confider and confidee. Confidential information is not a "monopoly" right. It requires the establishment of an obligation of confidence on the part of the recipient of the information. These obligations may arise as a matter of contract or as a matter of good faith — often referred to as "equity".

---

[19] [1967] R.P.C. 375 at 392.
[20] *Mustad v. Dosen* [1963] R.P.C. 41 and *Att.-Gen v. Guardian Newspapers Limited (Spycatcher)* [1988] 3 All E.R. 545 at 662–3.

In most cases, there is no problem about establishing the necessary relationship of confidence. It arises, as has been seen, in all employment contracts. It also arises between principal and agent, and between professional and client. In all these cases, there is an essential and obvious duty of good faith between the parties which is essential for the relationship to exist in the first place. Where material which has the necessary quality of confidence is disclosed in the context of such a relationship, the recipient is under an obvious obligation not to disclose it nor to use it for any purpose other than that for which it was disclosed.

The cases which cause difficulties in this area are where information is disclosed between companies or individuals not in an ongoing relationship of trust or good faith, but dealing with each other at arm's length or on a one-off basis. Some examples are discussed below.

Commercial dealings

10–23   Information is commonly disclosed in the course of ordinary business dealings governed by contract. A typical case would be the use of a manufacturing company to produce a novel component. The manufacturer is supplied with all the necessary information to make the component in the form of drawings and detailed specifications. These are inherently confidential.

It is relatively uncommon in such a case for the originating company to place specific restrictions on the use of information so supplied. Normally it would be operating under the manufacturer's standard terms and conditions. Nonetheless, the courts have no difficulty in implying a term into the contract that the information supplied be kept confidential and not be used save for the limited purpose of fulfilling the contract. A classic case of the implication of such a term is *Collins (Engineers) Limited v. Roberts & Co. Limited*[21] where the defendant was engaged by the claimant to manufacture bulk liquid containers to the claimant's own design. Detailed drawings were supplied for this purpose. When the bills were not paid, the defendant sought to manufacture on its own account. It was held that there was clearly an implied term in the contract that the drawings should be treated as confidential and not used for any other purpose. See also *Ackroyds v. Islington Plastics Limited*[22] and *Suhner v. Transradio.*[23]

The basis of the implied term in these cases is generally that it is necessary to imply the term in order to give "business efficacy" to the contract — *i.e.* the claimant would never have entered into the contract in the first place had it been contemplated that the defendant was free to go off and use the information on its own account.

In the alternative, the court may simply reduce the question to a matter of an obligation in equity (see discussion below), even though there was an existing contract between the parties.

Pre-contract negotiations

10–24   Valuable information is commonly disclosed in pre-contractual discussions, for instance where a potential licensor discloses information about his process to a licensee. Where a contract eventually arises, the information disclosed during negotiations will probably be subsumed into the general mass, and be protected by implied terms. A more difficult question arises where no contract is ever entered into.

---

[21] [1965] R.P.C. 429.
[22] [1962] R.P.C. 97.
[23] [1967] R.P.C. 329.

A classic case is *Seager v. Copydex*[24] where an inventor of a patented carpet grip tried to interest a manufacturer in making it. During the negotiations he revealed a number of pieces of useful information not in the public domain. No confidentiality undertaking was given, and no contract resulted. The defendant subsequently used some of the information in the development of a new grip of its own. The Court of Appeal held that it was bound by an equitable obligation not to use the information given to it by Mr Seager, and awarded damages.

The basis of the equitable obligation is that the recipient of the information knew (or ought to have known) that it was disclosed in confidence. The court will look at all the surrounding circumstances as to their effect on the conscience of a reasonable man. Whilst it is fair to say that the court will generally seek to imply an equitable obligation whenever obviously valuable information is handed over during contractual negotiations, this is not an invariable rule. See *Carflow Products (U.K.) Limited v. Linwood Securities (Birmingham) Limited*[25] where a prototype steering wheel lock was shown to a potential buyer at a meeting. Jacob J. held that the disclosure was not in confidence (thus invalidating a subsequent registered design). The buyer was not told that the disclosure was in confidence, and a reasonable man would not have thought so:

> People are habitually showing things they have designed or invented to would be manufacturers. If the latter are put under an obligation of confidence merely by seeing a prototype — which they would as reasonable men know could be the subject of independent protection — then they would be far less willing to look at things.

He specifically distinguished the case of disclosure of material (such as the shape of an article) which is protectable by design right or copyright, from the disclosure of material not so protectable (such as detailed plans). The latter are more likely to be confidential.

Nonetheless, *Carflow* shows that it is folly to rely entirely on equity to protect confidence in material freely handed over at meetings before any contract is entered into. It is plainly advisable for anyone going into such a meeting to extract confidentiality undertakings in advance.

Unsolicited information

10–25 There is no special rule exempting the recipient of unsolicited information from the obligation of confidence.[26] Again, the test is whether a reasonable man would as a matter of conscience believe that he ought to keep the material confidential.

Many companies receive a large number of unsolicited disclosures from individuals seeking to have their ideas exploited. Of course many of these consist of material already in the public domain. Others consist of trivial, vague or absurd ideas. As we have already seen, such information is not susceptible of protection by the law of confidence anyway. However, where genuine and substantial disclosure takes place, the circumstances are likely to support an enforceable obligation of confidence. This is because the recipient knows that the party disclosing the information intends to make money out of it. It would be utterly incompatible with this intention to hand the information over to a third party with no restriction on its use. The court is thus likely

---

[24] [1967] 2 All E.R. 415.
[25] [1996] F.S.R. 424.
[26] See for instance Lord Goff in *Att.-Gen v. Guardian Newspapers* [1988] 3 All E.R. (658–9).

to hold that the information was never intended to be made freely available to its recipients.

It is advisable for companies who do receive such unsolicited material from time to time to have a strategy for dealing with it. Once identified as something potentially valuable, it should not be disclosed within the company until a formal contract has been entered into with the confider. This may be in short form, but should lay down that the company will not make use of any of the information nor disclose it to third parties without the confider's consent unless and to the extent that the piece of information in question:

(a)   has entered the public domain other than through illicit disclosure by the recipient;

(b)   was already within the knowledge of the recipient, or has subsequently come to their knowledge in some other way, including disclosure by a third party or independent development;

(c)   was otherwise not capable of protection under the law of confidence, such as by reason of its trivial or vague character.

Only when the confider has signed such an agreement should the recipient proceed to consider the information further.

Information obtained by a third party in breach of confidence

10–26   The circumstances discussed so far have involved some kind of relationship between the confider of the information and its recipient. The law imposes obligations by way of contract and equity having regard to the nature of the relationship in question.

Trickier questions arise where there is no relationship. The most common case is where a third party obtains information by virtue of a disclosure which is itself a breach of confidence. The clearest example is the case of an employer which takes on an employee from a rival company. The employee brings with him trade secrets of his ex-employer and discloses them to his new employer in breach of his obligations of confidence.

Since the equitable basis of the breach of confidence action is founded on moral culpability, the courts distinguish between the innocent and the guilty recipients of information.

The rule is that the recipient of the information is bound by the obligation of confidence and is liable for breach if it knew, or ought to have known (in the sense that it deliberately closed its eyes to the idea) that the information was confidential.[27] This covers cases where the recipient was specifically told that the material was confidential, and cases where it was obvious to him that it was.

10–27   In many cases, there is no difficulty in establishing this. For instance, it is common enough for the new "employer" to be merely a corporate vehicle for the ex-employee's own business.[28] In most other cases, the new employer is unlikely to be so lacking in wordly wisdom as to be unaware of the origin of the valuable new information being made available to him.

---

[27]   Lord Goff, *ibid.*
[28]   Such a case was *Cranleigh Precision Engineering v. Bryant*, para. 10–21 above.

However, there are cases of genuinely innocent recipients (in some cases duped by their new employee). In such cases, the original receipt and use of the information will not be actionable. However, as soon as the recipient is put on notice of the obligation, he is bound by it, and cannot rely on his original innocence.

The potential problem so far as the originator of the confidential information is concerned is the case where the innocent recipient has published the information to the world before being put on notice. In such a case, the confidentiality of the information would have been destroyed, and no action could be taken against the recipient.

It will be apparent that it is crucially important for the originator of the confidential information to put the innocent recipient on notice as soon as possible.

Accidentally acquired information

10–28 Similar rules apply to the accidental recipient of confidential information. If he obtains information which he knows or ought to know is of a confidential nature which he was not meant to see, equity will impose on his conscience an obligation not to disclose that information nor to use it for his own benefit.

In *Att.-Gen v. Guardian Newspapers*[29] Lord Goff gave two examples:

> where an obviously confidential document is wafted by an electric fan out of a window into a crowded street, or when an obviously confidential document, such as a private diary, is dropped in a public place, and is then picked up by a passer-by.

Information stolen or obtained by underhand means

10–29 As a matter of common sense, if the courts will intervene to prevent use of confidential information by accidental recipients, then they will certainly do so in the case of those who obtain such information by theft, bugging or coercion. There has in the past been some suggestion[30] that, owing to the lack of a relationship between the thief and the victim in such a case, there was no cause of action for breach of confidence. This is obviously nonsense. Equity will intervene to impose obligations where it is just to do so as a matter of conscience.

In the few cases which have come before the courts raising this issue, judges have had no difficulty in recognising the existence of a cause of action.[31]

TO WHOM IS THE OBLIGATION OF CONFIDENCE OWED?

10–30 We have proceeded with the above analysis on the assumption that it was clear to whom the obligation of confidence in respect of the information was owed and who was entitled to take action to enforce it. In the vast majority of cases this is not a problem. In the case of a company whose vital secrets are created and developed by its employees, for example, there is no question but that the obligation is owed to the employer, as part of the general duty of good faith. Thus it is the employer who is entitled to take action to prevent the misuse of the information.

---

[29] See n.26 above.
[30] Based largely on *obiter* comments of *Megarry V-C* in *Malone v. MPC* [1979] 2 All E.R. 620 appearing to bemoan the lack of available relief in such a case.
[31] See *ITC Film Distributors v. Video Exchange Limited* [1982] 2 All E.R. 241; *Francome v. Mirror Group Newspapers* [1984] 2 All E.R. 417.

There are however cases in which the position is less clear. Since information is not property, but merely a set of obligations in contract and equity, we must examine the relationship between those responsible for generating the information.

Partnerships

10–31   The general rule is that partners owe each other a duty of trust and good faith. Where valuable information is generated for the purposes of the partnership, each partner owes the others a duty not to disclose the information save for the purposes of the partnership and not to use it on his own behalf. The rest of the partners have a right to take action against an errant partner who breaches this duty. While the partnership remains in existence, it would be a breach of an implied term of the partnership agreement.

The position after termination of the partnership is more difficult. It is hard to see how any assertion of breach of confidence could be made by one ex-partner against another unless by way of a breach of an implied term of the partnership agreement (or the agreement terminating the partnership). Whether such a term could be implied would depend on all the surrounding circumstances. The prudent course of action is plainly to ensure in the partnership agreement itself that the post-termination position of confidential information generated for the partnership is dealt with.[32]

Joint ventures

10–32   Information is commonly generated by two parties to a joint venture agreement. In the absence of an express term in that agreement, a term will generally be implied that information generated for the purposes of the joint venture may only be used by either party for those purposes. The position after termination is similar to that after termination of a partnership agreement (see paragraph 10–31 above).

Consultancy

10–33   So far as information provided to a consultant in the course of a contract is concerned, this will be protected in law by way of an implied term, and in equity[33]. If it is confidential in nature, it may not be used by the consultant for any purposes other than his performance of the contract.

The normal position after termination will be the same: the consultant must respect the confidentiality of the information with which he was provided. The only potential limitation on this concerns the position of full-time consultants on fixed term contracts whose work is more akin to that of employees. As we have seen, the law is concerned to protect an employee's ability to use his accrued skill and experience and provides a much more limited form of protection for an employer against use of confidential information after termination of employment. It would be perverse if this did not apply equally to those (as is common in, for example, the computer industry) who move from one full-time fixed term contract to another. Conversely it would be perverse to enable the employer to get around the usual restrictions on his right of action against employees by turning them all into "consultants".

---

[32] See *Murray v. Yorkshire Fund Managers Limited* [1998] 1 W.L.R. 951, CA, for a case where disclosure had been made to a "team" which did not amount to a partnership, and where it was held that there was no implied obligation of confidence to the Plaintiff (an ex-member of the team) after he had left the team.
[33] *Schering v. Falkman* [1981] 2 All E.R. 321.

One would expect the courts to treat such "consultants" in much the same way as they would ordinary employees. They should be free to use their accrued skill and experience and be restricted only in their use of "true trade secrets" (see paragraphs 10–12 to 10–19 above).

The position of information created *by* consultants in the course of their contracts is more complex. In the absence of contractual provisions to the contrary, the usual position will be that the information was created as part of services which were being carried out exclusively for the benefit of the party which was paying for the consultant's services. The value of the consultant's services would be markedly decreased if he could go off and use that information on his own account or for the benefit of a competitor. As a matter of business efficacy the court will therefore imply a term that any information generated in such a case which has the necessary quality of confidence is not to be used save by or for the benefit of the party paying for the consultant's services.

MISUSE OF CONFIDENTIAL INFORMATION

10–34 A cause of action for breach of confidence will arise when the following elements exist:

(a) Use or disclosure of confidential information beyond the scope of any authority to do so;

(b) Derivation of the information (directly or indirectly) from a person to whom an obligation of confidence is owed;

(c) Some degree of damage to the interests of the confider.

Use

10–35 Unlike true intellectual property rights, there is no "penumbra" of protection around confidential information. It is either used or it is not. Take as an example a case in which the defendant has had access to a confidential description of a manufacturing process. He then adopts a very similar manufacturing process in his own factory. The issue is not, as it would be in a copyright or design right case brought on the same facts, whether the process was sufficiently similar to the original. The only issue is whether the information was used in the course of developing it. This could be simply by replicating the process. Alternatively it could be by using the information as a starting point, and then "tweaking" various components of the process to suit his own ends.

Use may be proved by one of two approaches. Either direct evidence from witnesses and contemporary documents showing that the information was used. Or inference from the available facts: in the example given above the claimant might be able to tell by examining or testing the manufactured articles exactly how they were made.

Disclosure

10–36 Cases involving solely the wrongful disclosure of confidential information (and not its misuse) have been comparatively rare. In general, circumstances giving rise to an obligation not to use confidential information other than for specific purposes will give rise to a parallel obligation not to disclose the information to third parties.

As explained below, there is no *mens rea* in a breach of confidence action, so the employee who blurts out his company's trade secrets when under the influence of drink in the local pub is as liable as the one who sells those secrets to a competitor.

Derivation

10–37   Confidential information is not a monopoly right. In the example given in paragraph 10–35 above, the mere fact of access to the process information and the subsequent adoption of a similar process by the defendant is not sufficient of itself to establish the cause of action (although it may give rise to a strong inference of derivation).

If on the facts it emerges that the information in fact used by the defendant (although very similar to the information obtained from the claimant) was obtained from an unrelated source, then there is no cause for complaint.[34]

It should be borne in mind however, that there is no requirement for deliberate or malicious use of the information in question. In *Seager v. Copydex*[35] the use of the information was accepted by the Court of Appeal as having been inadvertent and "unconscious". Nonetheless, it was held to be an actionable breach of confidence.

Extent of use

10–38   The use does not need to be of the whole or even a "substantial part" of the confidential information. Use of any item of the information disclosed is enough to give rise to a cause of action, provided that the item in question has some value and would be protectable in its own right.

Damage

10–39   There is a great deal of doubt about the question of whether some degree of damage or detriment to the claimant is required for a breach of confidence action. Although a requirement of detriment has been stated by some courts in the past, the modern position would seem to be that the law will intervene to prevent potential damage in the future even where no existing damage can be shown, and may even give a remedy of nominal damages for past breaches.[36]

THE PUBLIC INTEREST DEFENCE

10–40   Since the law in this area is essentially an equitable doctrine, the courts will refuse to enforce confidences where to do so would be against the public interest.

It has often suggested that there may be two different defences here. The first being a general defence of "iniquity": that the nature of the information is so repugnant that the public interest demands that the courts should not enforce any obligation in relation to it. The second being a "whistle blower's" defence — the information itself may not be "iniquitous", but keeping it secret would cause grave damage to the interests of the public.

The better view now is that there is only one defence. It was stated very clearly by Lord Goff in *Att.-Gen v. Guardian Newspapers*[37]:

---

[34] See *Johnson v. Heat & Air Systems* [1941] R.P.C. 229.
[35] See n.24 above.
[36] See Lord Keith in *Att.-Gen v. Guardian Newspapers*, n.26 above, at 545.
[37] *ibid*. at 659.

although the basis of the law's protection of confidence is that there is a public interest that confidences should be preserved and protected by the law, nevertheless that public interest may be outweighed by some other countervailing public interest which favours disclosure. This limitation may apply, as the judge pointed out, to all types of confidential information. It is this limiting principle which may require a court to carry out a balancing operation, weighing the public interest in maintaining confidence against a countervailing public interest favouring disclosure.

He went on to say that this principle included the "so-called defence of iniquity". But it was now clear that the principle extended beyond that to other matters which it was in the public interest to disclose.

10–41 It will be clear from this that the "balancing operation" is between:

(a) The importance of enforcing the obligation of secrecy as a matter of public policy;

(b) The public interest in having the information disclosed.

Other questions such as the behaviour of either party, or the motives behind the disclosure (for money or out of genuine concern for the interests of the public) must be treated as essentially irrelevant.

In the commercial and industrial sphere, examples of a successful public interest defence have been relatively rare. The only clear cases are those concerning public health or safety. The most notable of these is *Lion Laboratories v. Evans*[38] where the claimant attempted to prevent the disclosure by the defendant of documents tending to show that the breathalysers produced by the claimant were defective and potentially causing wrongful convictions. At an interlocutory stage the Court of Appeal judged the public interest as favouring publication.

10–42 In *Lion Laboratories,* the defendants had sold the documents to a newspaper. On the facts of that case, the public interest was deemed to extend to publication *to the public,* given the potential importance of informing people under threat of conviction that the evidence against them might be defective. However, it should be borne in mind that the precise scope of the disclosure is something which the court will take into account in balancing the competing considerations. As Lord Goff put it[39]:

It does not follow however that the public interest will in such cases require disclosure to the media, or to the public by the media. There are cases in which a more limited disclosure is all that is required.

This places a serious limitation on the scope of the public interest defence, and demonstrates that there is no blanket protection for "whistle-blowers". In many cases, it may be that the public interest does not require general publication to the public, and is satisfied by some intermediate disclosure. In the case of a public safety issue, the courts may consider that the information should properly be disclosed (at least in the first place) to the Health and Safety Executive, or to Trading Standards Officers. It may be argued that going direct to the public before proper investigation by the

---

[38] [1984] 2 All E.R. 417.
[39] *Att.-Gen v. Guardian Newspapers*, n.26 above, at 545.

appropriate agencies is actually contrary to the public interest, given (a) that the information may turn out to be incorrect or ambiguous; and (b) that the public may misunderstand or overreact to the danger being disclosed.

In assessing whether the public interest requires disclosure in a case involving employee "whistle-blowers", the court may also investigate whether all appropriate avenues have been followed within the company itself. Depending on the urgency of the matters being revealed, it may be suggested that the public interest does not require disclosure outside the company unless and until it is clear that the company is not prepared to take the necessary steps either to correct the situation, or to reveal the information to the public itself.[40]

A company will be in a much better position to argue against public disclosure in such a case if it has in place complaints procedures for employees, preferably on a confidential basis.

REMEDIES

10–43 A broad range of equitable and legal remedies are available in an action for breach of confidence, just as for breaches of intellectual property rights. The details of such remedies and the procedures for obtaining them are not the province of this Chapter.[41] However, the breach of confidence action has certain peculiarities which need to be borne in mind.

Injunctions

Life-span

10–44 It is usually easy to identify the life span of a true intellectual property right. Patents last for 20 years from date of filing, registered designs 25 years from date of application, copyright 70 years from the death of the author, etc. This life span will equally apply to any final injunction granted to enforce the right.

The life span of an obligation of confidence is much more difficult to assess. Even in the case of a "real trade secret" such as the recipe for Coca Cola used as an example in paragraph 10–20 above, it is possible that at some point in the future the information in question may be made public, whether legally or illegally, such that the obligation of secrecy may no longer be enforced. In other cases, such as the "springboard" types of information dealt with in paragraph 10–21 above, the obligation will be held by the court to be inherently limited in time.

The upshot of this is that any injunctive relief in a breach of confidence case must be limited in some way. In the case of "springboard" relief, a specific time limit will be imposed — a matter of weeks or years, depending on the value of the information. In other cases, a condition will be imposed — namely that the injunction shall be discharged or cease to have effect upon the information coming into the public domain.

Damages in lieu of an injunction

10–45 An injunction being an equitable remedy, the court has a discretion whether or not to award it. In particular, whether to award damages instead of an injunction.[42]

---

[40] Though of course there will be cases where the employee is justified in believing that it would be counterproductive to approach his superiors about the matter
[41] See Chapter 24.
[42] s.50 of the Supreme Court Act 1981.

The general position is that an injunction will be granted wherever the confidential information remains confidential and there is an ongoing threat to use or disclose the information. However, damages alone may be regarded as an adequate remedy in the following cases.

Where the information is in the public domain

10–46   In many cases, the information in question will have ceased to be confidential by the time the action reaches trial. In such a case, the court will not award an injunction, for to do so would prevent the defendant from doing what he is now legally able to do. This is the case even if the sole reason the information has become public is the activities of the defendant which are being complained of. Once it is in the public domain, the court will not impose an obligation of secrecy on the defendant under any circumstances (save for the limited exception of "springboard" type injunction referred to paragraph 10–21 above). See *Att.-Gen v. Guardian Newspapers* (*"Spycatcher"*).[43]

A case such as *Spycatcher*, of course, where the defendant is himself responsible for the information coming into the public domain, is a classic instance of where it would be right for the court to award damages *in lieu* of injunctive relief. If the information is valuable, the release of it into the public domain will have destroyed its value entirely, and the defendant should compensate the claimant for the consequential damage to his business.

Where it would otherwise be disproportionate or oppressive to order an injunction

10–47   The courts will look at the overall commercial position between the parties to determine whether it is equitable to grant an injunction. There are many factors which may be taken into account. The primary one is the extent to which the defendant has established a business or a product line using the information in question. The courts may be unwilling to disrupt an ongoing commercial activity unless the use of the secret in the future can be shown to be likely to cause substantial difficulties for the claimant. Thus in *Seager v. Copydex*,[44] the Court of Appeal considered that damages were the only appropriate remedy where the defendant had subconsciously made use of the confidential information in developing its new product. It may also have been relevant in that case that the claimant had shown himself willing to licence the technology.[45]

Much may also depend on the extent of the use proved. Where the quantity of confidential information shown to be used by the defendant is relatively small, the courts will again tend to regard damages as the fairest remedy.[46] That may not be the case however where the use was deliberate and calculated to achieve a financial benefit.

It should be added that the above discussion concerns only injunctions against use of confidential information. The courts are much more willing to grant injunctions against disclosure to third parties (save of course where the information is in the public domain), since this is unlikely to cause any great inconvenience to the defendant, but will protect the claimant's interests for the future.[47]

---

[43] See n.26 above.
[44] See n.24 above.
[45] See also *Ocular Sciences v. Aspect Vision Care Limited* [1997] R.P.C. 289 at 406–407.
[46] *ibid.* at 407.
[47] *ibid.* at 409.

Delivery up or destruction on oath

10–48   This will tend to be automatic in the case of the grant of an injunction. The claimant is entitled to be protected against further disclosures of his secrets by an order which ensures the destruction or handing over of any materials which contain or embody the relevant information. Equally, if an injunction is not to be granted, delivery up will not be given either.[48]

Difficulties sometimes arise in relation to machinery or equipment which embodies the confidential information. Delivery up or destruction of an entire machine is likely to be disproportionate to the breach complained of. An alternative remedy may be granted which requires the alteration of the machine so that it no longer embodies the confidential information in question.

Assessment of damages

10–49   Damages in a confidential information case will be assessed on the same basis as any other case: the basic principle is that the claimant should be put in the position he would have been in had the breach not taken place. This is of course easier said than done. In a case where a defendant has taken the information and used it in developing a new product or process, it will be extremely difficult to assess how matters would have proceeded in the absence of the information, let alone what difference this would have made to the claimant.

The courts have established various methods of valuing information in such circumstances. The most useful discussion of this is the judgment of Lord Denning in *Seager v. Copydex No.2* [1969] R.P.C. 250 at 256. He divided information into three categories: "not very special"; "special" and "very special".

In the former category, where information could have been obtained by employing any competent consultant, the claimant would recover only the fee which a consultant would have charged for developing and providing the information. Where the information was "special", and could not have been obtained simply by going to a consultant, the claimant would recover the "market value" — namely the price which would be agreed for the information between a willing buyer and willing seller. Where the information was "very special" then damages should be awarded on the basis of a royalty for past and future use, capitalised as a lump sum.

These categories are somewhat arbitrary, and it is a little difficult to see why the different approaches should apply in each case. However, this judgment does at least identify three useful approaches which the courts may choose to adopt in any given case to do justice between the parties. As in all assessments of damages in intellectual property cases, the courts must recognise that there is no way of making any scientific measurement, or guaranteeing that the claimant has been adequately compensated. The courts inevitably have to adopt a "rough and ready" approach.[49]

Account of profits

Availability

10–50   Given the essentially equitable nature of the action for breach of confidence, an account of profits ought to be available as a remedy. The claimant is usually given a choice: he may choose to take damages or seek an account of profits, but not both. He is entitled to sufficient information before making this election for the choice to be

---

[48] *ibid.* at 410.
[49] Just as with patents — see *Gerber v. Lectra* [1995] R.P.C. 383.

an informed one.[50] However, since an account is an equitable remedy, it remains in the discretion of the court to refuse to order it, and to force the claimant to take damages instead.[51]

The circumstances in which an account may be refused are many and various, and similar factors are likely to be taken into account as are discussed above in relation to the availability of injunctive relief. In particular, the court is likely to be concerned with the innocence or otherwise of the defendant. It may be regarded as unfair to force an innocent defendant (such as the subconscious copyist in *Seager v. Copydex*) to disgorge all his profits.[52]

There is also the question of proportionality and practicality: if a deliberate breach of confidence has been only of minor benefit to the defendant in his business, it may be thought disproportionate to go through a complex and difficult accounting process. It may also be bordering on the impossible to value the effect of the breach on the defendant's profits.

Assessment

10–51 An account of profits does not necessarily entitle the claimant to all the profits made by the defendant from the business which misused the trade secret. This will only be the case where the secret was so central to the business in question, that it could not have been established or pursued without it.[53] In all other cases, an apportionment must be carried out to establish the value of the secret. This may be done by comparing the profits that were made from the business using the secret process with those which would have been had the closest available alternative process been used.[54]

The defendant is entitled to reduce his liability by taking account of the costs of setting up the business, his overheads, and the costs of sale and production. The difficulties and unpredictability of accounts of profits has resulted in very few claimants electing to pursue them in recent years.[55]

RESTRICTIVE COVENANTS

10–52 As we have seen, there are many restrictions placed by the law of confidence on the kind of information it will protect and the extent of the protection that the courts will give. In the context of contractual agreements, it is common to try to improve this protection by the use of express terms purporting to restrict the activities of one or both of the parties after the contract has come to an end. In particular, a term which prevents one party from competing with the other for a limited time, in order to stop him taking competitive advantage of what he has learnt during the course of the contract.

Such terms are known as "restrictive covenants". They are most commonly found in employment contracts, but may also be imposed in commercial contracts (for instance where trade secrets have been licensed).

Restrictive covenants need to be drafted with extreme care, because in principle

---

[50] See *Island Records v. Tring* [1995] F.S.R. 560.
[51] As was done in *Seager v. Copydex*, n.24 above.
[52] See also *Peter Pan v. Corsets Silhouette Limited* [1963] R.P.C. 349, where an account was ordered in the case of dishonest breach of confidence.
[53] This was the case in *Peter Pan, ibid.*
[54] *Siddell v. Vickers* [1892] R.P.C. 152 (a patent case); though *cf.* the more sophisticated way of getting to much the same result adopted in the recent case of *Celanese v. BP Chemicals* [1999] R.P.C. 203.
[55] A trend which is likely to continue given the result in *Celanese v. BP*, (*ibid.*) where the judge awarded even less than the Defendants were contending should be awarded against them!

they are in restraint of trade and therefore *prima facie* void. If litigated, it is up to the party seeking to enforce the restrictive covenant to justify it. Justification can only be on one of two grounds:

(a) The covenant was "reasonably necessary" to protect a trade secret;

(b) The covenant was "reasonably necessary" to prevent some personal influence over customers being abused in order to entice them away.[56]

The first of these grounds may justify certain restrictions on competitive trade as such. The second may justify specific restrictive covenants against the "solicitation" of customers. The following paragraphs discuss these two common types of restriction.

Restraints on competitive trade

10–53 In order to justify any restriction on competition, whether on ones own behalf or on behalf of someone else, it must first be shown that the party being restrained ("the covenanter") is in the possession of trade secrets. These must satisfy all the tests explained at the outset of this chapter. In particular, in the case of an employment contract, they must go beyond the general skill and knowledge obtained by the employee as a result of his employment. Thus, covenants in general restraint of trade will not be justified as against employees who are not (by reason of their job or their status) privy to the trade secret information of their employers. Covenants in such cases would clearly be mere attempts to prevent competition, which the law will not permit.

Having established the possession of trade secrets, the covenantee must justify the *scope* of the restriction on competition as being reasonable as between both parties in the light of the risks of disclosure or use.[57] The scope of the restriction must be no more than is adequate to protect the confidentiality having regard to those risks. It is important to remember that, this being a question of public policy, the courts will not take into account either the extent to which the covenanter entered the covenant with his eyes open (*i.e.* properly advised in law) or the "consideration" or remuneration given by the covenantee in return.[58]

10–54 When assessing the "reasonableness" of the restriction the courts must first have regard to the nature of the trade secrets which the covenanter is in possession of. To justify a restriction on competitive trade *per se,* it would seem that the trade secrets would have to go beyond mere knowledge of the identity of customers. Such information can adequately be protected by a mere non-solicitation clause.[59] They must be technical or business secrets. The more vital the secret information, the greater the restriction which will be justified.

The next question is the width of the definition of the trade which the covenanter is forbidden from engaging in. This must relate directly to the area of trade with which the particular secrets are concerned. It is most important to ensure that the two correlate precisely. A good example of the consequences of failure to do this is *Commercial Plastics v. Vincent* [1964] 3 All E.R. 546. There, an employee was engaged to make PVC calendered sheeting for adhesive tape. He had previously worked for three years generally in the field of calendering PVC. His restrictive

---

[56] *Herbert Morris v. Saxelby*, para. 10–14; *Faccenda Chicken,* n.11 above at 626.

[57] *Attwood v. Lamont* [1920] 3 K.B. 571 at 589.

[58] *Herbert Morris v. Saxelby*, para. 10–14 above at 707.

[59] See *Lucas v. Mitchell* [1974] Ch. 129.

covenant purported to prohibit him from working for any competitor in the PVC cal-endering field for one year after leaving his employment. The Court of Appeal held the restriction void, partly on the basis that the only trade secrets which the defen-dant could have had possession of related to the specific area of sheeting for adhesive tape. The extension of the clause to the whole field of calendering PVC was particu-larly unreasonable because of the defendant's established skills from his previous employment:

> He would be barred from making use of his skill and aptitude and general knowl-edge in that field, and that would be detrimental both to his interests and to the public interest.[60]

The courts will also be concerned to ensure that the length of time for which the restriction is imposed, and the geographical area over which it operates are reason-able in all the circumstances. Clearly the protection of a hugely important trade secret may justify a longer period of protection over a wider area than something of less significance.

10–55   A restriction which is open-ended in time will almost never be reasonable. In the case of an employee it may make it impossible for him to carry on the trade for which he is best qualified. It would have the effect of a contract of "serfdom". Simi-larly, a restriction which covers the whole of the United Kingdom for any substantial length of time will be hard to justify save in the most extreme case. The covenant in *Herbert Morris v. Saxelby* which covered the whole of the United Kingdom for a period of seven years was described in the following terms by Lord Shaw:

> From the point of view of the respondent it is, justly interpreted, a claim to put him in such a bondage in regard to his own labour that, if he seek to find employ-ment or advancement elsewhere, he must, for seven years of his life, become an exile.[61]

It is probably fair to say that longer periods of protection were thought justifiable in the early part of this century than would be acceptable now. Some early cases where genuine technical secrets were in issue held periods as long as five or seven years to be justified. In modern times, technology moves on at a much quicker pace. Matters which were secret when the contract was entered into may rapidly cease to be secret. Similarly, they may cease to have the commercial significance they once did. For this reason it is hard to see covenants in employment cases of more than one year being justified nowadays save in exceptional cases. On the other hand, the rapid improve-ments in world communications and the globalisation of the world economy may have made it more likely that the courts will recognise wider restrictions in terms of area.

Non-solicitation clauses

10–56   Where the trade secrets which a covenantee is concerned to protect concern the identity of his customers and their particular requirements, a non-solicitation clause ought to suffice to give him protection. Again, these must be justified as being no more than is adequate to provide reasonable protection against the "abuse" of per-sonal influence over customers in order to entice them away.

---

[60] See para. 10–14.
[61] *ibid.* at 718.

Although again *prima facie* in restraint of trade, such clauses have tended to be treated more generously from the covenantee's point of view than non-competition clauses *per se.* The following may be noted:

(a) The covenant may not be justified if it covers customers with whom the covenantee has not dealt;

(b) It should not extend to customers with whom the covenantee has only started to deal after termination of the contract;

(c) It is irrelevant that particular customers within the scope of the restriction did not propose to place any further business with the covenantee anyway[62];

(d) Covenants which would have the effect of preventing someone dealing with customers who were personally known to him *before* the commencement of the contract may be too wide[63];

Again, the extent of the clause in terms of time and in terms of area must be borne in mind when considering whether it is reasonable. If the area over which it operates is far beyond any area with which the covenanter has dealt, it may well be held unreasonable. Further, one can fairly assume that the ability to take advantage of a close relationship with customers to entice them away from is bound to diminish rapidly over time, as the customers build up a new relationship with the covenanter's successor. Thus any non-solicitation clause which extends beyond a year or so is likely to be held void.

Drafting and construction of restrictive covenants

10–57 It is important that restrictive covenants be drafted carefully so as to ensure that they do not cover areas over which the covenantee cannot reasonably claim protection. The basic rule is that if they do cover such areas, whether by accident or design, the courts cannot rewrite them. The only option is to strike them down as being invalid.

   Thus in *Commercial Plastics v. Vincent,* the plaintiff accepted in cross-examination that a narrower covenant could have given it adequate protection, but considered that it would have been difficult to draft it with sufficient precision. This was not regarded by the Court of Appeal as a satisfactory justification, though they went on to remark (at 555):

> The decision of this case against the plaintiffs is inevitable, but it is in a way regrettable, because the plaintiffs' case has underlying merits. They do seem to have important confidential information, for which they might reasonably claim protection by a suitably limited restrictive provision. The actual provision in this case can be described as "home made", that is to say not professionally drafted. It is unfortunate that a home-made provision, offered and accepted in good faith between commercial men and not in the least intended to be oppressive, has to be ruled out and declared void in a court of law for lack of the necessary limiting words. It would seem that a good deal of legal "know-how" is required for the successful drafting of a restrictive covenant.

---

[62] *Lawrence David v. Ashton* [1989] F.S.R. 87.
[63] *M&S Drapers v. Reynolds* [1956] 3 All E.R. 814.

10–58 It should however be noted that the courts have often attempted to do justice by taking a generous approach in cases where there was no obvious intention to be oppressive. It is often possible even in a reasonably well drafted restrictive covenant to show that it covers, on a literal construction, areas of activity which go beyond any reasonable restriction on trade. The courts in such cases have in the past shown themselves willing to engage in an "imaginative" approach to construction in order to avoid finding the covenant void. In *Littlewoods Organisation Limited v. Harris*[64] the question arose of the enforceability of a clause aimed at preventing a director of the Littlewoods mail order business from seeking employment with their main rival, Great Universal Stores, for a period of 12 months. The material part of the clause read:

> [Harris] shall not . . . enter into a Contract of Service or other Agreement of like nature with Great Universal Stores Limited or any company subsidiary thereto or be directly or indirectly engaged concerned or interested in the trading or business of the said Great Universal Stores Limited or any such company aforesaid.

Great Universal Stores Limited and their subsidiaries worldwide were engaged in a large range of different activities. Taken at its face value, the clause would have prevented Harris from working in fields far removed from his work for Littlewoods, and would thus have been void. However, the Court of Appeal were prepared to construe the clause as covering only the mail order part of GUS's business in the United Kingdom. So construed, the clause was reasonable to protect the confidential information Harris had gained in his employment by Littlewoods, and was enforceable. The court approved the words of Lindley L.J. in the old case of *Haynes v. Doman*[65]:

> the Court ought not to hold a just and honest agreement void, even when to enforce it would be just, simply because the agreement is so unskilfully worded as apparently, or even really, to cover some conceivable case not within the mischief sought to be guarded against. Public policy does not require so serious a consequence to be attached to a mere want of accuracy of expression.

10–59 This approach should be contrasted with a case in which the covenantee has not made any attempt to limit the terms of the restrictive covenant so that it is no wider than is reasonably necessary to avoid a real risk of misuse of confidential information. Thus the Court of Appeal in *Mont v. Mills*[66] struck down as void a clause which sought to restrain an employee for one year from joining "another company in the tissue industry". There was no attempt to limit this by reference to the nature of his employment or the particular nature of the business of the potential employer. Rejecting the *Littlewoods v. Harris* approach, they pointed out that if the courts were to ensure by an exercise of inventive construction that every restrictive covenant they were faced with was valid, there would be no incentive at all for employers to draft them carefully and in limited terms. And employees would be left at best in a state of total uncertainty, pending expensive litigation.

Mont v. Mills is a warning to those drafting restrictive covenants to ensure that they do not stray into unjustifiable areas. The *Littlewoods v. Harris* approach should be treated as the exception rather than the rule, and no reliance should be placed on it when drafting.

---

64 [1978] 1 All E.R. 1026.
65 [1899] 2 Ch. 13.
66 [1993] F.S.R. 577.

CHAPTER 11

THE LAW OF COPYRIGHT

Dr Uma Suthersanen, Fellow at Queen Mary Intellectual Property Research Institute and Lecturer at Centre for Commercial Law Studies, Queen Mary, University of London

INTRODUCTION

11–01   The area of copyright law has matured radically since it was introduced in fifteenth century Europe with the advent of printing. The earliest records of copyright-type laws were the royal privileges which were granted in response to the lobbying of book publishers for rights to protect their investments. However, the first real modern copyright law in the world was introduced in Britain with the Statute of Anne 1710, which granted a duration of protection of a mere 28 years and then, only to authors of books.[1] Since then, copyright law has had to adapt to successive technological changes in the last two centuries to extend its ambit to sound recordings, films, computer programs, electronic databases and finally, multimedia and Internet technology. The importance of copyright protection lies in this ability of the law to extend protection over an extensive range of works, from such diverse areas as fine and cultural arts to research and industrial production.[2]

"European copyright" law

Author's rights

11–02   The expression "copyright" is commonly used in reference to laws in United Kingdom and the United States, and also to the copyright laws in the European countries. However, readers should be aware that there is a demarcation between the Anglo-American and the European legal traditions, that is both academically and practically important. The European countries emphasise that copyright laws emanate from the need to protect the author or creator of the work. This emphasis is even reflected in the terminology employed by such countries — rather than "copyright", the rights are referred to as author's right (*droit d'auteur* — French, *diritto d'autore* — Italian, *Urheberrecht* — German). Furthermore, the author's rights are divided into two distinct categories — economic rights and moral rights. Economic rights protect the author's ability to exploit his work in the market by allowing him to prohibit others from reproducing or performing or transmitting his work; whereas, moral rights protect the author's personality and reputation.

In contrast, the United Kingdom and United States copyright systems place emphasis on the holder of the economic power, *i.e.* the entrepreneur or the producer of the work, as opposed to the creator or the author of the work. This emphasis has produced disparate consequences when compared with European systems, especially in respect of two areas: employee works (there are extremely few circumstances in which an employer can gain copyright *ab initio* under European Union Member States' national copyright laws, whereas the opposite is true in the United Kingdom and in the United States); and moral rights.[3]

---

[1] Copyright Act 1709 (8 Anne, c.19).
[2] For further reading, see W.R. Cornish, *Intellectual Property* (Sweet & Maxwell, 1999) (general introductory reading on United Kingdom copyright law); J.A.L. Sterling, *World Copyright Law* (Sweet & Maxwell, 1998) (for a global copyright perspective, concentrating on France, Germany, United Kingdom and United States, E.C. Directives and international copyright conventions).
[3] Moral rights are discussed in detail below.

Legislation

11–03 Unlike some other intellectual property laws, there is, as yet, no "European Community Copyright Law". However, while this Chapter will concentrate on United Kingdom copyright law, the nature of copyright law and other related areas is slowly evolving towards a more harmonised European standard, with the introduction of successive Directives in the following important areas:

- legal protection of computer programs[4];

- rental and lending right, and related rights[5];

- satellite broadcasting and cable retransmission[6];

- term of protection[7];

- legal protection of databases[8];

- legal protection of services based on conditional access[9];

- electronic signatures[10];

- electronic commerce.[10a]

In addition to these instruments, it is noteworthy that the European institutions are currently considering the following proposals which will have some effect on United Kingdom copyright law, if they are duly passed:

- Proposed Directive on artist's resale right (*droit de suite*)[11]

- Proposed Directive on the copyright in the information society[12]

Due to their growing influence, the discussion below will also encompass the more relevant E.C. legislation, with some reference to European Member States' national laws. Finally, we briefly examine the efforts to harmonise copyright law at the international level.

GENERAL COPYRIGHT PRINCIPLES

11–04 In general, copyright protection of a work grants the owner of the copyright a set of exclusive rights to do and to authorise others to do certain acts with respect to certain types of works for a limited period of time.

Specifically, a person concerned with copyright protection of his work should concentrate on five main issues: (a) does copyright subsist in the work? (b) if so, for how long? (c) who owns the copyright in the work? (d) has a third party committed any infringing acts? (e) what defences and remedies are available to the parties?

---

[4] Directive 91/250; [1991] O.J. L122/42.
[5] Directive 92/100, [1992] O.J. L346/61.
[6] Directive 93/83; [1993] O.J. L248/15.
[7] Directive 93/98; [1993] O.J. L290/9.
[8] Directive 96/9; [1996] O.J. L077/20.
[9] Directive 98/84; [1999] O.J. L320/54.
[10] Directive 99/93; [2000] O.J. L13/12.
[10a] Directive 2000/31; [2000] O.J. L178/1.
[11] Amended Proposal, [1998] O.J. C125/8.
[12] Amended Proposal, [1999] O.J. C180/6.

CONDITIONS OF PROTECTION

11–05 The main condition of protection, in relation to literary and artistic works, is originality. Related to this condition, is the important distinction between ideas and expression. A further criterion of protection within the United Kingdom is that some types of works must be physically fixed on a medium, before copyright can subsist in them.

Originality

11–06 All literary, dramatic, artistic and musical works must be original in order to qualify for copyright protection. The notion of originality does not import any meaning of quality or merit, and neither original thought or research is required. The standard for originality is minimal. First, the work must originate from the author and must not have been slavishly copied, directly or indirectly, from another source.[13] Thus, a drawing which is simply traced from another drawing cannot be considered an original artistic work, unless there is some alteration or addition by the subsequent artist which will convert the work into an original work. Secondly, it must be shown that the author expanded some "skill, labour and judgement" or "skill, labour and capital" in creating the work. Furthermore, the skill, labour and capital must be sufficient so as to "impart to the product some quality or character which the raw material did not possess, and which differentiates the product from the raw material."[14] As one court warned, "copying, *per se*, however much skill or labour is devoted to the process, cannot make an original work".[15]

Substantiality

11–07 Related to the criterion of originality is the further rule that the work must not be insubstantial or commonplace as this will necessarily imply a lack of originality. Therefore, in the past, copyright protection has been refused to single words (*e.g.* a name[16]), short titles (*e.g.* the title of a song or book[17]), and to diaries comprising commonplace information and material.[18]

European standard of "originality"

11–08 In contrast with the United Kingdom concept of originality whereby one has to show that skill, labour and effort were expanded to create a work, European civil law countries such as France and Germany employ a different concept. In these countries, it must be shown that the work exhibits the author's personality or individuality, or some "creative contribution"[19]; mere skill and labour are insufficient if absent the author's stamp of individuality.

There are provisions in three E.C. Directives on copyright law which offer a harmonised standard of "originality" throughout the European Union in relation to the following types of works: computer programs, photographs and databases. These

---

[13] *University of London Press v. University Tutorial Press* (1916) 2 Ch. 601; *Ladbroke (Football) Ltd v. William Hill (Football) Ltd* [1964] 1 W.L.R. 273.
[14] *Macmillan & Co. Ltd v. Cooper* (1923) 40 T.L.R. 186.
[15] *Interlego v. Tyco Industries* (1988) R.P.C. 343.
[16] *Exxon Corp. v. Exxon Insurance Consultants International Ltd* [1982] R.P.C. 69.
[17] *Francis Day & Hunter Ltd v. Twentieth Century Fox Corpn.* [1940] A.C. 112.
[18] *GA Cramp & Sons Ltd v. Frank Smythson Ltd.* [1944] A.C. 329.
[19] *Isermatic*, Cass. Civ., April 16, 1991; D. 1992, somm., at 13.

works will be considered original if it is the "author's own intellectual creation".[20] It is thought that the E.C. standard adopts the civil law rather than the common law approach. Until 1997, the United Kingdom withstood any attempt to incorporate the E.C. concept of originality. However, with the implementation of the Database Directive, the United Kingdom copyright law now specifically refers to the E.C. test of "author's own intellectual creation", but *only* in relation to databases.[21] It is uncertain at present whether the traditional test of "skill, labour and skill" will have to be abandoned in favour of this standard across other areas of copyright works.[22]

If the European standard is adopted, copyright protection for certain types of works may be more difficult to obtain. This is especially true for literary works such as routine reports, specifications, plans, compilations and lists, *i.e.* works which present factual data or information. Copyright could be denied unless it could be shown that the work exhibited the author's personality or individuality by the manner in which the data is arranged or presented. The different judicial attitudes in the United Kingdom and other European states may be illustrated in relation to the case of football pools coupons. In France, copyright protection was refused for tables of football matches with grids for entering the results on the ground that the work did not reach the rank of intellectual creation, either by choice or disposition of material; however, in the United Kingdom, the courts have given copyright protection in the past both to football coupons[23] and to grid patterns on lottery cards.[24]

Protection of ideas and concepts

11–09   An oft-repeated tenet under copyright law, which is not expressly stated in the copyright statute, is: copyright does not protect ideas, but the expression of ideas.[25] This is sometimes referred to as the idea-expression dichotomy. In other words, copyright protection gives you the right to prevent others from appropriating the *expression* of your ideas or the form in which your ideas are presented in the work. The rationale arises due to the long duration of protection given to certain types of copyright works and the public interest in the free exchange of ideas and concepts. Thus, in determining whether originality subsists in a work, the courts will apply this further qualification and look to see whether the originality of a work pertains to the expression of an idea, rather than the idea itself. How do courts make this distinction?

Some general rules can be observed. First, this principle is not applied with consistency across all types of works. For example, extremely simple diagrams or drawings can obtain copyright protection, although protection of such drawings may effectively lead to the protection of ideas. Instead, the idea-expression dichotomy is applied as a means by which the scope of protection can be measured, especially in relation to simple works. Thus, while a simplistic, single lined pencil drawing of a human hand may gain protection, copying the drawing may not necessarily lead to infringement unless

---

[20] Article 1(3), Computer Program Directive 91/250, (1991) O.J. L122/42; Recital 17 and Article 6 Term Directive 93/98, O.J. L290/9; Article 3(1) Database Directive 96/9, (1996) O.J. L077/20.

[21] See discussion below.

[22] This is made more probable as both the United States and Canada have recently adopted the European notion of originality. See *Feist Publications Inc. v. Rural Telephone Service Co. Inc.* 18 U.S.P.Q. 2d. 1275 (1991, United States Supreme Court); *Tele-Direct (Publications) Inc. v. American Business Information Inc.* 76 C.P.R. (3d) 296 (1998, Canadian Federal Court of Appeal).

[23] *Ladbroke (Football) Ltd v. William Hill (football) Ltd* (1964) 1 W.L.R. 273 ((on the grounds that the selection of bets, or the layout/mode of presentation of the bets were original).

[24] *Express Newspapers plc v. Liverpool Daily Post and Echo plc* [1985] F.S.R. 306.

[25] The principle is enshrined in other copyright instruments. See Art. 1(2), Computer Program Directive 91/250/EEC; Art. 9(2) TRIPS Agreement 1994; Art. 2, WIPO Copyright Treaty 1996.

the copy was an exact and identical facsimile copy of the drawing.[26] Similarly, copy-right protection will be granted to language or computer translations even though, strictly speaking, such translations are the conversion of ideas from one style or language to another.

11–10 Secondly, the notion that "copyright does not protect ideas" has been gener-ally interpreted as meaning that copyright subsists, not in ideas but in the form in which the ideas are expressed. Thus it is not an infringement of the copyright in the expression of the idea to take the idea and to apply it so long as that application does not involve copying the expression.[27] However, it should be noted that it is not an easy task to distinguish at any one time the demarcation between an idea and the expres-sion of it. It has been argued that during any creative process, there is a gradual pro-cess whereby a vague idea or suggestion gradually becomes more and more concrete, as an idea is gradually fleshed out, by incorporating sufficient elaboration and struc-ture. Thus, some courts have stated that it is only the "general idea or basic concept of the work" which is unprotected under copyright laws: but once the author sets out to transform the basic concept into a concrete form, he is taken to have "expressed" his idea, which will be protected. As one court said:

> "Anyone is free to use the basic idea — unless, of course, it is a novel invention which is protected by the grant of a patent. But no one can appropriate the forms or shapes evolved by the author in the process of giving expression to the basic idea."[28]

Formalities

11–11 There are no formalities required for copyright to subsist in a work such as registration or the affixation of a copyright notice or symbol. However, certain types of works must be fixed.

Fixation

11–12 A final criterion is that some types of works must be fixed in some medium. Copyright will not subsist in literary works unless such works are "recorded, in writ-ing or otherwise."[29] It should also be noted that it is immaterial whether the work is affixed by the author or with the permission of the author.[30] Thus, a person who gives an impromptu lecture or speech will not gain any copyright protection unless and until the words are physically affixed — either by the author himself, by writing or tape recording or filming himself, or by someone else who takes the speech in short-hand or records it. The criterion is unnecessary in relation to certain types of works which must necessarily be fixed to come into existence, *i.e.* artistic works, sound recordings and films.

---

[26] *Kenrick v. Lawrence* (1890) 25 Q.B.D. 99 (simple drawing of a hand and a ballot box granted copyright protection).

[27] *Designer's Guild Ltd v. Russell Williams (Textiles) Ltd* [2000] F.S.R. 121—overturned in HL, but not on this point.

[28] *Plix Products Limited v. Frank M Winstone (Merchants) and Others* (1986) F.S.R. 63 (High Court of New Zealand); affirmed in appeal (1986) F.S.R. 608 (Court of Appeal of New Zealand), at 66. See also *Ibcos Computers Ltd v. Barclays Finance Ltd* (1994) F.S.R. 275, Ch D.

[29] Copyright, Designs and Patents Act 1988, s. 3(2).

[30] *Ibid.*, s. 3(3).

TYPES OF WORKS

11–13 What types of works or products does copyright law protect? The United Kingdom copyright law sets out nine separate categories of works. The importance of these categories must be emphasised as any work or product which does not fall within these categories cannot claim copyright protection in the United Kingdom.[31] The nine categories are as follows:

(a) *original* literary, dramatic, musical or artistic works;

(b) sound recordings, films, broadcasts or cable programmes, and

(c) the typographical arrangement of published editions.[32]

Of the nine categories, the two most important categories of work in relation to industrial research and development are literary works and artistic works. It is within these two categories that one will find protection for research reports, developmental or manufacturing blueprints or drawings, computer-generated works, and computer software. There is also a growing significance of films and cable programmes, especially in relation to multi-media and Internet-related works.

Literary works

11–14 A "literary work" is defined in the Act as "any work, other than a dramatic or musical work, which is written, spoken or sung". "Writing" is further defined as: "any form of notation or code, whether by hand or otherwise and regardless of the method by which, or medium in which, it is recorded."[33] The law further states that the following will be considered as literary works:

(a) a table or compilation other than a database;

(b) a computer program;

(c) preparatory design material for a computer program, and

(d) a database."[34]

Besides these statutory definitions, there have been judicial attempts to define this concept. One judge likened a literary work to any work which is "expressed in print or writing, irrespective of the question whether the quality of style is high",[35] whilst another defined a literary work as "one intended to afford either information and instruction, or pleasure".[36] Thus, the definition of a literary work encompasses any subject matter which is expressed in words, mathematical notation or symbols. Accordingly, the courts have recognised the following as literary works: lectures,

---

[31] The position in the United Kingdom is different from that under the other E.C. Member States in that the latter do not require a strict observance of categories; in general, all works of literary, artistic and scientific nature are protectable, and it is left to the court's discretion to determine whether a particular type of work comes under the aegis of copyright protection.

[32] s. 1(1), Copyright, Designs and Patents Act 1988.

[33] *ibid.*, s. 178.

[34] *ibid.*, s. 3(1).

[35] *University of London Press Ltd v. University Tutorial Press Ltd* (1916) 2 Ch. 601.

[36] *Hollinrake v. Truswell* (1894) 3 Ch.D. 428; approved by the Court of Appeal in *Exxon Corp. v. Exxon Insurance Consultants International Ltd* [1982] R.P.C. 69.

poems, mathematical tables,[37] calculation charts,[38] codes and ciphers,[39] shorthand,[40] advertising jingles, time-tables, and directories. Furthermore, the notion of "literary work" is wide enough to extend to translations, editorial work, annotations, computer programs, electronic databases or compilations and, on occasion, to electronic circuit diagrams.[41]

As stated above, works will qualify for protection irrespective of the literary quality or style of the work;[42] however, there must be some minimal level of effort or labour or skill involved in creating the work, and the work must have some modicum of substantiality. This may be problematic in two types of literary works: short works; and compilations or databases. In respect of short works, protection has been refused, in the past, to insufficient or extremely short works such as names of persons[43] or companies[44], or short titles of books or films[45] or advertising jingles which consist of commonplace words.[46] The rationale is that literary works must give some "information, pleasure or instruction"; works such as innovative words or phrases do not exhibit the requisite skill or labour since the work itself is so minimal.[47]

Databases and compilations

11–15 Prior to 1996, the position on databases and compilations was this: such works could claim copyright protection as long as they exhibited a modicum of labour, skill and judgement.[48] However, with the introduction of the Database Directive, the law in the United Kingdom, as elsewhere in the European Union, has been changed. The result is that copyright protection of databases may be somewhat curtailed with the implementation of the European notion of originality, while a new database right has been introduced. The discussion on compilations and databases is dealt with at paragraphs 11–88, *et seq.* below.

Dramatic and musical works

11–16 The term "dramatic work" is defined vaguely as including a work of dance or mime.[49] It would thus cover activities such as ballet, plays, etc. A "musical work" is defined as meaning a "work consisting of music, exclusive of any words or action intended to be sung, spoken or performed with the music" (the words of a song would of course qualify as a literary work).[50]

---

[37] *Bailey v. Taylor* (1824) 3 L.J.O.S. 66 (mathematical tables); *Express Newspapers plc v. Liverpool Daily Post & Echo plc* [1985] F.S.R. 306 (a lottery card, which consisted of a grid of 25 letters).

[38] *Autospin (Oil Seals) Ltd v. Beehive Spinning* (1995) 23 R.P.C. 683, HC.

[39] *Anderson (D.P.) & Co. Ltd v. The Lieber Code Co.* [1917] 2 K.B. 469 (a telegraph code, consisting of 100,000 groups of five letters each).

[40] *Pitman v. Hine* (1884) 1 T.L.R. 39 (shorthand scheme).

[41] *Anacon Corpn Ltd v. Environmental Research Technology Ltd* [1994] F.S.R. 659; *Aubrey Max Sandman v. Panasonic U.K. Ltd & Matsushita Electric Industrial Co. Ltd* [1998] F.S.R. 651.

[42] *University of London Press Ltd, op. cit.* (examination papers).

[43] *Wombles v. Wombles Skips* [1975] F.S.R. 488 (copyright refused to word "Wombles").

[44] *Exxon Corp v. Exxon Insurance Consultants* [1982] R.P.C. 69 (no copyright protection for word "Exxon").

[45] *Francis Day v. Twentieth Century Fox* [1940] A.C. 112 (no infringement in taking title of film).

[46] *Kirk v. Fleming (J & R) Ltd* M.C.C. (1928–35) 44.

[47] *Exxon Corp v. Exxon Insurance Consultants* [1982] R.P.C. 69.

[48] *GA Cramp & Sons Ltd v. Frank Smythson Ltd* [1944] A.C. 329.

[49] s. 3(1), Copyright, Designs and Patents Act 1988.

[50] *ibid.*, s. 3(1).

Artistic works

11–17 This area of copyright law is closely related to design protection, and reference should be made to Chapter 12. The 1988 Act defines an artistic work as follows:[51]

(a)   a graphic work, photograph, sculpture or collage, irrespective of artistic quality;

(b)   a work of architecture being a building or a model for a building, or

(c)   a work of artistic craftsmanship.

The phrase "graphic works" is further defined to include the following types of art work[52]:

(a)   any painting, drawing, diagram, map, chart or plan; and

(b)   any engraving, etching, lithograph, woodcut or similar work.

Almost any type of two-dimensional drawing or design would come under the general category of graphic work, irrespective of their artistic merit or style or form. In the past, the courts have generously extended the notion of "graphic works" to protect simple drawings,[53] navigational charts,[54]; architectural plans,[55] technical or engineering drawings,[56] typefaces[57] and packaging labels.[58] Thus, the "artistic work" category is important in relation to blueprints, initial product design documentation, and organisational or developmental chart or plan. However, as explained below, the copyright owner of such drawings may not be able to rely on them to gain protection in respect of three-dimensional reproductions of the drawings.

Design for industrial products

11–18 What of three-dimensional products such as prototypes for furniture or toys? It will be extremely difficult for such works themselves to gain copyright protection unless they can come within the categories of architectural work, sculpture or work of artistic craftsmanship. The latter two categories have been narrowed to works which require artistic intention or merit and in the past, products such as baby rain-capes, modern furniture suites and toys have all been denied protection as these products could not come within the definition of artistic works. On rare occasions, it may be possible to argue that copyright protection subsists in the initial product mould. On one occasion, a court extended artistic copyright protection to the Frisbee toy and the mould utilised for making the Frisbee, rationalising that both these products could be

---

[51] s. 3(1), Copyright, Designs and Patents Act 1988, s. 4(1),(2).
[52] *ibid.*, s. 4(2).
[53] *Kenrick v. Lawrence* (1890) 25 Q.B.D. 93 (simple drawing of hand pointing to a square on an electoral voting paper was an artistic work).
[54] *Macmillan Publishers Ltd v. Thomas Reed Publications Ltd*, [1993] F.S.R. 455.
[55] *Chabot v. Davies* [1936] 3 All E.R. 221 (plans for a shop-front).
[56] *British Leyland Motor Corp v. Armstrong Patents Co.* [1986] R.P.C. 279, CA, HL (drawings of car exhaust pipe qualify as artistic works, but protection denied for other reasons); *Solar Thomson engineering Co. Ltd v. Barton* [1977] R.P.C. 537, CA (drawings of rubber "O" rings for pulley wheels granted copyright protection); *LB (Plastics) Ltd v. Swish Products Ltd* [1979] R.P.C. 551 (copyright protection granted for drawings of parts of drawers); *Canon Kabushiki Kaisha v. Green Cartridge Company (Hong Kong) Limited*, [1997] F.S.R. 817 (copyright held to subsist in drawings of parts for Canon's toner cartridges).
[57] *Stephenson, Blake & Co. v. Grant, Legros & Co.* [1916] 33 R.P.C. 406.
[58] *Charles Walker & Co. Ltd v. The British Picker Co. Ltd* [1961] R.P.C. 57.

considered as engravings under copyright law.[59] However, this decision was made during the pre-1988 copyright regime, and such stretches of imagination are highly unlikely under the current Act.

Even if such products did manage to come within the artistic works category, the scope of protection is curtailed upon the industrial application and marketing of the product.[60] The limitation sets in where certain types of artistic works have been commercially exploited by an industrial process and marketed. In such cases, copyright protection is effectively reduced to twenty five years from the date of such marketing activity. After this period, the work may be copied by making articles of any description, or doing anything for the purpose of making articles of any description, and anything may be done in relation to articles so made, without infringing copyright in the work. The underlying rationale of the provision is that copyright law is ill-designed to cope with artistic works which are destined for the industrial, mass-produced market; such works should receive protection comparable to that under the Registered Designs Act 1949.[61] In general, the only recourse for such products would be the registered design right or unregistered design right.

Functional drawings

11–19 Copyright protection will extend to all kinds of functional or technical drawings, especially in light of the low threshold of originality. Thus, even though an industrial product itself is denied copyright (see above), the drawings from which it is made will be protected though serious limitations are now placed on the enforcement of such copyright (see above).

Not only will such drawings be considered as artistic works, they may also, sometimes, be considered as literary works, especially where such drawings or documentation are accompanied by engineering notation and symbols which can be read by an engineer or skilled person. In one decision,[62] the plaintiffs claimed that they owned the copyright in engineering drawings and circuit diagrams in relation to an electronic dust meter analyser. The circuit diagram was of a conventional type showing various components (resistors, transistors, capacitors) utilising conventional symbols for these components, with a written piece of information indicating the appropriate rating on that particular component. The court accepted that the engineering drawings and circuit diagrams could be classified as both "artistic works" and "literary works" as they contained information in electrical engineer's notation.

11–20 The "section 51" defence However, irrespective of whether such drawings are considered as artistic or literary works, an extremely important limitation is placed on the scope of protection of such drawings under section 51 of the Copyright Designs and Patents Act:

---

[59] *Wham-O Manufacturing v. Lincoln Industries Ltd* [1985] R.P.C. 127; *cf. Davis (J&S) (Holdings) Ltd v. Wright Health Group Ltd* [1988] R.P.C. 403 (dental impression trays held not to be sculptures since they were merely steps in the overall process of producing dentures and were never intended to have permanent existence).
[60] Copyright, Designs and Patents Act 1988, s. 52.
[61] However, certain works do not come within the scope of section 52 including works of sculpture and printed matter primarily of a literary or artistic character. Copyright (Industrial Process and Excluded Articles) (No. 2) Order 1989 (S.I. 1989 No. 1070).
[62] *Anacon Corpn Ltd v. Environmental Research Technology Ltd* [1994] F.S.R. 659; see also *Aubrey Max Sandman v. Panasonic U.K. Ltd & Matsushita Electric Industrial Co. Ltd* [1998] F.S.R. 651.

"It is not an infringement of any copyright in a design document or model recording or embodying a design for anything other than an artistic work or a typeface to make an article to the design or to copy an article made to the design."[63]

This provision specifically states that the copyright owner of protected design documents or models cannot extend his scope of protection in drawings so as to prevent other parties from either making three-dimensional versions of his drawings or to copy articles which are made to such design documents or models — unless — the drawings or models are destined to be embodied in artistic works. The terms "design" and "design documents" are defined widely:

- design means the design of any aspect of the shape or configuration (whether internal or external) of the whole or part of an article, other than surface decoration;

- design document means any record of a design, whether in the form of a drawing, a written description, a photograph, data stored in a computer or otherwise.[64]

11–21 For example, the copyright owner of an architectural drawing will be able to prevent unauthorised copying of both two and three dimensional reproductions of his drawing since the drawing is eventually destined for an architectural work, which is an artistic work. However, the copyright owner of drawings depicting a car exhaust pipe or an electric terminal board will not be able to enforce his copyright against three dimensional reproductions of his drawing because the eventual product — the exhaust pipe or electric terminal board — will not be considered as an artistic work (see discussion above on industrial products).

This provision has the greatest impact on purely technical or functional drawings or other such documentation. Whilst the copyright owner of such works can use his copyright to prevent an unauthorised third party from copying such drawings or documentation in two-dimensional form (for example, by photocopying the work), he will not be able to prevent the unauthorised party from copying such works in a three-dimensional form, *i.e.* by making the product.[65] In such instances, the designer or copyright owner should look to any unregistered design right which he may have in the design of the article for protection.[66]

Computer-generated works

11–22 Computer-generated literary or artistic works are protectable under copyright law. A computer-generated work is defined as a work that is:

"generated by a computer in circumstances such that there is no human author of the work."[67]

---

[63] Copyright, Designs and Patents Act 1988, s. 51(1). Section 51(2) explains that this defence extends further to allow the copyist to communicate the copied design to the public by issuing it to the public, or including it in a film, broadcast or cable programme service.

[64] *ibid.*, s. 51(3).

[65] This is unlike the position prior to the Copyright, Designs and Patents Act 1988 when industrial copyright could be used for this purpose.

[66] See Chapter 12.

[67] Copyright, Designs & Patents Act 1988, s. 178.

11–23 Therefore, a work will qualify for protection, irrespective of whether the work is created by a human being or it is generated by a computer. Examples of computer-generated works would include satellite weather maps or navigational charts which are generated automatically from data acquired through instrument readings.[68] A distinction must be maintained between computer-assisted work and computer-generated work. This distinction may be important in relation to who will be considered the author or owner of the copyright.[69] Charts or drawings created with the assistance of computer software such as CAD (computer-assisted design) software would not be considered computer-generated work since the initial input of data and the subsequent selection of menu or drawing options would still be under the aegis of a human author. In such instances, the computer or the program is essentially a tool, with the author playing a determining role.

Other types of works

11–24 As indicated above, copyright law also recognises other types of works, namely sound recordings, films, broadcasts, cable-casts and typographical arrangements. These works are sometimes referred to as neighbouring or related rights. It should be noted that these works usually incorporate other types of copyright works. Thus, the filming of a presentation will give rise to several types of copyright works: the literary copyright in the speech, a sound recording copyright if the presentation is recorded, artistic copyright if the presentation includes the showing of any graphic slides, and the actual film copyright. The owner of the copyright in all these works may be the same person or different persons, depending on the circumstance in which the presentation took place.[70]

It was previously the case that the circumstances in which copyright in any of these other works would arise would have been negligible in a research or developmental context however, with the increased use of technology to create digital or electronic works such as multimedia encyclopaedias and dictionaries, some of these categories of works have gained a contemporary importance.[71]

Sound recordings[72]

11–25 This is defined as including the recording of any sound, including the reading or recitation of a literary work, on any medium, such as discs, tapes, compact discs and DAT, as long as the sound can be reproduced. Furthermore, there is no necessity that the sound be an intelligible or coherent sound. There is no copyright in a sound recording which is a copy of another sound recording.

Films[73]

11–26 The definition of a film means a recording on any medium provided a moving image can be produced from it. The phrase "any medium" would include celluloid, video tape and digital recordings of moving images; furthermore, if a sound track accompanies the film, it will be considered to be part of the film.[74] The wide definition can include audio-visual works and multimedia digital recordings. Thus, if

---

[68] H.Laddie *et al*, *The Modern Law of Copyright and Designs* (Butterworths, 1995), para. 3.18.
[69] See paras 11–34 *et seq.*
[70] *ibid.*
[71] See paras 11–71 *et seq.*
[72] Copyright, Designs and Patents Act 1988, s. 5A(1).
[73] *ibid.*, s. 5A(2).
[74] *ibid.*, s. 5(B)(2),(3), (5).

moving images are part of a web page, it can fall under the category of "film". However, separate copyrights would subsist in all the other individual elements which constitute the work in whole.

Broadcast[75]

11–27   A "broadcast" means a transmission by wireless telegraphy of visual images, sounds or other information which is capable of being lawfully received by the public or is transmitted for presentation to the public. Additionally, an "encrypted transmission" will be considered a broadcast, but only if decoding equipment is made available to the public. Thus, in a general sense, the notion of "broadcast" covers all transmissions of all types of information, whether by air or sound waves, or by terrestrial or satellite transmissions — as long as the transmission is by a wireless means, as opposed to transmission via a cable.

Cable programmes[76]

11–28   In contrast to the above, a cable-programme service means a service whereby information is sent via a telecommunications system for reception to the public. Certain types of services are excluded from the definition including interactive services such as services devoted solely to the running of a business or to private domestic use.[77]

Published edition[78]

11–29   This is a special type of copyright which subsists in the typographical format or arrangement of a published edition of a literary, dramatic or musical work.

FOREIGN AUTHORS OR WORKS

11–30   Works are entitled to copyright protection under United Kingdom law on two qualifying criteria: the nationality of the author, or the place of first publication of the work.[79] Both these categories are extremely wide due to the principle of national treatment imposed under the international copyright conventions such as the Berne Convention and the TRIPS Agreement. Furthermore, works will qualify for copyright protection in a majority of other countries on similar conditions.[80]

Nationality criterion

11–31   A work will qualify for protection if the author (not the copyright owner) of the work falls within one of the following categories:

- British citizen or domiciled person or resident in the United Kingdom; British Dependent Territories citizen, British National (Overseas), British Overseas citizen, British subject, British Protected person;

---

[75] Copyright, Designs and Patents Act 1988, s. 6 (1).
[76] *ibid.,* s. 7.
[77] *ibid.,* s. 11(2)
[78] *ibid.,* s. 8.
[79] *ibid.,* ss. 153–155.
[80] See para. 11–89, below.

- an incorporated body in the United Kingdom, if the author was a legal entity such as a company;

- a citizen, subject, domiciled person, resident or incorporated body in a Berne Convention state.

The above can be understood in relation to the following example. A work is commissioned by a United Kingdom company from a company incorporated in France. Under United Kingdom law, the owner of the copyright in the work will be the French company; the work will qualify for copyright protection under United Kingdom law since the French company is an incorporated body in a Berne Convention state (France is a signatory to the Convention). However, where a work is commissioned by a United Kingdom company from an Ethiopian author, the work will not qualify for protection under the nationality basis as Ethiopia is not a signatory of any international copyright convention. The work may nevertheless still qualify for protection on the publication basis.

Publication criterion

11–32   A work will be entitled to protection, regardless of the nationality of the author, if the work is first published in the following countries:

- United Kingdom;

- any Berne Convention country;

- or if first publication is in a non-Berne Convention country, when the second publication is made within 30 days in a Berne Convention country.

Thus, to return to our previous example, in the case of a work by an Ethiopian author, such a work will qualify for protection under United Kingdom law as long as the work was first published in the United Kingdom. Even if first publication is in Ethiopia, if a second publication of the work is made within 30 days in the United Kingdom, or indeed, in any other Berne Convention signatory country such as the United States, Canada or Germany, the work will still qualify for protection in the United Kingdom.

DURATION OF PROTECTION

11–33   Copyright in a work will arise automatically upon the work being created. There is no requirement of registration or of using the © symbol to attract copyright in the United Kingdom. The duration of protection varies depending on the type of work at hand as indicated in the table below.[81] For certain types of works, protection is dependent on the date of release or making of the work; whereas for other types of works, the duration of protection is anchored to the life of the author or a person closely related to the work (unless the identity of the author of the work is unknown).

---

[81] sections 12–15, of the 1988 Act.

Types of Work	Duration
Literary, dramatic, musical	Life of author plus 70 years after his death. If jointly authored work, life of author plus 70 years from the death of the last surviving author.
Artistic	Life of author plus 70 years after his death. If jointly authored work, life of author plus 70 years from the death of the last surviving author If the work is industrially manufactured and subsequently marketed, copyright protection is effective for 25 years from date of first marketing.
Anonymous/pseudonymous literary, dramatic, musical or artistic works	70 years from the date of creation or, if work made available to public, 70 years from the date of publication.
Films	Life of principal director or screenplay author or dialogue author or composer of music for film plus 70 years after the death of the last surviving person.
Anonymous films	70 years from the making of film or, if made available to public, 70 years from the date of publication.
Sound recordings	50 years from the making, or if released, 50 years from release.[82]
Broadcasts	50 years from the first transmission.
Published edition	25 years from first publication of new edition.
Computer-generated works	50 years from making the work.

Certain clarifications should be made. Although the duration of protection is tied to a particular person, this does *not* mean that the person is the copyright owner. For example, copyright protection of a literary work will last for the entire life of the author, and 70 years thereafter; however, the copyright owner may not be the author, but his employer. The rules for duration do not follow ownership in this particular instance.[83] Secondly, the table above is not exhaustive but concentrates on the most important categories of works. Different rules operate in respect of works which qualify for Parliamentary or Crown copyright. Finally, the duration of moral rights is considered separately below.

OWNERSHIP[84]

11–34 Ownership of copyright is considered more extensively in Chapter 14. However, it should be noted that the ownership of copyright in the work does not necessarily follow the ownership of the original or physical embodiment of the work. Thus, the recipient of a letter may hold physical ownership over the paper which will allow him to sell the letter, but not to make copies of the letter as the copyright in the letter will belong to the author of the letter.

---

[82] section 13A.
[83] See paras 11–34 to 11–35, below.
[84] sections 9, 11.

Prima facie, ownership of copyright is accorded to the author of the work, *i.e.* the person who creates the work (though slightly different rules exist in respect of films, sound recordings, etc.). An important exception exists in respect of certain types of works: if the work is made by an employee in the course of his employment, his employer is the owner of any copyright in the work, unless an agreement exists between them which specifies otherwise. The table below summarises the position in relation to the categories of works:

Types of Work	Ownership
Literary, dramatic, musical or artistic works	Author(s) of the work or employer
Films	Producer or principal director
Sound Recordings	Producer of sound recording
Broadcasts	Person making the broadcast
Cable programme	Person providing the cable programme service
Published edition	Publisher
Computer-generated works	Person who undertakes the arrangements necessary for the creation of the work

Joint authorship and ownership

11–35 Where two or more authors contribute to a work, both authors will be entitled to joint authorship of the copyright. The general rule is that authorship is determined with reference to the person who expands skill, labour and judgement in creating the work, as opposed to the person who merely supplies suggestions or ideas. The demarcation line between the two types of contribution, however, can be difficult to draw.[85]

INFRINGEMENT OF COPYRIGHT

11–36 The copyright owner has the right to prohibit certain types of acts if they are done by any person who is not licensed or authorised by the copyright owner. The prohibited acts are dependent on the type of work in question and are as follows:

- copy the work;

- issue copies of the work to the public;

- rent or lend the work to the public;

- perform, show or play the work in public;

- broadcast the work or include it in a cable programme service;

- make an adaptation of the work or do any of the above acts in relation to an adaptation.

---

[85] *ibid.*, s. 10(1). For more details, see Chapter 14 on ownership.

Primary and secondary infringement

11–37 The above acts are classified as acts of "primary infringement" — the copyright owner merely has to prove that a third party has committed one of the above acts without his authorisation. There is a second category of infringing acts termed collectively as "secondary infringement" — in the latter instance, the copyright owner will also have to prove that the offending party knew or had reason to believe he was dealing with an infringing copy. The requisite amount of knowledge will depend on the facts of each case and the court will look at the reasonable inferences which can be drawn from a concrete situation disclosed by the facts of the case.[86]

Copying the work

11–38 This is one of the most important rights conferred on the copyright owner and it is extensive in its breadth. The right to prohibit the copying of the work is available in respect of all categories of works.[87] Copyright should be distinguished from other exclusive rights such as under patent or registered design laws where the proprietor is granted a monopoly: copyright law does not prohibit third parties from producing works which are identical to the protected work provided they were made independently.

Courts look for two elements:

(a)   sufficient objective similarity between the two works;

(b)   causal connection or nexus between the two works.

Both must be satisfied to amount to infringement. Thus, even where a work is objectively identical to a protected work, this is not enough by itself for infringement of copyright. It must also be shown that the author had access to or was familiar with the protected work, and that he copied it, either directly or indirectly (even if it was done sub-consciously).

However, as a matter of evidence, if there is a striking similarity between the two works, and it is proved that the offender had access to the previous work, the courts will often be willing to infer that there has been copying.[88] In order to protect oneself from allegations of copying, especially in respect of functional works such as factual databases or computer programs where there may be a higher possibility that two independently created works will be substantially similar due to technical constraints, it may therefore be sensible to create works in a "clean room environment", where the designer can be shown to have had no access to existing copyright works.

Copying in any medium

11–39 The Act itself lays down certain guidelines: copying, for instance, includes the reproduction of a literary or artistic work in any material form, including storing the work in any medium by electronic means, or making copies of the work which are transient or incidental to other uses of the work.[89] Thus, the literary copyright in a report, which was originally produced in writing on paper, would be infringed if it was word-processed or scanned onto a computer without the author's permission. An

---

[86] *Sillitoe v. McGraw Hill* [1983] F.S.R. 545; see also *LA Gear Inc. v. Hi-Tec Sports plc* [1992] F.S.R. 121.
[87] See para. 11–13.
[88] *Francis Day & Hunter v. Bron* [1963] Ch. 587.
[89] Copyright, Designs and Patents Act 1988, s. 17

infringing act would have been committed irrespective of whether the unauthorised copy is permanent or is destroyed subsequently.

Substantial copying

11–40 It is an infringement to copy either the whole work or a "substantial" part of the work, either directly or indirectly.[90] Thus in the previous example, there is no need to prove that the whole report has been reproduced, as long as there has been reproduction of a substantial part of it. The question of whether a substantial amount of the work has been taken is dependent on several factors. As one court stated:

> ". . . the question whether he has copied a substantial part depends much more on the quality than on the quantity of what he has taken. One test may be whether the part which he has taken is novel or striking, or is merely a commonplace arrangement of ordinary words or well-known data. So it may sometimes be a convenient short-cut to ask whether the part taken could by itself be the subject of copyright. But in my view, that is only a short cut, and the more correct approach is first to determine whether the plaintiff's work as a whole is "original" and protected by copyright, and then to inquire whether the part taken by the defendant is substantial."[91]

A recent Court of Appeal decision has provided a set of guidelines in answering the question of substantial copying.[92] First, in considering the issue of substantiality, the part to be considered is the part of the copyright work which has been copied. Secondly, the copying must relate not to the idea, but to the expression of the idea: this principle applies the idea-expression dichotomy discussed above. Thirdly, substantiality is a qualitative not a quantitative test. For example, in our previous example, the written report is 50 pages but the vital statistics and information may only extend to 10 pages: if these 10 pages alone are taken, the court may conclude that a substantial part of the work has been appropriated, based on the quality of what is taken. The fourth guideline by the court was to point out that the antithesis of "substantial" is "insignificant". Fifthly, in considering whether the part which has been copied is substantial or not, no weight is to be attributed to elements within the work which are commonplace or well-known or derived from some other source. Finally, it should be borne in mind that the object of the law of copyright is to protect the product of the skill and labour of the maker and not to confer on him a monopoly in the idea that the work may express.

Free riding

11–41 Related to the above discussion is the issue of alteration of the original copyright work. The law is that infringement of copyright occurs when a defendant has taken a substantial amount of the plaintiff's work, irrespective of the fact that the former has expanded further effort and skill in changing or altering the original work so as to give rise to another original work. Thus, making a few changes to a work will not avoid infringement, unless those changes are so substantial that it can no longer be said that a substantial part of the original has been taken. Should a court be convinced that one has misappropriated a protected work so as to unfairly partake of the

---

[90] *ibid.*, s.16(3).
[91] *Ladbroke (Football) Ltd v. William Hill (Football) Ltd* [1964] 1 W.L.R. 273, at 283.
[92] *Designer's Guild Ltd v. Russell Williams (Textiles) Ltd* [2000] F.S.R. 121—overturned before going to press in HL. The fifth of these propositions may now be doubted.

author's fruits of labour, the court will not hesitate to hold that an infringement of copyright has occurred.[93]

Historical or factual works

11–42 Copyright protection is available for historical or factual works as long as it is shown that the author employed skill and labour in the selection and compilation of facts. However, the scope of protection given to such works can be smaller than that accorded to other types of literary works. Historical or factual works must necessarily rely on earlier sources or facts; thus, the law will allow a wider use to be made of a historical work than of a novel, so that subsequent works of knowledge can be built upon existing works of knowledge.[94] Having said this, it should be emphasised that this does not offer a *carte blanche* to subsequent users to copy substantial amounts from a historical or factual work.

Design documentation

11–43 A specific provision in the law is that the copyright owner of an artistic work will be able to prohibit the conversion of his two-dimensional artistic work into three dimensions, or *vice versa*. Thus, it will be an infringement of a protected architectural drawing to reproduce it as a three-dimensional building. However, there are limitations on the extent to which the copyright owner can enforce his rights in this respect in relation to functional or technical drawings, which we have discussed above (see paragraphs 11–13 *et seq.*, above).

Secondary infringement: importing, commercial dealing, etc.

11–44 The copyright owner's right to prevent copying of his work extends further to prevent certain dealings with infringing copies which constitute secondary infringement of the copyright.

First, he has the right to prevent *infringing copies* of his work from being imported into the United Kingdom.[95] Secondly, he has the right to prevent an unauthorised person from possessing, selling, letting, exhibiting or distributing infringing copies of the work. In general, the infringer must have committed the above acts in the course of business or for non-private or non-domestic use. However, an exception arises in the case of the distribution of an infringing work: if infringing copies of the work are distributed in non-commercial circumstances but to such an extent as to prejudice the copyright owner's interest, it will amount to infringement. Finally, the copyright owner can prevent any person who supplies or provides articles which are specifically designed or adapted for making infringing copies of the work. This provision will cover any person who attempts to make, import into the United Kingdom, possess or sell such articles. Since the acts of unauthorised importation of infringing works or of equipment and commercial dealing are categorised as secondary infringement, it must be shown that the offender knows or has reason to believe that he is importing or dealing with an infringing copy of the work.

---

[93] *Elanco Products Limited v. Mandops (Agrochemical Specialists) Limited* [1980] R.P.C. 213; *Ravenscroft v. Herbert* [1980] R.P.C. 193.
[94] *Ravenscroft v. Herbert* [1980] R.P.C. 193.
[95] Copyright, Designs and Patents Act 1988, s. 22. See also section 27 for a detailed definition of what constitutes "infringing copy".

Right to issue copies to the public[96]

11–45 The copyright owner has the right to decide when to release (or issue) to the public copies of his work. This means the act of putting a work into circulation in the European Economic Area by or with the consent of the copyright owner. The right only relates to putting copies of a work which had not previously been put into circulation in the EEA. Thus, once a copy of a book has been lawfully placed in circulation in the European Economic Area by the copyright owner or with his permission, he generally has no further rights over the subsequent distribution, sale or importation of *that* particular copy of the book — except where the rental and lending rights apply.

Furthermore, the right extends to the control of importation of copies of the work into the EEA. If the owner of a work has released the work *outside* the EEA, but not within it, he has a right to prevent an unauthorised person from importing the work into the EEA since this would constitute a release of a copy which had never been in circulation in the EEA. Thus, this provision gives extensive rights to the copyright owner against dealers in unauthorised copies of his work in that he can prevent them from bringing in such copies from outside the EEA as primary infringers. An exception would be where the copyright owner has circulated copies of the work outside the EEA in such manner as to amount to an express or implied consent to the importation of such works into the EEA.[97]

Rental and lending rights[98]

11–46 With the implementation of the E.C. Directive on rental and lending right, and related rights,[99] the United Kingdom copyright law now grants the copyright owner a right to control the rental or lending of legitimate copies of his work. The right is restricted to the following works:

- literary, dramatic, musical works;
- artistic works — but not works of architecture or works of applied art;
- films or sound recordings.

There is a slight distinction between "rental" and "lending". Rental means making a copy of the work available for use, on condition that it may or will be returned, for economic or commercial advantage — for instance, a store which makes a video available to anyone for a fee. Conversely, lending means a copy of the work is made available for use but for non-economic or non-commercial advantages and access is made through an establishment which is accessible to the public, *i.e.* a video made available through a public library.[1] It should be noted that unauthorised rental of copies of the work is an infringement irrespective of whether the rental is made by a business entity or an individual.

---

[96] *ibid.,* s. 18.
[97] See Chapters 16 and 23 for exhaustion.
[98] Copyright, Designs and Patents Act 1988, s. 18A.
[99] November 19, 1992 (92/100).
[1] Although a public library may charge a small fee, if the payment goes towards operating costs of the establishment, then this is not considered an economic or commercial advantage.

Performance, showing or playing[2]/Broadcasting or including it in a cable programme[3]

11–47 The performance of a literary, dramatic or musical work in public is a restricted act and requires the copyright owner's permission. The performance of a work includes delivering the work by means of a lecture or speech or audio-visual presentation, or even by means of a sound recording, film, broadcast or cable programme presentation. Similarly, it is an infringement to show or play a work in public without the copyright owner's consent — however, the right is only restricted to sound recordings, films, broadcasts or cable programs. Thus, for example, one can safely show artistic works in public without infringing (unless they are infringing copies of the work).

Secondary infringement: provision of apparatus and premises

11–48 The copyright owner's right extends not only to the persons who commit the offending act, but also to the following persons: (a) those who provide the copy of the sound recording or film; (b) those who provide apparatus necessary to play, show, perform or receive the work in an unlawful manner; (c) occupiers of premises who allow their premises to be used for the above infringing purposes.[4] However, the latter category of offenders are considered secondary infringers, and the copyright owner will have to show that they had knowledge or reason to believe that by providing copies or apparatus or premises, their activities would lead to infringing uses of the copyright work.

Adaptation[5]

11–49 It is an infringement to make an adaptation of a literary, dramatic or musical work without the copyright owner's consent. The term "adaptation" is strictly defined according to the type of work at hand. In relation to a literary or dramatic work, adaptation means the translation of the work, or converting the work into a non-dramatic work or vice versa, or making different versions of the work for example, converting a novel into a cartoon strip, etc. In respect of a computer program or a database, adaptation means the arrangement or altered version of the program or a translation of the program, *i.e.* converting the program from one computer language or code into different language or code.

Authorising an infringing act

11–50 Copyright infringement is not only committed by a person who does a restricted act without the licence of the copyright owner, but also by such persons who *authorise another* to do an infringing act. The term "authorise" not only covers those persons who expressly sanction or approve of an infringing act, but also those who are assumed to have authorised an act through their indifference or failure to act. In a landmark decision, it was held that a university had authorised the unlawful copying of books by their policy of installing photocopying machines in their library premises and of failing to place adequate notices on the machines to inform the library users as to their rights under the copyright law.[6]

This decision should be contrasted with a subsequent House of Lords decision whereby it was alleged that tape recorder manufacturers had authorised copyright

---

[2] Copyright, Designs and Patents Act 1988, s. 19.
[3] *ibid.*, s. 20.
[4] *ibid.*, ss. 25 and 26.
[5] *ibid.*, s. 21.
[6] *Moorhouse v. University of N.S.W.* [1976] R.P.C. 151.

infringement by the mere fact of their marketing and advertising a recorder which had the added feature of copying on to blank tapes records which are on discs and tape.[7] The court held that the manufacturers had not sanctioned, approved or countenanced infringement, explaining that the notion of "authorisation" within the copyright law meant a grant or purported grant, which may be express or implied, of the right to do the act complained of. Although the manufacturers had conferred on the purchaser of the tape recorder the power to copy, they had not granted or purported to grant him the right to copy. A further distinguishing feature from the university case discussed above, was the fact that the tape manufacturer had no control over the use of their models once they were sold.

PERMITTED ACTS[8]

11–51 The 1988 Act sets out a whole range of permitted activities which will not come within the scope of protection. The first general defence is that of fair dealing. Secondly, there are specific defences in relation to certain types of activities and works such as defences in relation to computer programs, functional drawings or educational reprography or use of a work for public administration or archival purposes.

Fair dealing

11–52 It is a defence to copyright infringement that the activity complained of amounts to "fair dealing" with the work for one of three purposes: research and private study; criticism and review; reporting current events.[9]

There is no statutory definition of the notion of "fair dealing", as such, though there are several statutory limitations as to what acts will constitute fair dealing. The guidelines furnished by the courts vary in relation to whether the fair dealing defence is raised in respect of research, criticism or reporting though as a general rule, courts will take into account the following: the motives of the user or the alleged infringer; the extent and purpose of the use; whether that extent was necessary for the purpose for which the fair dealing defence is claimed; and finally, whether the protected work had been published or circulated to the public.[10]

Research and private study

11–53 Fair dealing with a literary or artistic work for the purposes of research or private study is exempt from infringement. The notion of research includes both commercial and private research, but only in so far as one is not dealing with databases: in the latter case, the fair dealing defence does not permit one to use a database for commercial research.[11] In the case of copying a work, it is not an act of fair dealing if the copying is made by any person other than the researcher or student if the copier knows or has reason to believe that more copies of the work will be distributed subsequently.[12]

---

[7] *CBS Songs v. Amstrad* [1988] R.P.C. 567; *CBS Inc. v. Ames Records and Tapes Ltd* [1981] 2 All E.R. 812.
[8] Chapter III, Copyright, Designs and Patents Act 1988.
[9] *ibid.,* ss. 29–31.
[10] *Hyde Park Residence Ltd v. Yelland and others, The Times* February 16, 2000, CA.
[11] *ibid.,* s. 29(5).
[12] *ibid.,* s. 29(3).

Criticism and review

11–54   Fair dealing with a copyright work for the purposes of criticism or review is
not an infringing act as long as there is sufficient acknowledgement of the source and
author.[13] Criticism of a work need not be limited to criticism of style; it may extend
further to the ideas to be found in a work and its social or moral implications.[14] In one
case concerning the utilisation of passages from a book, the court held that the question of fair dealing was one of degree, of fact and impression to be determined by several factors, such as the number and extent of the quotations and extracts; the
proportions of the quotes: whether they are too many and too long; the use made of
them, *i.e.* whether they are used as a basis for comment, criticism or review; or are
they are used to convey the same information as the author, for a rival or competitive
purpose.[15] In another decision, the court held that too much weight should not be
placed on the expressed purpose, intention and motive of the plaintiff; consideration
must also be had as to the impact of the criticised work on the audience.[16] As the court
stated:

> "'Fair dealing' in its statutory context refers to the true purpose (that is, the
> good faith, the intention and the genuineness) of the critical work — is the pro-
> gramme incorporating the infringing material a genuine piece of criticism or
> review, or is it something else, such as an attempt to dress up the infringement
> of another's copyright in the guise of criticism, and so profit unfairly from
> another's work?"[17]

Reporting current events

11–55   Fair dealing with a copyright work for the purposes of reporting current
events is not an infringing act as long as there is sufficient acknowledgement of the
source and author.[18] Moreover, where photographs are concerned, it is not an act of
fair dealing to utilise a photograph for the purposes of reporting current events as its
whole commercial potential lies, in many instances, in it being a record of current or
newsworthy events. In respect of the defence of fair dealing for reporting, it should be
stressed that the use must have been necessary for the purpose of reporting current
events.

Incidental inclusion of work

11–56   The use of a work will not be infringing where it merely amounts to an incidental inclusion in an artistic work, sound recording, film, broadcast or cable programme.[19] Thus, the inclusion of a work in the background of a film will be accepted
as non-infringing if the use was "casual, inessential, subordinate or merely background", taking in account whether the use competes with or commercially prejudices
the exploitation of the copyright work.[20]

---

[13]   *Hyde Park Residence Ltd v. Yelland and others, The Times* February 16, 200, CA, ss. 29(1A), 30(1), 30(2).
[14]   *Time Warner Entertainments Co. v. Channel Four Television plc* [1994] E.M.L.R. 1.
[15]   *Hubbard v. Vosper* [1972] 2 Q.B. 84; *Beloff v. Pressdram Ltd* [1973] R.P. C. 765.
[16]   *Pro Sieben Media AG v. Carlton U.K. Television Ltd* [1999] F.S.R. 610.
[17]   *Pro Sieben Media AG v. Carlton U.K. Television Ltd* [1999] F.S.R. 610.
[18]   *ibid.,* ss. 29(1A), 30(1), 30(2).
[19]   *ibid.,* s. 31.
[20]   *IPC Magazines Ltd v. MGN Ltd* [1998] F.S.R. 431.

Miscellaneous statutory defences

11–57   There are a variety of educational or archival defences.[21] Thus, it is permissible to copy, show or play works for the purposes of education or instruction, as long as the activity is done by the person giving or receiving the instruction or education. In respect of educational establishments, such as universities, reprographic copies may be made for instruction but only to the extent that no more than one per cent of the whole work is made in any quarter of a year. This exception is aimed at educational establishments (*i.e.* those decreed to be so by the Secretary of State) and does not derogate from the rights of any individual to photocopy works as an act of fair dealing for research or private study purposes.

Defence of "public interest"

11–58   In some instances, the courts have condoned acts which technically infringe the copyright in a work but where such infringing acts are justified on public interest grounds.[22] In *Lion Laboratories Ltd v. Evans*[23], the plaintiffs were manufacturers of breathalysers which gauged the alcoholic content of motorists and which were being used by the police. The Court of Appeal held that the publication of internal research documentation by a reporter which evidenced that the equipment was faulty should be allowed in the public interest. However, in *Hyde Park Residence Ltd v Yelland and others*,[24] the Court of Appeal appeared to indicate that the defence of public interest would have a narrow application and that a court would only be entitled to refuse to enforce copyright if the work is:

- immoral, scandalous or contrary to family life;

- injurious to public life, public health and safety or the administration of justice;

- incites or encourages others to act in a way referred to above.

Design documentation

11–59   See paragraphs 11–20 *et seq.*

Computer programs

11–60   See paragraphs 11–72 *et seq.*

MORAL RIGHTS

11–61   Under the European copyright system, the copyright owner is granted two distinct sets of rights: economic rights (which protect his commercial interests in exploiting his work such as copying, issuing the work, etc., discussed above) and moral rights. The latter rights protect the author's personality and reputation by giving him the right to insist that his authorship be acknowledged or by making it unlawful to distort or mutilate the author's work. Historically, the United Kingdom and the United States have always been reluctant to adopt the latter category of rights into

---

[21] Copyright, Designs and Patents Act 1988 ss. 32–50.
[22] *ibid.,* s. 171(3). See also *Initial Services Ltd v. Putterill* [1968] 1 Q.B. 396 and *Hubbard v. Vosper* [1972] 2 Q.B. 84.
[23] *Lion Laboratories Ltd v. Evans* [1985] Q.B. 526.
[24] *Hyde Park Residence Ltd v. Yelland and Others, The Times,* February 16, 2000, CA.

copyright law. Several reasons were proffered. First, copyright law is concerned mainly with the economic concerns of the author, not his personality or integrity. Secondly, an author who wanted to reserve his right to claim authorship or the right to object to false attribution of authorship or to object to changes or alterations to his work, could always have recourse to other laws such as contract,[25] defamation and tort[26] laws. Nevertheless, due to growing international and European pressures, the United Kingdom finally introduced a limited set of moral rights under the 1988 Act.

Type of moral rights

11–62   There are four main moral rights[27]: (a) the right of paternity/attribution; (b) the right of integrity; (c) the right against false attribution; (d) the right of privacy in commissioned films and photographs.

Right of paternity[28]

11–63   This right allows an author the right to be identified as an author of a work or as the director of a film. However, in order to claim the right of paternity, the author must assert this right in writing, *i.e.* the author must proclaim in a visible and noticeable manner that he is the author of the work and that he claims the right to be identified as such. Note, however, that employees who create works in the course of their employment cannot claim this moral right against the owner of the copyright. There is also a further range of activities over which the author cannot exercise his moral rights for example, in respect of functional drawings or commercially exploited artistic works.

Right of integrity[29]

11–64   The right of integrity allows the author of a work to object to any derogatory treatment of his work. "Treatment" is defined to mean "any addition to, deletion from or alteration to or adaptation of the work", but not including translations of a literary work; while derogatory treatment refers to any treatment which amounts to distortion or mutilation of the work or is otherwise prejudicial to the honour or reputation of the author of the work or the director of the film. The test is an objective test. In one decision, drawings of dinosaurs by an artist were subsequently reduced in size when published in a book.[30] The artist complained that his right of integrity had been infringed; however, the court held that, from an objective and reasonable perspective, the reduction of size of his drawings did not amount to a distortion of his work. Employees are denied this moral right against the copyright owner unless the employee has been identified on the work. However, even where the moral right of integrity does apply, and an employer does change or alter the work, the employee may not have a course of action if his employer inserts a disclaimer on the

---

[25] *Frisby v. BBC* [1967] Ch. 932 (author objected to the cutting of one line in his play by BBC, even though the latter was given a licence to make minor alterations. The court held that the licence did not allow BBC to omit a line which the author considered important).
[26] *Clark v. Associated Newspapers Ltd* [1998] R.P.C. 261 (Spoof diaries written by an *Evening Standard* reporter were falsely attributed to Alan Clark; the latter's complaint of false attribution was upheld under both tort and copyright laws.)
[27] Chapter IV, ss. 77–89; Chapter V, ss. 94–95, Copyright, Designs and Patents Act 1988.
[28] *ibid.*, ss. 77–79
[29] *ibid.*, ss. 80–83.
[30] *Tidy v. Trustees of the Natural History Museum* (1996) E.I.P.R. D-86; (1998) 39 I.P.R. 501.

work to the effect that changes have been made without his employee's authorisation or consent.

Right against false attribution[31]

11–65 Any person has the right to object to any work being falsely attributed to him, *i.e.* if he is falsely named as an author of a work which he has not created. The right is infringed if copies of the work with the false description are issued to the public. Any person may claim this right — there is no need for the person to be an author himself.

Right of privacy in certain photographs and films[32]

11–66 This right does not actually benefit the author of a work, but is a right which is granted to the commissioner, and then, only in respect of photographs and films commissioned for private and domestic purposes.

Nature of moral rights

11–67 The right of paternity and the right of integrity are the two more important rights. There are several general points which should be noted. First, these two rights are only accorded in relation to certain types of works: literary, dramatic, musical and artistic works and films. Secondly, the rights are granted only to the authors of these works, or to the director of the film; therefore, the moral rights do not necessarily vest in the owner of the copyright.[33] Thirdly, there are many exceptions as to when and how the rights will be exercised. We have already mentioned that it is either impossible or difficult for an employee to enforce his moral rights. Another important exception is that the authors of computer programs and computer-generated works do not enjoy the rights of paternity and integrity. Fourthly, all moral rights will cease on the expiry of copyright protection of the work, *i.e.* life of the author plus 70 years thereafter. The right against false attribution only lasts for the life of the author plus 20 years thereafter. Finally, moral rights are not assignable but they can be subject to a waiver.[34]

The final attribute of moral rights is important as far as conflicting interests between the author or owner is concerned. For example, let us assume a company has commissioned a report from a third party consultancy. Under the normal rules, copyright ownership will vest in the author of the report, *i.e.* the consultant. This may be provided for in a specific contract whereby the consultant agrees to assign his copyright to the company. However, under normal rules, moral rights will also automatically vest in the consultant, and moral rights cannot be the subject of an assignment. In such instances, there should be a specific contractual clause whereby the consultant agrees to waive his moral rights. The waiver clause will apply to all four rights; the clause may be narrow or extremely wide; the author will be able to waive his moral rights in respect of a specific work or to several works or to all works generally; finally, he can waive rights in respect of existing works and in respect of future works.

---

[31] Copyright, Designs and Patents Act 1988, s. 84.
[32] *ibid.*, s. 85.
[33] See Chapters 12 and 15 on authorship and ownership.
[34] The provision on waiver is controversial, and is not generally accepted under the European copyright laws.

Remedies

11–68 The remedies for copyright infringement include civil and criminal penalties.

Civil remedies

11–69 In general, the copyright owner can ask for damages, injunctions, an account of profits, and delivery up.[35] Damages will be calculated so as to place the claimant in the position he would have been had the infringement not occurred, *i.e.* compensation for the actual losses suffered. This would usually be the amount of royalties the copyright owner would have obtained had a licence been obtained or the profit he is deprived of due to lost sales of his own products offered by the illicit competition. Damages are not payable if the offending party shows that he did not know or have any reason to believe that copyright subsisted in the work. If the defendant was an innocent infringer in these terms, the plaintiff's only recourse is to seek delivery up, an injunction and an account of profits.[36] In order to pre-empt a defence of innocent infringement, it is advisable to mark all copies of the work with a copyright notice or the symbol ©. In addition to normal damages, a copyright owner could go further and request additional damages though these are a form of punitive damages which are only awarded if the defendant has acted in a flagrant, deceitful or treacherous manner; nevertheless, they can be awarded where damages or an account of profits will not compensate the copyright owner.[37]

A request for damages should be contrasted with that for an account of profits whereby the court will award the plaintiff all profits or gains made by the defendant through his infringing activities. A final remedial measure is an order for delivery up of infringing copies or articles designed or adapted for making copies of the copyright owner's work. Related to the order of delivery up of copies or articles, is the court's power to order the disposal or destruction of such copies and articles.[38]

Criminal offences

11–70 The 1988 Act makes it an offence for any person to make, sell, import, or otherwise commercially deal with or distribute an article which is, and which the offender knows or has reason to believe, is an infringing copy of the copyright work.[39] It is necessary to show two circumstances: the offender had the necessary element of knowledge or belief; and that his activities were motivated on commercial grounds, and not on a private or domestic basis. It is a criminal offence to make an article specifically designed or adapted for making copies of a particular copyright article such as a plate for printing artistic works. If an offence committed by a corporate body such as a company is committed with the consent or connivance of a director, manager, secretary or other similar officer, then the latter person is also guilty of the offence.[40]

Copyright and modern technologies

11–71 The advent of digital technology has caused some consternation as to the appropriate form of intellectual property protection. The primary problem lies in the

---

[35] Copyright, Designs and Patents Act 1988, s. 96.
[36] *ibid.*, s.97(1).
[37] *ibid.*, s.97(2).
[38] *ibid.*, s. 114.
[39] *ibid.*, s. 107.
[40] *ibid.*, s. 110.

fact that products such as electronic databases, on-line products, multi-media and software packages, face classification difficulties under copyright law. For example, a software package may comprise the computer program, a built in dictionary or the-saurus and technical documentation — all of which would ordinarily be classified as literary works. However, the package may also contain digital artwork, graphic icons or photographs — which would be categorised as artistic works. Finally, the package may also contain sound recordings, musical works, or a sequence of images. If the package is available for downloading via the Internet, there may be a broadcast or cable program involved as well. As we noted above, different rights and terms of pro-tection arise for different types of works, and therein lies the dilemma. Should a sep-arate category be created for a multi-media work, as was done in the past for films or broadcasts? Or should such works be protected by intellectual property rights but not within the copyright arena? This was the partial solution to the protection of elec-tronic databases where a new database right was introduced, hovering at the edges of copyright law. The following sections seek to give a brief overview of copyright pro-tection in relation to two major areas: computer programs and electronic databases.

A final short section describes the potential copyright protection available for biotechnological products, such as DNA and protein sequences.

COMPUTER PROGRAMS

11–72   Computer programs are primarily protected under copyright law. They have been accepted as "literary works", both in the international and European legal arena.[41] As such, the law on literary works, as stated above, will ordinarily apply. How-ever, in an attempt to deal with the more difficult aspects of copyright protection of computer programs, the Directive on computer programs[42] has introduced several special provisions, which have been incorporated, to a certain extent, under United Kingdom copyright law.

Conditions for protection

11–73   There is no definition of computer programs, either under the E.C. or United Kingdom copyright laws, though the term includes preparatory design mate-rial leading to the development of a computer program provided that the nature of the preparatory work is such that a computer program can result from it at a later stage.[43] This would presumably include product specifications, flowcharts and dia-grams. Under the Directive, protection will be granted to a computer program which is original in the sense that it is the author's own intellectual creation — no further criteria as to the qualitative or aesthetic merit of the program will be applied.[44] Computer programs have, previously, in European Union Member States such as Germany[45] and France,[46] often failed this first hurdle due to the fact that courts required a higher threshold of originality for factual or technical works. Nevertheless,

---

[41]   Article 10(1), TRIPS Agreement; Article 4, WIPO Copyright Treaty.
[42]   Council Directive of May 14, 1991 on the legal protection of computer programs, 91/250; (1991) O.J. L122/42.
[43]   *ibid.,* Recital 7.
[44]   *ibid.,* Art. 1(3), Recital 8.
[45]   *Inkassoprogramm*, BGH May 9, 1985; (1986) 17 IIC 681. German Supreme Court.
[46]   *Pachot*, Cass. Ass. plén., March 7, 1986; (1986) 129 RIDA 136, French Supreme Court; *Isermatic*, Cass. 1re civ., April 16, 1991; D. 1992, somm. 13, French Supreme Court.

with the incorporation of the Directive, the standard of originality has been toned down in these countries.[47]

Idea/expression merger

11–74   The Directive clearly stipulates that copyright protection will only extend to the expression of a computer program, and that ideas and principles which underlie any element of the computer program, including its interfaces, will be excluded from protection.[48] Although most European Member States have incorporated these conditions, United Kingdom copyright law merely requires a computer program to be original; it is assumed that the exclusion as to ideas and principles will also apply though not specifically stated in the Act. As discussed above, the demarcation between an idea and its expression can be difficult.[49] In the case of computer programs, the difficulty is heightened by the fact that many programs operate under the same technical or functional constraints. Under such circumstances, two programmes may be identical due to the fact that there is only one way of expressing an idea *i.e.* the idea and expression has merged. In such an event, it may be decided that the computer program cannot enjoy copyright protection.[50]

Restricted acts

11–75   The Directive states that the copyright owner of a computer program has the right to authorise the following activities[51]:

- the permanent or temporary reproduction of a computer program by any means and in any form, in part or in whole: this includes the acts of loading, displaying, running, transmission or storage of the computer program which necessitate the reproduction of the program;

- the translation, adaptation, arrangement and any other alteration of a computer program and the reproduction of the results thereof;

- any form of distribution to the public, including the rental, of the original computer program or of copies thereof subject to the Community exhaustion rule that the first sale of a copy of a program by the rightholder or with his consent will exhaust his distribution right in relation to that copy.[52]

These provisions are implemented within the United Kingdom copyright law, albeit in slightly differing language. The owner of copyright in a computer program will have the same rights as accorded to other copyright owners and the discussion at para. 11–36 *et seq.* applies. Nevertheless, several problematic aspects are considered below.

Reproduction of a computer program

11–76   Infringement will occur if it can be shown that the whole or a substantial part of the program has been copied.[53] One difficulty lies in determining the scope of copy-

---

[47]   *Betriebssystem*, BGH 1991; (1991) 22 I.I.C. 723, German Supreme Court.
[48]   Art. 1(2), Recital 13, Computer Program Directive.
[49]   See paras 11–09 and 11–10.
[50]   *Total Information Processing Systems Ltd v. Daman Ltd* [1992] F.S.R. 171; *cf. Ibcos v. Barclay* [1994] F.S.R. 725.
[51]   Art. 4, Computer Program Directive.
[52]   Note that the exhaustion rule does not apply in relation to the right of rental, Art. 4(c), *ibid.*
[53]   s. 16(3), Copyright, Designs and Patents Act 1988.

right protection especially in relation to whether a program has been unlawfully repro-duced. For example, if a person is given a diskette containing a computer program, what is that person allowed to do under copyright law? First, does the mere use (*i.e.* loading or running) of a computer program constitute infringement? The answer is yes, under both E.C. copyright law (see above) and the United Kingdom law. Under the United Kingdom copyright statute, the act of reproduction includes copying a work in any material form, including the storage of a work in any medium by elec-tronic means.[54] Furthermore, any use of the computer program which results in tran-sient or incidental copies will be regarded as an act of reproduction.[55] Thus, the temporary reproduction of the computer program onto the offender's RAM or on his hard drive will be infringing, if it is done without the copyright owner's consent. How-ever, where a person is a lawful user of the program, there are certain permitted acts (see below).

A further difficulty is as to what is meant by substantial copying. In the case of writ-ten or printed literary works, this is a matter of comparing the two works, and judg-ing whether there is a similarity of expression between the two. However, in the case of computer programs, the courts have to face issues such as non-literal copying, *i.e.* where the exact program code is not copied but the end-result is that the offending program creates the same overall organisation, structure, user interface and screen display as the protected program. The offending program will have the same "look and feel" or "structure, sequence and organisation" of the protected program.

In one decision, the court issued the following guidelines as to how the question of substantial copying should be dealt with. First, a literal comparison between pro-grams is difficult as programs can be written in different computer languages which bear no literal similarity. Thus, non-literal elements such as structure, arrangement, menus, formats, etc. should be considered. Secondly, one should compare the pro-tected program with the offending program to see whether there are any similarities between the two works. If so, were such similarities due to copying? Finally, if some elements have been copied, are such elements a substantial part of the protected pro-gram or an insubstantial part of the protected program.[56] However, a subsequent deci-sion rejected these guidelines and held that the question of substantial copying in relation to computer programs was to be answered by the simple expedient of judging the degree of overborrowing by the defendant of the skill, labour and judgment which went into the copyright work; furthermore, in considering reproduction, one should not only compare the literal similarities between the protected program and the allegedly infringing program, but also the "program structure" and the "design fea-tures" of the programs.[57]

Adaptation

11–77 The copyright owner's permission should be obtained if one wishes to make an adaptation of a computer program. Adaptation in relation to a computer program is defined as "an arrangement or altered version of the program or a translation of it", where the notion of translation includes a version of the program which has been

---

[54] *ibid.,* s. 17(2). S. 178 defines electronic to mean: actuated by electric, magnetic, electromagnetic, electro-chemical or electromechanical energy; whereas, electronic form means in a form which is usable only by electronic means.

[55] *ibid.,* s. 17(6).

[56] *John Richardson Computers Ltd v. Flanders* [1993] F.S.R. 497. The test is a variation of the "abstraction, filtration, comparison" test applied in the United States, see *Computer Associates v. Altai* 23 U.S.P.Q. 2d 1241 (2nd Cir., 1992).

[57] *Ibcos v. Barclay* [1994] F.S.R. 725.

converted into or out of a computer language or code or into a different computer language or code.[58]

Permitted acts

11–78 In general, the copyright provisions which permit a person to do certain acts in relation to a copyright work will apply equally to computer programs.[59] In addition, the law accords the lawful user of a computer program additional privileges. A lawful user is any person who has a right to use the program, whether through lawful purchase of the program or under licence.[60] A lawful user will have the right to do the following activities:

- to make necessary back up copies of the program;[61]

- to copy or adapt the program if these acts are necessary for the lawful use of the computer program, especially if the lawful user needs to correct errors in the program: this permitted act is subject to any express prohibitions made by the copyright owner under any terms or conditions of agreement[62];

- to decompile the computer program (in certain cases).

The lawful user is allowed to do any of the above activities irrespective of any licence agreement which purports to state otherwise.[63] However, it should be noted that if the lawful purchaser of a copy of computer software has the right to copy or adapt the work in connection with his normal use of the work, he may lose this right if he transfers the purchased copy to a third party. In such instances, barring any contractual obligations to the contrary which may have been imposed by the copyright owner, the transferee will be allowed to do any act which the original purchaser was entitled to do — conversely, the original purchaser will no longer be entitled to do such acts after the transfer.[64] This rule is applicable not only to computer programs, but to all copyright works which are in an electronic form.

Decompilation of programs

11–79 One major issue in respect of copyright protection of computer programs is that of interoperability. A common goal of most application software programs is to interface successfully with another program or operating system so as to be compatible with the other program or system. Decompilation of the other program or system is a vital procedure in order to obtain interface details such as the source code. A second objective of decompilation however may also be for the decompiler to create a new competing program, to the detriment of the copyright owner of the first program. Any process of decompilation would ordinarily result in infringement. In order to address this problem, the European Union Member States, including the United Kingdom, have introduced a specific decompilation defence.

Under this defence, it is not an infringement for a lawful user to convert a computer program which is expressed in a low level language into a version expressed in a higher

---

[58] Copyright, Designs and Patents Act 1988, ss. 21(3)(ab), 21(4).
[59] See para. 11–51, above.
[60] Copyright, Designs and Patents Act 1988, s. 50(A)(2).
[61] *ibid.,* s. 50(A)(1).
[62] *ibid.,* s. 50(C).
[63] *ibid.,* s. 296A.
[64] *ibid.,* s. 56.

level language or to copy it while doing so. However, this is only allowed if the following conditions are fulfilled:

- it must be necessary to decompile the program to obtain the necessary information to create an independent program which can operate or interface with the decompiled program or with another program;

- the information obtained from decompilation must not be used for any other purpose;

- decompilation is not allowed if the information is readily available elsewhere to the user;

- the decompiler must not unnecessarily supply the information to any other person to whom it is not necessary to supply it in order to accomplish the decompilation;

- the decompiler must not use the information to create a program which is substantially similar in its expression to the decompiled program.[65]

DATABASES

11–80   A database is a collection of facts, data or independent works. It can be on an analogue medium such as paper, or on a digital medium, *i.e.* stored on the computer or CD-ROM. It can include such items as lists of clients, schedules of employees, etc.; anthologies; and on-line information services. In this decade, we have seen the growing economic importance of electronic databases which are created at the expense of much capital and time, and which are valuable commodities. In the past, copyright protection was easily available for such works in the United Kingdom as long as they satisfied the minimum threshold of originality, *i.e.* labour, skill and judgement.[66] A converse situation arose in Europe as databases or compilations of facts were considered unprotectable due to the absence of any creativity or personal intellectual contribution of the author.[67] The United States position was similar to that of the United Kingdom, until 1990, when the Supreme Court adopted a more European-based approach, *i.e.* no copyright protection for databases unless there was some creative element.[68]

In light of these diverse approaches, and of the economic significance of databases, an E.C. Directive[69] was introduced to harmonise the law, and United Kingdom copyright law has been amended accordingly. Now, a database owner has two potential means of protecting his investment: copyright and database right.

Legal definition of database

11–81   For the purposes of both copyright and database right, a database is defined as a:

---

[65] *ibid.,* s.50(B).
[66] *Ladbroke (Football) Ltd v. William Hill* [1964] 1 All E.R. 465; *BBC, ITP v. Time Out* [1984] F.S.R. 64; *Waterlow v. Rose* [1995] F.S.R. 207.
[67] *Coprosa,* (France), Cass. 1 civ., May 2, 1989, (1990) 143 R.I.D.A. 309 (compilations only protected if intellectual contribution by author); *Henault* (France), Cass. crim., June 2, 1983 (1983) 117 R.I.D.A. 85 (tables of football matches with grids not protected as it was not an intellectual creation).
[68] *Feist Publications v. Rural Telephone Service Co.* (1991) 499 United States; 111 S. Ct. 1282.
[69] European Parliament and Council Directive on Legal Protection of Databases, 96/9, March 11, 1996, O.J. L77/20, March 27, 1996.

"collection of independent works, data or other materials which are arranged in a systematic or methodical way, and are individually accessible by electronic or other means."[70]

The definition is wide enough to encapsulate many forms of collections, and the following guidelines as enumerated within the E.C. directive should be noted:

- the manner of accessing may be by electronic, electromagnetic or electro-optical processes or analogous processes;[71]

- both electronic (for example, CD-ROM and CD-i) and non-electronic databases are included;[72]

- literary, musical or artistic collections of works or collections of other material such as sound, images, numbers, facts and data are within the definition; however, a recording, or an audiovisual, cinematographic, literary or musical work *per se* will not be considered to be a database — only collections of such materials;[73]

- materials which are necessary for the operation or consultation of the database, such as a thesaurus or an indexation system can be termed databases;[74]

- it is not necessary for materials to have been physically stored in an organised manner.[75]

For example, a CD-ROM containing a selected collection of scientific papers, lectures and diagrams may have several copyrights subsisting in it: literary copyright in the individual papers and lectures, artistic copyright in the diagrams, and a database copyright and/or a database right in the whole collection.

Copyright protection

11–82 To gain copyright protection, a database must be original. However, the normal standard of originality is not applied. Instead, a database will only be considered original if and only, by reason of the selection or arrangement of the contents of the database, the database constitutes the author's own intellectual creation.[76] In other words, mere evidence of capital or investment or labour is insufficient to confer originality; instead, it must be shown that the author of the database exercised some criteria of judgement in selecting or arranging the data. If the database work satisfies this criterion, it will have copyright protection for the term of life of the author plus seventy years thereafter. All other rules regarding ownership and scope of protection which will normally vest in a copyright owner of a literary work,[77] will equally vest in the owner of database copyright.[78]

The only exception is that the range of non-infringing, permitted acts are widened

---

[70] Copyright, Designs and Patents Act 1988, s. 3A(1).
[71] Recital 13, Database Directive.
[72] *ibid.*, Recital 14.
[73] *ibid.*, Recital 17.
[74] *ibid.*, Recital 20.
[75] *ibid.*, Recital 21.
[76] Copyright, Designs and Patents Act 1988, s. 3(A)(2).
[77] *ibid.*, s. 3(1)(d) — for the purposes of copyright law, a database is considered a literary work.
[78] Note also s. 21(3)(ac), *ibid.*, where the owner of database copyright is entitled to prohibit others from adapting the protected database, *i.e.* from arranging or altering the version of the database.

in relation to a database. It is not an infringement of the database copyright for any person, who has a right to use the database, to do anything which is necessary for the purposes of access to and use of the contents of the database.[79] Moreover, the user is allowed to do the above acts irrespective of any licence agreement which purports to state otherwise.[80] It should be noted, nevertheless, that if the lawful purchaser of a copy of an electronic database has the right to copy or adapt the work in connection with his normal use of the work, he may lose this right if he transfers the purchased copy to a third party. In such instances, barring any contractual obligations to the contrary which may have been imposed by the copyright owner, the transferee will be allowed to do any act which the original purchaser was entitled to do — conversely, the original purchaser will no longer be entitled to do such acts after the transfer.[81]

Database right

Criterion of protection

11–83 The database right is a newly created right which exists independently of the database copyright.[82] The definition of a database is the same as that under copyright law, although the criterion for protection is markedly different. The database right will subsist in a database only if there is "substantial investment in obtaining, verifying or presenting the contents of the database."[83] Therefore, there is no requirement of originality or creativity, but instead it must be shown that capital and investment were present in creating the database. As the E.C. directive emphasises, the object of the *sui generis* database right is to ensure protection against misappropriation of the results of the financial and professional investment made in obtaining and collecting the contents of the database: as such, the nature of the protected investment need not necessarily consist merely of the deployment of financial resources but will also include the expending of time, effort and energy.[84] It should be further noted that the database right will subsist in the database, irrespective of whether the database itself or its underlying contents are protected by copyright.

Owner

11–84 The owner of the database right is the maker of the database, *i.e.* the person who takes the initiative in obtaining, verifying or presenting the contents of the database *and* assumes any investment risk involved. If such a person is an employee, who does it in the course of his employment, the database right will vest in the employer, unless there is an agreement which states otherwise.[85]

Duration of protection

11–85 The database right lasts for a period of 15 years from the year in which it was completed. But, if the database is made available to the public before the end of this 15 year period, the duration of protection is extended to 15 years from the year it was first made available to the public.[86] Thus, if a database was completed in December

---

[79] *ibid.*, s. 50D(1).
[80] *ibid.*, s. 296B.
[81] *ibid.*, s. 56.
[82] The database right is governed by the Copyright and Rights in Databases Regulations 1997, S.I. 1997 No. 3032. It came into effect on January 1, 1998 (U.K. Database Regulations).
[83] Article 13, U.K. Database Regulations.
[84] Recitals 39, 40, E.C. Database Directive.
[85] Articles 14, 15, U.K. Database Regulations.
[86] *ibid.*, Articles 17(1), 17(2).

1998, it will be protected until December 2013. However, if it is then made available in December 2010, then it will be protected until 2025. Finally, the duration of the database protection can be extended further if any substantial change is made to the contents of the database. In other words, if changes are made to the database (and such changes can include minor successive additions, deletions or alterations which over the years may accumulate to be considered a substantial change), then the database is considered to be a new product, and a new 15–year term of protection will apply to the changed database.[87]

This important provision can enable continuing protection of a database over years, as fresh input and investment alter or amend an existing database. A database which was completed on December 1998 would be protected until December 2013. However, if substantial changes are made to the database on December 2012, the term of protection will start again and the product will be protected until December 2027. If another substantial change is made on December 2025, the clock will be re-set yet again and a fresh 15–year term of protection will begin. The database can be protected continuously, and indefinitely, subject to substantial changes. The problem with the new provision is that there will be difficulty in assessing whether something constitutes a substantial change or not.

Infringement

11–86 The database right is infringed if any unauthorised person "extracts or re-utilises" all or a substantial part of the contents of the database. Moreover, even if a substantial part of the database is not taken, it is still an act of infringement to repeatedly and systematically extract or re-utilise insubstantial parts of the database if this amounts, over the period, to a substantial appropriation of the database contents.[88] However, the statute provides for certain permitted acts:

• a lawful user may extract or re-utilise insubstantial parts of the contents of the database for any purpose;[89]

• fair dealing with a substantial part of a database is permissible if the part is extracted by a lawful user, and it has been extracted for the purpose of illustration for teaching or research (as opposed for any commercial purpose), and the source is indicated;[90]

COPYRIGHT AND GENES

11–87 Ordinarily, patent law will be the primary means of securing protection for most biotechnological products, save for the peripheral copyright protection available to laboratory results, reports, etc. Nevertheless, one suggestion is that copyright and the database right may subsist in the DNA sequence.[91] If one views the sequence as being a message written in a chemical alphabet, comprising of permutations of the four letters A,G,C, and T, it is arguable that the DNA or amino acid sequence is an original literary work. One can compare this stance with the early copyright cases on

[87] Article 17(3), U.K. Database Regulations.
[88] *ibid.,* Article 16.
[89] *ibid.,* Article 19.
[90] *ibid.,* Article 20.
[91] A final possibility for supplementary protection of molecular sequences is the unregistered design right; however, this poses a very slim opportunity of protection and is discussed in the following Chapter.

telegraph codes where it was held that copyright can subsist in arbitrary sequences of letters which of themselves impart no understandable message to the lay reader but nevertheless held some meaning for the accomplished expert.[92]

Furthermore, there may not be any difficulty in demonstrating that the work is an original work in that much skill, labour and judgement had gone into configuring the genetic sequence. If this is so, then copyright will subsist in a scientific record which sets out the structure of DNA and other similar subject matter. The unauthorised reconstruction of the protein or DNA molecule from the published sequence could be an infringement of the copyright that subsists in that literary work.[93]

A second possibility is that the *sui generis* database right could subsist in the sequence. As noted above, the threshold of protection is the evidence that the maker invested "substantial investment in obtaining, verifying or presenting the contents of the database."[94] The string of letters representing the DNA sequence would also have no trouble in coming within the definition of "database", *i.e.* "a collection of independent works, data or other materials which are arranged in a systematic or methodical way, and are individually accessible by electronic or other means."[95]

International copyright

11–88   In addition to national copyright laws, there are also international copyright treaties whereby co-operation on a global scale is achieved in certain areas such as the nature of protected subject matter, qualifications for protection and reciprocal treatment of authors or works emanating from different countries. We briefly list the main objectives of the major international copyright conventions.

Berne Convention for the Protection of Literary and Artistic Works 1886

11–89   This was the first major international copyright treaty and is still the major copyright convention. There are currently 127 countries which are members of the Convention, including all the E.C. Member States, most of Eastern and Central Europe, United States, Canada, Australia, and China.[96] The Berne Convention adopts two major principles, which have become the basic tenets of international copyright law:

- principle of national treatment: each member of the Union grants nationals of other Member States the same treatment as it grants its own nationals;

- the principle of minimum rights: certain minimum rights are granted to authors protected under the Conventions.

We have already come across the principle of national treatment under United Kingdom copyright law. Anyone can qualify for protection in the United Kingdom if he is a national, etc. of a Berne Convention Member State; likewise, any work can qualify for protection in the United Kingdom if the work is first published in a Berne

---

[92] *Anderson & Co. Ltd v. Lieber Code Co* [1917] 2 K.B. 469.

[93] For a wider discussion on this area, see Laddie, *et al*, *op. cit.*, Chapter 21; A. Speck, "Genetic Copyright", (1995) E.I.P.R. 171; J. Stanley & D.C.Ince, "Copyright Law in Biotechnology: A View from the Formalist Camp", (1997) 3 E.I.P.R. 142.

[94] Article 13, U.K. Database Regulations.

[95] Copyright, Designs and Patents Act 1988, s. 3A(1).

[96] A current list of Berne Convention members is available at http: //www.wipo.org.

Convention country.[97] Herein lies the importance of the Berne Convention: any author who is a national of a Berne country, can go to any other country of the Convention and claim national treatment and the minimum guaranteed rights. The minimum rights include the right to authorise copying, public performance, translation, broadcasting of works, and moral rights. The Convention covers literary, dramatic, musical and artistic works, sound recordings, films, and broadcasts.

Trade related aspects of intellectual property rights (TRIPS)

11–90 The TRIPS Agreement is an annex to the World Trade Organisation Agreement (1994) and it establishes international standards for all intellectual property rights, including copyright.[98] Every country which is a member of WTO must incorporate the minimum TRIPS standards into their national laws. The TRIPS Agreement, in respect of copyright, basically adopts the Berne Convention. In short, with the TRIPS Agreement, the Berne Convention principles apply to *all* members of the WTO.[99]

CONCLUSION

11–91 There is a tendency to regard copyright law as being relevant only to works of fine and cultural arts such as books, prints, paintings and music. However, as the above discussion shows, copyright protection is more prevalent than this, extending to protect technical documentation, graphs, database collections, computer programs and multimedia works. The importance of copyright protection is three-fold. First, it offers cheap and quick protection as there are no formalities to be complied with. Secondly, copyright material will invariably subsist in much of the preparatory work that accompanies any project, whether it is scientific research, product innovation or development. Thirdly, the term of protection for most copyright works is much longer than that offered under patent, registered design and unregistered design laws. Nevertheless, despite the advantages of copyright protection, its major shortcoming is that it is difficult to predict in advance whether a work will be protected under the law, and if so, how far the scope of that protection will be.

---

[97] See para. 11–30, above.
[98] A current list of WTO members is available at http: //www. wto.org.
[99] For a further discussion on protection under the Berne and TRIPS instruments, see Sterling, *World Copyright Law*, *op. cit.*, Chapters 18 and 21.

CHAPTER 12

THE LAW OF DESIGNS

Dr Uma Suthersanen, Fellow at Queen Mary Intellectual Property Research Institute and Lecturer at Centre for Commercial Law Studies, Queen Mary, University of London[1]

INTRODUCTION

12–01   In ordinary language, the definition and concept of "design" is difficult to grasp. It can equally apply to industrial design, functional design, ornamental or artistic design. To some extent, it can encompass paintings, architectural buildings and works of sculpture. The term has also been increasingly applied to computer software structure and even genetic structure. The concept of design is also dependent, partly, in its mode of production. The width of the concept is the first indication of the problem intellectual property law faces in relation to the appropriate nature of protection.

Thus, the question of legal protection of designs cannot be answered easily. Inventive or functional designs may be protected under patent laws, or in some countries, under utility model or petty patents law. If the design is extremely artistic in nature, it can be protected under artistic copyright law; and if the design is distinctive enough, it can even be protected under trade mark laws. In between the these laws, most countries have a "registered design law" or "registered industrial design" law. In addition to this, some countries, such as the United Kingdom and Hong Kong, have a further layer of protection: the unregistered design right.

Legislation

12–02   This Chapter concentrates on the two specific design rights provided by the United Kingdom law. The registered design right is governed by the Registered Designs Act 1949 and the role of the registered design law is to offer protection to new, eye-catching, aesthetic industrial designs. The registered design right operates rather like the patent regime: an application for registration is required whereupon a 25–year monopoly is accorded to the design owner. On the other hand, the unregistered design right offers protection to original three-dimensional designs, irrespective of their visual or artistic quality and is governed by the Copyright, Designs and Patents Act 1988. The right is automatic upon recording the design and is similar to copyright in that the design proprietor only has the right to prohibit the copying of the design. In addition to this, we refer briefly to the European harmonised design law, the Design Directive 1998, which has not been implemented into the domestic law.

PROTECTABLE DESIGNS

12–03   Different types of designs are protected under the registered design law and under the unregistered design right.

Registered design law

12–04   The Act defines a design as follows:

---

[1] For further reading, see U. Suthersanen, *Design Law in Europe* (Sweet & Maxwell, 2000).

Features of shape, configuration, pattern or ornament applied to an article by any industrial process, being features which in the finished article appeal to and are judged by the eye, but does not include:

(a) a method or principle of construction, or

(b) features of shape or configuration of an article which:

    (i)   are dictated solely by the function which the article has to perform, or
    (ii) are dependent upon the appearance of another article of which the article is intended by the author of the design to form an integral part.

Eye-appealing designs

12–05 From the definition, it is clear that both two-dimensional and three-dimensional designs may be registered. The criterion of industrial process merely means that the design must be of the nature that is capable of being industrially multiplied or replicated. Secondly, the Act stipulates that protection is only conferred on such features of design which "appeal to and are judged by the eye".[1a] The test of eye-appeal does not mean that the design must have an aesthetic or artistic appeal, but merely that the features of the design must "attract the attention of the beholder" or be "special, peculiar, distinctive, significant or striking" in appearance. If the appeal of the designed product is on the basis of its functionality or fitness for its purpose, it may not be protectable since it is not being judged solely by the eye. In the words of Lord Reid:

> "the words "judged solely by the eye" must be intended to exclude cases where a customer might choose an article of that shape not because of its appearance but because he thought that the shape made it more useful to him."[2]

The features are to be judged from the perspective of the customer, and not that of the court or designer familiar with articles of the type. The eye-appealing features do not need to be outwardly visible in the finished article. Protection can be gained for internal features of an article which are not visible at the point of sale. In one decision, registration was sought for the interior design of a chocolate egg consisting of the tonal contrast between two layers of differently coloured chocolate. An objection was made on the grounds that the design had no eye appeal since it could not be seen at the point of sale, but rather at the point of consumption. This argument was rejected: as long as an eye-appealing design is present at all times in the finished article, whether at the point of manufacture, sale or consumption, the design is registrable.[3]

Aesthetic designs

12–06 The design must have some aesthetic content. This is a separate and distinct test from the "eye–appeal" test. The appearance of the design must be a material consideration to would-be purchasers or users of the designed article. The appearance of a design will only be considered material if aesthetic considerations are normally taken into account to a material extent by would-be purchasers or users of article of that description (or would be taken into account if the design were to be applied to the article).

The provision was introduced in the 1988 amendments to the Registered Designs Act 1949 and has not had the benefit of much case-law. In *Goodyear Tire & Rubber Co.*,[4] the

---

[1a] Registered Designs Act 1949 s. 1(1).
[2] *Amp v. Utilux* (1972) R.P.C. 103, at 108.
[3] *Ferrero's Application* (1978) R.P.C. 473.
[4] *Goodyear Tire & Rubber Co.* [1995] I.P.D. 18052; SRIS 0/93/94.

design in question was for the tread of a motor vehicle tyre. The applicant tendered proof, by affidavit, that customers had taken into account the aesthetic nature of the product when they bought tyres; this evidence was accepted by the Registry. Thus, despite the convoluted nature of the test, it would appear that it places little additional burden on the applicant. This accords with the opinion of the House of Lords in *Amp v. Utilux*, in relation to the "eye-appeal" criterion:

> "The onus is on the person who attacks the validity of the registration of a design. So he would have to show on a balance of probability that an article with the design would have no greater appeal by reason of its appearance to any member of the public than an article which did not have this design. Looking to the great variety of popular tastes this would seem an almost impossible burden to discharge."[5]

Designs applied to an article

12–07   The design must be applied to an article which is defined as "any article of manufacture", including any part of an article if that part is made and sold separately.[6] Unlike other E.C. Member States where protection is focused on the design only, United Kingdom design law is product specific and is granted to a design as applied to a specific article. The Act also protects designs as applied to sets of articles. A set of article refers to a number of articles which have the same general character and which are ordinarily on sale together or are intended to be used together, for example, cutlery sets. Each article must bear the same design or a design with modifications or variations not sufficient to alter the character or substantially to affect the identity of the design. The definition of an article is wide and the Registry has accepted applications for designs for paper,[7] rugs, portable buildings or structures (though not for ones that are erected on the site), and for an article in a kit form (*i.e.* articles not manufactured or sold in a finished state).

As stated in the definition, one can either register the design for a whole article or part of an article. In the latter case, the part must be made and sold separately. This requirement has been recently the subject of discussion, especially in relation to spare parts of motor vehicles, or indeed of any other product. Can one register parts of a product such as the bumper or panel of a motor vehicle or the outer casing of a computer? One court held that the words "made and sold separately" mean that the manufacturer or proprietor intended the article to be put on the market and sold separately, such as a hammer handle or the bit of a bradawl.[8] However, a recent ruling has stated that, to be registrable as part of an article, that part must have an independent commercial life at the time of manufacture.[9] This decision involved a design application to register individual body parts of a motor vehicle including body panels, bumpers, and grills. It was held that items such as body panels and grills had no independent life at the time of manufacture as people would not go into a shop to buy a body panel, independent of a car, for its own sake, but only for replacement purposes. However, other items of the car, such as steering wheels, lamps and wheel covers, were held to be registrable, since they could be sold and bought independently of the motor vehicle, for use with other motor vehicles of different makes.

---

[5] *Amp v. Utilux* (1972) R.P.C. 103.
[6] Registered Designs Act 1949, s. 44(1).
[7] *Lamson Industries Application* (1978) R.P.C. 1.
[8] *Sifam Electrical v. Sangamo Weston* (1973) R.P.C. 899.
[9] *Ford/Fiat* (1993) R.P.C. 399, Ch. D.; (1994) R.P.C. 545, CA; (1995) R.P.C. 167, HL.

Excluded designs

12–08 Several types of designs or design features are not protected under the registered design law. First, the following articles are specifically excluded from registration[10]:

- works of sculpture (but the Act will protect casts or models used or intended to
  be used as models or patterns to be multiplied by any industrial process);

- wall plaques, medals and medallions;

- printed matter primarily of a literary or artistic character, including book jack-
  ets, calendars, certificates, coupons, dress-making patterns, greeting cards, labels,
  leaflets, maps, plans, playing cards, postcards, stamps, trade advertisements,
  trade forms and cards, transfers and similar articles;

- designs which incorporate a portrait of Her Majesty the Queen or any other
  member of the Royal Family; a reproduction of the armorial bearings, insignia,
  orders of chivalry, decorations or flags of any country, city, borough, town,
  place, society, body corporate, institution or person; or the name or portrait of
  a living person or a person recently dead — unless consent has been obtained.

Secondly, features which constitute a "method or principle of construction" are
excluded. There is no definition of this phrase but it is accepted to exclude any features
of the design which do no more than convey the idea of a general shape appropriate
to the function to which the article is intended to perform. Thirdly, design features
which are dictated solely by the function which the article has to perform are excluded
— *i.e.* functional features. The functionality exclusion only applies to three-
dimensional designs, and not to patterns or ornamentation. In the leading case in this
area, the design of an electrical terminal to be used in a washing machine was rejected
on the grounds that the design was dictated by its function. The test is whether every
single feature of the design is dictated by or "attributable to or caused or prompted
by" the article's function, taking into account the designer's intention. The fact that
there are other designs of different shapes which can perform the same function is not
sufficient to escape this exclusion.[11]

12–09 Fourthly, "must-match" features of the design are excluded. Again, this exception only applies to three-dimensional designs, and not to patterns or ornamentation.
Such features are those which are dependent upon the appearance of another article,
of which the article is intended by the designer to form an integral part. This provision
was introduced specifically to exclude protection for spare parts of products, and to
prevent the designer gaining a monopoly over his product by tying up the spare parts
market, especially in relation to motor vehicles. In the landmark decision on this
point, the court held that the issue had to be determined from a practical and realistic commercial perspective. Designs of articles such as seats, wheel covers, steering
wheels and wing mirrors are not caught by the "must match" clause because they can
be fitted to other vehicles for sportier appearance, or greater comfort or for a variety
of other reasons. Such items are not dependent on the appearance of a vehicle. However, with respect to articles such as body panels, instrument panels, grilles, bumpers,
such designs were meant by the designer to form an integral and essential part of the

---

[10] Registered Designs Rules, rules 24– 26, 1989.
[11] *Amp v. Utilux* (1972) R.P.C. 103.

motor vehicle. Thus, the latter group of articles would be caught by the "must-match" exclusion clause.[12]

Finally, the registrar has a general discretion to refuse the registration of any design, including such designs which would, in his opinion, be contrary to law or morality.[13] Therefore, the registrar may refuse a design on the grounds that it would offend accepted social conventions, but not merely if it was vulgar or unpalatable. Thus, an application for a anatomically correct male doll wearing a kilt, which could be raised, was held not to have come within this provision.[14]

Unregistered design right

Definition of design

12–10   An unregistered design right will be conferred on "any aspect of the shape or configuration (whether internal or external) of the whole or part of an article",[15] including computer-generated designs. Unlike the registered design law, only three-dimensional designs will benefit from unregistered design protection. Furthermore, there is no further criteria of eye-appeal or aesthetic consideration. Protection will be granted to any three-dimensional design, whether it is visible to the eye or not, and irrespective of whether it has any aesthetic appeal. Thus, in a decision involving the design of a contact lens, the court held that microscopic design features, such as the lens size and dimension were protectable subject matter under the unregistered design right scheme,[16] thus widening the potential of this right to protect nanotechnology designs, and perhaps, genetic structures. As speculated in the previous Chapter, copyright protection of DNA and other molecular sequences poses a viable means of supplementary protection. In relation to the unregistered design right, however, although the definition of "design" and recent jurisprudence on this matter appears to indicate the possibility of unregistered design right protection for genetic sequences, one should note that the unregistered design right is subject to a wide category of excluded subject matter: thus, potential hurdles for attempting design protection of sequences would be the criterion of originality whereby commonplace configurations would be excluded, and the interface exclusion (both discussed below).

An important point to remember is that the unregistered design right can subsist in the whole article, *and* in the individual components of the article.[17]

Excluded designs

12–11   The following features are excluded from unregistered design right protection[18]:

(a)   a method or principle of construction,

(b)   features of shape or configuration of an article which are dependent upon the appearance of another article of which the article is intended by the designer to form an integral part (*must–match features*); or

(c)   features of shape or configuration of an article which enable the article to be

---

[12]   *Ford/Fiat* (1993) R.P.C. 399 (Chancery Div.); (1994) R.P.C. 545 CA; (1995) R.P.C. 167, HL.
[13]   Registered Designs Act 1949, s. 43(1).
[14]   *Re Masterman's Application* (1991) R.P.C. 89.
[15]   Copyright, Designs and Patents Act 1988, s. 213 (2).
[16]   *Ocular Sciences v. AVCL*, (1997) R.P.C. 289, at 345.
[17]   *Farmers Build Ltd v. Carier Bulk Materials* (1999) R.P.C. 461.
[18]   Copyright, Designs and Patents Act 1988, s. 213 (3).

connected to, or placed in, around or against, another article so that either article may perform its function (*interface features*), or

(d)   surface decoration.

The first two exclusions, relating to method or principles, and must-match features, are identical to those under the registered design law, and the principles enunciated above apply. The third exclusion clause is often referred to as the "must-fit" provision or the "interface" provision. The provision is similar in spirit, if not in statutory language, to the "solely dictated by function" exclusion under the registered design law.

The court in *Ocular Sciences v. AVCL* offered the following three part test as to the application of the interface exclusion.[19] The first step is to determine whether one design has interfaced (*i.e.* connect to, placed in, around or against) with another article. The other article need not be inanimate, but can be a part of the human body. Thus, in the case of contact lenses, some features of the lens have to be placed against the human eyeball. Once interface features are identified, the next step is to determine the function of each design product. The function of the lens is to provide corrective vision, to allow oxygen to pass through, and to fit the eyeball; whereas, the function of the eye is to see. The last step of the test is to see whether any of the interface features perform these purposes. In *Ocular*, it was held that the lens diameter had been specifically chosen to enable the soft lens to fit onto the eyeball — this was thus excluded from protection.

Finally, surface decoration is excluded from unregistered design right protection. Surface decoration relates to patterns or ornamentation — protection for such designs must be sought under artistic copyright or registered design law. However, the clause is not limited to two–dimensional features; raised three–dimensional patterns or decoration can be caught under the exclusion if the three–dimensional decoration is on the surface of an article or if its main purpose is to decorate the surface of an article.[20]

CONDITIONS OF PROTECTION

12–12   The main differences between the two regimes relate to issues of novelty and registration (registered design law), and of originality (unregistered design right).

Registered design

Registration

12–13   Registration is a condition for protection under the Registered Designs Act 1949.[21] The application for registration must be made by the proprietor of the design; if unregistered design right subsists in the design, the applicant must declare that he is also the unregistered design right owner.[22] The application is filed in the Designs Registry division of the Patent Office. The application will consist of the following: (a) the prescribed application form (Designs Form 2A); (b) a statement of novelty indicating the features in which novelty is claimed; (c) representations or specimens of the design; (d) priority documents stating the first application number, the convention country,

---

[19]   *Ocular Sciences v. AVCL* (1997) R.P.C. 289.
[20]   *Mark Wilkinson v. Woodcraft Designs* (1998) F.S.R. 63.
[21]   Registration of a design in the U.K. automatically extends protection in the following regions: England, Scotland, Wales, Northern Ireland, Isle of Man, Brunei, Fiji, Gibraltar, Malaysia, Singapore, The Turks and Caicos Islands, Tuvalu and Vanuatu.
[22]   Design Form 2A.

the official date and a copy of the representation of the design filed in the convention country. All documents must be filed in the English language.

Novelty

12–14 A design will not be registered, or if registered may be held invalid if it is not new. A design will be considered not new if it is identical or substantially similar to a design which has previously been registered or published in the United Kingdom, in respect of the same article or any other article, before the date of application for registration.[23] A design will be considered substantially similar to another design if it only differs from the other design in immaterial details or in features which are variants commonly used in the trade. It is not necessary that the whole design or every part of the design be new. A design is still valid even if the majority of parts of the design are old as long as some part of it imparts a novel character.[24] Furthermore, a design may still be considered novel, despite all its parts being old, if the combination of the parts or the appearance of the combination as a whole is new.[25]

The test of novelty is less severe than that under patent law as only disclosures made within the United Kingdom will destroy the novelty of the design. It should be emphasised that any potential designer must consider the possibilities of registration before releasing or marketing the design in the United Kingdom, as once a design product has been released on to the market in the United Kingdom either by way of sale or any other type of disclosure, novelty will be lost.

Prior art

12–15 The design is tested for novelty against the prior art, *i.e.* all designs published in the United Kingdom before the priority date of registration. There is no necessity for every member of the public to have knowledge of the design: the design will be deemed to be published as long as it is made available to any person in the United Kingdom who is free in law and equity to use or disclose the material.[26] Thus, in the past it has been held that disclosure to one person is sufficient to constitute publication and therefore is capable of destroying the novelty of the design.[27]

In respect of prior documents, there is no restriction as to the type of document which will be considered, *i.e.* it can be a trade catalogue, newspaper, patent specifications, or written description, which discloses a design identical or similar to the design being registered. The design can either be depicted as a picture, photograph or drawing or it can be described in words as long as an ordinary competent member of public can visualise the design from the words used.[28] Alternatively, the prior art can consist of any design which has been released to the public by being on display or by being sold.

Novelty of a design may be destroyed due to a previously published similar design which has been applied to a different type of article. For example, an applicant might apply to register his design consisting of animal shapes as applied to book-cases; a search then reveals that similar animal shaped designs are already available in the public but as applied to CD stacking racks. The law is that novelty can be destroyed if a similar design has already been applied to any article; however, it is possible that a

[23] Registered Designs Act 1949, s. 1(4).
[24] *Walker & Co. v. A.G. Scott & Co.* (1892) 9 R.P.C. 482.
[25] *Re Clarke's Design* (1896) 13 R.P.C. 351.
[26] *Sommer Allibert (U.K.) Ltd v. Flair Plastics* (1987) R.P.C. 599, *Re Vredenburg's Design* (1934) 53 R.P.C. 7.
[27] *Humpherson v. Syer* (1887) 4 R.P.C. 407.
[28] *Rosedale Associated Manufacturers Ltd v. Airfix Products Ltd* (1957) R.P.C. 239.

design which is exactly the same as a prior design but applied to another type of article, may still be held to be novel, if the combination (the old design + a new substrate) gives the design something different which can be considered to be novel.[29]

Excluded from prior art

12–16 Several types of publications, uses or disclosures will not be considered as being part of the prior art.

(i) any disclosure of the design by the proprietor in circumstances of confidence
12–17 Prior publication or disclosure of a design will not destroy its novelty if the disclosure or publication was made in secret or in confidence. Thus, if the proprietor of the design needs to show the design to several parties, in order to gauge the marketability or commercial viability of the design, the design will not inadvertently lose its novelty provided that the proprietor imposes conditions of confidentiality on all such parties. However, this exception is only applicable in relation to disclosures by the proprietor (though see (ii) below).

(ii) disclosure in breach of good faith[30]
12–18 If the proprietor shows the design in secret to one person, and that recipient subsequently publicises the design, the novelty of the design is still preserved as the recipient's disclosure was in breach of the confidential obligation owed to the proprietor of the design.

(iii) publication of the design by placing orders for textile designs[31]
12–19 In the case of a new or original textile design which is intended for registration, the acceptance of a first and confidential order for goods bearing the design will not invalidate a subsequent application to register the design. The provision is limited to textile designs; thus, proprietors of non-textile designs must either register the design first, or impost a confidentiality clause on any buyers.

(iv) any publication of the design at a certified exhibition[32]
12–20 Designs which are displayed at exhibitions which are certified by the Secretary of State will be exempt from the novelty provisions. However, the design must be submitted for registration not later than six months after the opening of the exhibition; failure to do so will destroy the novelty of the design. Furthermore, it must be stressed that the exhibition exception will only apply to those exhibitions which are certified exhibitions at the time when it takes place, and it is not open for the Secretary of State to issue a retrospectively effective certificate after the exhibition in question has taken place.[33]

(v) the publication of the parent design[34]
12–21 In many design areas, such as textile, fashion or household goods, designers will not only produce one article which is made to a design, but will produce several types of articles which may incorporate the design, or a slightly modified design. The proprietor may have registered the main design (*i.e.* the parent design) in respect of one type of product; if the design is commercially successful on other types of products, the proprietor may subsequently register the design (dependent designs or designs of

---

[29] Laddie, *et al*, *op. cit.*, p. 205, para. 31.18 ; and Suthersanen, *op. cit.* para. 16–228.
[30] Registered Designs Act 1949, ss. 6(1)(a)–(b).
[31] Registered Designs Act 1949, s. 6(1)(c).
[32] *ibid.*, 6(2).
[33] *Mod-Tap W. Corporation v. B.I. Communications Plc* (1999) R.P.C. 333.
[34] Registered Designs Act 1949, 4(1).

addition) in respect of other articles. In such an instance, the law will not take into account the prior registration of the parent design as part of the prior art; thus, the novelty of the dependent design is preserved. This provision is reserved to cases where the dependent design is identical to the parent design, or where the dependent design has modifications or variations but which are not sufficient enough to alter the character or substantially affect the identity of the parent design. The term of protection of dependent design is anchored to the registration of the parent design: when the latter registration expires, the registration of dependent designs will similarly expire.

(vi) any prior use of a copyright artistic work for a corresponding design[35]

12–22 This provision is of assistance in the case where a copyright artistic work, such as a cartoon drawing, sculpture or a work of artistic craftsmanship, is subsequently industrially exploited, and the proprietor of the copyright work wishes to claim registered design protection. The prior existence or publication of the design, as a copyright artistic work, does not necessarily affect the novelty of the design upon application for registered protection. However, this grace provision is limited. The subsequent application for registration of the copyright work can only be made by the copyright owner or with his consent. Secondly, and very importantly, this concession is given to the copyright owner only if he has not previously industrially exploited his copyright work. Exploitation of the work includes sale, letting for hire, offering or exposing for sale or hire any article to which the design or copyright work has been industrially applied, *i.e.* by applying the design to 50 articles or more.[36]

Objective test

12–23 The test of novelty is objective. Once the prior art has been identified, the design for which registration is sought is compared objectively with the prior art to see how close visually the two designs are. The visual test is usually from the perspective of a customer viewing the articles from a normal distance. Evidence can be accepted from members of the public as to their confusion between the two designs — this must mean that the designs are visually similar. However, if very close scrutiny is required to tell the two designs apart, then they will be considered to be identical or substantially similar. It is the overall impact of the design which is important — if the overall impact of the design as applied to one article is visually different to the overall impact of the design as applied to another article, then the second design is likely to be found to be novel.

There must be substantial differences between the two designs before they will be considered dissimilar. Furthermore, a design will not be new if it only differs from a prior design in "immaterial details or in features which are variants commonly used in the trade".[37] Immaterial details mean detailed differences which have no visual effect on the identity or character of the design. A commonly used trade variant means features or elements (including colours) of design which have been used before, as opposed to merely having been published before; the use must be common so use by a single trader is unlikely to be sufficient. Finally, the publication must have occurred within the United Kingdom. Coupled with this test is the principle that functional features are usually ignored when judging the novelty of the design. Therefore, if the similarities between the design in question and the prior art are due to functional constraints or due to the choice of materials, etc., such similarities are to be ignored.

---

[35] Registered Designs Act 1949, s. 6(4), 6(5).
[36] Registered Design Rules, r. 35.
[37] Registered Designs Act 1949, s. 1(4).

The date of application

12–24 A design must be new at the date of application for registration or at the date of priority. Therefore, the design for which registration is sought is compared with prior designs which have been published at the date on which the application for registration is made at the Designs Registry. In the alternative, if an international application is made under an international convention,[38] the date at which novelty is judged is the "priority date", *i.e.* the date on which the first foreign application is made.

Unregistered design right

Originality

12–25 The main condition of protection is that the design must be original. The Act states that a design will not be considered original if it is commonplace in the design field in question at the time of its creation.[39] The test of originality is thus more restrictive than under copyright law.

The leading case on this is the Court of Appeal's decision in *Farmers Build Ltd v. Carier Bulk Materials*[40] whereby Mummery L.J. concluded that the proper approach to the question of originality was as follows. First, the court must determine whether the design of the article was copied from earlier designs, in the same field, or whether the design originated from the designer. In doing so, the court must be satisfied that the design for which protection is claimed has not simply been copied (*e.g.* like a photocopy) from the design of an earlier article. The court should further note that, in the field of designs of functional articles, one design may be very similar to, or even identical with, another design and yet not be a copy: It may be an original and independent shape and configuration which is coincidentally the same or similar. If the court is satisfied that it has been slavishly copied from an earlier design, it is not an "original" design in the "copyright sense" and the "commonplace" issue does not arise. If, on the other hand, the court is satisfied that the design has not been copied from an earlier design, then it is "original" in the "copyright sense".

12–26 Secondly, the court has to decide whether the design is "commonplace". For this, it is necessary to ascertain the similarity of the design in question to the design of similar articles in the same field of design. The closer the similarity of the various designs to each other, the more likely it is that the designs are commonplace, especially if there is no evidence of copying, which would account for the resemblance of the compared designs. Furthermore, if a number of designers working independently of one another in the same field produce very similar designs by coincidence the most likely explanation of the similarities is that there is only one way of designing that article. In those circumstances the design in question can fairly and reasonably be described as "commonplace". In such instances, the unregistered design right would be withheld from a design that, whether it has been copied or not, is bound to be substantially similar to other designs in the same field. If, on the other hand, there are aspects of the plaintiff's design of the article which are not to be found in any other design in the field in question, and those aspects are found in the defendant's design, the court would then be entitled to conclude that the design in question was not "commonplace".

This comparative exercise must be conducted objectively and in the light of the

---

[38] Paris Convention/Hague Agreement.
[39] Copyright, Designs & Patents Act 1988, s. 213 (4).
[40] *Farmers Build Ltd v. Carier Bulk Materials*, (1999) R.P.C. 461.

evidence, including evidence from experts in the relevant field pointing out the similarities and the differences, and explaining the significance of them.

Fixation

12–27 No registration is required but the design must be recorded in a design document or in an article made to the design.[41] A design document is defined as any record of the design, whether it is in the form of a drawing, a written description, a photograph, data stored in a computer or any other form.[42]

Qualification

12–28 There is no reciprocity provision under the unregistered design right provisions such as those available under copyright law. Therefore, not everyone will be entitled to an unregistered design right and much depends on the nationality of the first owner of the design right. The following rules apply[43]:

- **designer:** the designer can claim protection if he/it is the owner, and if the designer is a citizen or a subject or a habitual resident in the United Kingdom, or another Member State of the E.C.; or if the designer is a body corporate formed under United Kingdom or E.C. law and has a place of business in that country;

- **employer/commissioner:** the employer or commissioner can claim protection if he/it is the owner, and if the employer or commissioner is a citizen or a subject or a habitual resident in the United Kingdom, or another Member State of the E.C.; or if the employer or commissioner is a body corporate formed under United Kingdom or E.C. law and has a place of business in that country;

- **first marketer:** Any person or company can claim protection, if he/it first markets articles incorporating the design in the United Kingdom or the E.C., and if he/it is a citizen or a subject or a habitual resident in the United Kingdom, or another Member State of the E.C.; or if the first marketer is a body corporate formed under United Kingdom or E.C. law and has a place of business in that country. Marketing means selling, letting for hire, or offering or exposing for sale or hire, in the course of business.

From the above provisions, it is clear that a design created and owned by (for instance) United States nationals or United States corporations will not qualify for unregistered design right protection; save in the usual case that it qualifies under the "first marketing" rules (see also 12–30 below).

OWNERSHIP[44]

Registered design

12–29 The author (*i.e.* creator) of the design is the original proprietor of the design. If a design is generated by computer in circumstances where there is no human author,

---

[41] Copyright, Designs and Patents Act 1988, s. 213 (6).
[42] *ibid.*, s. 263.
[43] *ibid.*, s. 217–220.
[44] See Chapter 14 for more details.

the author will be the person by whom the arrangements necessary for the creation of the design are made. If the design is created by an employee in the course of employment, the employer will be considered the proprietor of the design. Similarly, where the design is created under commission for money or money's worth, the commissioner is the design proprietor. An important qualification is that an application for registration of a design in which the unregistered design right subsists will not be entertained unless the applicant is also the unregistered design right proprietor: in practice, the applicant makes this declaration in the official application form for registration.[45]

Unregistered design right

12–30 The unregistered design right will belong to the designer in the first instance. The designer is the creator of the design or, in the case of computer-generated designs, the person who makes the necessary arrangements for the creation of the design. However, if the design is created by an employee in the course of his employment, or is created pursuant to a commission, the employer or the commissioner will own the design right, respectively.[46] A final unusual provision is that in relation to the first marketer. If the design has qualified for design right protection only because of first marketing of articles made to the design, the first owner of the design right will be that person who markets such articles, irrespective of the designer, employer or commissioner.[47] Thus, assuming a design is created by an American citizen or an American corporation, but articles made to the design are subsequently imported into France and first marketed there by a French company, only the French company is entitled to claim unregistered design right protection.

INFRINGEMENT

Registered design

12–31 Registration confers on the proprietor of a registered design an exclusive right to do the following acts[48]:

- to make or import for sale, hire or for use for the purposes of a trade or business such an article;

- to sell, hire or offer or expose for sale or hire such an article;

- to make anything for enabling such an article to be made in the United Kingdom or elsewhere;

- to do any of the above acts in relation to a kit or make anything for enabling such a kit to be made or assembled in the United Kingdom or elsewhere; a kit means a complete or substantially complete set of components intended to be assembled into an article.

The right in a registered design is infringed by a person who, without license, does any of the above acts.

---

[45] *ibid.*, s. 3(2); Designs Form 2A, Registered Design Rules 1989 (S.I. 1989 No. 1105).
[46] Copyright, Designs and Patents Act 1988, ss. 214–215.
[47] *ibid.*, s. 215(4).
[48] Registered Designs Act 1949, s. 7.

Kits

12–32 A kit is a "complete or substantially complete set of components intended to be assembled into an article".[49] This will apply to articles which are delivered in separate components to be assembled or finished by the purchaser such as DIY furniture or toy kits. However, reference to a "set" indicates that the components must be suitable and intended for assembly into a specific article, not merely susceptible for assembly into something — therefore, LEGO bricks may not necessarily be considered as kits unless they are clearly intended to be assembled into one particular article. In respect of kits, not only is the manufacturer who makes the kit without authorisation liable for infringement, but liability is also imposed on the supplier, importer and retailers of the kit; however, the end-user or consumer will not normally be part of the infringement chain unless they make the article from the kit for commercial purposes (see below).

No proceedings can be taken in respect of an infringement committed before the date on which the certificate of registration of the design is granted.

Act of infringement

12–33 Strictly speaking, registered design protection confers an exclusive right or monopoly on the design proprietor; therefore, there is no need to prove any causal connection or act of copying.

The infringing article must have been made or imported for commercial purposes. Therefore any act which is done for domestic or private purposes does not constitute infringement. Similarly, the infringing article must comprise of a design applied industrially, and a single, hand-made article incorporating the protected design would not infringe the design owner's rights.

Infringing article

12–34 The infringing article must incorporate the protected design, or a design not substantially different from it. This is a visual test applied from the perspective of the customer by comparing the protected article and the allegedly infringing article. The court tends to look for the essential features of the protected design, and determine whether these have been reproduced by the offender. However, even if the essential elements have been taken by the offender, but the overall visual effect of the offending article is markedly different from the protected article, it may not be an infringement. If consumer confusion is proven, it will go towards proving that the confusion arises due to the fact that consumers retain visually striking features of the design, and the confusion must be ascribed to the substantial similarity of the protected and offending designs. A related test is the "imperfect recollection" test where the courts assume that the average consumer cannot recollect detailed features of the design.[50,51]

Scope of protection

12–35 Protection of the design is product or article specific. The scope of protection of the design only extends to the type of articles for which the design is registered, and an infringement may not have occurred if someone applies the protected design to a different type of article. For example, a design proprietor may have a registered design

---

[49] *ibid.*, s. 7(4).
[50] *Sommer Allibert (U.K.) Ltd v. Flair Plastics* (1987) R.P.C. 599.
[51] *ibid.*

in relation to the shape of a crocodile as applied to tea-pots; he will not be able to stop a third party from applying the identical design to ink stands.

Additionally, the scope of protection does not extend to the features of the design which are excluded as non-registrable due to their functionality or due to their dependence on the appearance of another article (the must-match provision). Finally, the scope of protection of the protected design will depend on the prior art. If the registered design represents a substantial leap or breakthrough from the prior art (consisting of the same genre of designs as applied to similar articles), then it will have a wider scope of protection. Conversely, if the registered design is derived from commonplace or well-known designs in that field, and its novelty is only to be found in its combination, that design will be considered to be very close to the prior art, and accordingly, will have a small scope of protection. In the latter case, the design will probably belong to a crowded field, where the monopoly conferred to the proprietor must be necessarily narrow, and any small deviation or difference in the competing article will be sufficient to avoid infringement.

Permitted acts

12–36   There are no specific defences available to the infringer.[52] However, it should be noted that in the case of innocent infringement, the defendant is not liable for damages. To prove innocent infringement, the infringer must show that he was not aware and had no reasonable grounds for supposing that the design was registered.[53] Thus, as a precaution, a design proprietor should mark all articles which incorporate the registered design with a registration notice (either the word "registered" or ®), followed by the design registration number. Although this is not a legal requirement, it can pre-empt any defence of innocent infringement.[54]

Unregistered design right

12–37   The owner of the design right has the right to reproduce the design for commercial purposes, by making articles to that design, or by making a design document recording the design for the purpose of enabling such articles to be made.[55] The right is infringed if anyone does the above-stated acts without the permission or license of the design right owner, or if anyone authorises some other person to do such unlawful acts.[56] Private, non-commercial or domestic reproduction of the design is allowed.

Reproduction

12–38   As with copyright law (and not the registered design law), there must be an act of copying. Thus there must be some causal connection between the protected design and the allegedly infringing design, such that the latter is not an independent work. Reproduction is defined by statute. Reproducing a design by making articles to that design means "copying the design so as to produce articles exactly or substantially to that design."[57] This means that the allegedly infringing article must be identical or

---

[52] Note however the exceptions for Crown use where any Government department and any other person authorised by the Government may use a registered design for the services of the Crown. See Registered Designs Act 1949, s. 12 and Schedule 1.
[53] *ibid.*, s. 9.
[54] See below permitted acts.
[55] Copyright, Designs and Patents Act 1988, s. 226.
[56] *ibid.* s. 226(3).
[57] *ibid.* s. 226(2).

substantially similar to the article incorporating the protected design. The question of similarity is an objective test to be judged through the eyes of a potential user or customer.[58] If the overall similarity between the two articles is such that a potential customer would conclude that they are made to the same design, then infringement has occurred.[59]

The unregistered design right is not limited to the article to which it is applied. As one court held, it is not necessary that the infringing design be used on exactly the same type of article as the protected design — infringement can occur if a design, as applied to coffee pots, is copied but applied to wine bottles. However, it may be that a design applied to a certain class of articles has a different impact from the same design as applied to other types of articles. Thus, it may well be that an offender has copied the protected design, but because it has been applied in a different context, the copied design is visually different. In such a case, there may be no infringement.[60]

Secondary infringement[61]

12–39   In addition to the above, the following acts are prohibited, if done without the authorisation of the design right owner:

- importing an infringing article into the United Kingdom for commercial purposes;

- possessing an infringing article for commercial purposes;

- selling, letting for hire, offering or exposing for sale or hire in the course of a business an infringing article.

However, the above require knowledge on the part of the offender in that he knew or had reason to believe he was dealing with an infringing article.

Permitted acts

12–40   There are no specific defences available. However, if the infringer is innocent, the design right owner cannot claim damages against him. An infringer will be innocent if he can show that at the time of the infringement, he did not know and had no reason to believe that design right subsisted in the design.[62]

Licence of right

12–41   Any person wishing to reproduce the design or make articles to the design can apply to the unregistered design right proprietor for an appropriate licence during the last five years of the unregistered design right term.[63] Such a person will be automatically entitled as of a right to a licence, on terms to be settled either by the parties or, in default, by the Comptroller-General of Patents, Designs and Trade Marks.[64] Furthermore, in the same period, it is open to the defendant to undertake a licence of right in the course of infringement proceedings. This is an attractive option for an infringer as the undertaking to take a licence of right will preclude the following remedies being

---

[58]   *Mark Wilkinson v. Woodcraft Designs*, (1998) F.S.R. 63.
[59]   *Parker v. Tidball* (1997) F.S.R. 680.
[60]   *Electronic Techniques v. Critchley Components*, (1997) F.S.R. 401.
[61]   Copyright, Designs and Patents Act 1988, s. 227.
[62]   *ibid.*, s. 233.
[63]   *ibid.*, s. 237(1).
[64]   *ibid.*, s. 237 (2) (*i.e.* by application to the Patent Office).

granted against him: injunction; order for delivery up; damages or an account of profits in excess of double the amount which would have been payable by him as a licensee
if a licence was granted before the earliest infringement. Such an undertaking is available to the defendant at any time before the final order in infringement proceedings,
without any admission of liability.[65]

DURATION OF PROTECTION

Registered design

12–42   The initial duration of protection is five years from the date of registration of
the design. This period may be renewed every five years, with the maximum term of
protection extending to 25 years from the date of registration. An application to renew
the term of protection must be made within three months prior to the expiration of
the prior five year period, and a renewal fee is payable.[66] However, two exceptions
should be noted. First, the registration term of dependent designs will terminate at the
end of the protection period of the parent design. Secondly, if the registered design is
based on an existing artistic work in which copyright subsists, the registered design
protection will expire when the copyright in the artistic work expires (given the much
longer period of copyright protection, this is unlikely to be a common scenario).

Unregistered design right

12–43   The unregistered design right lasts between ten and sixteen years, depending
on the speed with which the designed article is brought into the market. Prima facie,
it expires 15 years from the end of the year in which the design was first recorded in a
design document or an article was first made to the design. However, if the articles
made to the design are made available for sale or hire anywhere in the world by the
design right owner (or with his permission), and this exploitation occurs within
the first five years of recording the design, the design right will expire 10 years from
the end of the year in which this commercial exploitation first occurred.[67] Therefore, if
a design is first recorded in January 2000 and never commercially exploited: protection
will cease on December 31, 2015. If the designer records the design in January 2000,
and starts to sell articles made to this design in January 2003, protection will cease on
December 31, 2013. If the designer records the design in January 2000, but starts to
sell them in January 2010, protection will cease on December 31, 2015.

REMEDIES

Registered design right

12–44   The following types of remedies are available: injunction; delivery up; damages; account of profits, as an alternative to damages. In special circumstances, the
claimant may also ask for the following relief: freezing of assets, within the jurisdiction of the courts; seizure of relevant documents or products; seizure of infringing
products or design documents.

---

[65] *ibid.*, s. 239.
[66] Designs Form 9A.
[67] Copyright, Designs and Patents Act 1988, s. 216.

Threats

12–45   There is statutory protection in respect of groundless threats of proceedings for registered design right infringement. If any person threatens by circulars, advertisements or otherwise any other person with proceedings for infringement, the aggrieved party may bring an action. Available remedies include a declaration to the effect that the threats are unjustifiable; an injunction against the continuance of the threats; and damages for losses sustained by the aggrieved party. However, this provision is not available against threats to bring legal proceedings simply for making or importing articles. The defendant may avoid liability by proving that the threats were not groundless because the claimant had or would have committed infringing acts. A mere notification that a design is registered does not constitute a threat.[68]

Criminal Offences

12–46   In addition to civil actions, the following acts constitute a criminal offence: failure to comply with the registrar's directions in respect of designs relevant for national defence purposes; making or causing a false entry in the register; making or causing or producing or tendering a document falsely purporting to be a copy of a register entry; falsely representing that a design is registered when it is not or when the right has expired (including persons who sell articles marked "registered").[69] Criminal proceedings may be brought in the Magistrates Courts or, if a serious offence, may be brought in the Crown Court. Such prosecutions can be brought either privately or by the police, Trading Standards or Customs & Excise. The offender is liable for certain specific offences to a term of imprisonment not exceeding two years or a fine not exceeding a certain level in the standard scale depending on the offence committed. If the offence is committed by a corporation, then the director, manager, secretary or other similar officer will be guilty of the offence if they gave their consent.

Unregistered design right

12–47   The unregistered design right owner has the following civil remedies: injunction; damages, (though these are not available against an innocent infringer); account of profits, as an alternative to damages; additional or punitive damages where the infringing act is flagrant or the defendant has acquired a disproportionate profit; delivery up of infringing articles and anything designed or adapted for making article as to a particular design; these goods may be subsequently subject to forfeiture or destruction.[70] Similar provisions exist in respect of "threats" to those discussed above for registered designs.[71]

SEMICONDUCTOR TOPOGRAPHIES

12–48   The unregistered design right also provides protection for the topography of semiconductor products, including the design of integrated circuits or chips.[72] The law regarding the unregistered design right, as discussed above, applies to semiconductor topographies, save in respect of some issues, which are discussed below. In addition to

---

[68] Registered Designs Act 1949, s. 26.
[69] *ibid.*, s. 5, 33–35a.
[70] s. 229 *et seq*, Copyright, Designs and Patents Act 1988.
[71] *ibid.*, s. 253.
[72] Copyright, Designs and Patents Act 1988, as modified by the Design Right (Semiconductor Topographies) Regulations 1989 (S.I. 1989, No. 1100) of June 29, 1989.

unregistered design right protection, it may also be possible to claim copyright protection in the drawings or photographs of the topography design.

Definition[73]

12–49 A semiconductor product is defined as:

- an article the purpose, or one of the purposes, of which is the performance of an electronic function and which consists of two or more layers, at least one of which is composed of semiconducting material and in or upon one or more of which is fixed a pattern appertaining to that or another function.

A semiconductor topography, in turn, is defined as the design of:

- the pattern fixed, or intended to be fixed, in or upon a layer of a semiconductor product, or a layer of material in the course of and for the purpose of the manufacture of a semiconductor product; or

- the arrangement of the patterns fixed, or intended to be fixed, in or upon the layers of a semiconductor product in relation to one another.

Analogous to the unregistered design right, protection will not extend to a method or principle of construction; features of shape or configuration of an article which enable the article to be connected to, or placed in, around or against, another article so that either article may perform its function (interface provisions); features of shape or configuration of an article which are dependent upon the appearance of another article of which the article is intended by the designer to form an integral part (must-match provision).

Ownership[74]

12–50 The designer is the first owner of any design right in the topography design, unless the design is created in the course of employment or in pursuance of a commission or in the course of employment. In the latter cases, the commissioner or the employer is the owner of the design right, unless there is an agreement in writing stating otherwise. However, where the topography design qualifies for design right protection by virtue of first marketing, the person by whom the articles are marketed is the first owner of the design right.

Qualification[75]

12–51 The category of persons who can qualify for protection are slightly different from that under the unregistered design right. A qualifying person is:

- a citizen or subject of, or an individual habitually resident in the United Kingdom, or another Member State of the European Economic Community, Australia, Japan, Switzerland, the United States, Finland, French overseas territories, Iceland, Norway; and British Dependent Territory citizens;

- a body corporate or other body having legal personality which has a place of business at which substantial business activity is carried on in the United Kingdom, or another Member State of the European Economic Community or in Gibraltar;

---

[73] 1989 Regulations, Reg. 2.
[74] 1989 Regulations, Reg. 5.
[75] 1989 Regulations, Reg. 4.

Australia, Japan, Switzerland, the United States, Finland, French overseas
territories;

- any qualified person who first markets the semiconductor product, and
  who is exclusively authorised to put the product on the market in every
  E.C. Member State, where the marketing takes place within the territory of any
  Member State; marketing means selling, letting for hire, or offering or exposing
  for sale or hire, in the course of business.

Rights conferred[76]

12–52   The owner of a topography design right has the exclusive right to reproduce
the design by making articles to that design or by making a design document recording the design for the purpose of enabling such articles to be made. The topography
design right owner can sue for secondary infringement if any of the following acts
occur in relation to an article which the offender knows or has reason to believe is an
infringing article:

- importing the infringing article into the United Kingdom for commercial
  purposes;

- having the infringing article in his possession for commercial purposes;

- selling, letting for hire, offering or exposing for sale or hire in the course of a
  business the infringing article.

Permitted acts

12–53   There are several exceptions to the topography design right. First, reproduction is allowed if it is done privately and for non-commercial aims. Secondly, anyone
can reproduce the topography design for the purpose of analysing or evaluating the
design or analysing, evaluating or teaching the concepts, processes, systems or
techniques embodied in it. This is the reverse engineering defence. Furthermore, the
creation or reproduction of another original semiconductor topography as a result of
an analysis or evaluation of the first topography or of the concepts, processes, systems
or techniques embodied in it is not an infringing act. This provision emphasises the
copyright idea/expression dichotomy — it is acceptable to analyse the topography
design and to use the ideas underlying it, but not to copy the manner in which these
concepts and ideas have been expressed.

Finally, it should be noted that the licence of right, which is available under the normal unregistered design right, is not available in the case of the topography design
right.

Duration[77]

12–54   The duration of protection of a topography design right is slightly different
from that of a normal unregistered design right. Protection commences automatically
upon the topography being recorded in a design document or from the moment an
article is first made to the topography. If the topography or articles made to the
topography are not made available for sale or hire anywhere in the world, protection
will expire 15 years from the time of first recording or first making. However, if the
topography or articles made to the topography are made available for sale or hire

---

anywhere in the world, protection will expire 10 years from the end of that calendar year. Thus, the maximum duration of protection can be for 26 years.

E.C. DESIGN DIRECTIVE

12–55 A new E.C. Design Directive was passed by the European Parliament and Council on October 13, 1998.[78] No Member State, as yet, has implemented this legislation. The following discussion briefly looks at the more salient points of the Directive.

Non-Harmonisation in Europe

12–56 Currently, all the E.C. Member States protect designs under *sui generis* design law, with the exception of Greece. The European Union institutions have been attempting, since 1991, to formulate a harmonised approach to the protection of designs in Europe. As a result of this effort, two pieces of legislation were proposed: a Community Design Regulation; and a Design Directive. The Regulation seeks to achieve a unitary protection system throughout the Community by creating a two-tier system of rights: the Registered Community Design Right (an exclusive 25 year right based on registration); and the Unregistered Community Design Right (a three year quasi-copyright, not dissimilar to the United Kingdom unregistered design right). The progress of the Community Design Regulation has halted due to technical reasons.

The purpose of the second legislation, the Design Directive, was to harmonise several aspects of Member States' national laws. This Directive was passed on October 14, 1998, with the result that each Member State will retain its own national-based registration system, but it will be brought into line with the Directive. The Directive legislates *only* on the registered design law of each Member State; it does not interfere with any other right such as the unregistered design right or copyright law.

Definition[79]

12–57 The definition of "design" has been widened by the Directive, which will entail some changes within the United Kingdom registered design law. Design means "the appearance of the whole or a part of a product resulting from the features of, in particular, the lines, contours, colours, shape, texture and/or materials of the product itself and/or its ornamentation." Furthermore, there is no need for industrial application or an industrial product. Instead, design protection is to extend to both industrial and handicraft items, including parts intended to be assembled into a complex product, packaging, get-up, graphic symbols and typographic typefaces. This widened protection regime will also encompass computer graphics and icons, product packaging and get-up. However, computer programs are clearly excluded.

Excluded subject matter[80]

12–58 Design protection will not be granted to the following subject matter:

- features of appearance of a product which are solely dictated by its technical function;
- "must-fit" features, *i.e.* features of appearance of a product which must necessarily be reproduced in their exact form and dimensions in order to permit the

---

[78] Directive 98/71 of the European Parliament and of the Council of October 13, 1998 on the legal protection of designs, [1998] O.J. L289/28. For a more detailed discussion, see Suthersanen, *op. cit.* Chapters 5 to 7.
[79] *ibid.*, Article 1.
[80] *ibid.*, Article 7.

designed product to be mechanically connected to or placed in, around or against another product so that either product may perform its function;

- designs which are contrary to public policy or to accepted principles of morality.

There is an exception in respect of "must-fit" features: designs which serve the purpose of allowing multiple assembly or connection of mutually interchangeable products within a modular system will be protected. This has been taken to refer to designs destined for modular systems, and which by their nature must fit or assemble with each other. Examples of this include modular toy systems such as LEGO bricks or modular furniture systems.

Conditions of protection

12–59 It will be recalled that, the United Kingdom registered design law confers protection on new, eye-appealing, and aesthetic designs. The new Directive eschews all requirements of eye-appeal or aesthetic merit or quality. Instead, the design must be new and have individual character.

Novelty

12–60 A design will be considered new if no identical design (designs will be deemed to be identical if their features differ only in immaterial details) has been made available to the public before the date of filing of the application for registration or, if priority is claimed, the date of priority.[81]

Individual character[82]

12–61 A design will have individual character if the overall impression it produces on the informed user differs from the overall impression produced on that user by any design which has been made available to the public before the date of filing of the application for registration or the priority date. In assessing individual character, the degree of freedom of the designer in developing the design shall be taken into consideration.

Prior art[83]

12–62 Both the requirements of novelty and individual character will be gauged by reference to prior art designs "made available to the public" before the date of application for registration or the priority date. The phrase in quotation marks is further defined within the Directive to refer to any design which has been published following registration or otherwise, or exhibited, used in trade or otherwise disclosed. However, the prior art will not include any designs which have been disclosed in circumstances which could not reasonably have become known in the normal course of business to the specialised circles in the design area in question, within the Community, before the date of filing of the application for registration or the date of priority.

Furthermore, a design will not have entered the prior art if it has been disclosed to a third person under explicit or implicit conditions of confidentiality; or if it has been disclosed as a consequence of an abuse in relation to the designer or his successor in title. Finally, the Directive sets down a 12–month grace period during which any disclosure

---

[81] *ibid.*, Article 4.
[82] *ibid.*, Article 5.
[83] *ibid.*, Article 6.

of the design by the designer (or another party with his authority) will be ignored for the purposes of gauging the novelty and individual character of the design.

Visibility

12–63　It is assumed that the design must be visible to the naked eye since protection is conferred on the "appearance" of a product; however, the design may be internal or external and need not be visible during normal use of the product, with one exception. This is in relation to component parts, and is ostensibly to exclude internal parts or spare parts of products. The exception applies to component parts of a complex product. A complex product, in turn, is defined as "a product which is composed of multiple components which can be replaced permitting disassembly and reassembly of the product."[84] If the design under consideration is one which is applied to or incorporated in a product which constitutes a component part of a complex product, such a design will only be considered new and to have individual character:

- if the component part, once it has been incorporated into the complex product, remains visible during normal use of the latter, and

- to the extent that those visible features of the component part fulfil in themselves the requirements as to novelty and individual character.

Normal use is defined to mean use by the end user excluding maintenance, servicing or repair work. In other words, internal parts of larger products will only be considered if they remain visible during the use of the larger product.

Rights conferred

Rights[85]

12–64　As stated above, the Directive's purpose is to harmonise the registered design laws of E.C. Member States. The actual procedure as to application and registration will be governed by the national design law. Upon registering the design in any Member State, the Directive states that registered owners should obtain the following rights: the exclusive right to use the design and to prevent any third party not having his consent from using a design which is identical to his own or one which will not produce a different overall impression on the informed user. Such use will also cover the acts of making, offering, putting on the market, importing, exporting, stocking or using a product which incorporates the protected design. However, the scope of protection will not cover acts which are done privately and for non-commercial purposes; or which are done for experimental or educational purposes.[86]

Duration[87]

12–65　Upon registration, the design will be protected for an initial term of five years from the date of filing, which can be renewed for four further periods, up to a maximum of 25 years.

Spare parts

12–66　One of the original objectives of the design harmonisation programme was to solve the spare parts problem — *i.e.* to curtail protection of such products. However,

---

[84]　Directive 98/71 of the European Parliament, Article 1.
[85]　*ibid.*, Articles 9, 12.
[86]　*ibid.*, Article 13.
[87]　*ibid.*, Article 10.

after three different versions of the solution were presented and rejected, the final Directive has no special provision dealing with this issue. The only means by which spare parts will be excluded are under the "functionality" and "must–fit" provisions, the latter having been influenced by the United Kingdom law.

Relationship with copyright

12–67 The Directive states that a design which is protected by a registered design right should also be eligible for protection under the law of copyright of individual Member States. However, the extent to which, and the conditions under which, such protection is conferred, including the level of originality required, will be determined by each Member State.

At present, most Member States provide for protection of designs under their copyright laws. However, the requirements and scope of copyright protection for designs vary from Member State to Member State. Thus, France and Belgium offer very generous copyright protection to designs, both artistic and functional; whereas, it is difficult to obtain copyright protection for industrial designs in Germany and the Scandinavian countries. Finally, it is almost impossible to obtain copyright protection for designs in Italy.

COMPARISON OF PROTECTION

Type of product

12–68 Depending on the type of product at hand, the creator has a variety of intellectual property protection available to him. The following table offers a quick perspective of the types of protection under copyright and design laws available to different types of products.

Function Designs	Copyright	Unregistered design right	Registered design right
Heavy or light engineering-based goods, mechanical, transportation	Possible protection for drawings, plans, tables. However, very unlikely to obtain protection for three-dimensional versions of design documents	Possible, if original, non-commonplace, three dimensional design. No protection for interface or must-match designs	Unlikely unless eye-appealing, aesthetic, new design. No protection for functional or must-match designs
Artistic Designs Furniture design; fashion and textile design; exhibition and display	Possible protection for two dimensional versions of work, so long as works exhibit skill, labour and judgement; unlikely for three dimensional works unless some artistic merit or quality	No protection for two dimensional or surface designs. However, likely protection for any three-dimensional design	Possible protection if design is eye-appealing, aesthetic and novel

Industrial designs	Copyright	Unregistered design right	Registered design right
Furniture design; automotive design; building materials; pre-fabricated or pre-assembled building parts. Examples include jewellery, chairs, furniture cabinets, lighting apparatus, kitchen appliances	Possible protection for design documentation, but not three dimensional designs unless work exhibits some artistic merit or quality. See above	Likely protection for three-dimensional designs	Possible protection if design is eye-appealing, aesthetic and novel

Nature of Protection

12–69 This table gives a brief comparison of the nature and extent of protection afforded by each type of intellectual property right.

	Copyright	Unregistered design right	Registered design right
Legislation	Copyright, Designs and Patents Act 1988	Copyright, Designs and Patents Act 1988	Registered Designs Act 1949
Protected subject matter	Literary and artistic works, including compilation, database, preparatory design material for a computer program, plan, drawing, diagram, engraving, photograph, sculpture, collage, work of artistic craftsmanship, architectural building and model	Three dimensional designs *i.e.* any aspect of the shape or configuration (whether internal or external) of an article	Two or three dimensional designs *i.e.* features of shape, configuration, pattern or ornament as applied to an article
Excluded subject matter	Ideas, copied works	Methods or principles of construction, must-fit and must-match features, surface decoration	Functionally dictated, features, must-match features, methods or principles of construction
Criteria for protection	Original *i.e.* not copied. For some types of works, artistic merit is required	Original and not commonplace	Novel, eye-appealing and aesthetic, industrially applied

	Copyright	Unregistered design right	Registered design right
Formalities for protection	None, but work must be fixed on a medium, including digital fixation	None, but design should be recorded or an article must be made to the design	Registration
Duration of protection	Life of the author plus 70 years; if design is industrially applied, 25 years from industrial application	15 years from making, or 10 years from marketing	25 years from registration
Ownership	Creator of work. If made in the course of employment, employer. Note computer-generated designs	Creator of design, or if in course of employment or commision, employer or commissioner of work	Creator of design, or if in course of employment or commission, employer or commissioner of work
Rights	A right against copying. Limits on scope of protection of functional drawings. Moral rights	Anti-copying right	Exclusive monopoly

Can all three rights subsist together?

12–70   All three rights can co-exist with each other in the same article. Thus, the preparatory drawings and plans of any project would be initially protected by copyright. However, it may not be able to rely on copyright protection for the final product unless the product itself qualified as a sculpture or a work of artistic craftsmanship. This may be difficult as courts look for artistic merit. Neither can indirect protection be sought for the product through the drawings. Section 51 of the Copyright, Designs and Patents Act 1988 cuts down the extent to which one can rely on drawings as a means to protect non-artistic works. Finally, the product may also benefit from protection under the unregistered design right if it is not commonplace in that particular product industry.

The overlapping areas of copyright and unregistered design right are illustrated in the decision of *Electronic Techniques (Anglia) Ltd v. Critchley Components Ltd.*[88] The product in question here was a transformer. ETAL, the plaintiff, designed, manufactured and sold transformers for insertion inside electronic equipment such as computers and telephone modems. The plaintiff also, in support of its business, distributed data sheets which described the specification and performance characteristics of a particular model of transformer. Critchley, the defendant, sold about 300 different transformers, of which nine or 10 were supplied in direct competition to the plaintiff's transformers. The defendant also supplied data sheets with its products. The plaintiff instituted proceedings in which it claimed an infringement of copyright and unregistered design right.

Copyright

12–71   Copyright was claimed in the data sheets. The data sheet contained a narrative in highly condensed and technical language and notation, and essential information

---
[88] *Electronic Techniques v. Critchley Components*, [1997] F.S.R. 401.

relating to the transformer. The data sheet also contained circuit diagrams which were recommendations by ETAL as to how the client should align his circuitry when using ETAL transformers. The plaintiff claimed literary copyright in the narratives in the data sheets; this was accepted by the court. The plaintiff also claimed that the diagrams could be both artistic and literary copyright works. However, the court was not enthusiastic as to this approach: different copyrights can protect a particular product, or an author can produce more than one copyright work during the course of a single episode of creative effort; but a single piece of work by an author cannot give rise to two or more copyrights in respect of the same creative effort. In other words, the diagrams had to be either literary copyright works or artistic copyright works, but not both.

Unregistered design right

12–72   The transformer was made up of the following elements: a magnetic core, a plastic bobbin, wire which is wound around the bobbin, a plastic box into which the assembled core, bobbin and windings are placed, and finally, insulating resin. Unregistered design right was claimed in relation to the shape and configuration of the whole transformer. The issue was whether the must-fit provision applied. The must-fit provision should be applied to the claimed design. If individual components of the transformer had been claimed (for example, the interface between the core, bobbin and winding), then the interface features between these internal components of the transformer could be excluded from protection under the must-fit provisions. However, this would only occur if the unregistered design right was claimed in each individual internal component. Here, the unregistered design right was claimed in the whole transformer *i.e.* in the design of the combination of all the internal components.

In the above instance, would it have been possible to register the design? This would appear to be highly unlikely as the conditions of protection are: novelty, eye-appeal, and the purchaser of the product must acquire it on the basis of aesthetic considerations.

CONCLUSION

12–73   The most suitable type of protection for a product design is very much dependent on the nature of the product, and the design. It will be extremely rare for copyright protection to be available for three–dimensional, utilitarian designs unless protection is only sought for the design documentation, drawings or reports. A more suitable means of protection is unregistered design right. This right is attractive in that the protection will arise automatically, without the need to seek registration. However, the right is limited to preventing others from reproducing the design. In contrast, the registered design law offers, at the price of registration, a monopoly over the protected design. In addition to the above, it should also be noted that some product designs will be protectable under patent laws, or utility model laws. The latter type of protection is aimed at minor process or product innovations, which exhibit a lower threshold of novelty and inventiveness than that under patent law. Although not available under the United Kingdom, other European Union Member States such as Germany, Austria, Spain and Italy offer this umbrella of protection. A final option for design creators and producers is to seek trade mark protection for the shape of the product.[89]

---

[89] See Chapter 13.

CHAPTER 13

THE LAW OF TRADE MARKS AND PASSING OFF

Iain Purvis
Barrister, 11, South Square, Gray's Inn

INTRODUCTION

13–01 The subject matter of trade marks — the protection of the distinctive names, get-up and livery of businesses and their products — is not of central relevance to a book concerned with the protection of technology. However, trade marks are vital components of the intellectual property armoury of most businesses, and registered marks are often included in transactions involving the acquisition or licensing of technology rights. This short Chapter dealing with registered trade marks (with a brief reference to passing off) has thus been included for those involved in such transactions, and for the sake of completeness. It should not be treated as a comprehensive guide, but merely as a brief introduction to the main aspects of the law in this area. The one exception to this is the subject of the registration of the shapes of articles as trade marks, which is given more detailed treatment because of its potential importance as a new source of intellectual property protection for the fruits of technology.

REGISTERED TRADE MARKS

System of registration

13–02 The United Kingdom (in common with most industrialised countries) operates a system of registration of trade marks. A register of trade marks is maintained by a branch of the Patent Office called the Trade Marks Registry. The law of registered trade marks in the United Kingdom is governed by the Trade Marks Act 1994, which in turn enacts the European Trade Marks Directive,[1] the aim of which is to harmonise the law of registered trade marks in Member States. This harmonisation is only partial, since some of its provisions are only optional and have been enacted by some countries and not others.

A parallel system of registration is available by way of a "community trade mark". The community trade mark and the office which maintains the register ("OHIM")[2] was established by E.U. Council Regulation of December 20, 1993.

Organisations can choose whether to apply for a United Kingdom registered mark (and other marks in other E.U. countries) or a community mark, or both. The rules governing whether a mark can be registered and the effects of a registered mark in the United Kingdom are the same whether it is a United Kingdom or community mark.[3] When an application for registration is made, the relevant office considers whether it satisfies the requirements of the Trade Marks Act (U.K. office) or the Directive (OHIM).

There is also a system for "international registration" established under the "Madrid Protocol".[4] The idea is to reduce the cost and complexity of applying for

---

[1] Council Directive 89/104.
[2] Based in Alicante, Spain.
[3] Though, obviously, the community mark gives protection throughout the E.U., whereas the United Kingdom mark only gives protection in the United Kingdom.
[4] The Protocol relating to the Madrid Agreement concerning the International Registration of Marks (1989).

marks in different countries by enabling a single international application and filing which will achieve the effect of applying for a registered mark in each of the ratifying countries. The United Kingdom is one of the countries to have ratified the Protocol. The effect of an international registration designating the United Kingdom is in all important respects the same as that of a national United Kingdom trade mark.[5]

A registered trade mark is granted in respect of specific goods or services. For administrative purposes, these are defined by reference to international "classes". Registration gives a monopoly in the use of the mark in relation to the specific goods or services for which it is granted, and extends to the use of similar marks for similar goods or services where the similarity is likely to cause confusion. In certain cases it even extends to the use of the mark or similar marks for dissimilar goods and services.[6]

Length of protection

13–03   Unlike other intellectual property rights, there is no maximum period of protection for a trade mark. Once registered, it lasts for an initial period of 10 years, which may be extended by the proprietor on payment of a fee for another 10 years, and so on indefinitely.[7] The mark may be revoked or declared invalid by the Trade Marks Registry or by the court (either because it was registered in breach of the Directive, because of non-use, or because it has ceased to fulfil the basic requirements of a trade mark[8]), but will otherwise only cease to exist in the event of non-renewal.

REGISTRABILITY

13–04   The Directive and the Trade Marks Act lay down various requirements which a trade mark must satisfy in order to be validly registered. If it does not fulfil those requirements, then the application to register should be refused by the relevant office. Even if the application is granted, failure to comply with these requirements may be raised as grounds for revoking the mark at any time after grant.

Definition of a trade mark

13–05   The first requirement which a mark must satisfy before it can be registered is that it must fall within the definition of a trade mark established by law. Under Article 2 of the Trade Marks Directive a "trade mark" is a "sign" which can be represented graphically and is capable of distinguishing the goods or services of one undertaking from those of another. In particular, it *may* consist of "words (including personal names), designs, letters, numerals or the shape of goods or their packaging".[9]

This is a very broad definition. The list of matters which may constitute a trade mark is non-limiting. There is therefore on the face of it nothing to prevent the registration of musical jingles or even smells, provided that the technical difficulties of representing such a mark graphically can be overcome. The principal limitation placed on registrability by the definition of Article 2 is that it must be "capable of distinguishing".

---

[5] The complexities of the provisions of the Trade Marks (International Registration) Order 1996 are beyond the scope of this book.
[6] The question of infringement is dealt with in more detail below at 13–24 to 13–29.
[7] Trade Marks Act 1994, ss.42–3.
[8] See below.
[9] Article 2 of the Directive.

Capable of distinguishing

13–06 As we have seen, any "sign" which is not capable of distinguishing the goods or services of one undertaking from those of another is not permitted to be registered because it is not a "trade mark".[10]

This definition is a statutory recognition of what has long been held by the courts in the United Kingdom and in Europe: that the purpose of a trade mark is as an indication of origin.[11] Put in its crudest terms, the mark must be sufficiently distinctive that the public can rely on it as a guarantee that the goods on which it is used are from the same source as other goods they have bought in the past bearing the mark. If the sign in question is not capable of doing this, then it is not a trade mark which can be registered under the Directive.

So far as ordinary "word" trade marks are concerned, this restriction is very much a "first filter". It would preclude registration of utterly descriptive words such as "soap" for soap. [12] The word "soap" could never serve, in the public mind, to distinguish one company's soap from that of another. However, it would not preclude "soap" written in a unique stylised script.

GROUNDS FOR REFUSAL OF AN APPLICATION TO REGISTER

13–07 Even where a sign is a "trade mark" as defined by Article 2, it may be excluded from registration by other provisions in the Directive. These provisions are divided into two: "absolute grounds for refusal" and "relative grounds for refusal".

"Absolute grounds for refusal" arise simply because of the nature of the mark itself. "Relative grounds" arise out of third party rights.

ABSOLUTE GROUNDS

Devoid of any distinctive character[13]

13–08 This is a similar objection to "incapable of distinguishing", which is part of the definition of a trade mark — see 13–06 above. However, there is an important difference. The objection that a mark is devoid of distinctive character may be overcome by showing that the mark has in fact acquired a distinctive character as a result of the use made of it.[14] Since such a mark could not by definition be "incapable of distinguishing" it is clear that the phrase "devoid of distinctive character" is meant to be an objection of wider effect.

In *Philips v. Remington* [1999] R.P.C. 809, Aldous L.J. took as an example the two marks "weldmesh" and "welded mesh". An application to register the former for welded mesh products would be rejected on the ground that it was devoid of distinctive character, since it is plainly descriptive of the products for which it is intended to be used. However, it is possible that substantial use of the trade mark would cause the mark to become associated in the minds of the public not with welded mesh products generally, but with welded mesh products from a particular source. At that point, the mark would have acquired a "secondary meaning" as a guarantee of origin, and would be able to overcome the objection that it was devoid of distinctive character.

---

[10] Article 3(1)(a).
[11] See, for instance, the Preamble to the Directive.
[12] The example given by Jacob J. in *Philips v. Remington* [1998] R.P.C. 283.
[13] Article 3(1)(b) of the Directive.
[14] Article 3(1).

This is a difficult hurdle to overcome, and the trade mark applicant would have to show that a "significant proportion" of the relevant public (*i.e.* customers, dealers and producers of the product or services) had come to see the mark as identifying goods from a particular source.[15]

The words "welded mesh" however, are so inherently descriptive that they could never become a guarantee of origin. No amount of use would ever cause them to be distinctive of any particular supplier. They are "incapable of distinguishing" and therefore are unregistrable in any circumstances.

Descriptive, laudatory and geographical marks

13–09   Article 3(1)(c) lists a number of types of descriptive marks all of which are barred from registration unless they are shown to have acquired a distinctive character through use. These are marks which consist exclusively of signs or indications which may serve in trade to designate the kind, quality, quantity, intended purpose, value, geographical origin, time of production of the goods or the rendering of the service, or other characteristics of the goods or service.

The idea is to protect marks which other traders might legitimately wish to use either now or in the future to make a factual statement about their products or services. The test is not whether the marks are actually being used now for that purpose by other people. It is enough that it is a reasonable assumption that the mark may be used for that purpose in the future.[16]

Thus the mark "Bootle" for bicycles would be refused (in the absence of distinctiveness acquired through use), whether or not anyone at present made bicycles in Bootle. It is reasonable to expect that this may happen in the future, and the public would therefore see the word Bootle as an indication of geographic origin. However, the mark "North Pole" for bicycles should be accepted since there is no reasonable chance of bicycle manufacture at the North Pole, and the public would not see the mark as an indication of geographical origin.

It should be noted that this exclusion only applies to marks which consist "exclusively" of the excluded matter. Thus if "Bootle" were only part of a mark consisting of a picture of a bicycle outlined against the moon, there would be no objection under this provision.

Customary in the trade

13–10   A mark shall not be registered if it consists exclusively of signs or indications which have become customary in the current language or in the bona fide and established practices of the trade.[17] Again this is subject to the mark having acquired distinctiveness through use.

There is obviously a great deal of overlap between this objection and that under Article 3(1)(c) — see paragraph 13–09 above. It is rather hard to imagine a mark which fell within this exclusion but was not excluded on the grounds that it may serve in trade to designate a characteristic of goods or services.

---

[15] See the judgment of the ECJ in the *WSC Windsurfing Chiemsee* case [1999] E.T.M.R. 585. The Advocate-General suggested "over 50 per cent" distinctiveness amongst the relevant public was necessary.
[16] *Chiemsee* case *ibid.*
[17] Article 3(1)(d).

The "shape exclusions"

13–11   Signs which consist exclusively of "the shape which results from the nature of the goods themselves"; "the shape of goods which is necessary to obtain a technical result"; or "the shape which gives substantial value to the goods" are not registrable, even upon acquired distinctiveness being shown.[18]

The exact scope of these limitations is not easy to understand, and has been the subject of some controversy, at present unresolved. Registering the shape of goods as a trade mark also causes particular difficulties with the various "distinctiveness" exclusions discussed above. The current position on the registration of shapes of goods as trade marks under United Kingdom law is discussed below.

"Shape marks" generally

13–12   The leading United Kingdom authority on the registration of shape marks is the decision of the Court of Appeal in *Philips v. Remington* [1999] R.P.C. 809. This was an infringement action in which the validity of the mark was challenged by the defendant. The claimant had registered a picture of the head of its well-known three-headed rotary shaver as a trade mark for electric shavers. There was evidence that the picture was recognised by a substantial proportion of the public in the United Kingdom as being the claimant's shaver. If the public saw a shaver with the head shown in the picture, they would believe, in the absence of a statement to the contrary, that it came from the claimant. The Court of Appeal held that the mark was invalid on a number of grounds.

First of all they held that that it was not a trade mark within the meaning of the Act because it was not capable of distinguishing the products of the claimant from those of other traders. The reason given for this was that the trade mark was purely a pictorial description of a three headed rotary shaver. It could never be "capable of distinguishing" since it could not distinguish three headed rotary shavers made by Philips from those made by other companies. The fact that there were no other products actually on the market was irrelevant, since the question of capability of distinguishing was a matter of law.

Aldous L.J. summarised his judgment on the point as follows (at 818):

> "In my view the shape of an article cannot be registered in respect of goods of that shape unless it contains some addition to the shape of the article which has trade mark significance. It is that addition which makes it capable of distinguishing the trade mark owner's goods from the same sort of goods sold by another trader."

By "addition ... which has trade mark significance", it may be presumed that the court had in mind the depiction of the shape of an article with the name of the trade mark owner or his logo branded onto it. It is of course a little difficult to see what the point of such a registration would be, as opposed to a registration of the name and logo on its own.

13–13   The Court also held that the mark was devoid of distinctive character. Philips sought to argue that it had acquired distinctiveness through use, but this argument was rejected. This is perhaps a little surprising, since there was evidence of very substantial use of the mark, not only in the form of the shaver itself, but also in two-

---

[18] Article 3(1)(e).

dimensional form on the packaging and in advertising. Furthermore, there was a finding of fact that the mark had become associated with Philips in the mind of the public. This was held to be irrelevant on the basis that such association was due merely to the fact that the Philips shaver was the only three headed rotary shaver on the market.

Thirdly the mark was held invalid on the ground that it consisted exclusively of an indication of the kind and intended purpose of the goods for which it was registered: *i.e.* the depiction of the shape merely showed a three headed shaver and nothing more. This finding necessarily followed from the decision that some "addition" to the mere shape of the article was necessary to achieve distinctiveness.

Finally, the mark was held invalid on the basis of one of the specific "shape" exclusions discussed above: namely that the shape depicted in the mark was "necessary to achieve a technical result". This was the case even though there were many other shapes which could have achieved the same result. The Court held that the wording of this exclusion simply meant that "the essential features of the shape are attributable only to the technical result". Because the design was dictated purely by technical considerations, it was therefore not registrable.

The objections based on the other "shape" exclusions failed:

(a)   The shape did not result from the nature of the goods themselves. For this purpose, the "goods themselves" should not be regarded as "three headed shavers of the shape used by Philips", but rather as the goods for which the mark was registered (in this case "electric shavers"). There are of course many other shapes of electric shavers, and therefore the Philips shape cannot have resulted from the nature of the goods.

(b)   The shape did not "give substantial value to the goods". It was held that this phrase was intended to exclude "aesthetic-type shapes". It required that the shape chosen by the proprietor was substantially more valuable than other shapes available for use. In this case, that had not been shown, and the objection therefore failed.

It will be appreciated that the approach taken by the Court of Appeal was highly restrictive. In particular, the establishment of a requirement that something more than a shape which has achieved general recognition by the public as denoting the goods of a particular manufacturer seems to set up a new requirement for registrability in the case of shapes which does not apply to word or device marks. If this approach is adopted generally, it is hard to see how more than a handful of shapes could ever be accepted for registration. However, this is an area of active disagreement, not only amongst practitioners and academics, but also between the courts of different member states applying the same rules under the Directive.[19] There are obvious policy issues here: the economic power granted by a mark granted for a useful shape is potentially very great, particularly since (unlike other monopoly rights) a trade mark has no time limit. On the other hand, if a particular shape is indeed distinctive of goods from a particular business, the adoption of that shape by others may lead to deception of the public, which is undesirable. A number of fundamental questions arising out of the Philips decision have been referred by the Court of Appeal to the European Court of Justice. Until their decision on those questions is given, this whole area must be treated as one of great uncertainty.

---

[19] The Swedish courts came to the opposite decision from that of the United Kingdom courts in the *Philips* case.

Public policy

13–14 Trade marks which are contrary to public policy or to accepted principles of morality may not be registered.[20]

Deceptive marks and bad faith

13–15 No mark may be registered if it is of such a nature as to deceive the public, for instance as to the nature, quality or geographical origin of the goods or service. Nor can a mark be registered in "bad faith". The exact scope of this objection remains unclear, but the test generally adopted in the United Kingdom courts is that put forward by Lindsay J. in *Gromax v. Don & Low Nonwovens,*[21] namely that it includes dishonesty and "dealings which fall short of the standards of acceptable commercial behaviour observed by reasonable and experienced men in the particular area being examined". Under previous statutes, similar provisions have been held to prevent the deliberate "stealing" of a mark used abroad by a third party who was known to the applicant to be about to launch under the mark in the United Kingdom. It may also prevent the registration of marks for commercial advantage where the applicant has no intention whatsoever of using them.

RELATIVE GROUNDS

13–16 These grounds require refusal of registration where a mark conflicts with an "earlier trade mark", or other "earlier right".

Earlier trade mark

13–17 An "earlier trade mark" for the purposes of a "relative grounds" objection may be one of the following:

(a) a community trade mark;

(b) a United Kingdom registered trade mark;

(c) an international trade mark (U.K.);
(or an application for any of the above marks)

(d) a trade mark "well known" in the United Kingdom and entitled to protection under the Paris Convention.[22]

Whether it is "earlier" is judged by comparing its date of application with that of the application in issue, taking into account any applications from which either claims priority.

"Well known marks" under the Paris Convention

13–18 This is a mark which is well known in the United Kingdom as being the trade mark of a foreign person or business enterprise from a country which is a signatory to the Paris Convention for the Protection of Intellectual Property (1883) or the more

---

[20] Article 3(1)(f).
[21] [1999] R.P.C. 367 at 379.
[22] Article 4(2); Trade Marks Act, ss. 5 and 6.

recent World Trade Organisation agreement (1994). It applies whether or not the person or business has any business or "goodwill" in the United Kingdom.[23]

Grounds for refusal based on earlier marks

(a) *Identical goods/services and identical marks*

13–19 An application will be refused by reason of an earlier mark where the marks are identical and the goods and services which it is applied for are identical to those for which the earlier mark is registered (or has been applied for, or, in the case of a Paris Convention mark, has acquired a reputation).

(b) *Similar goods/services and similar marks*

13–20 Registration will also be refused in circumstances where the marks are similar and the goods or services are identical or similar, provided that the similarity is such that there exists a likelihood of confusion on the part of the public[24]. Confusion includes the likelihood of association.

The test is not to be applied in two stages (*i.e.* "are the marks/goods similar?" and "is there a likelihood of confusion?"). Rather, there is a single test: by reason of the similarity of the marks/goods, is there a likelihood of confusion?[25] The court is entitled to take into account the distinctiveness of the earlier mark — the public are more likely to take account of small differences where the earlier mark is descriptive, than where the earlier mark is highly distinctive.[26] It is also entitled to take into account the actual use and reputation of the earlier mark. The general rule is that the more famous the earlier mark, the more likely it is that confusion will be caused by a non-identical later mark, or a later mark used for different goods.[27] However, the fame of the earlier mark is only one factor in a global assessment of likelihood of confusion.[28]

The reference to "likelihood of association" should not be taken too broadly. It is not enough that the mark being applied for "brings the earlier mark to mind". It must go further than that and suggest a connection between the two.[29]

(c) *Similar marks, dissimilar goods/services*

13–21 Registration will be refused where the mark is identical or similar but the goods are dissimilar from those of the earlier mark if two conditions are satisfied:

(a)   the earlier mark has a reputation in the United Kingdom (or in the case of a community mark in the European Community);

(b)   the use of the later mark without due cause would take unfair advantage of, or be detrimental to, the distinctive character or the repute of the earlier trade mark.[30]

---

[23] *ibid.*, s.56.
[24] Article 5(2).
[25] *Sabel v. Puma* [1998] R.P.C. 199, ECJ.
[26] *ibid.*
[27] *Canon v. Metro Goldwyn Meyer* [1999] R.P.C. 117, ECJ.
[28] *Lloyd Schuhfabrik* [1999] All E.R. (E.C.) 587, ECJ.
[29] *Sabel*, n.25 above.
[30] Article 4(4)(a) of the Directive (voluntary); TMA 1994, s.5(3).

This provision is intended to protect marks which are known to a "significant part of the relevant sectors of the public",[31] against abuse by traders in unrelated fields.

The question of when use of a mark on unrelated goods will be deemed to have taken unfair advantage of or be detrimental to an earlier mark is highly subjective. Thus the owners of the EVER READY trade mark for batteries failed to prevent the registration of the same mark by a different company in respect of condoms. The owners of the VISA mark for financial services, however, succeeded in preventing registration of their mark in respect of condoms.

(d) *Other earlier rights*

13–22 So far, we have considered only objections based on earlier trade marks which are protected by statute or international treaty. Section 5(4)(a) of the Act prohibits registration where the use of the trade mark is "liable to be prevented" by a third party by virtue of the law of passing off. This protects the interests of those who have an actionable goodwill in the United Kingdom based on a mark which is not registered.[32]

Under section 5(4)(b), registration is also prohibited where the use of the mark is liable to be prevented by virtue of another legal right such as copyright or design right. Thus, for instance, no design can be registered as a trade mark if it is an unlawful copy of someone else's copyright work.

Consent

13–23 In the case of the "relative grounds" for refusal of registration dealt with above, they may be overcome by obtaining the consent of the owner of the earlier right to the registration of the mark.[33]

INFRINGEMENT

13–24 A registered trade mark is a monopoly right. This is to say that it grants "exclusive rights" to the owner over a specifically defined area. The question of infringement therefore involves no element of intention. It is a purely objective test based on the similarity between the registered mark and the mark alleged to infringe.

The scope of the exclusive rights granted to the trade mark owner is set out in Article 5 of the Directive (TMA 1994, s. 10). The wording almost exactly parallels that used by the Directive in respect of refusal of marks based on "earlier rights".[34] Thus, a trade mark will be infringed by the use of a sign identical to the registered mark in relation to goods or services identical to those for which it is registered. This is regardless of any question of confusion or likelihood of confusion.[35]

A trade mark will also be infringed by the use of an identical sign in respect of goods or services similar to those for which the mark is registered or by the use of a sign similar to the registered mark in respect of goods or services identical or similar to those for which the mark is registered. However, here there will only be infringement where, because of the relevant identity and similarity of marks and goods/services, there exists a likelihood of confusion on the part of the public between the sign and the registered mark.[36]

---

[31] *General Motors v. Yplon* [1999] E.T.M.R. 122 (A-G's opinion).
[32] See section on passing off paras 13–39 to 13–43 below.
[33] TMA, 1994, s.5(6).
[34] See paras 13–16 to 13–23 above.
[35] Article 5(1)(a), section 10(1).
[36] Article 5(1)(b), section 10(2).

Finally, a trade mark will be infringed by the use of an identical or similar sign in respect of *dissimilar* goods, where the mark has a reputation in the United Kingdom and where use of the sign without due cause takes unfair advantage or and/or is detrimental to the distinctive character of the mark.

All the points made above in relation to the equivalent provisions on "earlier rights" apply equally when considering the comparison to be made between marks for the purposes of infringement.

Infringing use

13–25 Infringement only takes place when the sign complained of is "used". Section 10(4) gives a non-exhaustive definition of acts which constitute "use" for the purposes of infringement. Most of these are fairly obvious: affixing the sign to the goods or to packaging; selling or putting goods or services on the market under the sign; importing or exporting or stocking goods under the sign; using the sign in advertising. Because the section is non-exhaustive it would appear that the oral use of a trade mark may also infringe.

The Act provides specifically for "secondary" infringement by packagers or printers who apply the infringing mark to material intended to be used for labelling or packaging goods, or advertising, if they knew or had reason to believe that the application of the mark was not duly authorised by the proprietor or a licensee. [37] This would apply equally to goods intended for sale abroad.

Comparative advertising

13–26 Under the old law, it was an infringement to engage in comparative advertising which used the registered trade mark of a rival in the course of identifying the advantages of the product being advertised. Honesty and accuracy was no defence. This is no longer the case. Under section 10(6) of the Act, the use of a trade mark for the purpose of identifying goods or services as those of the proprietor of the mark or a licensee is permitted provided that it is not "otherwise than in accordance with honest business practices in industrial or commercial matters" and "takes unfair advantage of, or is detrimental to, the distinctive character or repute of the trade mark".

A number of cases on "honest practices" have come before the courts, in which trade mark owners sought a finding of infringement on the basis that comparisons being made in advertising material were not fair. The approach taken has been quite robust, reflecting the intention of the statute to permit comparisons between products and services to be made, save where those comparisons are clearly deceptive or misleading. The courts will not engage in minute linguistic analysis or seek to resolve honest disputes of fact or opinion between traders, most of whom can use their own advertising to set the record straight.[38]

DEFENCES

13–27 The Act provides for specific defences against an action for infringement in three particular cases as set out below.

---

[37] section 10(5).
[38] see for instance *Barclays Bank v. Advanta* [1996] R.P.C. 307.

Use of another registered mark

13–28   A registered trade mark cannot be infringed by the use of another registered mark.[39] However, this defence may be defeated if it is shown that the defendant's registered mark was invalid at the time.[40]

Use of a mark other than as a trade mark

13–29   As we have seen, the purpose of a registered trade mark is as an indication or guarantee of the origin of goods or services. The Act provides for various defences where the mark is being used for other purposes, provided that these uses are in accordance with honest practices in industrial or commercial matters.[41] These are as follows:

(a)   Use by a person[42] of his own name and address;

(b)   Use of indications concerning the kind, quality, quantity, intended purpose, value, geographical origin, time of production or other characteristics of goods or services.
       We have seen (see paragraph 13–09 above) that marks which may serve to indicate these features of goods or services are not registrable save where they have become distinctive through use. This defence serves to protect other traders in the event that such acquired distinctiveness is shown and the mark has been validly registered. Thus, taking our earlier example, if a bicycle manufacturer trading under the name "Bootle" acquired sufficient distinctiveness in the name to enable registration, he would not be able to use that registration to prevent a rival manufacturer from telling the public that his bicycles were made in Bootle (unless this information was untrue or was given for the purpose of causing deception, in which case it would fall foul of the "honest practices" proviso).

(c)   Use of the mark where it is necessary to indicate the intended purpose of a product or service (in particular as accessories or spare parts).
       This is fairly straightforward. A motor accessory manufacturer may sell his spark plugs as "suitable for Ford Orions" without risk of infringement of Ford's registered mark.

EXHAUSTION OF RIGHTS

13–30   A trade mark cannot be infringed by use in relation to goods which have already been put on the market in the European Economic Area by the trade mark owner or with his consent, unless there exist "legitimate reasons" for the owner to oppose further dealings with the goods. This is particularly where the condition of the goods has been changed or impaired since they have been put on the market.[43]

This rule embodies the doctrine of "exhaustion of rights". A trade mark owner's interest in controlling the use of a mark is deemed to be exhausted with regard to particular goods at the point when he permits them to enter circulation in the market.

---

[39] section 11(1).
[40] section 47(6).
[41] section 11(2).
[42] This probably includes companies, though the point remains undecided.
[43] section 12.

From then on, he is not entitled to any say in how or where they are sold within that market. For these purposes the market is the free market in Europe created by the treaties establishing the European Economic Area. Thus, if Nike sell a pair of shoes under the Nike mark anywhere within the EEA, they may be purchased by anyone, and may be subsequently sold anywhere else in the EEA at any price and from any outlet without interference. However, this would not apply to such shoes if they had been put on the market in the United States.[44] In principle the sale of such shoes under the Nike mark in Europe would be an infringement of trade mark. It could only escape were the court to find that the circumstances of the sale were such that Nike were deemed to have consented to the sale in Europe under the mark, within the meaning of section 9(1) of the Act.[45]

INVALIDITY AND REVOCATION

13–31 A trade mark may be attacked after grant in two ways: by an application to revoke, or by an application for a declaration of invalidity.

Declarations of invalidity

13–32 An application for such a declaration seeks a review of the question of whether the mark should have been granted in the first place. It is common for a defendant to make such an application after being sued for infringement. However, it may be made at any time to the High Court or to the Trade Marks Registry, or OHIM in the case of a community trade mark.

All the grounds for refusal of a mark, both absolute and relative, considered above, may be relied on in such an application. Validity is judged as of the date of the application, with the important proviso that a mark which ought to have been refused on the basis of lack of inherent distinctiveness but which has subsequently obtained the necessary distinctiveness will not be declared invalid[46].

A declaration of invalidity need not be made in relation to the entirety of the goods and services for which the mark was registered. It may be made in respect of some of them only.[47] Thus the mark "Fish" registered in relation to a range of cooking utensils could be declared invalid in relation to fish kettles on the ground that it was devoid of distinctive character but could survive in relation to toasters.

Where a mark is declared invalid, it is deemed never to have been made. However, this does not destroy the effect of any past transactions involving the mark.[48]

Revocation

13–33 A mark may be revoked after grant on one of three grounds. Such revocation takes effect from the date of the application to revoke, or from the date when the grounds of the application started to exist.[49]

---

[44] The European Court have considered this matter in two cases: *Silhouette v. Hartlauer* [1998] F.S.R. 729; and *Sebago v. GB-UNIC* [1999] E.T.M.R. 467.
[45] See the finding in *Zino Davidoff v. A&G Imports* [1999] R.P.C. 631. The issue of "implied consent" is a matter of hot debate at the moment, and is the subject of several references to the European Court. Detailed consideration of this issue is beyond the scope of this short chapter, but further details may be found in Chapter 23 on "Licensing".
[46] section 47(1).
[47] section 47(5).
[48] section 47(6).
[49] section 46(6).

Non-use

13–34 The most common ground for revocation is where the mark has not been used by the proprietor or with his consent in relation to the goods or services for which it is registered for a period of five years or more. Any uninterrupted period of five years without use, when there are no "proper reasons for non-use", will give grounds for revocation, starting with the period of five years following the date of completion of the registration procedure.[50] Thus the trade mark proprietor is effectively given an initial five years "grace" period in which to commence using his mark.

"Proper reasons for non-use" cover the case where the proprietor genuinely wished and intended to use his mark, but because of unexpected production problems or a temporary collapse in the relevant market has been prevented from doing so.[51] They do not include the case where the proprietor simply changes his mind about adopting the mark.

Any use of the mark (or another version of the mark not differing from the registered mark in its distinctive character) in relation to goods within the scope of the registration and within the period of five years prior to the application to revoke will defeat the application. This includes applying the mark to goods intended solely for export.[52] However, such use must be "genuine" — merely placing the mark on a one-off basis on a single article every few years simply to avoid the risk of revocation would not count. Furthermore, proprietors should be aware that a mark may be revoked for non-use in relation to some of the goods or services for which it registered, even where it has been used in relation to others.[53]

Use of the mark in the period of three months immediately before the application to revoke, where the trade mark proprietor was aware that such an application was to be made, does not count.[54]

Common name

13–35 If through the acts or inactivity of the proprietor, a mark becomes the common name in the trade for a particular product or service, then it may be revoked.[55] The reference to "inactivity" here shows how important it is for trade mark owners to control and to monitor the use of their mark. It should where possible be identified as a trade mark (by use of the "®" or "™" devices[56]). In the case of marks which are evidently in danger of becoming "generic", active steps should be taken to point out to those who misuse the mark that it is a trade mark which indicates the goods or services of a particular supplier. The proprietor should also be careful in all his advertising and other descriptions of his goods or services to ensure that the mark is not used as a common noun or adjective, so as to suggest that it is a general descriptive term.

Liable to mislead the public

13–36 Similarly a mark may be revoked if, in consequence of the use made of it by the proprietor or with his consent, it has become liable to mislead the public,

---

[50] section 46(1)(a) and (b).
[51] See for instance *Magic Ball Trade Mark* [2000] R.P.C. 439.
[52] section 46(2).
[53] section 46(5).
[54] section 46(3).
[55] section 46(1)(c).
[56] There should be no use of the "®" device unless the mark has actually been registered: this is a criminal offence under s.95.

particularly as to the nature, quality or geographical origin of the goods or services.[57] Again, this emphasises the importance, particularly in licence agreements, of proprietors retaining active control over the way in which a mark is used.

ACQUIESCENCE

13–37 The Act provides a statutory cut off period for objections to trade marks by the owners of earlier rights. This is a period of five years provided the mark was being used with the knowledge of the owner of the earlier right.[58] If he makes no objection during this period, the owner of the earlier right not only loses his ability to apply for a declaration that the registration of the trade mark was invalid, but also may no longer oppose the use of the trade mark for the goods or services in relation to which it has been used. It is therefore obviously important for trade mark owners to be diligent in following up any information about conflicting marks.

REMEDIES

13–38 The same remedies are available in an action for infringement of a trade mark as are available in actions for infringement of other intellectual property rights (injunctions, delivery up of infringing articles, damages, accounts of profits, etc.). There is also a specific remedy of obliteration of the mark from the goods in question.

There are specific provisions in the Act concerning counterfeit goods. These give powers to the Customs and Excise to seize goods as they arrive in the United Kingdom from outside the European Economic Area.[59] Furthermore, infringement of trade mark can in certain circumstances be a criminal offence.[60]

PASSING OFF

13–39 Passing off is a non-statutory tort which protects the goodwill of a business insofar as that goodwill is associated with a particular mark, brand, device or get-up. There is no requirement of registration. The following is a very brief summary of the law in this area.

The elements of the tort are essentially three-fold.[61] First, it must be shown that the claimant has acquired a goodwill or reputation in the United Kingdom in the name, mark or other indicia which he is seeking to protect. Secondly, there must have been a misrepresentation by the defendant leading or likely to lead to confusion or deception. Thirdly, this misrepresentation must have caused damage to the claimant. These elements are dealt with below.

Goodwill

13–40 Goodwill has been described as "the attractive force which brings in custom". It must be shown that the claimant has a business in the United Kingdom operating under or by reference to the mark or get-up sought to be protected. If the business has

---

[57] section 46(1)(d).
[58] section 48 (Directive Article 9).
[59] sections 89–91: the details of these rather complex provisions are outside the scope of this Chapter.
[60] section 92.
[61] Referred to as the "classical trinity" by Nourse L.J. in his summary of the law in *Consorzio del Prosciutto di Parma v. Marks & Spencer Plc* [1991] R.P.C. 351 at 368.

no actual presence or base in the United Kingdom, then it must be shown that there are customers here.[62] The mark or get-up must be recognised by the claimant's customers as indicating his goods or his business. This does not mean that the customers need necessarily know the actual identity of the claimant: it is enough that they associate the mark or get-up as denoting goods from a particular source. Sometimes, issues arise as to the true owner of the goodwill in a mark, particularly where goods are associated with more than one business (for instance goods produced by a foreign manufacturer but sold in the United Kingdom by an exclusive distributor). Here, the goodwill will generally vest in the person responsible for controlling the way the mark is used and the quality of the goods sold under the mark.[63]

Just as with registered trade marks, there can be no protectable goodwill unless the mark or get-up is capable of distinguishing the goods of one trader from those of another. Thus purely descriptive or geographical marks will not be protected in passing off, unless the claimant can show that they have acquired through use a "secondary meaning" indicating that the goods in question are connected with him.[64]

Misrepresentation

13–41 The tort of passing off depends on showing that the defendant has adopted a mark or get-up which is so similar to the claimant's mark or get-up as to be likely to cause confusion amongst the claimant's customers or potential customers. The court will not require actual evidence of confusion amongst such customers, but such evidence can be decisive if it can be found. The sort of confusion required is not limited to actual confusion between the claimant's and defendant's businesses or products. It also covers a confused belief that the defendant's business has been authorised or licensed by the claimant. A passing off action can arise even where there is no actual "common field of activity" between claimant and defendant (though it will be more difficult to establish likelihood of confusion in such a case).[65]

There is no need to establish a deliberate intention to deceive, or even reckless behaviour, on the part of the defendant. Fraud is not an element in the tort of passing off. However, the courts have stated on a number of occasions that evidence of deliberate intent will make a finding of passing off more likely (the reasoning being that if someone has set out to deceive, the court should not readily find that he has failed to achieve his objective).

Damage

13–42 The claimant must show that it has suffered (or is likely to suffer) damage to its business by reason of the misrepresentation. This normally follows as a matter of course once misrepresentation is established, in that confusion amongst customers will almost inevitably result in loss of custom, or at least "dilution" of the value of the brand.

Ownership of goodwill

13–43 The goodwill associated with a particular mark or get-up is treated as a "property right". It is generally owned by the person who exercises control over the

---

[62] See for instance *Anheuser-Busch v. Budejovicky Budvar* [1984] F.S.R. 413.
[63] See *A/B Manus v. RJ Fullwood & Bland* [1949] Ch. 208.
[64] *Reddaway v. Banham* [1896] A.C. 199; *Cellular Clothing v. Maxton* [1899] R.P.C. 326.
[65] See *Eastman v. John Griffiths* [1898] R.P.C. 105 and *Lego System v. Lego M Lemelstrich* [1983] F.S.R. 155 ("Kodak" for bicycles and "Lego" for garden sprinklers: actions succeeded); *Rolls Razors v. Rolls (Lighters)* [1949] R.P.C. 137 ("Rolls" for lighters: action failed).

use of the mark, and in particular over the nature and quality of the products sold under it. It may be transferred by assignment.[66] However, care needs to be taken when assigning goodwill without the rest of the business with which the mark has always been associated. There is a risk in some cases that the public will continue to associate the mark with the old business, at least until they are educated to the contrary.

The law of passing off recognises the possibility of "shared goodwill" where a particular brand is used by a number of independent companies to define a particular quality of goods or a particular geographical origin. A number of successful cases have been brought by the Champagne manufacturers to protect their shared goodwill in the Champagne brand against dilution of that brand by others.[67]

---

[66] See Chapter 22 on Assignments.
[67] The most recent example is the "elderflower champagne" case of *Taitinger v. Allbev* [1993] F.S.R. 641.

CHAPTER 14

OWNERSHIP OF INTELLECTUAL PROPERTY

Iain Purvis
Barrister, 11, South Square, Gray's Inn

INTRODUCTION

14–01   This Chapter is concerned solely with the question of "first ownership" of intellectual property rights. The legal aspects of the transfer of intellectual property rights are dealt with in Chapter 22 on assignments.

This Chapter deals with issues relating to the ownership of patents, registered designs, unregistered designs and copyright, with a brief reference to database and topography rights.

THE BASIC RULES

14–02   The basic rules as to ownership of the principal statutory intellectual property rights are, unsurprisingly, set out in the relevant statutes. They are generally straightforward, recognising that the person responsible for creating the subject matter of the right in question is entitled to ownership of it. However, this is subject to a number of qualifications. In particular, where the right in question is generated by an employee in the course of his employment, or pursuant to a commission. Furthermore, the statutory rules commonly give way to the rules of equity: where the courts step in to reflect the true intentions of contracting parties or to enforce "fairness". The position in equity is dealt with as a separate section, since it applies equally to all intellectual property rights.

PATENTS

Introduction

14–03   The primary entitlement to be granted a patent is that of the inventor or inventors.[1]

Inventorship

The invention

14–04   Before deciding on the identity of the inventor, the "invention" must first of all be defined. In the vast majority of cases in which the question of inventorship arises, the invention is defined by the "claims" of the patent or the application.[2] However, if the question of inventorship arises before a patent has actually been applied for, some attempt must be made to define the nature of the invention before embarking on the enquiry.

Who made the invention?

14–05   It may be more difficult than it seems to identify the inventor simply by considering the claims of the patent. First of all, the claims are likely to contain a number

---

[1] Patents Act 1977, s.7(2)(a).
[2] *ibid.* s.125 — see Patents Chapter 8.

of different elements. Merely contributing an element to be found in the claim does not of itself give rise to a claim to be one of the inventors.[3] For instance, many of the elements of the claim are likely to be commonplace in the field. Their presence in the claim may be simply to define the known art upon which the patent seeks to improve.[4] Or they may be obvious in themselves.

In order to identify the inventor, one must therefore go behind the wording of the claims to find the "inventive concept" of the patent and the person responsible for it.[5] This amounts to deciding what gives the patent its inventive improvement over the prior art. Not everyone who was involved in the process which led to the making of the invention can claim responsibility for this. Many people may have made contributions which were invaluable — for instance the researcher who discovers a long forgotten prior art process in an obscure document, which the patented invention then improves upon. However, they cannot be described as "inventors".

Joint inventorship

14–06 Even once the inventive concept is established, it is commonly the case that responsibility for it cannot be claimed by a single individual. The invention may have arisen out of the process of "bouncing ideas around". Alternatively there may be two or more different inventive concepts within the patent specification.[6] In either of these cases, there are joint inventors, with joint prima facie entitlement to apply for a patent.

Employment

14–07 The Patents Act contains detailed provisions for inventions made in the course of employment. They apply only to employees employed mainly in the United Kingdom, and employees with no main place of employment or with no ascertainable place of employment whose employer has a place of business in the United Kingdom to which the employee is attached[7]. They apply to any inventions whether or not they result in a patent.

No definition of "employment" is given, but it may be assumed that the same approach will be taken in this area as in other areas of the law.[8] Two common problems arise: the status of directors, and whether an individual is an employee or an "independent contractor".

Directors

14–08 The mere existence of a directorship is not an indication that a person is employed by a company. Many directors are not employed in the legal sense, even though they may work full time in the company's interests. The question of employment falls to be decided by considering various factors. In particular:

(a) Is there a written contract setting out the director's duties and remuneration?

---

[3] *Henry Brothers (Magherafelt) Limited v. Ministry of Defence* [1997] RPC 693.
[4] The "pre-characterising portion" of the claim — see Patents Chapter 8 at 8–10.
[5] See *Henry Brothers*, n.3 above.
[6] For instance in different claims.
[7] s.43(2).
[8] In a number of authorities, generally concerned with tax law and employment protection legislation, the courts have established the factors which will be taken into account in determining whether someone is employed or not.

(b) Is he obliged to attend the workplace at any particular time, or to work any particular number of hours per week?

(c) Is he paid a fixed salary, or is he paid simply according to the hours he works, or paid a percentage of the profits of the company?

It should be borne in mind that even if a director is not employed by his company, an invention made by the director may nonetheless belong to the company as a matter of equity (see separate section on equitable title at paragraph 14–42 below).

Independent contractors

14–09 The other position which gives rise to difficulties is that of the independent contractor. The mere description of someone as a "contractor" rather than an "employee" is not decisive. The position needs to be looked at as a matter of substance rather than form.

The old established test for distinguishing between contractors and employees used to be whether the company was entitled to dictate to the individual in question not only what he was to do but how he was to do it, sometimes called the "control test". However, this is generally regarded as somewhat outdated. The courts will nowadays consider the more general question of whether the individual can be seen as an integral part of the business of the employer, or whether he was working on his own account.[9] The following factors are often considered to be relevant:

(a) Does the company pay N.I. contributions or deduct PAYE on behalf of the individual?

(b) Does the company provide the tools of the individual's trade, and those who assist him in doing his job?

(c) Is the individual obliged to work any particular number of hours per week, or restricted in his place of work?

(d) Is he paid a fixed salary as opposed to an hourly rate?

(e) Will he share in the profit made by the company from the job he is doing?

(f) Is he closely managed as to how he works, and generally subject to the "discipline" of the company?

The general rule

14–10 The general rule is that an invention made by an employee shall be taken to belong to the employee save in the cases set out below.[10]

Normal or specifically assigned duties of the employee

14–11 Under section 39(1)(a), where an invention is made in the course of the normal duties of the employee or duties specifically assigned to him, the invention will belong to the employer *provided that* an invention "might reasonably be expected to result" from the carrying out of those duties.

---

[9] See *Lee Ting Sang v. Chung Chi-Keung* [1990] 2 A.C. 374.
[10] Patents Act, s.39(2).

The key questions here are the type of duties which the employee was carrying out when the invention was made, and the position or status of the employee.

It is fairly easy to identify types of employee or types of duties which would usually satisfy section 39(1)(a) — researchers, engineers, or anyone whose job involves the resolution of technical problems. There are more difficulties with those working in technical departments whose job would not normally give rise to an invention: draftsmen for instance, working to detailed instructions from their superiors. Each case must be treated on its own facts.

Special obligations

14–12 The other circumstance in which an invention will belong to the employer is set out in section 39(1)(b): where it is made in the course of the duties, of the employee and those duties are such that the employee had "a special obligation to further the interests of the employer's undertaking".

This is intended to catch the senior employee whose invention cannot be said to have been made in the course of his "normal" duties, or duties "specifically assigned to him", or where it would not be reasonable to expect him to make inventions. If the employee's position is nonetheless so senior that he is under a fiduciary obligation to the company, requiring him not to profit from his duties at the company's expense, the law will reflect this obligation by vesting the invention in the company. This subsection will obviously apply to company directors, and also to those in very senior management positions.[11]

Contracting out

14–13 Under section 42 of the Patents Act, any contract which purports to diminish the rights of an employee (to whom the above rules apply) in an invention is unenforceable to that extent. Thus there is little point in making any provision in contracts of employment with such an employee as to ownership of inventions, unless it is intended to *increase* the rights of the employee.

Disputes about ownership

14–14 Any dispute about the ownership of an invention or an application for a patent or a granted patent may be brought before the Patent Office under sections 8 or 37 of the Patents Act 1977. Alternatively it may be brought before the High Court, which has an inherent jurisdiction to resolve such issues. Disputes about entitlement to European and foreign patents may (assuming that the United Kingdom has jurisdiction under the Brussels Convention) be brought under section 12.

In all such proceedings, there is a presumption (which may be rebutted) that the applicant for a patent is the true owner[12]. Thus the burden of proof is on the person seeking to rectify the position.

Powers of the Patent Office

14–15 The Patent Office has a number of different powers which it may exercise in the event of such a dispute. These are quite complex, and the details are beyond the scope of this book. However, in summary, where a patent has been applied for and not granted it may:

---

[11] See *Staeng's Patent* [1996] R.P.C. 183.
[12] s.7(4).

(a) order that the existing application proceed in the name of the true owner; or

(b) reject the application altogether.

Where a patent has been granted, it may:

(a) transfer the patent to the true proprietor and rectify the Register accordingly; or

(b) revoke the patent[13]

There are also powers to enable the true proprietor to make his own application for a patent, claiming the benefit of the filing date of the application by the wrong party. The court may also exercise these powers, or direct the Patent Office to do so.[14]

REGISTERED DESIGNS

14–16 The proprietor of a design (and thus the person entitled to register it) under the Registered Designs Act is the "author" of that design.[15] This is subject to two exceptions: commissioned works and works created in the course of employment.

Authorship

14–17 The author of a design is defined in the RDA[16] as the person who creates it. Obviously, in the usual case, this will be the person who created the original embodiment of the design, whether in the form of drawings or a prototype article made to the design. However, there are exceptions to this.

Amanuensis

14–18 A draftsman who responds to very detailed verbal or written instructions to produce a design and has no creative input of his own can be viewed as a mere amanuensis — no more an author of the design than a playwright's secretary is the author of his plays. In such a case, the real author is the person providing the detailed instructions.

Perhaps more commonly, this situation may establish a case of joint authorship: where the generality of the design is created by the person providing the instructions, but the details are those of the draftsman.[17]

Disputed authorship

14–19 In determining whether any particular person qualifies as an author of a design, the courts will consider the exact nature of his contribution. Disputes of this kind tend to arise more in copyright than registered design cases, and this issue is discussed in detail in the section on copyright below at 14–33.

---

[13] Under s.72(1)(b).
[14] Although the court may not exercise its jurisdiction more than two years after the grant of the patent, unless the patentee knew at the time of grant or at the time the patent was transferred to him that he was not entitled — s.37(9).
[15] RDA 1949 (as amended), s.2(1).
[16] s.2(3).
[17] See *Cala Homes v. Alfred McAlpine* [1995] F.S.R. 818.

EMPLOYMENT

14–20 The employment provisions of the RDA are much less complex than those of the Patents Act and much more favourable to the employer. The RDA simply recites that where a design is created by an employee in the course of his employment, his employer shall be treated as the original proprietor of the design. There is no provision (as is found in the CDPA for copyright — though not for design right) that this rule is subject to agreement to the contrary.

Questions of whether or not someone is employed fall to be resolved according to the same principles discussed above in relation to patent law (paragraphs 14–07 to 14–09).

Course of employment

14–21 Work can be within the course of employment even where it falls outside the scope of the normal duties of the employee. Modern employment relationships tend to be fairly flexible, and an employee will expect to be asked to perform tasks outside his usual run of activities. The test for judging whether any particular activity was outside the course of employment has been said to be whether, if the employee had refused to carry out that activity, he would have been in breach of his contract of employment.[18]

So far as the law of registered designs is concerned, this issue is unlikely to arise very often.

Work done outside working hours

14–22 The most common problem in this area is with designs created outside the office on the employee's own initiative and in his "spare time".

It may be difficult for an employee to establish that a work was entirely created in "spare time": not least because his contract of employment may not lay down any specific working times, and he may be expected to work where necessary outside normal office hours. For similar reasons, the mere fact that a work was created at home rather than in the office may not be decisive.

The determinative factor here is likely to be the nature of the design in question. If it is an improvement on something which the employer already produces, or is directly relevant to a project which the employee is working on for his employer, the courts are likely to find that the work was done in the course of employment. Similarly, if it is obviously directly competitive with an existing product of the employer. In these cases, were the employee to keep the design for himself and seek to exploit it for his own ends, he would be acting in breach of his duty of fidelity to his employer. The courts are unlikely to be prepared to listen to an argument from the employee which would inevitably lead to that conclusion.

On the other hand, there is nothing to prevent an employee from pursuing at home his own projects unrelated to his employer's business, and to keep the proceeds of that work for himself.

---

[18] *Copinger and Skone James on Copyright*, para. 5–16, citing *Stevenson Jordan v. McDonald* [1952] 1 T.L.R. 101.

COMMISSIONED WORKS

14–23 A design created in pursuance of a commission for money or moneys worth belongs to the commissioner.[19]

Commission

14–24 Commissioning requires a specific order or instruction to create the work in issue.[20] The classic case is where a manufacturing company goes to an independent designer to create a new design for a product. The designer is requested to produce the design and is paid for it.

There is an important distinction between this situation and the case where a company orders a product from a manufacturer to meet certain specific requirements. This is likely to require the manufacturer to engage in some design work. But the design work itself is not commissioned or paid for. The only payment is for the finished goods. In such circumstances, the design is not a commissioned work and (subject to the intervention of equity, and implied licences) the rights in any registered design belong to the designer.

Money or money's worth

14–25 This simply excludes the case where an order is placed for a design but no consideration is provided. This would be an unusual case, since even if no sum was actually agreed the law would commonly intervene to fix a reasonable remuneration on a *quantum meruit* basis.

Works generated by computer

14–26 Under section 2(4) of the RDA specific provision is made for works generated by computer in circumstances such that there is no human author. This must be fairly rare. It would clearly not apply to designs produced on a CAD system for instance, since in those cases there clearly is a human author.

For works falling within this sub-section, the author is taken to be the person by whom the arrangements necessary for the creation of the design are made.

UNREGISTERED DESIGN RIGHT

14–27 The rules are almost identical to those set out above in relation to registered designs. The first owner of an unregistered design is the "designer", defined as "the person who creates the design".[21]

The position as to commissioned works and works created in the course of employment is the same as set out above in relation to registered designs. It should be noted that the question of employment and commissioning can assume an increased significance in the field of design right. This is because a work can qualify for protection by reference to the nationality or habitual residence either of the designer, or (in the case of works created in the course of employment) the employer, or (in the case of commissioned works) the commissioner. Thus, for instance, a work which would otherwise be excluded from protection because the designer was American may nonetheless

---

[19] RDA, s.2(1A). Slightly different wording applies to designs created prior to August 1, 1989.
[20] See *Plix Products v. Winstone* [1986] F.S.R. 63.
[21] CDPA 1988, s.214(1).

qualify if it can be shown that the design was commissioned by a United Kingdom company.

First marketing

14–28   The only significant difference from the rules relating to Registered Designs is in the case of designs which do not qualify for protection on the basis of nationality or habitual residence of the designer, the employer or the commissioner, but only on the basis of the nationality or habitual residence of the person who first markets the articles made to the design (section 220).

In such a case, ownership of the design vests solely with the person who first markets the articles.

Joint designs

14–29   The CDPA, unlike the RDA, makes specific provision for the case of joint designers. A "joint design" is defined as a design produced by the collaboration of two designers in which the contribution of each is not distinct from that of the other or others.[22]

This test is the same as that for joint authorship of copyright works (see below). If the contribution to the design is distinct, such that one can divide the overall design into two or more, each created by a different person, then the design must be treated as two or more separate designs.

Prospective ownership

14–30   The CDPA makes specific provision for the case where a design is "assigned in advance" by agreement with the prospective owner of the right.[23] Thus if it is anticipated that a work protectable by design right may come into being in the course of development of a product for a particular customer, and the circumstances are not sufficient to amount to "commissioning", the parties may agree in advance that any design right which comes into being will belong to the customer. This agreement must be in writing and signed by the prospective owner.

The effect of section 223 is essentially to give legal recognition to what would have been the position in equity anyway. The right is deemed to have been assigned at the moment it came into existence, *provided that* the assignee would have been entitled to require the prospective owner to assign it to him (*i.e.* by obtaining an order for specific enforcement of the agreement). It will thus take effect against any successor in title to the prospective owner with the exception of a bona fide purchaser without notice.

COPYRIGHT

14–31   The law of ownership of copyright follows very much the same principles we have seen above in the case of registered and unregistered designs.

Authorship

14–32   As with registered designs, the first owner of the copyright in a work is the "author" of that work, author being defined as the person who creates the work.[24]

---

[22]   CDPA, s.259(1).
[23]   CDPA, s.223.
[24]   CDPA, s.9(1).

Degree of involvement required to be an author

14–33 Copyright cases commonly give rise to disputes about authorship. A typical scenario is where a work is created by a company with a particular customer in mind. That customer may have been closely involved in the development process, and made numerous suggestions as to the design. At what point is this contribution enough to give rise to a claim to be a joint author?

The key question to be considered in any assessment of whether a contribution is sufficient to amount to authorship is the nature of the skill and labour being contributed. This must be the kind of skill and labour which the copyright in question is intended to protect. A good example of a contribution which was not of the right kind is provided by the modern case of *Fylde Microsystems v. Key Radio Systems* [1998] F.S.R. 449, concerning copyright in a computer program. There, the defendant, who was the intended customer for the program, claimed joint authorship with the writer of the software on the basis of a close involvement at a technically advanced level during the course of its development. There was no doubt that the defendant's contribution had been very valuable. It consisted of specifying what the software had to do, providing data to enable the software to communicate with the relevant hardware, and reporting faults (including identifying the causes of some of them). However, despite the importance of this contribution, it was held not to be the kind of skill and labour which is protected by copyright in computer programs. This is the skill and labour involved in actually writing the source code. All the skill and labour in writing the source code (including creating the structure of the programs) was provided by the claimant, who was thus held to be the sole author of the copyright work.

In the case of a conventional literary work such as a book, copyright will protect the skill and labour which went into writing it. This includes constructing the plot and structure of the book. However, copyright gives no protection to mere ideas, so merely contributing the basic idea for "a book about the crusades, concentrating on the siege of Acre" would not give rise to a claim of joint authorship in the book which resulted. Nor would the contribution of a proof reader in checking the work for errors. In the case of an artistic work such as a drawing, one can be a joint author without actually setting pencil to paper,[25] but again the contribution must go beyond the mere provision of ideas, and must actually affect the way the ideas are expressed on the paper.

In summary, an author must have actually created something protected by copyright which found its way into the finished work. Joint authors "share responsibility" for the expression of the ideas embodied in the finished work.[26]

Special rules as to authorship

14–34 Section 9(2) lays down certain specific rules as to who qualifies as the author in the case of particular categories of work: sound recordings and films, broadcasts, cable programmes and typographical arrangements. These are summarised in the copyright Chapter 11 at 11–34.

Computer generated works

14–35 The same rules apply as for unregistered design right (see above).[27]

---

[25] See *Cala Homes*, n.17 above.
[26] See *Robin Ray v. Classic FM* [1998] F.S.R. 622.
[27] CDPA, s.9(3).

Joint authorship

14–36   Again, the position is the same as for design right: the contribution of each author must be "not distinct" from that of the other. If it is distinct, then there must be two separate works.[28]

Employment

14–37   The position is the same as for registered designs, save that ownership by the employer is specifically stated in the Act to be "subject to any agreement to the contrary".[29]

Commissioned works

14–38   Unlike registered designs and unregistered design right, there is no provision in the CDPA that the commissioner is the first owner of copyright in a work. His rights of ownership, if any, will arise entirely as a matter of equity.

Prospective ownership

14–39   The same rules apply in relation to prospective ownership of future copyright as apply in relation to design right (see paragraph 14–30 above).[30]

DATABASE RIGHT

14–40   The owner of a database right is the "maker" of the database.[31]

"Maker" is defined as the "person who takes the initiative in obtaining, verifying or presenting the contents of a database and assumes the risks of investing in that obtaining, verification or presentation".[32]

The Regulations provide for ownership by employers in the case of databases made in the course of employment in very similar terms to those used for copyright and unregistered designs under the CDPA.[33] There is no provision for ownership by commissioning, so the commissioner is left to his remedies in equity (see paragraphs 14–42 to 14–50 below).

TOPOGRAPHY RIGHTS

14–41   The statutory rules applying to ownership of topography rights differ from those applying to design rights generally in that the rules vesting design right in commissioners and employers are expressed to be subject to "any agreement in writing to the contrary".[34]

OWNERSHIP IN EQUITY

14–42   The circumstances in which equitable ownership may become separated from legal ownership are many and various, but can usefully be divided into three. First, the

---

[28] CDPA, s.10(1).
[29] CDPA, s.11(2).
[30] CDPA, s.91.
[31] Regulation 15.
[32] Regulation 14(1).
[33] Regulation 14(2).
[34] Design Right (Semiconductor Topographies) Regulations 1989, reg. 5.

deliberate creation of an equitable interest by the parties. Second, the imposition of a trust by the court by reason of a fiduciary relationship between the parties. Third, the implication of a trust by the courts to reflect the intention of the parties.

The first of these is very unusual in the case of intellectual property rights and is outside the scope of this Chapter. The second and third commonly arise in commercial situations.

FIDUCIARY RELATIONSHIP BETWEEN THE PARTIES

14–43   Where an intellectual property right is created by a trustee in the context of a fiduciary relationship, the court will generally imply a trust over the asset on the basis that it will not permit a trustee to profit from his position personally.

Directorship

14–44   The most common commercial situation in which this applies is that of the company director. We have seen above that where a director is not strictly an "employee" of his company, the statutory rules do not give legal title to an intellectual property right created by him to the company, even where it was clearly created in the context of and in the course of his directorship. English law deals with this anomaly by establishing an implied trust over the intellectual property right in favour of the company.[35] As explained above, the rationale for this is that personal ownership of the right would amount to an impermissible profit for the director at the expense of the company to which he owes a fiduciary duty. The effect of the implied trust is to leave the "legal" title in the hands of the director, whilst giving the company "equitable" title. The practical impact of this is explained in paragraph 14–50 below.

Partnership

14–45   The same applies in the case of a work created by a partner in the course of a partnership business.[36]

INTENTION OF THE PARTIES

14–46   Equity will often imply a trust over an intellectual property right in order to give effect to the intentions of the parties where the formalities required to transfer the intellectual property right at law have not been complied with. For instance, it will usually treat an agreement to assign as having transferred the beneficial title to the intellectual property right to the assignee even before the formal assignment of the legal interest has taken place. More significantly, it will intervene in the same way to give effect to the unexpressed intentions of the parties where an intellectual property right has been created in the course of a transaction between them, as explained below.

Commissioned works and works created by independent contractors

14–47   We have seen how that in the case of registered designs and unregistered design rights the relevant statute grants title to the commissioner, but that there is no such provision in the case of patents or copyright. Obviously, if there is an express

---

[35] *Antocks Lairn v. Bloohn* [1971] F.S.R. 490.
[36] See for instance *Ibcos Computers* [1994] F.S.R. 275.

term in the commissioning contract dealing with the question of title to intellectual property rights created in the course of performing it, then this is usually determinative. If it provides that title to intellectual property rights will vest in the commissioner, then copyright will probably pass automatically (by virtue of the "assignment of future copyright" provisions in the CDPA) and equity will give effect to the obligation by imposing a trust over any resulting patent rights or inventions.[37] The same applies in the case of intellectual property rights created by an independent contractor in the course of his work for a particular client.[38]

Where (as is very common) there is no such express term, the question is whether a term vesting the intellectual property rights in the commissioner should be implied: see paragraph 14-48 below.

When an agreement to assign will be implied

14-48 Such a term can only be implied where the parties must be taken by their conduct and the circumstances of the case to have intended that all relevant rights would be owned by the commissioner/client, or where the term is "necessary" to give business efficacy to the contract.[39]

There tend to be two principal factors:

(a) whether the commissioner/client paid for the creation of the work which gave rise to the intellectual property right;

(b) whether it was the intention of the parties that the commissioner/client would be exploiting the work in such a way that he would need all relevant rights to be vested in his own name: for instance to take action against infringers, or to license third parties.

Where both these factors apply, the courts have tended to imply a term that the intellectual property right be assigned to the commissioner/client. The necessary consequence of this is that the commissioner/client owns the right in equity.

An alternative: the implied licence

14-49 It should be emphasised that the commissioning relationship is not an "all or nothing" situation. The courts have a range of options available by way of implied terms, and do not necessarily need to go so far as implying an agreement to assign.

The most common alternative is an implied royalty-free licence. This may be exclusive or non-exclusive depending on all the facts of the case. In the recent case of *Robin Ray v. Classic FM*,[40] the court emphasised that terms should only be implied where they were "necessary" to give effect to the intentions of the parties, or to give business efficacy to the contract. Since the licence solution was a less drastic one than a complete assignment of all relevant rights, it would be preferred unless for some reason a complete assignment was "necessary". However, this was an unusual case, where there was no obvious need for the commissioner to exclude the author from exploiting the work or to be able to enforce the copyright against third parties. In the normal case, where it is intended that the commissioner/client will need to do both these things, an implied assignment is the more likely result.

---

[37] According to the maxim "equity treats as being done that which ought to be done".
[38] See for example *Massine v. De Basil* [1936–45] M.C.C. 223.
[39] See *Robin Ray v. Classic FM*, n.26 above.
[40] *ibid.*

Effect of equitable assignment

14–50 The effect of an implied term of this kind is technically that the ownership of the right is split between law and equity. One party owns the property at law (since the statutory requirements for assignment — requiring a document in writing signed by the assignor — have not been complied with) but the other owns the property in equity.

This conflict is, however, more apparent than real. In practice, all the useful rights belong to the party who owns the right in equity (the "beneficial owner"). In particular, it has the right to call upon the legal owner to transfer his legal title by formal assignment, and to force him to do so if he refuses. In the case of registered rights, this should be done as soon as possible in order that the Register reflects the true position.[41]

The legal owner may take proceedings for infringement, but, in the event of successful litigation, any damages recovered would be held on trust for the beneficial owner.

The beneficial owner may commence proceedings in its own name even before perfecting its rights to the legal title, and will, where appropriate, be granted interim relief.[42] However, it needs to join the legal owner in the action at some stage before final judgment (or else perfect its own title by assignment).[43]

It is unclear whether the legal owner has a defence against the beneficial owner to an action for infringement in connection with exploitation of the intellectual property right before assignment. However, the point may be academic, since he would have to account for any profits he made and pay for any damage caused to the beneficial owner as a matter of equity.[44]

---

[41] Neither the Register of Patents nor the Register of Designs is permitted to bear any notice of the existence of a trust (PA 1977, s.32(3) and RDA 1949, s.17(2)), and in the case of patents no monetary relief is available unless the change of ownership has been registered (with a six month grace period).

[42] This applies to all intellectual property rights, including patents, except for registered designs where the only party entitled to take action under the RDA is the "registered proprietor" — see s.7.

[43] See *Performing Right Society v. London Theatre of Varieties* [1924] A.C. 1. It should also be borne in mind that in the case of registered designs, only the registered owner is permitted to take action.

[44] See *Ibcos*, n.36 above where no argument was raised by the legal owner on this point.

CHAPTER 15

INTRODUCTION TO COMPETITION LAW

Mark Brealey, Barrister, Brick Court Chambers and Kelyn Bacon, Barrister, Brick Court Chambers

E.C. COMPETITION LAW

15–01 The application of E.C. competition law and the exercise of intellectual property rights in the European Union give rise to various competing interests. An awareness of the relevant objectives is important, as they guide the application of the legal rules and determine which rule takes precedence.

15–02 The objectives of E.C. competition law are not expressed as often as might be expected. It is, however, possible to discern three main objectives. The first is to *maximise consumer welfare* by achieving productive and allocative efficiency. Productive efficiency aims to ensure that companies innovate and reduce cost. Allocative efficiency aims to ensure the most efficient allocation of resources. The objective seeks to prevent companies, usually dominant ones, from restricting output, but maintaining profit by increasing prices. If output is restricted (and the monopolist makes the same profit by producing less), consumers are deprived of goods that they want, and the economy does not perform efficiently. Maximising consumer welfare is not concerned with the allocation of wealth, but with its creation. By contrast, the second objective, the *protection of the consumer* is concerned with the allocation of wealth. It is concerned to ensure that consumers are not exploited, for example, by excessive prices or disadvantaged by discriminatory prices. Included in this second objective is the protection of traders who are in weaker competitive positions. In order to protect the consumer, E.C. competition law often protects the small and medium sized company. For example, a policy of predation by a dominant company may act to the advantage of the dominant company, but not in the long term interests of the consumer who finds that the choice of supplier has been restricted or that prices have increased following the exit of the smaller competitor.

15–03 The twin objectives of maximising consumer welfare and protecting the consumer are classic objectives of most competition rules in market economies, although the extent to which competition law interferes with the business affairs of companies to achieve these objectives is a difficult balancing act. As the E.C. Commission has acknowledged: "Competition policy must strike a balance between reliance on market forces to maintain competition and selective intervention where necessary."[1]

15–04 The degree of intervention may differ depending on the national jurisdiction or on the mood prevailing in the jurisdiction at the time. For example, since the 1970s, the Chicago School of economists in the United States has advocated a more Darwinian approach.[2] This policy of "let the strongest survive" recognises the first objective of maximising consumer welfare, but largely ignores the second, par-

---

[1] 14th Competition Report on Competition Policy, p. 1.

[2] Perhaps the best known advocate was Bork "The Antitrust Paradox: a Policy at war with itself."

ticularly the protection of smaller competitors. The acceptance of this policy by the courts in effect changed the application of the competition rules. Vertical restraints suddenly received a far more relaxed treatment. The *per se* illegality approach to restrictions of intra-brand competition adopted by the United States Supreme Court in *Schwinn*[2a] was abandoned in *Sylvania*.[2b] Although E.C. competition law does not adopt such a laissez-faire approach, there has been an undoubted shift along the United States lines. Until the late 1990s, vertical restraints were regarded with suspicion and tightly regulated by block exemptions. This policy has changed and, absent certain hard core restrictions, E.C. competition law will not interfere unless the supplier has a fair degree of market power. The competition enforcement policy prevailing in Brussels at present is aimed squarely at traditional cartels (market sharing; price fixing, etc.).

15–05 The fact that E.C. competition policy still regards export bans as a hard core restriction capable of unlawfully distorting competition reflects the third and last objective of the E.C. competition objective: *integration*. Over 100 years ago this was a loose objective of United States antitrust law when the formation of trusts between competitors, particularly the railway companies, to hike prices affected the transport of goods from east to west. It is no longer an objective in that jurisdiction. By contrast, the formation of a single market remains high on the European agenda and has on occasions taken precedence over the first two objectives.[3] For how long this objective will be given star billing is questionable. As more companies operate on a pan-European basis, as more consumers accept without question goods manufactured in other Member States, as more non-economic barriers fall, there will be less need to ensure that companies do not erect economic barriers. Nevertheless the principle expressed by the Court of Justice in 1966 in *Consten & Grundig*[4] will remain valid for the foreseeable future.

> "An agreement . . . which might tend to restore the national divisions in trade between Member States might be such as to frustrate the most fundamental object of the Community. The Treaty, whose preamble and content aim at abolishing the barriers between States . . . could not allow undertakings to reconstruct such barriers. Article 85(1) is designed to pursue this aim . . ."

15–06 The Treaty establishing the European Union impacts on the exercise of intellectual property rights in two principal ways. First, it subjects their exercise to the E.C. Competition rules. If the exercise of the right is consensual, Article 81 applies. If the right is exercised by a dominant company, Article 82 applies. In both instances there may be a conflict between the objective of protecting intellectual property rights (*e.g.*, a reward for creative effort) on the one hand and the need to maximise consumer welfare, the protection of the consumer and the integration of the European market on the other. At the expense of generalising, any exercise of an intellectual property right which is contrary to the first objective (maximising consumer welfare[5]) and the third objective (integration) is likely to be condemned. The outcome of any conflict with the second objective (protection of the consumer)

[2a] *United States v. Arnold Schwinn & Co.* 388 U.S. 365, 87 S. Ct. 1856 (1967).
[2b] *Continental T.V. Inc v. GTE Sylvania Inc*, 433 U.S. 36, 97 S. Ct. 2549 (1977).
[3] *e.g. Distillers* [1980] E.C.R. 2229.
[4] [1966] E.C.R 299, 340.
[5] Which is primarily concerned with dominance.

is always difficult to predict, but there is now a substantial body of case law that assists. Much of the case law and legislation referred to in Section 1 of Chapter 16 is relevant to this area of possible conflict.

15–07 The second major impact that the Treaty has on intellectual property rights concerns the extent to which reliance may be placed on such rights to restrict imports. Articles 28 and 30 of the Treaty provide for the free movement of goods between Member States. As will be seen in Section 2 of Chapter 16 these articles introduce the principle of exhaustion of intellectual property rights within the European Union.

UNITED KINGDOM COMPETITION LAW

15–08 The Competition Act 1998, Chapters I and II, mirror the prohibitions contained in Articles 81 and 82 respectively. The application of the United Kingdom rules on competition will inevitably pursue the first two objectives of the E.C. rules; namely, *maximising consumer welfare* and *protecting the consumer*. The objective of integration is, however, absent in the rules. There is no need to integrate the United Kindgom. Although the Act refers to an effect on trade within the United Kingdom, this difference may have an important impact in practice as to how the Act is applied. For example, an export ban on goods being transported from London to the Midlands will not be condemned as conflicting with an objective relating to integration.

UNITED STATES ANTITRUST LAW

15–09 In the United States, competition law is known as "antitrust law". As mentioned above, it covers similar subject matter, but currently places different emphasis on the protection of the consumer. In common with United Kingdom competition law, integration is not a policy objective.

CHAPTER 16

E.C. COMPETITION LAW AND INTELLECTUAL PROPERTY[1]

Mark Brealey, Barrister, Brick Court Chambers, and Kelyn Bacon, Barrister, Brick Court Chambers

SECTION 1: COMPETITION LAW

INTRODUCTION

16–01 The Court of Justice of the European Communities has consistently stated that the competition rules of the Treaty will not affect the *existence* of intellectual property rights and that the securing of such rights, in itself, cannot be regarded as an elimination of competition.[2] Neither will the exercise of intellectual property rights, *per se*, constitute an infringement of the competition rules.[3]

On the other hand, the Court has continued to maintain that the Community competition rules may not be frustrated by the improper use of intellectual property.[4] Accordingly, the Treaty may prohibit the use of intellectual property as the "instrument" for the abuse of a dominant position,[5] or the "purpose, the means or the result" of a restrictive agreement.[6]

The following sections will therefore consider the impact of the Community competition rules on the use of intellectual property. Three principal issues are relevant: Agreements under Article 81 of the Treaty[7] (Section A), the rules on mergers and joint ventures (Section B), and the abuse of a dominant position contrary to Article 82[8] by exercising intellectual property (Section C). Section D will address the issue of competition procedure.

A. ARTICLE 81 AND IP AGREEMENTS

Prohibited agreements

Prohibition

16–02 Article 81(1) contains the basic prohibition on restrictive agreements, and provides that:

"The following shall be prohibited as incompatible with the common market: all agreements between undertakings, decisions by associations of undertakings and

---

[1] Regard should also be had to the following works: Bellamy & Child, *Common Market Law of Competition* (Sweet & Maxwell); Tritton, *Intellectual property in Europe* (Sweet & Maxwell); Green & Robertson, *Commercial Agreements and Competition Law* (Kluwer); Butterworths, *Competition Law*, R. Whish and P. Freeman, eds; and Valentine Korah, *Technology Transfer Agreements and the E.C. Competition Rules* (Clarendon Press).
[2] Case 53/87 *CICRA et Maxicar v. Renault* [1988] E.C.R. 6039.
[3] Case 24/67 *Parke Davis v. Probel* [1968] E.C.R. 55.
[4] Cases 56 & 58/64 *Consten & Grundig v. Commission* [1966] E.C.R. 299; *Parke Davis*, above n. 3.
[5] Case 102/77 *Hoffmann-La Roche v. Centrafarm* [1978] E.C.R. 1139.
[6] Case 262/81 *"Coditel II"* [1982] E.C.R. 3381.
[7] Ex Article 85.
[8] Ex Article 86.

concerted practices which may affect trade between Member States and which have as their object or effect the prevention, restriction or distortion of competition within the common market, and in particular those which:

(a)   directly or indirectly fix purchase or selling prices or any other trading conditions;
(b)   limit or control production, markets, technical development, or investment;
(c)   share markets or sources of supply;
(d)   apply dissimilar conditions to equivalent transactions with other trading parties, thereby placing them at a competitive disadvantage;
(e)   make the conclusion of contracts subject to acceptance by the other parties of supplementary obligations which, by their nature or according to commercial useage, have no connection with the subject of such contracts."

Extra-territorial effect

16–03   The prohibition applies not only to agreements between undertakings where one or more of the parties is established within a Community Member State, but extends to agreements between undertakings established in third countries which have the object or effect of restricting competition within the Community. Otherwise, the Court held in the *Woodpulp* cases, if the applicability of the competition law prohibitions depended on the place where the agreement, decision or concerted practice was formed, the undertakings would have an easy means of evading those prohibitions. The decisive factor is thus the place where the agreement is *implemented*.[9] In that case, the agreement was implemented in the Community since the undertakings sold directly to purchasers in the Community; the competitive activities of those undertakings thus affected the Community market.

Agreements between undertakings

Agreements

16–04   The reference in Article 81(1) to agreements between undertakings and decisions by associations of undertakings has been interpreted broadly. These terms include not only contractual agreements and binding decisions, but also "gentleman's agreements" outside the express contractual terms,[10] non-binding recommendations by trade associations,[11] and even unilateral conduct where supported by a policy adopted or accepted by a network of undertakings, for example the exclusion by a supplier of a certain distributor from its distribution network.[12]

This broad definition of agreement covers nearly every conceivable form of exploitation of intellectual property rights, *e.g.* vertical licences, horizontal and reciprocal licences, and patent pools. The ownership of an intellectual property right, however, also gives rise to special considerations. Although the Treaty does not interfere with "property ownership,"[13] nevertheless, the competition rules clearly apply to the purchase or assignment of intellectual property rights. As the Court of First Instance stated in *Tetra-Pak v. Commission*:

---

[9]   Cases 89/85, etc. *Åhlström and others v. Commission* [1988] E.C.R. 5193.
[10]   Case 41/69 *ACF Chemiefarma v. Commission* [1970] E.C.R. 661.
[11]   Case 8/72 *Vereeniging van Cementhandelaren v. Commission* [1972] E.C.R. 977; Case 123/83 *BNIC v. Clair* [1985] E.C.R. 391.
[12]   Case 107/82 *AEG Telefunken v. Commission* [1983] E.C.R. 3151; although on unilateral see now Case T-41/96 *Bayer AG v. Commission*, October 26, 2000 not yet reported. The CFI held there that, contrary to the view of the Commission, the fact that distributors did not interrupt their commercial relations with Bayer after Bayer took action to restrict exports did not amount to an agreement within Article 81(1).
[13]   Article 295 (ex Article 222).

"... the mere fact that an undertaking in a dominant position acquires an exclusive licence does not per se constitute abuse within the meaning of Article [82]. For the purposes of applying Article [82], the circumstances surrounding the acquisition, and in particular its effects on the structure of competition in the relevant market must be taken into account."[14]

An assignment of an intellectual property right will be treated as akin to a licence and thus within Article 81 where the risk of exploitation remains with the assignor. This is particularly the case where the sum payable in consideration of the assignment is dependent on the turnover obtained by the assignee, the products manufactured or the number of operations carried out employing the assigned intellectual property.[15]

Concerted practice

16–05 The concept of a concerted practice is more difficult to define than an agreement. The court held in *Dyestuffs* that it refers to "a form of co-ordination between undertakings which, without having reached the stage where an agreement properly so-called has been concluded, knowingly substitutes practical co-operation between them for the risks of competition."[16] Consciously parallel conduct does not, in itself, amount to a concerted practice, though it may be strong evidence of such a practice if it leads to conditions of competition which do not correspond to the normal conditions of the market,[17] and the court has been ready to infer a concerted practice from parallel conduct where there is evidence of communication between the parties. The inference will be displaced, however, where there is a legitimate explanation for the parallel conduct. In the *SACEM* case, a number of national copyright-management societies refused to grant access to their repertoires to users established in other Member States. There was a reasonable explanation for this, since to allow access by users abroad would require them to organise monitoring systems in other countries. The conduct, therefore, did not amount to a concerted practice.[18]

Undertaking

16–06 An undertaking, for the purposes of Article 81, includes a company, a partnership, a sole trader,[19] and professionals.[20] The fact that the entity is non-profit making is irrelevant. The test is whether the entity carries out an economic activity. Employees are not regarded as undertakings.[21]

Economic Unit

16–07 E.C. competition law examines an agreement between undertakings from an economic perspective, not an overly legalistic one. Although an agreement may be concluded between two separate legal entities, the agreement will not be an agreement between undertakings if the undertakings constitute one economic unit. The

---

[14] Case T-51/89 [1990] E.C.R. II-309, para. 23.
[15] Article 6(2) of Regulation 240/1996 [1996] O.J. L31/2.
[16] Cases 48, 49, 51–57/69 *ICI v. Commission* [1972] E.C.R. 619, para. 64.
[17] *ibid.*, para. 66.
[18] Cases 110/88, 241/88 and 242/88 *Lucazeau v. SACEM* [1989] E.C.R. 2811.
[19] Case 258/78 *Nungesser v. Commission* [1982] E.C.R. 2015 (independent plant breeder).
[20] *COAPI*, [1995] O.J. L122/37, [1995] 5 C.M.L.R. 468 (patent agents); Case C-35/96 *Commission v. Italy* [1998] E.C.R. I-3851 (customs agents); Cases C-180–184/98 *Pavlov*, September 12, 2000 n.y.r. (self-employed medical specialists).
[21] Case 40/73 and others *Suiker Unie v. Commission* [1975] E.C.R. 1663 at para. 539.

overriding criterion is whether both entities have the independence to determine their own commercial course of action. Thus, a licence agreement between a parent and subsidiary will not be regarded as concluded between two "undertakings" if the companies form a single economic unit, where the subsidiary has no real freedom to determine its course of action.[22] Similarly, true agency agreements (*e.g.* where the agent does not take title to the goods and assumes no risk) often escape the prohibition contained in Article 81 on the basis that the principal and agency constitute one economic unit.[23]

Effect on inter-state trade

Effect on the pattern of trade

16–08   The agreement, decision or concerted practice must be "capable of constituting a threat, either direct or indirect, actual or potential" to the pattern of trade between Member States.[24] The test is a very wide one, and catches agreements involving products which are not yet, but which might in the future be traded between Member States, as well as products which are not exported themselves, but which are used to create a secondary product which *is* exported.[25]

National agreements

16–09   An agreement whose effects extend to the entire territory of a Member State will be regarded as affecting inter-State trade, since its effect will be to create a barrier to entry for potential producers or suppliers from other Member States.[26]

Effects of the agreement as a whole

16–10   Article 81(1) does not require that each individual clause in an agreement should be capable of affecting trade between Member States. Only if the agreement as a whole is capable of affecting trade is it then necessary to examine which are the clauses of the agreement which have as their object or effect a restriction or distortion of competition.[27] Thus, although a no-challenge clause might not on the facts affect trade between Member States, it might still infringe Article 81(1) if it distorted competition and the agreement as a whole affected trade between Member States (*e.g.* because it restricted parallel imports).

Networks of Agreements

16–11   Trade between Member States can be affected by a network of similar agreements concluded by the same licensor. Although each individual agreement may not have an impact taken in isolation, the agreement when considered in conjunction with the other similar agreements may impact on the market, for example by foreclosing the market to competitors.[28]

---

[22] Case 15/74 *Centrafarm v. Sterling* [1974] E.C.R. 1147.
[23] *Suiker Unie* above n. 21; Commission Notice on Agency O.J. 1962 p. 2921.
[24] *Consten & Grundig*, above n. 4.
[25] *BNIC v. Clair*, above n. 11, which concerned agreements on the price of potable spirits (domestic trade only) which were used to manufacture cognac (marketed throughout the Community).
[26] *Vereeniging van Cementhandelaren*, above n. 11.
[27] Case 193/83 *Windsurfing International v. Commission* [1986] E.C.R. 611 at 95.
[28] Case C-234/89 *Delimitis* [1991] E.C.R. 935.

Distortion of competition: relevant considerations

16–12 Not all restrictions on conduct lead to restrictions of competition prohibited by Article 81(1). In practice, whether a restrictive clause in an agreement relating to intellectual property distorts competition within the meaning of Article 81(1) can be examined in three stages:

(a) whether the restriction forms part of the existence of the intellectual property right;

(b) whether the restriction has as its object or effect the distortion of competition;

(c) whether the restriction constitutes an ancillary restraint.

(a) *Does the restriction form part of the existence of the intellectual property right?*

16–13 Existence versus exercise: The Court of Justice has consistently stated that the competition rules of the Treaty will not affect the *existence* of intellectual property.[29] In other words, if the restriction contained in the agreement does no more than what the Community perceives to be the essential function of the right, competition law will not interfere. A summary of the essential function of patents, trade marks and copyright is contained in the section dealing with the free movement of goods.[30] The Commission decision in *Bayer* illustrates the principle.[31] Here the German manufacturer of dental products restricted purchasers of its products from repackaging them. The Commission (applying the Court of Justice's jurisprudence on essential function of trade marks[32]) considered that the restriction would not have restricted competition had it been limited to repackaging that altered the original state of the product. The restriction was, however, too wide since it covered all repackaging. The restriction accordingly infringed Article 81(1). Similarly, it is part of the essential function of a patent that the patentee require a royalty for the exploitation of the patent.[33] On this basis, any obligation on a licensee to continue paying royalties beyond the duration of the licensed patents in order to facilitate payment would not restrict competition within the meaning of Article 81(1).[34] Likewise, it would not be a restriction of competition to ensure that the licensee refrain from divulging know-how which is secret since such an obligation would form part of the essential function of know-how.[35]

(b) *Does the restriction have as its object or effect the prevention, restriction or distortion of competition?*

16–14 Relevant market[36]: Any examination of whether certain conduct has an anti-competitive object or effect must involve an identification of the relevant market. The main purpose of market definition is to identify the competitive restraints on any given undertaking. The objective of defining a market in both product and geographic dimensions is to determine the market power of the undertaking concerned, in particular by reference to its market shares, its competitors, and whether those competitors are capable of constraining its behaviour.

---

[29] *Parke Davis v. Probel* above n. 3.
[30] See below at para. 16–109, 16–112 and 16–120 respectively.
[31] [1992] 4 C.M.L.R. 54.
[32] See below at para. 16–112.
[33] *cf.* Case 262/81 *Coditel v. Cine-Vog Films* [1982] E.C.R. 3381 at 12 (copyright).
[34] Case 320/87 *Ottung v. Klee* [1989] E.C.R. 1177
[35] *Delta Chemie/DDD* [1989] C.M.L.R. 535, para. 31.
[36] See generally the E.C. Commission's 1997 Market Definition Notice: [1997] O.J. C372/5; [1998] 4 C.M.L.R. 177.

16–15   Product market: In defining the relevant product (or service) market regard should be had to demand substitution and supply substitution. Demand substitution constitutes the most effective disciplinary force on the supplier of a given product. Demand substitution between two products exists where the products:

> "are regarded as interchangeable or substitutable by the consumer, by reason of the products' characteristics, their price and their intended use."[37]

In practice, it is necessary first to consider whether the two products that are being compared are substitutable as regards their intended use. If the use of the products is not interchangeable, they are most unlikely to be found in the same market. In *Renault/Volvo*,[38] the Commission defined separate markets for trucks of 5–16 tons and 16 tons and over since they had different uses: the larger trucks were used for long distance.

If the two products are similar as regards their characteristics and use, considerations of price become critical. Where the two products being compared have substantially different prices, they are likely to be in different markets. In *Digital/Kienzle*[39] the Commission ruled that mini-computers priced $100,000–$1million were in a different market to computers priced over $1million. In *Nestle/Perrier*[40] the Commission found that bottled water was in a different market from soft drinks, because *inter alia* soft drinks were twice as expensive. In defining the relevant market, the so-called "SSNIP test" (Small but Significant and Non-transitory Increase in Price) is often applied. The question to be answered is whether the parties' customers would switch to other products in response to a hypothetical, small (5–10 per cent) but permanent price increase in the products. If substitutability is enough to make the price increase unprofitable because of the resulting loss of sales, additional substitutes are included in the relevant market.

Supply-side substitution should also be taken into account.[41] The relevance of this is, again, to ascertain the competitive restraints on the undertaking in question. The issue to be considered is to what extent companies manufacturing other products would or could switch production and manufacture the product in question. The Commission distinguishes between short term and longer term switching. Where suppliers are able to switch production to the relevant products and market them in the short term, all the products are contained in the same market. For example, two different grades of paper (*e.g.* high and low quality) may not be substitutes because art books cannot be published with low quality paper. But if a paper mill manufacturing low quality paper is able to switch to manufacturing high quality paper with negligible cost in a short time frame, both the high and low quality papers will be included in the relevant market. Where, however, the manufacturer of the low quality paper needs to adjust significantly (in terms of capital cost, investment, distribution), the two grades of paper would not be in the same market.

16–16   Geographic market: In addition to the relevant product market, the relevant geographic market must be ascertained. The relevant geographic market may be defined as an area in which the conditions of competition are sufficiently homoge-

---

[37] *ibid.*
[38] Case IV/M04 [1991] 4 C.M.L.R. M297.
[39] Case IV/M57 [1992] 4 C.M.L.R. M99.
[40] Case IV/M190 [1993] 4 C.M.L.R. M17.
[41] Failure to do so has resulted in the annulment of E.C. Commission decisions: *e.g.* Case 6/72 *Europemballage and Continental Can v. Commission* [1973] E.C.R. 215.

neous and which can be distinguished from neighbouring areas because the conditions of competition are appreciably different in those areas.[42] For example, the United Kingdom may be a separate geographical market because of the existence of high transport costs, brand loyalty and the need for a national distribution network.

16–17 Object: The intellectual property agreement must be examined to determine whether it has as its object the prevention, restriction, or distortion of competition. The court held in *Consten & Grundig* that it is sufficient if the object of the agreement is to restrict competition, regardless of the effects.[43] Price fixing and market sharing agreements are generally regarded as having distortion of competition as an object. This comes close, therefore, to a *per se* rule of illegality.

16–18 Effect: If the agreement does not have an anti-competitive object it will be necessary to analyse its effects on the market. This may involve a detailed economic analysis, particularly where the agreement is a vertical one.[44] Vertical agreements are those agreements between traders at different levels of trade. Horizontal agreements, *i.e.* agreements between traders at the same level of trade are inherently more likely to distort competition since any collusion on price or markets is likely to have a greater impact (albeit not always adverse) on efficiency and ultimately on consumer choice. Non-reciprocal intellectual property licences can be something of a hybrid. They have the appearance of vertical arrangements (the grant by the licensor to the licensee), but may have horizontal implications as both licensor and licensee may be actual or potential competitors. It is for this reason that non-compete restrictions are treated more favourably in distribution agreements (exempted subject to certain qualifications) than in intellectual property licences (black-listed).[45]

16–19 Appreciability: Article 81 only prohibits agreements that appreciably distort competition in the Common market: *i.e.* are not *de minimis*. In its Notice on Agreements of Minor Importance,[46] the Commission has taken the view that Article 81(1) will not be infringed if the aggregate market shares held by the participating undertakings do not exceed, on any of the relevant markets, five per cent in the case of horizontal agreements (undertakings operating at the same level of production or marketing) or 10 per cent for vertical agreements (undertakings operating at different economic levels). This rule is, however, subject to a proviso: the application of Article 81(1) will not be ruled out for (a) horizontal agreements whose object is to fix prices, limit production or sales, or share markets or sources of supply; or (b) vertical agreements whose object is to fix resale prices or confer territorial protection. Mixed agreements (*i.e.* agreements that are both horizontal and vertical) are subject to the five per cent threshold. The Notice on Minor Agreements does not apply where in a relevant market competition is restricted by the cumulative effects of parallel networks of similar agreements established by several manufacturers or dealers.[47]
It is not the case that an agreement between undertakings, whose market share exceeds these thresholds, will inevitably distort competition. The effect of exceeding the thresholds is that there can be no presumption against infringement.

---

[42] Commission Market Definition Notice, above n. 36.
[43] Above n. 4.
[44] See *e.g. Delimitis* n. 28 above.
[45] See below at para. 16–45 (intellectual property licence) and paras 16–64 to 16–66 (vertical agreements).
[46] [1997] O.J. C372/13.
[47] *ibid.*, para. 10.

The Commission takes the view that an agreement remains outside the prohibition contained in Article 81(1) if the market shares are exceeded by no more than one tenth during two successive years. Otherwise, it is clear that an agreement can float in and out of the prohibition. Companies must, therefore, be vigilant not only of their market share, but of the market in general and the performance of their competitors.

(c)   *Does the restriction constitute an ancillary restraint?*

16–20   The concept of ancillary restraints: Even if a restriction distorts competition, it may escape the prohibition in Article 81 through the doctrine of "ancillary restraints" as developed by the Court of Justice. This concept is similar to (some may say borrowed from) antitrust jurisprudence in the United States, where a distinction is made between an ancillary restraint which is necessary to make a pro-competitive agreement work and a naked restraint which has no purpose but to restrict competition and which is then subject to the test of reasonableness.

16–21   The test: In E.C. law, while the test for determining whether a restraint is ancillary (and thus outside the prohibition altogether) is not entirely clear, it appears to be similar to that applied in the United States. From the cases decided by the Court of Justice and by the Commission the question to be determined is whether the restriction is essential to the working of an otherwise pro-competitive agreement.

16–22   Examples: In *Pronuptia*,[48] the court accepted as permissible restrictions in a franchise agreement which imposed obligations on the franchisee not to open a shop of the same or similar nature in an area where he could compete with a member of the franchise network, during the validity of the contract and for one year after its expiry; and not to transfer his shop to another party without the franchisor's prior approval. These restrictions, the court considered, were essential in order to enable the franchisor to provide know-how and assistance to the franchisee, without running the risk that this might benefit his competitors. Further restrictions, such as a ban on relocation without the franchisor's approval, an obligation to obtain approval of advertising, and an obligation to sell only products supplied by the franchisor or by approved suppliers, were likewise essential for the franchisor to ensure the identity and reputation of the franchise network.

Although each case must viewed in the light of its own facts, licences of intellectual property rights are particularly apt to fall within the ancillary restraint doctrine because of the considerable financial investment often associated with the intellectual property right. For example, in *Erauw-Jacquery* the Court of Justice ruled that a provision in a licence prohibiting the licensee from selling and exporting basic seeds fell outside Article 81 altogether. It was accepted by the court that for a plant breeder's right to remain uniform and stable, it was essential for the holder to be able to control those who propagated the seed. The Court stated:

"... the development of the basic lines may involve considerable financial commitment. Consequently, a person who has made considerable efforts to develop varieties of basic seed which may be the subject-matter of plant breeders' rights must be allowed to protect himself against any improper handling of those varieties of seeds. To that end, the breeder must be entitled to restrict propagation to the growers which he has selected as licensees. To that extent, the provision pro-

---

[48] Case 161/84 [1986] E.C.R. 353.

hibiting the licensee from selling and exporting basic seed falls outside the prohibition contained in Article [81(1)]."[49]

Whilst this is undoubtedly a special case given the nature of the right in issue (plant
breeders rights over basic seeds), it shows how flexible the courts are prepared to be
when applying the doctrine of ancillary restraints, given that export bans are generally
regarded as serious infringements of Article 81(1). This flexibility, however, gives rise
to difficulties when applying the doctrine in practice. Questions relating to how much
investment is necessary or how restrictive the restriction can be are often difficult to
answer with any precision. For example, the case of *Erauw-Jacquery* may be compared
with *Nungesser*[50] where the court condemned the conferral of absolute territorial protection on a licensee of plant breeders rights (not of the basic seed), but held that an
exclusive licence to exploit the right fell outside Article 81(1) on the basis of the "new
technology exception." These cases are discussed further below.

Distortion of competition: examples

Exclusivity/territorial restrictions

16–23   One of the main objectives of European Union is the creation of a single market. Member States should not impose restrictions on the free movement of goods. It
should be as easy to sell goods across national borders as within a single Member
State. Article 81 continues with this theme. Export bans and other types of territorial
restriction are viewed as potentially serious infringements of the competition rules. It
is, however, accepted that licensees may need appropriate protection from competition
in certain circumstances. A distinction must be drawn between:

(a)   open exclusivity for new technology;

(b)   open exclusivity for existing technology; and

(c)   exclusivity conferring absolute territorial protection.

Open exclusivity for new technology

16–24   In *Nungesser* the court held that an "open" exclusive licence relating to new
technology falls outside Article 81. Open exclusivity means a licence whereby the
licensor agrees not to licence anyone else for the same territory and not to compete
with the licensee (but the agreement does not prevent imports by parallel importers
and licensees in other territories). Without the exclusive right the licensee would be
deterred from accepting the risk of cultivating and marketing the product.[51] The
court only gave limited guidance as to what is meant by "new technology" referring
to "years of research and experimentation."[52] It is not clear whether the technology
must be brand new or new to the territory in question. Recital 10 to Regulation
240/96 (the block exemption for technology transfer licences) appears to suggest that
the technology must be new to the territory. This would appear to be justified, as the
rationale for this rule is a reward to the licensee for accepting the risk in marketing
the product.[53]

---

[49] Case C-27/87 [1988] E.C.R. 1919, para. 10.
[50] Above n. 19.
[51] *ibid.*, paras 57–58.
[52] *ibid.*, para. 56.
[53] The view of the Commission in *Knoll/Hill-Form*, 13th Competition Report, para. 144.

Given its uncertain ambit, there is a reluctance to rely on the new technology exception in practice. However, when notifying an agreement to the E.C. Commission for negative clearance or an exemption, the exception should be referred to if the facts appear to support it. The exception is often invoked in civil litigation where one party is challenging the enforceability of the licence. In such cases the exception can be a particularly useful shield.[54]

Open exclusivity where the technology is not new

16–25 If the technology cannot be regarded as new, or if the period of the agreement or the relevant provision exceeds the period which can reasonably be justified for rewarding the licensee for accepting a risky technology, the exclusivity cannot be regarded as ancillary to an otherwise pro-competitive agreement. The agreement must be viewed in the context of the overall market (*e.g.* effect on competitors: market position of the parties to the agreement). If Article 81 is infringed, regard should be had to the block exemption, Regulation 240/96.[55]

Absolute territorial protection

16–26 A "protected" exclusive licence, whereby the parties take measures to prevent competition from parallel importers or other licensees, is, according to the court in *Nungesser,* contrary to Article 81(1) since it would enable the licensee artificially to partition the common market[56]; and in *Pronuptia* territorial exclusivity was not permitted in the context of a well-known franchise brand, for the same reasons.[57] The exceptions are limited: for example, where restrictions on unfettered dissemination or distribution are required either by the particular characteristics of the market, or the particular characteristics of the product. Thus, in *Erauw-Jacquéry*, a clause prohibiting a licensee of plant breeders' rights fell outside the Article 81(1) prohibition on the grounds that the holder of those rights had to retain protection against improper handling of its seed varieties, and therefore had to have the right to reserve propagation for its chosen licensed propagating establishments.[58]

Monopoly leveraging

16–27 This can be done in at least two ways: first by tying the protected product to the sale of another product; secondly, by charging royalties on unprotected products.

Ties

16–28 Article 81(1) does not apply to tying arrangements that are necessary for the exploitation of the intellectual property right. For example, a patentee may require that the licensee obtain products from the licensor for reasons of quality. Otherwise Article 81(1)(e) expressly prohibits an obligation on the licensee to acquire other products which "by their nature or according to commercial useage have no connection with the protected product."[59] Whether a tie in fact distorts competition depends largely on the market power of the licensor and the foreclosure of the tie on third party suppliers.[60]

---

[54] See section D below for further details.
[55] [1996] O.J. L32/2, see below at paras 16–40 *et seq.*
[56] *Nungesser*, above n. 19, paras 60–61.
[57] Above n. 48.
[58] Above n. 49.
[59] As to the conditions for exemption under Regulation 240/96 see Article 2(1)(5) and below at para. 16–48.
[60] *Vassen/Moris* [1979] 1 C.M.L.R. 511.

Royalties

16–29   In principle, Article 81 will prohibit an obligation to pay royalties on unprotected items. As with tying restrictions, competition law acts to prevent the intellectual property right owner extending his monopoly to unprotected products. In practice, however, a wide margin is left to the parties to decide on the most appropriate method of payment for the right to exploit the intellectual property. If, for example, the nature of the finished product makes it difficult to charge royalties on the patented product separately, the parties may agree on a royalty on the finished product. The parties are also left free to spread royalty payments so that the licensee remains obliged to pay royalties even after expiry of the patent.[61]

Assignment of improvements

16–30   An obligation on the licensee to assign in whole or in part improvements to the licensed technology is likely to infringe Article 81(1). In contrast, an obligation to grant the licensor a non-exclusive licence relating to improvements is not generally restrictive of competition. In other words the grantback obligation should not be too strong. The justification is probably twofold: first, to prevent the licensor from extending his monopoly otherwise than by his own creativity; secondly, to reward the licensee for improvements and, thus, to encourage improvements to the licensed technology.

The Commission is particularly vigilant to ensure that at the end of the licence term the licensee does not find himself in a position where he is unable to use his own improvements when they cannot be used independently of the licensor's technology, which he no longer has the right to exploit whereas the licensor continues to be able to use the licensee's improvements. The Commission has therefore insisted that the licensor should be unable to use non-severable improvements at the end of the licence, leaving it to the parties to re-negotiate in the event that they wish to continue to use the initial know-how as improved by the licensee.[62]

Customer/field of use restrictions

16–31   Any obligation on the licensee to restrict his exploitation of the licensed technology to one or more fields of application or product markets is not generally restrictive of competition.[63] By contrast, a customer restriction is generally of the type that restricts competition. Where, however, a licence is granted to give a single customer a second source of supply, a prohibition on the second licensee from supplying other customers is justified.[64]

The rationale for the distinction between customer restriction and field of use restriction is not always easy to follow or apply in practice. A field of use restriction benefits from a less strict application of the competition rules on the basis that, if the licensor is entitled to transfer the technology at all, he must be able to license for a limited purpose. In other words, the field of use restriction forms part of the existence of intellectual property rights, whereas a customer restriction forms part of their exercise.[65]

---

[61] See *AIOP/Beyrard* [1976] 1 C.M.L.R. D14; *Rich Products/Jus-Rol* [1988] 4 C.M.L.R. 527; Case 380/27 *Ottung v. Klee* [1989] E.C.R. 1177 and the conditions for exemption under Regulation 240/96, Recital 21 and Article 2(1)(7) of Regulation 240/96 and below at para. 16–52.
[62] See *Delta Chemie/DDD* n. 35 above; as to the conditions for exemption under Regulation 240/96 see Article 2(1)(4) and below at para. 16–46.
[63] See Recital 22 of Regulation 240/96.
[64] It does not benefit from the automatic block exemption, although it benefits from the opposition procedure.
[65] As to the distinction between existence and exercise see para. 16–13 above.

Non-competition obligations

16–32   Any restriction on either party competing with the other is likely to fall within the prohibition.[66] A provision that allows the licensor to terminate the exclusive nature of the agreement or to licence improvements in the event that the licensee deals in competing technology is generally permitted.[67]

Sub-contracting

16–33   A person may choose to sub-contract the manufacturing of certain products using his technology. The Commission views sub-contracting in a favourable light. Guidance is obtained from its Notice on Sub-contracting.[68] The Commission considers that Article 81(1) will not be infringed by the following restrictions in sub-contracting agreements:

(a)   technology or equipment provided by the contractor may only be used for the purpose of the sub-contracting agreement;

(b)   the technology or equipment should not be made available to third parties;

(c)   the goods, services or work resulting from the use of such technology or equipment may be supplied only to the contractor or performed on his behalf;

(d)   the sub-contractor assigns or licences on an exclusive basis all improvements that are incapable of being used independently from the contractor's patents or know-how.

However, it is essential that these restrictions are part of an agreement where the contractor makes available to the sub-contractor intellectual property or plans, studies or equipment that is distinctively the contractor's. These restrictions are likely to fall within the prohibition contained in Article 81(1) if the sub-contractor has ready access to the relevant technology or equipment himself.

Patent pools and joint licensing

16–34   An aggregation of patent rights for the purpose of joint licensing (commonly called a "patent pool") and joint licensing of patents may provide competitive benefits by integrating complementary technologies, reducing transaction costs, clearing blocking positions and avoiding costly infringement litigation. By promoting the dissemination of technology, cross licences and pooling arrangements can be pro-competitive.[69] Nevertheless, some patent pools can restrict competition, whether among intellectual property rights within the pool, in downstream products incorporating the pooled patents, or in innovation among parties to the pool.

Patent pools as defined above may be distinguished from reciprocal licensing under which one person grants another a patent and/or a know-how licence in exchange for the grant of a patent and/or know-how licence. Patent pools may also be distinguished from joint ventures, the purpose of which is for the joint venture to carry out further research or development, or exploit the intellectual property rights licensed to it by the parent companies.[70]

---

[66] And is not block exempted: Article 3 (2).
[67] Article 2(1)(18) of Regulation 240/96.
[68] [1979] O.J. C1/2.
[69] For a useful guide see U.S. Department of Justice Federal Trade Commission Anti-Trust Guidelines for the Licensing of Intellectual Property, section 5.5.
[70] This distinction is made at Article 5 of Regulation 240/96, [1996] O.J. L31/2.

16–35 Contents of pool: A licensing scheme premised on invalid patents is unlikely to withstand competition law scrutiny. This is because the licensor will be claiming a monopoly in respect of an unprotected product. Furthermore, a patent pool that aggregates competing technologies and sets a single price for them will give rise to competition concerns. By contrast, a pool of complementary intellectual property rights can be an efficient and pro-competitive method of disseminating those rights to would be end users. The appointment of an expert to assess that the patents are complementary and not substitutes is regarded as important.

16–36 Clauses: Even if the pool itself is regarded as pro-competitive, specific terms may nevertheless restrict competition. The terms and conditions of the collective licence should be non-discriminatory. An open pool will tend to militate against any allegation of discrimination. Royalty rates should generally be non-discriminatory and should not be used as a device to co-ordinate prices of downstream products. Any grant back of improvements must not be such as to reduce a party's incentive to innovate. For example, an obligation to license each other all future technology at minimal cost could reduce the members' incentives to engage in research and development.

Other restrictions

16–37 It is important to recognise that E.C. competition law recognises few *per se* rules of infringement. Export bans, market sharing and price fixing are hardcore restrictions which are likely to infringe Article 81(1). The technology transfer block exemption, Regulation 240/96, contains references to other types of clauses that may distort competition (*e.g.* restrictions on production; prohibitions on challenging intellectual property); or that do not generally distort competition (minimum quantities; most favoured licensee).

    Whether competition is distorted in fact depends on a proper market analysis being carried out: the existence of intra-brand competition (*i.e.* competition in the same product); the existence of inter-brand competition (*i.e.* competition from competing products) that might act as a counterweight in the event of there being limited intra-brand competition; the foreclosure effect of the particular clause on other suppliers; the nature of the intellectual property right (*e.g.* whether it is new; the scale of research and development) and the risk associated with exploiting it.

Exemptions

Article 81(3) exemption

16–38 Agreements which fall, *prima facie*, within Article 81(1) may nevertheless be exempted by the Commission under Article 81(3). This provides:

> "The provisions of paragraph 1 may, however, be declared inapplicable in the case of:
>
> — any agreement or category of agreements between undertakings;
> — any decision or category of decisions by associations of undertakings;
> — any concerted practice or category of concerted practices;
>
> which contributes to improving the production or distribution of goods or to promoting technical or economic progress, while allowing consumers a fair share of the resulting benefit, and which does not:

(a) impose on the undertakings concerned restrictions which are not indispensable to the attainment of those objectives;

(b) afford such undertakings the possibility of eliminating competition in
respect of a substantial part of the products in question."

Only the Commission has the power to grant exemptions under Article 81(3). It may
grant individual exemptions where the above criteria are satisfied though in practice
these are extremely rare. Individual cases are more usually dealt with by means of
"comfort letter" indicating that the criteria for exemption would be met. It may also
adopt block exemptions for certain categories of agreements.

Block exemptions

16–39 The Community's approach to block exemptions has changed from a rigid
application in the 1980s and 1990s to a more relaxed regime in the 2000s. In the case of
vertical agreements relating to the re-sale of goods or the provision of services, the
former block exemptions on exclusive distribution,[71] exclusive purchasing[72] and on franchising[73] have all been replaced by a single block exemption for vertical agreements,
Regulation 2790/1999.[74] The draftsmen have moved away from listing en masse the type
of clauses that are exempted, to listing certain hardcore restrictions that are not presumed beneficial.

The main block exemption relating to intellectual property is Regulation 240/96 on
technology transfer agreements.[75] This follows the old approach of listing the various
types of restriction that merit automatic exemption. Of limited practical use is the
Regulation 418/85 R&D block exemption.[76] There are two important aspects to note
about Regulations 240/96 and 418/85. First, if an agreement contains a restrictive
clause that is not envisaged by the block exemption, the whole block exemption
ceases to apply. In other words, if the agreement contains further restrictions of competition that are not expressly dealt with in the block exemption, the exemption does
not apply in its entirety. Secondly, the fact that the block exemption does not apply does
not mean that Article 81 is in some way infringed. It is still necessary to prove that the
agreement appreciably distorts competition.[77]

Technology transfer block exemption

16–40 Format of the exemption: An agreement which falls within the scope of the
block exemption, which contains any of the exempted terms (set out in Articles 1 and
2), which does not contain any blacklisted clauses and which does not restrict competition in any other way is automatically exempted. It does not have to be notified for
individual exemption. If the technology transfer agreement contains anti-competitive
clauses which do are not listed in Articles 1 and 2 then (provided again that it does not
contain blacklisted clauses) it should be notified to the Commission, but may be
exempted under the "opposition procedure" set out in Article 4 of the block exemption
("the grey list"). The agreement is then regarded as falling within the block exemption
if the Commission does not oppose its exemption within a period of four months.

---

[71] Regulation 1983/83.
[72] Regulation 1984/83.
[73] Regulation 4087/88.
[74] [1999] O.J. L336/21.
[75] [1996] O.J. L32/2.
[76] [1985] O.J. L53/5; also Regulation 417/85 on specialisation agreements. Both regulations are due to be
replaced: [2000] O.J. C118/3.
[77] *Courage v. Crehan* [1999] Eu.L.R. 834.

16–41 Scope of the exemption: The block exemption covers licences[78] and certain assignments[79] of (1) patents (2) know-how and (3) a mixture of the two.

The definition of patents includes patent applications, supplementary protection certificates, utility models, semiconductor topographies and plant breeders' certificates.[80] In addition, Recital 4 states that the exemption applies to patents of Member States, Community patents and European patents.

Know-how must be secret, substantial and identified.[81] The threshold for the substantiality test, in particular, is relatively high: the know-how must be such that it "can reasonably be expected at the date of conclusion of the agreement to be capable of improving the competitive position of the licensee, for example by helping him to enter a new market or giving him an advantage in competition with other manufacturers". The practical effect of the definition is to exclude flimsy know-how that is used as a vehicle for substantiating a restrictive provision. In order to be identified, the information must "either be set out in the licence agreement or in a separate document or recorded in any other appropriate form at the latest when the know-how is transferred or shortly thereafter."

The exemption also covers intellectual property rights other than patents, provided that they are "ancillary".[82]

The block exemption generally does not apply to patent or know-how pools, certain licences in the context of a joint venture, or certain reciprocal licensing between competitors. In each case, there is an increased risk that the agreement will have adverse effects on competition. There is an exception, in the case of patent pools and reciprocal licensing, for agreements where the parties are not subject to certain territorial restrictions in the common market. There is a further exception for certain licences between parent undertakings and joint ventures. Two further categories of agreement are outside the exemption: agreements relating to sales (*i.e.* distribution agreements) and agreements containing provisions relating to intellectual property rights other than patents which are not ancillary.[83]

16–42 Basic exemption: The basic exemption is contained in Article 1(1) of the block exemption. Within the scope of the block exemption, Article 81(1) will not apply to agreements to which only two undertakings are party, and which include one of more of the following obligations:

"(1) an obligation on the licensor not to license other undertakings to exploit the licensed technology in the licensed territory ['sole licence'];

(2) an obligation on the licensor not to exploit the licensed technology in the licensed territory himself ['exclusivity if also sole licence'];

(3) an obligation on the licensee not to exploit the licensed technology in the territory of the licensor within the common market ['protection for licensor'];

(4) an obligation on the licensee not to manufacture or use the licensed product, or use the licensed process, in territories within the common market which are licensed to other licensees ['protection for other licensees'];

---

[78] Including sub-licences: Recital 9.
[79] Where the risk associated with exploitation remains with the assignor - see Recital 9 and Article 6.
[80] Article 8.
[81] As defined in Article 10(2)-(4).
[82] *i.e.* contain no anti-competitive provisions other than those attached to the exempted know-how and patents: Article 10(15).
[83] Article 5.

(5)   an obligation on the licensee not to pursue an active policy of putting the licensed product on the market in the territories within the common market which are licensed to other licensees, and in particular not to engage in advertising specifically aimed at those territories or to establish any branch or maintain a distribution depot there ['ban on active selling'];

(6)   an obligation on the licensee not to put the licensed product on the market in the territories licensed to other licensees within the common market in response to unsolicited orders ['ban on passive selling'];

(7)   an obligation on the licensee to use only the licensor's trademark or get up to distinguish the licensed product during the term of the agreement, provided that the licensee is not prevented from identifying himself as the manufacturer of the licensed products ['use of trade mark'];

(8)   an obligation on the licensee to limit his production of the licensed product to the quantities he requires in manufacturing his own products and to sell the licensed product only as an integral part of or a replacement part for his own products, provided that such quantities are freely determined by the licensee ['use only licence']".

16–43   Duration: The exemptions apply for varying periods depending on the type of restriction (as set out in Article 1(1)) and the type of agreement (whether patent, know-how or mixed). In summary:

(a)   In a pure patent licensing agreement, the exemption for the obligations in (1) (2) (7) and (8) above only lasts for as long as there are parallel patents in the territory of the licencee. The exemption for the obligation in (3) above only lasts for as long as the licensor is protected by a patent. A ban on active selling and protection for other licensees (paragraphs 4 and 5) lasts for the duration of the licence provided that there is still patent protection in the other licensees' member state(s). A ban on passive selling in a pure patent licence (paragraph 6 above) lasts for a period not exceeding five years from the date when the licensed product is first marketed in the common market by one of the licensees and the product remains patented in the other territory .

(b)   Where the licence is a pure know-how licensing agreement, the exemption for the obligations in (1)–(5) above cannot exceed ten years from the date when the licensed product is first marketed in the Common market by one of the licensees. The same five year period applies for restrictions on passive sales.

(c)   Where the licence is mixed the exemption applies for as long as there is patent protection: the ten year rule does not apply to obligations in paragraphs (1) – (5). The five year rule for restrictions on passive sales applies.

In the case of improvements, it is possible to extend the term of the block exemption to cover them, subject to conditions, as described below.

There is an obvious distinction between the duration of the agreement and the duration of permitted post-term restrictions. Following termination of the agreement, the licensor may maintain the confidentiality obligations for so long as the know-how remains secret but not necessarily substantial. Similarly, a post-term use restriction is permitted for patents, for so long as the patents are in force.[84] Post-term royalties are dealt with below.

---

[84]   Article 2(1)(3).

16–44   White, black and grey lists: Article 2 sets out the "white list" of obligations which are not generally regarded as restrictive of competition, but which are exempted just in case in the particular circumstances they do fall within Article 81(1). Article 3 sets out the "black list" of provisions which, if included, take the *entire agreement* outside the block exemption. Finally, the exemption provides in Article 4 that anti-competitive clauses which are not listed in Articles 1 and 2, and which do not fall under the black list, may be exempted through the opposition procedure. In such a case, the agreement should be notified to the Commission, but is regarded as falling within the block exemption if the Commission does not oppose exemption within a period of four months. Examples of such clauses are given in Article 4: the "grey list". In addition, the exemption may be withdrawn in the circumstances listed in Article 7. In considering the effect of the block exemption on various types of agreements, it is necessary to consider the provisions of Articles 2–4, and 7 as a whole.

16–45   Restrictive covenants[85]:   Covenants restricting disclosure of know-how, prohibiting sub-licensing and assignment, and post-term use of the technology, are on the white list. Similarly, a prohibition on the licensee constructing production facilities for third parties is white-listed, provided that it does not restrict the licensee's own production. By contrast, non-competition covenants are on the black list.

16–46   Grant-back[86]:   Provisions for the grant of a licence of the licensee's improvements to, and new applications of, the licensed technology are white-listed, provided that the licence in respect of severable improvements is non-exclusive, and that the licensor reciprocally undertakes to grant a licence of his own improvements. However, an obligation for the licensee to *assign* any rights to improvements to, or new applications of, the licensed technology falls under the black list.

It should be noted that the wording of improvements clauses is always problematic. In this context, it may be difficult to determine which improvements are severable. For example, it would be possible to define severable improvements as being those that may be exploited independently. The problem with this sort of definition is that technology which is incapable of independent exploitation at the outset may become exploitable independently during the course of the contract. If on the other hand the definition of severability were loose, this would leave open crucial questions of interpretation.

The position of post-termination grant-back provisions is unclear. Some licensors provide for post-term use of the licensee's improvements by the licensor, subject to royalty payments but with no reciprocal licence. The old know-how block exemption provided for payment of reasonable royalties in these circumstances, but the current block exemption does not contain explicit provisions on this point.[87]

16–47   Extending the licence to cover improvements: The block exemption covers agreements whose initial duration is automatically prolonged by the inclusion of new improvements by the licensor. This is subject to the proviso that the licensee has the right to refuse the improvements, or each party has the right to terminate the agreement at the expiry of the initial term and at least every three years after that.[88]

Automatic prolongation of the agreement by the inclusion of new improvements may not, under the block exemption, extend the licensor's obligation not to license

---

[85]   Article 2(1)(1), (2), (3), (12) and Article 3(2).
[86]   Article 2(1)(4) and Article 3(6).
[87]   See *Delta Chemie/DDD* n. 35 above.
[88]   Article 8(3).

other undertakings in the licensed territory beyond the permitted durations set out in Article 1(2)–(4). However, as stated in Recital 14, in circumstances where the improvements are "distinct from the licensed technology" it is possible to conclude a fresh agreement.

16–48 Tie-ins[89]:   Tie-ins of goods or services are white-listed insofar as they are necessary for the technically proper exploitation of the licensed technology and to meet minimum quality specifications that are applicable to the licensor and to all other licensees. Other forms of tying are covered by the grey list and will therefore be considered by the Commission. The acceptability of such a "grey" clause may well depend upon the licensor's market power.

It is not clear whether the tie must remain technically necessary for the duration of the agreement in order to qualify for the white list, or whether the relevant time is the date of entering into the agreement, which is referred to in the context of the grey list tying example.

16–49 Field of use restrictions and market sharing[90]:   One of the most efficient ways of licensing technology is by field of use. This is reflected in the block exemption, where field of use restrictions relating to technical fields of application and product markets appear on the white list.

However, with the exception of arrangements relating to a second source of supply for a customer, which are on the white list, market-sharing agreements between competing manufacturers, which restrict customers which may be served within the same field of use or same product market, are black-listed. The black list will probably catch not only direct customer restrictions (*e.g.* specific named customers) but also more subtle restrictions aimed at the same objective. Therefore, if the restriction cannot be rationalised on the grounds that customers are buying different products, or for different uses, it may be safest to use the opposition procedure.

16–50 Incentives for the licensee: provisions that require licensees to pay minimum royalties, to have a minimum throughput,[91] and to use best endeavours to manufacture and market products,[92] are all on the white list as they are a good means of ensuring that the licensee works the technology. But the use of these provisions to prevent a competing manufacturer from using competing technology may lead to the withdrawal of the exemption.[93] Similarly, Recital 21 states that the setting of royalty rates to achieve a black-listed objective (such as non-competition covenants in R&D, production, and the use or distribution of competing products) renders the agreement ineligible for exemption.

16–51 Restrictions on the licensee's production: Included with the restrictions covered by the basic exemption in Article 1(1)(8) is an exemption for clauses requiring the licensee to limit production to the quantities required for making the licensee's products, and not to sell the licensed products independently of the licensee's own products   —   provided that the licensee is free to determine the quantity of products. Added to this provision is an item on the white list exempting quantitative limitations on the licensee's production where the licence is granted to provide a second source of

---

[89] Article 2(1)(5) and Article 4(2)(a).
[90] Article 2(1)(8) and Article 3(4).
[91] Article 2(1)(9).
[92] Article 2(1)(17).
[93] Article 7(4).

supply.[94] However, all other quantitative restrictions on the licensee's output fall under the black list.[95]

16–52 Royalty Scheduling: Some licensees may not be able to pay the full economic cost of the licence whilst the licensed rights subsist. The payment of royalties after the expiry of the licensed patents is therefore included in Article 2 if it is to facilitate payment. The same applies to payments during the unexpired term of a know-how licence if the know-how becomes publicly known, provided that the disclosure is not by the licensor. [96]

16–53 No challenge clauses: Licensing is often more than a theoretical alternative to litigation. Many licensees take licences as a precautionary measure, with the aim of reviewing the need for the licence at a later date, and many licensors grant licences to settle actual or potential infringement actions. In the block exemption, clauses enabling the licensor to terminate the licence if the licensee challenges the licensor's rights (and those of connected undertakings) in the common market are white-listed.[97]Similarly, termination is permitted in respect of the licence of a *patent* if the licensee claims that the patent is not necessary. This is balanced by the appearance on the grey list of an outright prohibition on the licensee challenging the substantiality of the licensed know-how or the validity of the patents licensed in the common market, which belong to the licensor (or a connected undertaking).[98]

16–54 Blocking parallel trade: The provisions of the exemption must not be used to prevent parallel trade. Amongst the black-listed provisions is an obligation to refuse orders from those who would market in other territories, or to frustrate parallel trade without justified reason.[99]

16–55 Price fixing: Direct or indirect price fixing agreements are regarded as almost *per se* anti-competitive. Thus restrictions on the prices which a party may charge for the licensed products are black-listed, and individual exemption is extremely unlikely.[1]

R&D block exemption

16–56 Basic exemption: The R&D block exemption, in Article 1, exempts joint venture agreements entered into for the purposes of joint R&D and/or joint production resulting from prior joint R&D. The exemption is subject to conditions as to market shares (Article 3) and a number of further conditions (Article 2). The regulation has not had much practical effect; it has been perceived as too inflexible since the conditions for exemption have proved too narrow. It is due to be replaced by a block exemption of much wider ambit and flexibility.

16–57 Agreements exempted: Article 1 of the block exemption covers joint R&D and its joint exploitation. The obligation to exploit may be in the same agreement or arising out of a prior agreement on R&D. Pure joint R&D agreements, excluding

---

[94] Article 2(1)(13).
[95] Article 3(5).
[96] Article 2(1)(7).
[97] Article 2(1)(15).
[96] Article 4(2)(b).
[99] Article 3(3).
[1] Article 3(1).

joint exploitation, do not usually fall within Article 81, but are covered by the regulation in any event. R&D includes experimental production, technical testing, the "obtaining" of intellectual property in respect of the results of the research. Exploitation means the manufacture of the goods or the assignment or licensing of the intellectual property.

16–58   Condition: Article 2 provides that the block exemption should apply upon certain conditions being met. The most important of these are that all parties must have access to the results (Article 2(b)), and agreements that are purely for joint research must allow each party independently to exploit the results and any pre-existing technical knowledge (Article 2(c)). Also, joint production is permitted only in relation to results which are protected by intellectual property or innovative know-how (Article 2(d)). The latter criterion ensures that the R&D is not a cosmetic reason for a patent pool, cross licensing or joint distribution.

16–59   Market shares & duration: Article 3 sets out certain market share thresholds and time limits. These may be summarised as follows:

(a)   where the parties are not competing manufacturers of products capable of being improved or replaced by the contract products, the exemption applies for the duration of the agreement. If the agreement extends to joint exploitation, the exemption lasts for five years from the time when the contract products are first marketed in the Common Market. Thereafter, the exemption will continue so long as the market share of the contract goods (together with such other goods produced by the parties which are in the same market as the contract goods) does not exceed 20 per cent. The exemption continues to apply where the market share is exceeded during any period of two consecutive financial years by not more than one tenth (Article 3(4)).

(b)   where the parties are competing manufacturers, the exemption lasts for five years from the time when the contract products are first marketed in the Common Market, but only if the parties share of the market at the date of the agreement does not exceed 20 per cent.

(c)   where one of the parties (or another appointee) is entrusted with the distribution of the contract goods, the market share must not exceed 10 per cent.

16–60   Restrictions of competition: Article 4 sets out a list of further restrictions, to which the exemption extends. These include restrictions on independent R&D in the same or closely connected fields during the execution of the programme; obligations to purchase the contract products only from each other, the joint venture or third parties jointly charged to make them; the reservation to certain parties of rights to manufacture the products in specified territories; confining manufacture to technical fields of use except if the parties were competitors when the agreement was entered into; territorial restrictions on marketing, for up to five years from the time the products are first put on the market; and the grant of exclusive distribution rights to one of the parties or the joint venture.

16–61   White list: Further provisions fall under the white list in Article 5: they are not regarded as restrictive of competition, but are exempted just in case. Similar to the technology transfer block exemption, the white list here includes confidentiality obligations, obligations to pay or share royalties, and obligations to observe minimum quality standards.

16–62 Black list: Article 6 sets out clauses which take the agreement outside the block exemption, including restrictions on R&D in unrelated fields during and after the research project or in the same field after the completion of the joint programme; restrictions on the quantity of contract products which the parties may make; restrictions on prices and customers; territorial restrictions after five years from the time the products are first marketed (*i.e.* the territorial protection exceeds that permitted under Article 4); requirements not to grant production licences to third parties in the absence of joint manufacture by the parties themselves; and contractual measures to prevent cross-supplies between dealers, in particular by the exercise of intellectual property; certain "no challenge" clauses relating to intellectual property.

16–63 Opposition procedure: The block exemption provides in Article 7 for an opposition procedure leading to exemption if the Commission does not indicate opposition within six months of notification.

Vertical agreements block exemption

16–64 Basic exemption: The new block exemption for vertical agreements replaces the previous separate exemptions for exclusive distribution, exclusive purchasing, and franchise agreements. The basic exemption, contained in Article 2 of the block exemption, provides that Article 81(1) shall not apply to:

> "agreements or concerted practices entered into between two or more undertakings each of which operates, for the purposes of the agreement, at a different level of the production or distribution chain, and relating to the conditions under which the parties may purchase sell or resell certain goods or services ('vertical agreements')."

The exemption is subject to the proviso that the market share of the supplier (and, in the case of exclusive supply agreements, the buyer) does not exceed 30 per cent of the relevant market. There is a further market share requirement in circumstances where there is a parallel network of similar vertical restraints.

16–65 Non-severable black list: Article 4 sets out the clauses which may not be included in the agreement: resale price maintenance; restriction of the territory into which, or the customers to whom, the buyer may sell the goods or services (with a few specified exceptions); restriction of active or passive sales to end users by retailers in a selective distribution system; restriction of cross-supplies between distributors in a selective distribution system; and restriction on access to spare parts by service providers.

16–66 Severable black list: The block exemption also contains, in Article 5, a further list of clauses which may not be exempted, but which do not take the remainder of the agreement outside the exemption: any non-compete obligation exceeding five years; post termination non-compete obligations, limited to a maximum of one year after termination unless necessary to protect know-how transferred by the supplier to the buyer; and restriction on members of a selective distribution system selling brands of particular competing suppliers.[2-3]

[2-3] See Article 5 for the full details.

B. Mergers and joint ventures

Joint ventures

16–67   A joint venture (JV) is more than a simple co-operation agreement but less than a full merger. A JV is usually jointly controlled by other firms, although it can take the form of loose co-operation between two or more firms. Control means the possibility of exercising, directly or indirectly, a decisive influence on the activities of the JV. Other forms of association (*e.g.* the acquisition of minority holdings or representation on the board) which do not involve joint control but which enable a firm to influence appreciably the activities of another may have effects similar to those of JVs.

Different rules apply to "co-operative JVs" and "full-function JVs". Co-operative JVs fall to be considered under Articles 81 and 82, whereas full-function JVs fall to be considered primarily under the Merger Regulation 4064/89. Essentially, a full-function JV is one that forms on a lasting basis all the functions of an autonomous economic entity which does not give rise to any co-ordination of the competitive behaviour of the parent companies amongst themselves or between them and the joint venture. For example, there is normally no risk of co-ordination where the parent companies transfer the whole of certain business activities to the joint venture and withdraw permanently from the JV's market. If the JV operates in a market neighbouring that of the parent companies, the risk of co-ordination depends on the technical and economic relation between the products involved.

Co-operative joint ventures

16–68   Examination of co-operative joint ventures: Like patent pools, so-called co-operative joint ventures, whereby the parent undertakings remain independent, may have many benefits to competition and technological progress, by allowing companies to share risks, save costs, pool know-how and develop innovative products and services. However, JVs may also have anti-competitive effects, which may trigger Article 81(1).[4] Competition may be restricted either through the formation of the JV *per se* (the paradigm case being where the parent companies are competitors in the field) or through the specific clauses of the agreement, for example, cross-licensing of intellectual property, sharing of know-how, or assignment to the JV of the rights to exploit any intellectual property arising from collaboration or joint research.

A joint venture may be examined from three perspectives: (i) competition between parent companies; (ii) competition between parent companies and the JV; (iii) effects on the position of third parties.

Competition between parent companies

16–69   Competition between parent companies will be restricted if those companies are actual or potential competitors. Where there is no actual or potential competition, the creation of the joint venture has a positive effect since it creates a new competitor. The issue is whether the parent companies could enter the relevant market alone (in which case they are potential competitors) or whether they need each other to enter that market (in which case they are not). In its thirteenth report on competition policy (1983),[5] the Commission sets out a checklist of questions with respect to potential competition between the parent companies.

---

[4] See Commission Notice concerning the assessment of co-operative joint ventures pursuant to Article 85 of the EEC Treaty, [1993] O.J. C43/2; and new draft guidelines on the applicability of Article 81 to horizontal co-operation, [2000] O.J. C118/14.

[5] Paragraph 55; regard should also be had to the draft guidelines on horizontal co-operation, *ibid.*

- **Contribution to the joint venture:** Does the investment involved substantially exceed the financing capacity of each party? Does each partner have the necessary sources of supply of input products? Does each parent company have sufficiently qualified managerial capacity?

- **Production of the joint venture:** Is each parent company familiar with the technology being applied? Does each parent company manufacture the products upstream or downstream and have access to the necessary production facilities?

- **Sales by the joint venture:** Is the actual or potential demand such as to enable each parent company to manufacture the product on its own? Does each parent company have access to the necessary distribution channels for the joint venture's product?

- **Risk factor:** Can each partner bear the technical and financial risks associated with the production operations of the joint venture alone?

- **Barriers to market access:** Is each parent company on its own capable of entering the geographical market concerned? Is access to that market impeded by artificial tariff or non-tariff barriers? Can each parent company overcome those barriers without undue effort or cost?

Competition between the parent companies and the JV

16–70 Competition between the parent companies and the JV may be restricted where the JV operates on the same markets as its parents. The main risks that must be assessed are whether the parent companies and the JV have the possibility of sharing geographical or product markets or customers.

Effects on the position of third parties

16–71 Third parties who are customers may find that a JV reduces the number of suppliers which in turn reduces the scope for price competition.[6] A JV which makes a product which is supplied to the parents may affect third party suppliers because the parents are likely to cease to buy from them in the future. The very creation of the joint venture as a strong competitor may constitute a barrier to entry into the market by third parties.

Joint ventures which do not usually fall within Article 81(1)

16–72 In its Notice of 1968 concerning co-operation between enterprises[7] the Commission listed certain forms of co-operation which are neutral of competition. These include the following types of JV:

- JVs which perform certain internal organisational tasks on behalf of their parent companies, for example analysing and processing data, tax or business consultancy services or gathering information on credit guarantees or debt collection.

- JVs which deal solely with research and development up to the stage prior to industrial application of the invention, even where the parents compete with each other. There should, however, be no express or implied restriction on the parties for independent research and development.

---

[6] *e.g. GEC/Weir* [1978] 1 C.M.L.R. D42.
[7] [1968] O.J. C75/3.

- JVs which are formed by non-competing firms and which are designed to provide a joint after-sales or repair service.

- JVs set up to create a common quality label.

Joint ventures likely to fall under Article 81(1)

16–73   Competition is likely to be restricted and exemption necessary in the following cases:

- Where the parents of a JV are regarded as potential competitors and can reasonably be expected to enter the JV's market individually.

- Where the parents agree not to compete with each other or with the JV (*e.g.* in independent research and development in the field covered by the agreement). Although the Commission has on occasions taken the view that it is inherent in the formation of a joint venture that the parents will not compete with the joint venture or with each other in the field of the joint venture,[8] this is by no means a hard and fast rule.

- Where the co-operation extends to exploitation of the results of any research and development and the subsequent sale or distribution of the relevant product.

- Where the JV operates on a market neighbouring that of its parents, but where the two markets are interdependent. This may be the case where the JV manufactures products which are complementary to those of the parents.

- Where the JV has a spill-over effect on the activities of the parents. For example, the supply by the JV to its parents of an intermediate product may restrict competition between the parents in the supply of the finished products made from the intermediate product: in *Vacuum Interrupters* an agreement between switchgear suppliers to buy vacuum interrupters and components from the JV was found to restrict competition in the supply of switchgear.[9] In *WANO*, ICI and Wasag wanted to create a JV in black powder. They had interests also in other explosives and safety fuses. The Commission considered that the JV would give "opportunities and strong inducements" for market sharing in safety fuses.[10]

- Where there is a series of interlocking joint ventures. In this case even where the creation of a manufacturing JV does not restrict competition because the parties are not actual or potential competitors, competition may be restricted downstream between the various joint ventures themselves because they agree not to compete with each other. [11]

16–74   Ancillary restraints: Even if the JV does give rise to a restriction of competition (which is significant, and affects inter-State trade), it will not necessarily infringe Article 81(1). The doctrine of ancillary restraints discussed above for licence agreements also applies to JVs: if the formation of the JV itself does not restrict competition, then clauses necessary for that JV will not fall within Article 81(1). In its *Odin* decision, the Commission held that a JV for the development of a food carton did not infringe Article 81(1). The formation of the JV did not restrict competition, since the

---

[8] *GEC v. Weir*, n. 5 above.
[9] [1977] 1 C.M.L.R. D67 (No. 1); [1981] 2 C.M.L.R. 217 (No. 2).
[10] *WANO Schwarzpulver* [1979] 1 C.M.L.R. 403.
[11] *Optical fibres* [1986] O.J. L236/30.

parents were not actual or potential competitors and it did not lead to foreclosure of the market; and restrictive clauses such as the grant to the JV of the exclusive right to exploit proprietary know-how were essential for its successful functioning.[12]

16–75 Exemption: Moreover, restrictive provisions in JVs, just as in any other agreement, may be exempted under Article 81(3), either by way of individual exemptions or under block exemptions, in particular the block exemptions for technology transfer agreements[13] and R&D agreements.[14]

In its fifteenth report on competition policy (1986)[15] the Commission listed a number of general economic objectives and benefits of joint ventures: integration of the internal market, especially by means of cross-border co-operation; facilitation of risky investments; promotion of innovation and transfer of technology; development of new markets; improvement of the competitiveness of Community industry; strengthening the competitive position of small and medium-sized firms; and eliminating structural over-capacity.

Full-function joint ventures

16–76 Definition: Different rules, however, apply to a full-function joint venture, defined as a joint venture "performing on a lasting basis all the functions of an autonomous economic entity" — in other words where it results in a permanent structural change in the participating undertakings. Such a joint venture is deemed to constitute a concentration, within the meaning of the "Merger Regulation": Council Regulation 4064/89 on the control of concentrations between undertakings.[16]

16–77 Turnover thresholds: The Merger Regulation applies to concentrations which have certain turnover thresholds: either

(a) combined aggregate world-wide turnover of all undertakings greater than 5 billion Euros plus aggregate Community-wide turnover of each of at least two of the undertakings greater than 250 million Euros, *unless* each of the undertakings achieves more than two thirds of its aggregate Community-wide turnover within one and the same Member State; or

(b) combined aggregate world-wide turnover of all undertakings greater than 2.5 billion Euros plus, in each of at least three Member States, combined aggregate turnover of all undertakings concerned greater than 100 million Euros, plus, in each of at least three of those Member States, aggregate turnover of at least two of the undertakings concerned greater than 25 million Euros, plus aggregate Community-wide turnover of each of at least two of the undertakings concerned greater than 100 million Euros, *unless* each of the undertakings does not achieve more than two-thirds of its E.C.-wide turnover within the same Member State.

---

[12] *Elopak/Metal Box—Odin* [1990] O.J. L209/15; also *Mitchell Cotts/Sofiltra* [1988] 4 C.M.L.R. 111.
[13] As to which see above at paras 16–40 *et seq*; see particularly Articles 5(1)(2) and 5(2)(1) of the block exemption.
[14] As to which see above at paras 16–56 *et seq.*
[15] At para. 26.
[16] [1990] O.J. L257/14, as amended by Council Regulation 1310/97, [1997] O.J. L180/1; Article 3(2) and see for guidance on the interpretation of this provision the Commission Notice on the concept of full-function joint ventures [1998] O.J. C66/1.

16–78   Test for prohibition: Where the Merger Regulation applies, it supersedes control under Articles 81 and 82, and the concentration must be notified to the Commission within a week after conclusion of the relevant agreement, announcement of the public bid or acquisition of a controlling interest. The Commission will then evaluate the concentration and will declare it incompatible with the Common Market and prohibited under Article 2(3) of the Regulation if it creates or strengthens a dominant position, as a result of which effective competition would be significantly impeded in the common market or in a substantial part of it.

C. Abuse of a dominant position

Prohibition

16–79   Article 82 provides:

> "Any abuse by one or more undertakings of a dominant position within the common market or in a substantial part of it shall be prohibited as incompatible with the common market in so far as it may affect trade between Member States. Such abuse may, in particular, consist in:
>
> (a)   directly or indirectly imposing unfair purchase or selling prices or other unfair trading conditions;
>
> (b)   limiting production, markets or technical development to the prejudice of consumers;
>
> (c)   applying dissimilar conditions to equivalent transactions with other trading parties, thereby placing them at a competitive disadvantage;
>
> (d)   making the conclusion of contracts subject to acceptance by the other parties of supplementary obligations which, by their nature or according to commercial useage, have no connection with the subject of such contracts."

The requirements for an infringement of Article 82 are therefore (1) dominance and (2) abusive conduct and (3) effect on trade between Member States.

Dominance

16–80   Definition: In order to determine dominance it is necessary to determine (1) the relevant market; and (2) the power of the undertaking on that market.

Relevant market

16–81   Product market: Assessment of the relevant market under Article 82 involves determining the relevant product market as well as the relevant geographic market.[17] The primary consideration is demand substitution and in this respect the relevant product market is defined as comprising "all those products and/or services which are regarded as interchangeable or substitutable by the consumer, by reason of the products' characteristics, their prices and their intended use".[18]

Intellectual property, like any other kind of property, could in theory itself be regarded as a "product", in which the holder of the intellectual property enjoyed a

---

[17]   See section on market definition above at paras 16–14 to 16–16.
[18]   Commission Notice on the definition of the relevant market for the purposes of Community competition law, [1997] O.J. C372.

monopoly. However E.C. competition law does not recognise a market for the licensing or use of an intellectual property right itself (see *Volvo v. Veng*[19]). Rather, the relevant product market is the market for the products protected by the intellectual property. In *Magill*, the television broadcasters ITV and the BBC held copyright in programme listings which were the basic information necessary to publish television listings magazines. The relevant question was not their position on the market for access to the copyright information (in which they had a monopoly) but rather their position on the market for television magazines produced using that information.[20]

This approach was adopted by the English High Court in *Philips v. Ingman*.[21] Laddie J., applying the judgment of the Court of Justice in *Volvo v. Veng*, held that a market in which a dominant position is held had to be defined by reference to particular products or services. A market could not be defined by reference to an intellectual property right itself. The case of *Magill*[22] did not support "the suggestion that an intellectual property right gives rise to a per se dominant position in the area of technology covered by the right."[23] Nor did an intellectual property right constitute an essential facility so as to make access mandatory. The Commission's *Port of Rødby* decision[24] did not support compulsory licensing: otherwise "[o]wnership of an intellectual property right would, per se, mean the ownership of a dominant position or an essential facility and the refusal of a licence would, per se, amount to an abuse of it."[25] In short, therefore, dominance cannot be defined by the patented technology since this would destroy the very existence of the patent, which is to reward the inventor by giving him the exclusive right to exploit the invention (or as Laddie J. stated to prevent someone else from exploiting the patent).

Laddie J. was prepared to admit that there may be exceptions to this rule. However, it is incumbent to plead "*the exceptional features which take the case outside* Volvo v. Veng."[26] One such exceptional case (that was not fully explored in *Philips v. Ingman* but is one that is generally thought to be favoured by the E.C. Commission) concerns the situation where the intellectual property right has become a standard. Once an essential technology is included in a product that has become an industry standard, with the agreement of the proprietor of the underlying rights to that technology, particularly a technology that is made mandatory pursuant to Community legislation, the owner of the underlying intellectual property might occupy a dominant position *vis-à-vis* manufacturers requiring licences of the intellectual property in order to be able to participate in the market for the equipment in question. Thus, if a patentee (a) chooses to "lend" his technology to a standards authority so as to gain greater acceptance for that technology and (b) the standard becomes universally accepted so that it is impossible for a manufacturer to do business unless he meets that standard, the essential function (*i.e.* the specific subject matter) of the patent is not undermined by compulsory licensing. The compulsory licensing is the *quid pro quo* of having the technology accepted as the industry standard.

---

[19] Case C-238/87 [1988] E.C.R 6211.
[20] Cases C-241/91P and C-242/91P *RTE and ITP v. Commission* [1995] E.C.R. I-743, para. 47.
[21] [1998] Eu.L.R. 666.
[22] Above n. 20.
[23] *ibid.* at 684 E.
[24] [1994] 5 C.M.L.R. 457.
[25] Above, n. 21 at 683 A.
[26] *ibid.* at 686.

16–82 Geographic market: The relevant geographic market "comprises the area in which the undertakings concerned are involved in the supply and demand of products or services, in which the conditions of competition are sufficiently homogeneous and which can be distinguished from neighbouring areas because the conditions of competition are appreciably different in those areas".[27] Under this test, the geographic market may be as large as the whole world, or as small as a single port. In intellectual property cases the relevant geographic market will often be defined as the area covered by the intellectual property.

Market power

16–83 A dominant position has been defined as "a position of economic strength enjoyed by an undertaking which enables it to prevent effective competition being maintained on the relevant market by giving it the power to behave to an appreciable extent independently of its competitors, its customers and ultimately of its consumers."[28]

The holder of an intellectual property right such as a patent or design right has a legal "monopoly" in any product that is protected by the relevant right. But this is not sufficient for dominance under Article 82. Dominance must be examined in the context of a market. This may include products that compete with the patented product. In *Parke Davis v. Probel*, Advocate General Roemer stated that to prove that a patent holder was dominant "it was not enough to prove the existence of a legal monopoly (like that enjoyed by the owner of a patent) but that it was necessary to determine the position which the holder of the patent occupies on the market, whether the product which he manufactures is of its nature subject to competition, or at least does not meet substantial competition, and whether, therefore the person enjoying the monopoly is in a position to fix prices and terms".[29] This approach has been endorsed by the Court in relation to other intellectual property rights.[30]

Thus under Community competition law mere ownership of intellectual property cannot confer a dominant position. Rather, a person owning intellectual property will only be dominant where it enjoys economic strength on the market for the products protected by those rights. Hence in *Magill*, the dominant position of the television companies derived not from their copyright *per se*, but because their *de facto* monopoly in that information enabled them to prevent effective competition in the market for weekly television magazines.[31]

Abuse

16–84 Examples of abuse: Abuse may be exploitative or exclusionary. Examples are listed in Article 82 itself and include excessive pricing, predatory pricing, discrimination (including, for example, price discrimination), fidelity rebates, refusal to supply, tying, and applying unequal conditions to equivalent transactions with other trading parties. Article 82 also covers in principle abusive use of an intellectual property right, particularly where the object or effect is to divide markets along national lines.

---

[27] Commission Notice on the definition of the relevant market, above n. 18.
[28] Case 27/76 *United Brands v. Commission* [1978] E.C.R. 207, para. 65.
[29] Case 24/67 [1968] E.C.R. 55 at 78.
[30] See *e.g.* Case 40/70 *Sirena v. EDA* [1971] E.C.R. 69; Case 51/75 *EMI v. CBS* [1976] E.C.R 811; and Case 102/77 *Hoffmann-La Roche v. Centrafarm* [1978] E.C.R. 1139 (trademarks); and Case 78/80 *Deutsche Grammophon v. Metro* [1971] E.C.R. 487 (copyright).
[31] Above n. 20, para. 47. Note that the Court of Justice, unlike the Court of First Instance and Commission, was careful *not* to suggest that the dominance of the television companies was based on their monopoly in a market for copyright information itself.

16–85 Refusal to license: This has given rise to particular problems in the field of intellectual property. The refusal to license intellectual property does not, itself, infringe Article 82. So the court held in *Volvo v. Veng*, when considering Volvo's refusal to license independent repairers to make spare body parts for Volvo cars. The right of the proprietor of a protected design to prevent third parties from manufacturing, selling and importing without his consent products incorporating that design constituted the subject matter of the intellectual property right. A refusal to license the right could not, therefore, constitute an abuse of a dominant position.[32]

However, the exercise of the exclusive right might in "exceptional circumstances" involve abusive conduct which infringes Article 82.[33] In *Volvo v. Veng* the court gave as examples the arbitrary refusal to supply spare parts to independent repairers, the fixing of prices for spare parts, or a decision to stop producing spare parts for a model which was still in circulation. In *Magill* the refusal to licence copyright information prevented the appearance of a new product for which there was a potential consumer demand, which constituted abuse under Article 82(b).[34] A refusal to license might also infringe Article 82 where it has the effect of partitioning the market.[35]

D. PRACTICE AND PROCEDURE

Civil proceedings in national courts

16–86 Nullity under Article 81(2): An agreement which falls within Article 81(1), which is not covered by a block exemption, and which has not been exempted individually is automatically void under Article 81(2). Under United Kingdom law, this has two consequences.

16–87 Severance: First, the agreement may not be enforced unless the offending provisions may be severed leaving the remainder of the agreement intact. The test for severance has been variously stated in the case law,[36] but the underlying principle is that severance will only be prevented where the restrictive provision forms substantially the *whole* or *principal* consideration for the agreement, such that its deletion entirely alters and negates the main purpose of the agreement.

16–88 Damages/injunction: Secondly, breach of Article 81 may give rise to a cause of action for damages and/or injunctive relief.[37] However recent case law of the English Court of Appeal has confined this right to third parties which are harmed by the agreement. Parties to a prohibited agreement, themselves, are prevented under the principle of *ex turpi causa non oritur actio*[38] from raising the illegality of the agreement by way of defence of set off arising out of a counterclaim to an action for

---

[32] Above n. 19, para. 8.

[33] *Magill* case, above n. 20, para. 49.

[34] *ibid.*, para. 54.

[35] Case 22/79 *Greenwich Film Production v. SACEM* [1979] E.C.R. 3275, para. 12 (copyright).

[36] See *Crehan v. Courage* [1999] Eu.L.R. 834, with references to, *inter alia, Chemidus Wavin v. TERI* [1978] 3 C.M.L.R. 514, *Amoco Australia v. Rocca Bros Motor Engineering* [1975] A.C. 561 and *Marshall v. NM Financial Management* [1997] 1 W.L.R. 1527.

[37] See *Garden Cottage Foods v. Milk Marketing Board* [1984] 1 A.C. 130, which concerned Article 82 but which has been applied to the competition rules of the Treaty in general in *e.g. An Bord Bainne v. MMB* [1984] 1 C.M.L.R.519; *Bourgoin v. MAFF* [1986] 1 Q.B. 716; and *Heathrow Airport v. Forte* [1998] Eu.L.R. 98. The right to damages and injunctive relief in a national court for breach of Articles 81 and 82 has been established and applied in nearly all other Member States of the E.C.

[38] See *Tinsley v. Milligan* [1994] 1 A.C. 340.

breach of contract,[39] or *a fortiori* from bringing a new action for damages for breach of Article 81.

**16–89	Breach of Article 82:**	This may, like Article 81, give rise to a cause of action for damages and/or injunctive relief in national courts.

Powers of the Commission

16–90	Breach of Articles 81 and 82 also has administrative law consequences. Under Regulation 17, the Commission has wide powers to enforce the competition rules. In particular, it is the only body which may grant an exemption under Article 81(3). It may also investigate infringements of Articles 81 and 82, issue formal decisions in respect of infringements, and impose various sanctions including interim measures, orders to terminate infringements, and fines.

Notification

**16–91	Purpose of notification:**	An agreement which is covered by a block exemption is, as discussed above, automatically exempt under Article 81(3). Parties to an agreement or practice not covered by a block exemption may notify the agreement to the Commission, on the standard form known as Form A/B, to obtain either an individual exemption under Article 81(3), or to obtain a decision granting "negative clearance": *i.e.* a decision that the agreement or practice does not fall within Articles 81(1) or 82.

Due to the large number of notifications, however, the vast majority of cases are dealt with not by means of a formal exemption or negative clearance decision, but rather by means of a "comfort letter" indicating that the Commission does not object to the notified agreement or practice.

Benefits of Notification

**16–92	Consequences:**	Notification has several benefits. First, it protects the parties from fines. The Commission cannot impose fines in respect of acts taking place between notification and issuing the decision; or after the Commission has issued a negative clearance decision or comfort letter. Secondly, if an individual exemption is granted under Article 81(3), the Commission may grant the exemption with retroactive effect to the date of notification. Thirdly, national courts must avoid taking decisions which may contradict Commission decisions. Therefore if an agreement has been notified, and it is not obvious that the agreement either falls outside Article 81(1), or that it falls within Article 81(1) and would not qualify for exemption, then the national court should generally stay any proceedings on the validity of the agreement.[40]

Drawbacks of Notification

16–93	Filling in form A/B can be time-consuming, especially with regard to analysis of the relevant market. When the form is submitted, the Commission then becomes aware of the agreement, and may undertake an investigation, possibly identifying

---

[39] *Gibbs Mew v. Gemmell* [1998] Eu.L.R. 588 (CA) and *Crehan v. Courage*, above n. 36. However at the time of writing a reference to the ECJ is pending in *Crehan*.
[40] Case C-234/89 *Delimitis* [1991] E.C.R. 935.

other potential infringements. It is possible that the Commission may require amendments to be made to the agreement, in order to issue a comfort letter or to grant an exemption, and may attach conditions to an exemption. This gives the other party the chance to try to renegotiate the agreement, perhaps when the other is in a weaker position than when it was first negotiated.

16–94   Non-notifiable agreements: Article 4(2) of Regulation 17[41] sets out certain, exceptional categories of case in which the Commission may grant exemption without notification. These include, most importantly, all vertical agreements relating to the conditions under which the parties may purchase, sell or resell goods or services. Non-notifiable agreements also include agreements between undertakings from one Member State, which do not relate either to imports or exports between Member States, and agreements made between two undertakings which only impose restrictions on the exercise of the rights of the assignee or user of industrial property rights (in particular patents, utility models, designs and trademarks) or of the person entitled under a contract to the assignment or grant of certain rights relating to industrial processes.

This Article provides an exception from the requirement to notify for certain categories of agreement thought to have little anti-competitive effect. However an agreement which falls within Article 4(2) may still be found to infringe Article 81(1) and, if it does, may be the subject of fines.

Investigations and decisions

16–95   Provision of information: The Commission may investigate agreements or practices either following a notification, or following a complaint by a third party, or on its own initiative. Regulation 17 gives it powers to compel undertakings to provide necessary information, and provides for fines where undertakings intentionally or negligently supply incorrect information, as well as for failing to supply information at all pursuant to a decision requiring the information to be supplied.

16–96   Dawn raids: In addition to requesting or requiring information, Commission officials may carry out investigations of undertakings either with prior notice, or without any warning — the so called "dawn raids". Undertakings may be fined for refusing to submit to an investigation ordered by formal decision of the Commission. Article 14 of Regulation 17 provides that Commission officials are empowered:

> "to examine the books and other business records, to take copies of or extracts from the books and business records, to ask for oral explanation on the spot and to enter any premises . . ."

The power to ask questions is subject to the right against self-incrimination. In other words, the Commission can attempt to ascertain relevant facts ("where was X on Y date?"), but cannot elicit incriminating answers ("did X fix the price with Z on Y date?")[42]

16–97   Legal professional privilege: Legal professional privilege extends to documents made for the purpose and in the interest of the undertakings right of defence.

---

[41] As amended by Regulation 1216/1999, [1999] O.J. L148/5.
[42] See generally Case 374/87 *Orkem v. Commission* [1989] E.C.R. 3283

The document may come into existence prior to the initiation of any Commission investigation, but must relate to the subject matter of that investigation. An important limitation is that the privilege does not extend to documents emanating from in-house lawyers, unless the document is summarising advice from independent or out-house lawyers.[43]

16–98   Statement of objections and hearing: Prior to issuing a decision finding an infringement of Article 81 or 82, the Commission must issue a statement of objections, with a time limit for the parties to reply. Any party which the Commission proposes to fine or penalise may require the holding of an oral hearing. Proceedings at the hearing — as all other Commission proceedings under Regulation 17 — are administrative, rather than judicial in nature. However the Commission must respect the rights of defence and accordingly must conduct its procedure fairly.

16–99   Adoption of decision: After hearing the views of the parties, and prior to taking any adverse decision under Article 81 or 82, the Commission must consult the Advisory Committee on Restrictive Practices and Monopolies. The Commission may then adopt and authenticate its final decision.

Sanctions

16–100   Interim measures: Prior to adoption of a formal decision, the Commission may adopt interim measures in cases of urgency where the practices in issue are likely to cause serious and irreparable damage to a party, or damage which would be intolerable for the public interest.[44]

16–101   Termination: A decision finding an infringement of Article 81 or 82 may order termination of the infringement, which may include positive action to be taken by the undertaking. In the *Magill* case, for example, the Commission ordered the television companies to supply weekly programme listings to each other and to third parties on request and on a non-discriminatory basis; and if this was done by way of licence the royalties had to be reasonable.[45]

16–102   Fines: Finally, where undertakings intentionally or negligently breach Articles 81 or 82, the Commission may impose fines of up to 1 million Euros or 10 per cent of the turnover of the undertaking concerned in the previous business year, whichever is the greater. The Commission may impose fines even where an agreement with the object of restricting competition has never in fact been implemented, and even if the infringing conduct has ceased. However, in fixing the amount of the fine, the Commission must have regard to the gravity and duration of the infringement.

---

[43] Case 155/79 *AM & S v. Commission* [1982] E.C.R. 1575.
[44] Case T-44/90 *La Cinq v. Commission* [1992] E.C.R. II-1.
[45] [1989] O.J. L78/43, [1989] 4 C.M.L.R. 757.

SECTION 2: FREE MOVEMENT OF GOODS

Introduction

The Treaty provisions

16–103 The creation of an internal market constitutes a fundamental principle of European Community Law. The internal market is characterised by the abolition as between Member States of obstacles to the free movement of goods, persons, services and capital.[46] National intellectual property rights are by their very nature liable to destroy such a market since each national intellectual property right bestows on the owner a degree of exclusivity within the national territory.

The Treaty attempts to balance these conflicting interests in Articles 28 and 30 (ex Articles 30 and 36). They provide:

> "Quantitative restrictions on imports and all measures having equivalent effect shall be prohibited between Member States." [Article 28]

> "The provision of Article 28 shall not preclude prohibitions or restrictions on imports . . . justified on grounds of . . . protection of industrial and commercial property. Such prohibition or restriction shall not, however, constitute a means of arbitrary discrimination or a disguised restriction on trade between Member States." [Article 30]

The effect of these Treaty Articles is that reliance on national intellectual property rights to prevent the importation of goods (*e.g.* by injunctive relief) from other Member States constitutes a quantitative restriction on imports within the meaning of Article 28 of the Treaty.[47] The first sentence of Article 30 provides that restrictions on imports may be justified in order to protect intellectual property rights, but the second sentence contains an important proviso; reliance on such rights will not be justified if it constitutes a means of arbitrary discrimination or a disguised restriction on trade between Member States.[48]

Specific subject matter

16–104 The Court of Justice has interpreted the first sentence of Article 30 so as to give greater effect to the principle of a single market. The court has held that reliance on intellectual property rights to prevent importation or the marketing of imported goods can only be "justified" under Article 30 if such reliance is for the purpose of safeguarding the rights which constitute "the specific subject matter" or the essential function of the intellectual property right. This means that the parallel importer might not infringe intellectual property law applicable in the importing Member State. For example, the packaging of imported products may need to be modified so that they can be lawfully sold in the Member State of importation. This repackaging may infringe the manufacturer's trade mark. Despite the infringement, however, Community law allows the repackaging since, if the manufacturer of the product could rely on

---

[46] Article 3(c).
[47] Case 6/81 *Industrie Diensten Groep v Beele* [1982] E.C.R. 707.
[48] The same principles apply to trade within the European economic Area (EEA). The EEA Agreement applies to the European Union, Iceland, Liechtenstein and Norway.

domestic trade mark rights to prevent such re-packaging, the single market could be artificially partitioned along national markets.[49]

This dilution of the protection afforded by national law is justified as follows. In the absence of harmonisation of national intellectual property law at Community level (*i.e.* all national laws afford the same protection), the requirement that the national right must comply with the overriding Community condition of essential function is clearly necessary. There needs to be some Community benchmark by which the national right can be tested so that the Common Market is not compartmentalised. Thus, if the national right does not conform to what the Community perceives to be the proper function of the intellectual property, the competing interest of free trade must take precedence. By contrast, if the national right does conform to what the Community perceives to be the proper function of the intellectual property, the competing interest of free trade is secondary to the intellectual property right's protection.

This balancing exercise between protection under national law and the essential function of the right recognised by Community law is well illustrated by *Warner Brothers v. Christiansen.*[50] The owner of a Danish video shop, Christiansen, purchased the video of "Never Say Never Again" in the United Kingdom, rented it out in Denmark but refused to pay any royalty to Warner Brothers, the owners of the copyright, on each rental as required by Danish law. Christiansen argued that the United Kingdom law did not recognise rental rights and that, since Warner Brothers had consented to the marketing of the video in the United Kingdom, they must be taken to have exhausted copyright, including rental rights, in Denmark. The court rejected this argument. Although the Danish law constituted a restriction on trade contrary to Article 28 because it was liable to hinder trade in video cassettes,[51] the Danish recognition of rental rights formed part of the specific subject matter of copyright. Thus, the protection of national intellectual property rights took precedence over the Community principle of free trade. The court stated:

"... where national legislation confers on authors a specific right to hire out video-cassettes, that right would be rendered worthless if its owner were not in a position to authorise the operations for doing so. It cannot therefore be accepted that the marketing by a film-maker of a video-cassette containing one of his works, in a Member State which does not provide specific protection for a right to hire it out, should have repercussions on the right conferred on that said film-maker by the legislation of another Member State to restrain, in that State, the hiring-out of that video cassette."[52]

A. HARMONISATION

Protection under national law

16–105 In the absence of harmonisation at Community level, national law continues to apply as regards the conditions and procedures under which protection is afforded. Thus in *Keurkoop v. Nancy Keen Gifts,*[53] Nancy Keen Gifts sought an interlocutory

---

[49] For the circumstances where the repackaging is permitted see para. 16–114.
[50] Case 158/86 [1988] E.C.R. 2605. See also Case 42/85 *Bassett v. SACEM* [1987] E.C.R. 1747.
[51] Paragraph 10. Compare Case 187/80 *Merck v. Stephar* [1981] E.C.R. 2063, where a Dutch patentee could not prevent re-importation into Netherlands of goods previously marketed with consent in Italy where no patent protection existed.
[52] Paragraph 18.
[53] Case 144/81 [1982] E.C.R. 2853.

injunction against Keurkoop to prevent the latter from selling a handbag, the appearance of which was virtually identical with Nancy Keen's registered design of the bag. The defendant could not challenge Nancy Keen's exclusive rights as the Benelux law of designs granted the exclusive right to the first person to file the design in spite of the fact that it had previously been filed elsewhere. The European Court of Justice ruled that Article 36 (now Article 30) did not preclude the application of such a national law. The Court stated in (paragraph 18):

"on that issue the Court can only state that in the present state of Community law and in the absence of community standardisation or of harmonisation of laws the determination of the conditions and procedures under which protection of designs is granted is a matter for national rules and, in this instance, for the common legislation established under the regional union between Belgium, Luxembourg and the Netherlands referred to in Article 233 of the treaty".

This principle has been consistently applied by the Court of Justice and has been extended to patents,[54] copyright,[55] trademarks.[56]

No discrimination

16–106 Although the conditions and procedures under which protection is granted is a matter of national rules in the absence of Community harmonisation, nevertheless such conditions or protection cannot discriminate against nationals of other Member States by virtue of Article 12 (ex Article 6) of the Treaty.[57]

Harmonisation of national laws

16–107 Although the protection of intellectual property rights remains largely subject to the laws of the Member States, some progress has been made by the Community in harmonising national laws. In respect of patents, the Community has adopted Regulation 1768/92 concerning the creation of a supplementary protection certificate for medicinal products, Regulation 1610/96 concerning the creation of a supplementary protection certificate for plant protection products and Directive 98/44 on the legal protection of biotechnological inventions. It has also signed the Second Community Patent Convention, but this is not yet in force, and it is unlikely that it will be in the short term. As regards copyright and related rights, the following directives have been adopted: Directive 91/250 on the legal protection of computer programmes, Directive 93/98 harmonising the term of protection of copyright, and Directive 96/9 on the legal protection of databases, which also introduced a new database right. In the field of designs, there is Directive 98/71 on the legal protection of designs, and Directive 87/54 on the legal protection of topographies of semiconductor products. As regards trade marks, Directive 89/104 to approximate the laws of the Member States relating to trade marks and Regulation 2868/95/E.C. implementing Regulation 40/94, which established the Community Trade Mark, have been adopted. There are other Community measures relating to intellectual property, but those listed above are of particular relevance to technology.

---

[54] Case 35/87 *Thetford Corporation v. Fiamma SpA* [1988] E.C.R. 3585.
[55] Case 53/87 *CICRA et Maxicar v. Renault* [1988] E.C.R. 6039.
[56] Case C-317/91 *Deutsche Renault v. Audi* [1993] E.C.R. I-6227. There is now harmonisation of the national rules relating to trade marks.
[57] Cases C-92/92 and 326/92 *Phil Collins v. Imtrat* [1993] E.C.R. I-5145.

Certain of these measures (and legislation under United Kingdom national law) contain provisions that reflect community laws on the free movement of goods.[58]

B. Exhaustion of rights

16–108 In its attempt to regulate the tensions between the single market and the protection of national intellectual property law, the Court of Justice has applied the principle of exhaustion of rights. This principle applies to all forms of intellectual property rights. The effect of the principle is that the owner of the intellectual property right has the exclusive right to place the product concerned on the common market (and the EEA) for the first time and prevent infringement of that right. However, the act of first marketing by the owner or with his consent in Member State A thereafter exhausts his right to object to the product being imported into Member State B. Put simply, if Kodak markets Kodak film in France it cannot object to the importation into the United Kingdom of the French film.

Whilst this book does not cover trade marks in depth, it is worth considering the trade mark cases more fully in this section as trade marks have been the subject of many of the seminal cases.

C. Patents

Essential function

16–109 The essential function of a patent ensures that the creative effort of the inventor is respected by guaranteeing the inventor the exclusive right to use the invention for the purpose of manufacturing the product or putting it into circulation for the first time either directly or by the grant of licences. It is not part of the specific subject matter of the patent to rely on a national patent to prevent the importation of the product which has been marketed in another Member State by the patentee or with his consent. In *Centrafarm v. Sterling Drug*, the court stated:

> "In as much as it provides an exception to one of the fundamental principles of the Common Market, Article 36 in fact only admits derogations from the free movement of goods where such derogations are justified for the safeguarding rights which constitute the specific subject matter of this property.
>
>   In relation to patents, the specific subject matter of the industrial property is the guarantee that the patentee to reward the creative effort of the inventor, has the exclusive right to use the invention with a view to manufacturing industrial products and putting them into circulation for the first time, either directly or by the grant of licenses to third parties as well as the right to oppose infringements.
>
>   An obstacle to the free movement of goods may arise out of the existence, within a national legislation concerning industrial and commercial property, of provisions laying down that a patentee's right is not exhausted when the product protected by the patent is marketed in another Member State, with the result that the patentee can prevent importation of the product into his own Member State when it has been marketed in another Member State."[59]

---

[58] See in particular Directive 98/44 (biotechnology) — Article 10; Directive 91/250 (computer programmes) — Article 4(c); Directive 98/71 (designs) — Article 15; Directive 87/54 (semiconductor topographies) — Article 5(5); Directive 89/104 (trade marks) — Article 7; Directive 40/94 (community trade marks) — Article 13.

[59] Case 15/74 [1974] E.C.R. 1147, paras 8–10.

Consent

16–110 Consent will be inferred where the imported product has been previously marketed in another Member State by a parallel patentee, its subsidiary, or its licensee. It is probable that the licensee and subsidiary could not object to parallel imports of products first marketed in another Member State by the licensor or by the parent company, since they must bear the consequences of their choice.[60] The crucial test is whether the marketing has been carried out by a person who is legally and economically independent from the person entitled to sue on the patent. In *Centrafarm* the court continued:

> "Whereas an obstacle to the free movement of goods of this kind may be justified on the ground of protection of industrial property, where such protection is invoked against a product coming from a Member State where it is not patentable and has been manufactured by third parties without the consent of the patentee and in cases where there exist patents, the original proprietors of which are legally and economically independent, a derogation from the principle of the free movement of goods is not, however, justified where the product has been put onto the market in a legal manner by the patentee himself or with his consent in the Member State from which it has been imported, in particular in the case of a proprietor of parallel patents.
>
> In fact, if a patentee could prevent the import of protected products marketed by him or with his consent in another Member State, he would be able to partition off national markets and thereby restrict trade between Member States, in a situation where no such restriction was necessary to guarantee the essence of the exclusive rights flowing from the parallel patents."[61]

The marketing in the other Member State must, however, be truly consensual. Thus the Court of Justice has held that a product has not been marketed with the consent of the patentee where the product has been manufactured in another Member State under a compulsory licence.[62] A legal obligation to manufacture or market the product in another Member State would vitiate consent for the purposes of the exhaustion of rights principle. However, ethical obligations cannot be a basis for derogating from the rule on the free movement of goods. Consequently, where a company markets a product in one Member State where there is no patent protection, that company cannot thereafter object to parallel imports from that Member State to Member States where there is patent protection.[63]

Disguised restriction on trade

16–111 Even if reliance on national patent law seeks to protect the essential function of the patent, the restriction on imports will not be justified if reliance constitutes a discriminatory or disguised restriction on trade.[64] In *Commission v. United Kingdom* the Court of Justice considered that section 48 of the Patents Act 1977 constituted a disguised restriction on trade. The Act provided that a compulsory patent licence could be granted even though the domestic demand was being met by imports;

---

[60] By analogy with Case C-9/93 *Ideal Standard* [1994] E.C.R. I-2789.

[61] Above n. 59, paras 11–12.

[62] Case 19/84 *Pharmon v. Hoechst* [1985] E.C.R. 2281.

[63] Joined Cases C-27/95 and C-268/95 *Merck v. Primecrown* [1996] E.C.R. I-6285.

[64] *e.g.* Case C-30/90 *Commission v. United Kingdom* [1992] E.C.R. I-829 (compulsory patent licence could be granted unless patent was being worked by manufacturer within the United Kingdom). Also Case 434/85 *Allen and Hanburys v. Generics* [1988] E.C.R. I-245.

whereas, by contrast, a compulsory patent licence would not be granted if demand was being met by domestic production. The court considered that this encouraged the patentee to manufacture in the United Kingdom and amounted to a disguised restriction on trade.[65]

D. TRADE MARKS

Essential function

16–112   The early case law of the Court of Justice regarded trade marks as a lower class of intellectual property right (see *Sirena v. EDA*[66]). This view was severely criticised by Advocate General Jacobs in *HAG II* who stated:

> "the truth is that, in economic terms, and perhaps also from the "human point of view", trade marks are no less important, and no less deserving of protection, than any other form of intellectual property. They are in the words of one author "nothing more nor less than the fundament of most market based competition"[67] . . . Like patents, trade marks find their justification in a harmonious dovetailing between public and private interests. Whereas patents reward the creativity of the inventor and thus stimulate scientific progress, trade marks reward the manufacturer who consistently produces high quality goods and they thus stimulate economic progress. Without trade mark protection there would be little incentive for manufacturers to develop new products or to maintain the quality of existing ones."[68]

Although the Court of Justice had subsequently qualified its approach in *Sirena*,[69] in *HAG II* it clearly endorsed the view of its Advocate General, defining the specific subject matter of trade marks as follows:

> "Trade mark rights are, it should be noted, an essential element in the system of undistorted competition which the Treaty seeks to establish and maintain. Under such a system, an undertaking must be in a position to keep its customers by virtue of the quality of its products and services, something which is only possible if there are distinctive marks which enable customers to identify those products and services. For the trade mark to be able to fulfil this role, it must offer a guarantee that all goods bearing it have been produced under the control of the single undertaking which is accountable for their quality.
>
>   Consequently, as the Court has ruled on numerous occasions, the specific subject matter of the trade marks is in particular the guarantee to the proprietor of

---

[65] *ibid.* See now section 48B of the 1977 Act as inserted by Regulation 5 of the Patents and Trade Marks (World Trade Organisation) Regulations 1999.

[66] Above n. 30, where the court stated at para. 7: "the exercise of a trade mark right is particularly apt to lead to a partitioning of the markets, and thus to impair the free movement of goods between States which is essential to the Common Market. Moreover a trade mark is distinguishable in this context from other rights of industrial or commercial property, in as much as the interest protected by the latter are usually more important, and merit the higher degree of protection, than the interest protected by an ordinary trade mark."

[67] W.R. Cornish, *Intellectual Property: Patent Copyright, Trademarks and Allied Rights* (2nd ed., 1989) p. 393 [now in 4th edition].

[68] [1990] E.C.R. I-3711, paras 17–18.

[69] In Case 16/74 *Centrafarm v. Winthrop* [1974] E.C.R. 1183; and *Hoffmann-La Roche v. Centrafarm*, above n. 30.

the trade mark that he has the right to use that trade mark for the purpose of putting a product into circulation for the first time and therefore to protect him against competitors wishing to take advantage of the status and reputation of the trade mark by selling products illegally bearing that mark. In order to determine the exact scope of this right exclusively conferred on the owner of the trade mark, regard must be had to the essential function of the trade mark, which is to guarantee the identity of the origin of the marked product to the consumer or ultimate user by enabling him without any possibility of confusion to distinguish that product from products which have another origin."[70]

Consent

16–113 The principle of exhaustion of rights applies to trade marks. The Court of Justice in *Centrafarm v. Winthrop* held that:

"The exercise, by the owner of a trade mark, of the right which he enjoys under the legislation of a Member State to restrict the sale, in that State, of a product which has been marketed under the trade mark in another Member State by the trade mark owner or with his consent is incompatible with the rules of the EEC Treaty concerning the free movement of goods within the Common Market."[71]

As in the case of patents, consent is not implied where the first marketing has been carried out by an undertaking which is "economically and legally independent" of the trade mark proprietor.[72] Thus, Community law will not imply consent where the goods were produced by an assignee which has no economic link with the owner of the mark in the imported Member State. Consent will be implied where the goods are marketed by a licensee or a subsidiary of the owner. Although a sister company does not have control over the quality of goods produced by its sister company in another Member State, there are sufficient legal and economic links between the two to justify the implication of consent. As the Court of Justice stated in the *Ideal Standard* case, Articles 28 and 30 "require the group to bear the consequences of its choice."[73]

Disguised restriction on trade

16–114 A trade mark owner cannot prevent an importer from replacing the packaging or labelling or re-branding a product if this would lead to an artificial partitioning of the market within the meaning of the second sentence of Article 30. Such a disguised restriction will arise where four conditions are satisfied[74]:

(i)   the restriction on the parallel importer artificially partitions the Common Market;

(ii)   the repackaging, re-branding or re-labelling does not affect the original condition of the product;

(iii) the repackaging, relabelling or re-branding is not such as to damage the reputation of the trade mark and its owner;

---

[70] Above n. 68, judgment paras 13–14.
[71] Above n. 69, para. 12.
[72] See *HAG II*, above n. 68, para. 15.
[73] Above n. 60, para. 38.
[74] Cases C-427, 429 & 436/93 *Bristol-Myers Squibb v Paranova ("Paranova")* [1996] E.C.R. I-3457, paras 49–50; Case C-349/95 *Loendersloot v. Ballantine ("Ballantine")* [1997] E.C.R. I-6227, para. 29; and Case C-379/97 *Pharmacia & Upjohn v. Paranova ("Pharmacia")*, [1999] E.C.R. I-6927 para. 19.

(iv) the parallel importer provides the trade mark owner with a specimen of the product as repackaged, relabelled or re-branded and the repackaging etc identifies the parallel importer as having carried it out.

(i) *Artificial partitioning of the markets between Member States*

16–115 The words "artificial partitioning of the markets" do not imply that the importer must demonstrate that, by putting an identical product on the market in various forms of packaging in different Member States, the trade mark owner deliberately sought to partition the markets between Member States. By using the word "artificial", the court stresses that the action taken by the trade mark owner must be justified by the need to safeguard the essential function of the trade mark.[75] Nevertheless, the parallel importer is only entitled to repackage, re-label or re-brand so far as it is "necessary in order to market the product in the Member State of importation." As the court stated in *Paranova*, "the power of the owner of trade mark rights protected in a Member State to oppose the marketing of repackaged products under the trade mark should be limited only insofar as the re-packaging undertaken by the importer is necessary in order to market the product in a Member State of importation."[76] This principle of necessity carries with it the principle of proportionality. Thus the trade mark owner may object to the repackaging if, for example, fixing new labels would suffice. As the court indicated in *Ballantine*:

> "The person carrying out the relabelling must however use the means which make parallel trade feasible while causing as little prejudice as possible to the specific subject matter of the trade mark right. Thus, if the statements on the original labels comply with the rules on labelling in force in the Member State of destination, that those rules require additional information to be given, it is not necessary to remove and reaffix or replace the original labels, since the mere application to the bottles in question of a sticker with the addition information may suffice."[77]

Furthermore, the Court of Justice in *Pharmacia* stated that "the condition of necessity will not be satisfied if replacement of the trade mark is explicable solely by the parallel importer's attempt to secure a commercial advantage".[78] By contrast, the condition of necessity is satisfied if, in a specific case, the prohibition imposed on the importer against replacing a trade mark or repackaging hinders effective access to the markets of the importing Member States.[79] In *Glaxo Group v. Dowelhurst*, Laddie J. considered that this principle of necessity should not be interpreted restrictively, and that necessity would be met "when it is shown that the use is reasonably required to overcome actual or potential hindrance to further commercialisation of the products." This would be the case where a large number of customers would not accept re-stickering as opposed to re-boxing.[80] However the judge did not regard the issue as *acte clair* and has referred a question to the Court of Justice as to the meaning of "effective market access."

---

[75] *Paranova*, para. 57.
[76] para. 56. See also *Pharmacia* paras 39–46 and *Ballantine* para. 35.
[77] para. 46. See also para. 55 of *Paranova*.
[78] para. 44.
[79] para. 43.
[80] Judgment of February 28, 2000, Ch.D, n.y.r., para. 104.

(ii) *Re-packaging does not adversely affect the condition of the product*

16–116 This condition concerns the original condition of the product inside the packaging. For example, the mere fact that the parallel importer has grouped together in single external packaging blister packs that may have come from different production batches with different use-by dates is not such as to adversely affect the original condition of the product. The court, however, has recognised that the original condition of the product inside the packaging might be indirectly affected where the new packaging omits certain important information, or gives inaccurate information concerning the nature, composition, effect, use or storage of the product.

(iii) *No damage to the reputation of the trade mark*

16–117 The presentation of the re-labelled or re-packaged product must not be such as to damage the reputation of the trade mark and its owner.[81] Account must be taken of the nature of the product and the market for which it is intended. In the case of pharmaceutical products it is recognised by the court as a sensitive area where the quality and integrity of the product are important criteria. The trade mark owner may in principle object to defective, poor quality or untidy packaging. Nevertheless, the court recognises that in the case of pharmaceutical products the requirements to be met by the presentation of a re-packaged pharmaceutical product may vary according to whether the product is sold to hospitals or through pharmacies to consumers. In the case of hospitals, the products are administered to patients by professionals for whom the presentation of a product is of little importance. In the latter case the presentation of the product is of greater importance for the consumer.

The law is currently in a state of uncertainty as to this third condition. A conceptual difficulty has arisen as to whether repackaging that does not affect the reputation of the trade mark:

(a) precludes justification to prevent importation within the meaning of Article 30 first sentence; or

(b) merely constitutes one of the four conditions that must be satisfied under Article 30, second sentence.

This distinction has enormous practical implications. The approach in (a) clearly favours parallel imports while the approach in (b) favours the enforcement of national intellectual property rights. In the *Glaxo* case Laddie J. inclined to the approach in (a) but referred the issue to the Court of Justice for a preliminary ruling.

(iv) *The owner of the mark receives a sample of the re-packaging*

16–118 The trade mark owner must be given advance notice of the re-packaged product being put on sale. The owner may also require the importer to supply him with a specimen of the re-packaged product before it goes on sale, to enable him to check that the re-packaging is not carried out in such a way as directly or indirectly to affect the original condition of the product and that the presentation after re-packaging is not likely to damage the reputation of the trade mark. Similarly, such a requirement affords the trade mark owner a better possibility of protecting himself against counterfeiting.[82] In *Glaxo*, Laddie J. held the sanction for marketing the product without

---

[81] para. 75 of *Paranova*; para. 50 of *Ballantine*.
[82] para. 78 of *Paranova*; see also *Hoffmann-La Roche*, above n. 30.

notice was an actionable infringement. He has referred questions to the Court of Justice seeking to determine the consequences of failing to give notice, how much notice to be given and the means by which notice is given.

Since it is in the trade mark owner's interest that the consumer or end user should not be led to believe that the owner is responsible for the re-packaging, an indication must be given on the packaging of who re-packaged the product. That implies that the national court must assess whether it is printed in such a way as to be understood by a person with normal eyesight, exercising a normal degree of attentiveness. Further, a clear indication may be required on the external packaging as to who manufactured the product, since it may indeed be in the manufacturer's interest that the consumer or end user should not be led to believe that the importer is the owner of the trade mark, and that the product was manufactured under his supervision.[83]

International exhaustion of trade mark rights

16–119 A trade mark owner remains entitled to prevent the marketing of products coming from a country outside the European Economic Area under an identical mark to that registered in a Member State. Any infringement action does not affect the free movement of goods between Member States, and does not come under the prohibition set out in Article 28 of the Treaty.[84] The position is different once the goods are lawfully in free circulation within the Community. The question then arises whether the goods have been marketed in another Member State by an undertaking which is economically and legally independent of the trade mark owner.

Community law precludes international exhaustion of trade mark rights. Since Directive 89/104 embodies a complete harmonisation of national rules relating to the trade mark rights (in Articles 5–7 of the Directive), the Court of Justice in *Silhouette*[85] held that the Directive could not be interpreted as leaving open to Member States the possibility of allowing international exhaustion (*i.e.* exhaustion of the national rights in respect of products imported from non-Member States marketed there with the owner's consent). The court considered that a situation in which some Member States could provide for international exhaustion while others provide for Community exhaustion, would inevitably give rise to barriers to free movement of goods and the freedom to provide services.

However, whereas Community exhaustion of rights is indefeasible in the sense that the trade mark owner cannot deny consent by imposing restrictions on the [territorial] destination of the goods, it appears that notwithstanding *Silhouette*, a trade mark owner may be considered to have internationally exhausted his rights where he has consented either expressly or implicitly to the importation of the goods from outside the EEA to within the EEA. There is no rule of Community law providing a presumption against consent to further exploitation in trade mark cases. Thus, where the goods in issue have been placed on the market in circumstances where the trade mark owner could have placed, but did not place, an effective restraint on their further sale and movement, the trade mark owner may be treated as having consented to such marketing.[86]

---

[83] See *Paranova* paras 67 to 77.
[84] *EMI v. CBS*, above n. 30.
[85] Case C-355/96 [1998] E.C.R. I-4799.
[86] See Laddie J. in *Zino Davidoff S.A. v. A & G Imports Limited* [1999] R.P.C. 631. A request for a preliminary ruling was made by the High Court on June 25, 1999. The question seeks to determine whether consent must be express or implicit before trade mark rights are exhausted on an international level.

E. Copyright

Specific subject matter

16–120   The specific subject matter of copyright is the exclusive right to reproduce a protected work to protect the moral rights in the work and to ensure a reward for the creative effort.[87] It is not part of the specific subject matter of copyright to prevent the importation of a copy of a literary, dramatic, music or artistic work which has been marketed by the copyright owner or with his consent. As with patents and trade marks, the acid test is whether the copyright work has been marketed in the other Member State by a legally and economically independent person. As the Court of Justice stated in *Deutsche Grammophon*:

> "If a right related to copyright is relied upon to prevent the marketing in a Member State of products distributed by the holder of the right or with his consent on the territory of another Member State on the sole ground that such distribution did not take place on the national territory, such a prohibition, which would legitimise the isolation of national markets, would be repugnant to the essential purpose of the Treaty, which is to unite national markets into a single market."[88]

International Exhaustion

16–121   Articles 28–30 do not preclude the copyright owner preventing the importation of works protected by copyright into the EEA where those goods have been put on the market outside the EEA by him or with his consent.[89]

F. Design rights

16–122   The Community rules on exhaustion of intellectual property rights apply equally to registered designs.[90] *Keurkoop v. Nancy Keen Gifts* is described in paragraph 16–105 above.

G. Passing off

16–123   Articles 28–30 do not prevent national laws on passing off.[91]

---

[87] CFI in *Magill*, Case T-69189 *RTE and others v. Commission* [1991] E.C.R. II-485 paras 70–71. Note that under United Kingdom copyright legislation moral rights are distinct from copyright — see Chapters 11 and 12.
[88] Case 78/70 [1971] E.C.R. 487, para. 12.
[89] Case 270/80 *Polydor v. Harlequin Record Shops* [1982] E.C.R. 329.
[90] *Keurkoop v. Nancy Keen*, above n. 53.
[91] *Beele*, above n. 47.

CHAPTER 17

COMPETITION LAW IN THE UNITED KINGDOM

Mark Brealey, Barrister, Brick Court Chambers and Kelyn Bacon, Barrister,
Brick Court Chambers

A. RELEVANT LEGISLATION

The old rules — the RTPA

17–01 The scrutiny of intellectual property rights from a competition perspective
radically changed on March 1, 2000, when the 1998 Competition Act ("CA") came
into force.[1] The previous competition legislation, the Restrictive Trade Practices Act
1976 ("RTPA"), rarely applied to vertical agreements relating to intellectual property
rights. First, it was based on "the open and shut door principle" according to which a
limited grant of intellectual property rights could not be said to be restrictive since the
grantee had started with no rights to begin with. This excluded exclusive intellectual
property licences and any resultant distortion of intra-brand competition. Secondly,
the RTPA only applied where at least two parties accepted a restriction as to the supply of goods or services. Thus, a patent licence that contained restrictions on the
licensee, but no restrictions on the patentee, fell outside the RTPA.[2]

The new rules — the CA

17–02 Exclusive intellectual property licences and licences that impose restrictions
only on one party are now subject to competition scrutiny under the CA, the provisions of which mirror the prohibitions contained in Articles 81 and 82 of the E.C.
Treaty. Chapter I of the CA (equivalent to Article 81) prohibits agreements which have
as their object or effect the distortion of competition within the United Kingdom. The
CA provides for exemptions, both individual and block, similar to those provided for
in E.C. competition law.

Abuse of market power stemming from intellectual property rights is also dealt
with by the CA. Chapter II of the CA (equivalent to Article 82) prohibits the abuse
by an undertaking in a dominant position which may affect trade within the United
Kingdom. The anti-monopoly provisions of the Fair Trading Act 1973 ("FTA") have
been retained. So too have the provisions of the FTA regulating mergers.

The CA also makes provision for enforcement powers and rules of procedure.
These also reflect E.C. competition rules, particularly those laid down by Regulation
17 (enforcement) and Regulation 3385/94 (notifications). The Director General of
Fair Trading ("DGFT") is charged with ensuring compliance with the CA in much
the same way as the Commission is charged with supervising compliance with the
E.C. competition rules. Appeals against decisions of the DG are to a new Competition Tribunal.

---

[1] See Flynn & Stratford, *Competition Understanding the 1998 Act* (Palladian Law Publishing); Freeman &
Whish, *a Guide to the Competition Act 1998* (Butterworths); Singleton, *Competition Act 1998*
(Blackstone's); *Butterworths Competition Law*, ed. Whish & Freeman.
[2] For an analysis of the previous law see *Butterworth Competition Law*: Green & Robertson, *Commercial
Agreements and Competition Law (*"Kluwer").

Relevance of E.C. competition law

17–03 Reference to E.C. competition law is of critical importance. Not only is the terminology of Chapter I and Chapter II broadly similar to that of Articles 81 and 82 of the E.C. treaty, but the CA makes express reference to the need to ensure consistency between the CA and the Treaty.

By virtue of section 60(1), the CA must be interpreted in "a manner which is consistent with the treatment of corresponding questions arising in Community law in relation to competition within the Community." Sections 60(2) & (4) impose an express obligation on a court (which includes tribunals) and the DGFT to ensure consistency with the Treaty and the case law of the ECJ and CFI. Section 60(3) imposes a lesser duty to "have regard" to the E.C. Commission, its decisions and notices.

As part of its duty to ensure consistency between the CA and E.C. law, it is likely that the courts and the Competition tribunal could make a reference to the Court of Justice of the European Communities pursuant to Article 234 (ex Article 177).[3]

B. RESTRICTIVE AGREEMENTS

The Chapter I prohibition

Transitional arrangements

17–04 Section 1 of the Act provides that the RTPA 1976 shall cease to have effect. This needs to be read with Schedule 13 and the transitional provisions. In summary:

- The RTPA, etc. will not apply to agreements made after March 1, 2000 (the starting date).

- Agreements made before the date when the Act was passed (November 9, 1998) had to comply with the previous rules. If particulars of the restrictive agreement were not notified, the agreement is void in respect of restrictions accepted as to the supply of goods or services.

- Agreements made in the interim period (November 6, 1998 — March 1, 2000) were subject to the old rules but relaxed. Only price fixing agreements were notifiable. Other restrictive agreements were deemed non-notifiable and failure to comply with the procedural requirements did not render the restrictions void.

- The Act provides for transitional periods (after March 1, 2000):

  — There is no transitional period for agreements made after March 1, 2000.

  — Agreements made before March 1, generally have a one year transitional period if they were non-notifiable agreements.

  — If the Restrictive Practices Court has ruled that an agreement is not contrary to the public interest, the agreement benefits from a five year transitional period.

  — Agreements which benefited from section 21(2) clearance (agreements with no significant anti-competitive effect) are exempted for their duration unless there is a material change to the agreements which may affect competition.

---

[3] See by way of analogy Case C-7/97 *Bronner* [1998] E.C.R. I-7791. This is also the view of the Government: see Lord Simon, *Hansard*, HL, col. 963, November 25, 1997.

The basic prohibition

17–05 Section 2, CA provides that agreements between undertakings, decisions by associations of undertakings or concerted practices which may affect trade within the United Kingdom and have as their object or effect the prevention restriction or distortion of competition within the United Kingdom are prohibited.

Section 2(2) provides five examples of such practices:

(a) fixing purchase or selling prices or other trading conditions;

(b) limiting production, markets, technical development or investment;

(c) sharing markets or sources of supply;

(d) discrimination;

(e) tying.

Since this Chapter reflects Article 81, regard must be had to the latter in interpreting the former. Of particular note are the following Community law concepts; concept of undertaking, concerted practice, economic unit, effect on pattern of trade, existence and exercise of intellectual property rights, definition of relevant market, ancillary restraints, open exclusivity, monopoly leveraging, assignment of improvements, customer and field of use restrictions, patent pools and joint venturers.[4]

Extra-territorial effect

17–06 Section 2(3) provides that the Chapter I prohibition applies "if the agreement . . . is or is intended to be, implemented in the United Kingdom." This follows the Court of Justice in *Woodpulp*.[5]

Voidness

17–07 Section 2(4) provides that any agreement which is prohibited by the Chapter I prohibition is void. This does not mean what it says. It means that the *restriction* in the agreement is void. Furthermore, E.C. cases on severance and damages[6] are all relevant.

Appreciability

17–08 The Chapter I prohibition will only apply where an agreement brings about an appreciable distortion of competition. This requirement again does not appear in the text of section 2 itself, but is borrowed from E.C. law. The issue was the subject of debate in the House of Lords. Lord Simon rejected the proposal that there should be an express reference to appreciability since the relevant E.C. jurisprudence would apply by virtue of section 60.[7] So far as E.C. law is concerned, the Commission's 1997 notice on minor agreements provides that vertical agreements are presumed not to appreciably distort competition if the market share of the parties concerned does not exceed 10 per cent and horizontal agreements are presumed not to have an appreciable effect on competition if the market share does not exceed five per cent. No such presumptions exist if the agreement fixes prices or imposes territorial restrictions.

---

[4] See Chapter 16 for an explanation of these concepts.
[5] Cases 89/85, etc., *Åhlström and others v. Commission* [1988] E.C.R 5193. See also above para. 16–03.
[6] See 16–87 and 16–88.
[7] *Hansard*, HL, col. 259, November 13, 1997.

The Director General in guidelines on the Chapter I prohibition has taken a view that an agreement would generally have no appreciable effect on competition if the parties combined share of the relevant market does not exceed 25 per cent. Such a presumption will not apply to agreements which fix prices or constitute one of a network of similar agreements which has a cumulative effect on the market in question. Although this share of the market seems high, section 2 will apply very much to localised markets where the market share of the parties may be higher.

Exclusions

17–09 Section 3 provides that the Chapter I prohibition does not apply to certain agreements: namely mergers and concentrations (Schedule 1); competition scrutiny under other enactments (Schedule 2) for example, the Financial Services Act 1986; steel agreements and agricultural products which are separately regulated under the E.C. Treaty (Schedule 3); and Professional Rules (Schedule 4). Under the RTPA, the services of solicitors, barristers and accountants, etc., were excluded. The 1998 Act only provides for the exclusion of professional rules which are notified to the director and designated. Thus, price fixing between professionals would not be excluded.

An important exclusion relates to vertical agreements. An Exclusion Order made under section 50 of the Act excludes vertical agreements from the application of the Chapter I Prohibition.[8] Vertical agreement is defined as:

> "an agreement between undertakings, each of which operates, for the purposes of the agreement, at a different level of the production or distribution chain and relating to the conditions under which the parties may purchase, sell or re-sell certain goods or services and includes provisions contained in such agreements which relate to the assignment to the buyer or use by the buyer of intellectual property rights, provided that those provisions do not constitute the primary object of the agreement and are directly related to the use, sale or resale of goods or services by the buyer or its customers."

The exclusion does not apply where the vertical agreement has the object or effect of restricting the buyer's ability to determine its sale price. This is without prejudice to the possibility of the supplier imposing a maximum sale price or recommending a sale price, provided that these do not amount to a fixed or minimum sale price as a result of pressure from or incentives by any of the parties.[9]

The exclusion of vertical agreements for the sale of goods or services brings the United Kingdom Act into line with E.C. Competition law. Commission Regulation 2790/1999 has exempted from the prohibition contained in Article 81 all vertical agreements unless they have certain hard core restrictions (*e.g.* price fixing).[10] The rationale for the exemption is that vertical agreements concluded between parties that have little market power are unlikely to distort the competitive process. The intra–brand distortion will not significantly affect the market. It is only when the parties to the agreement have significant market power (so that inter–brand competition is limited) that the effects of the vertical agreement may be significant.

---

[8] The Competition Act 1998 (Land and Vertical Agreements Exclusion) Order 2000 (S.I. 2000 No. 310).
[9] Article 4.
[10] [1999] O.J. L336/21.

Intellectual property licences

17–10   Importantly, the Exclusion Order only exempts the licensing of intellectual property rights if they are ancillary to the vertical agreement. Otherwise intellectual property licenses are not excluded. The effect is therefore that the Chapter I prohibition applies to all intellectual property licences. Regard should be had to the E.C. rules on competition to determine the circumstances where an infringement of Chapter I is likely,[11] to the *de minimis* exception that is understood to form part of Chapter I,[12] and to the provisions relating to parallel exemption.[13-14]

Exemptions

Individual exemptions

17–11   Section 4 of the Act provides that the DGFT may grant an individual exemption if:

(a)   a party has requested it pursuant to section 14;

(b)   the criteria laid down in section 9 are satisfied. These broadly mirror Article 81(3). The agreement must contribute to improving the production or distribution of goods or promoting technical or economic progress and allow the consumers a fair share of the resulting benefit: it must not impose on the undertakings concerned unnecessary restrictions or eliminate competition in a substantial respect (the two positive and two negative conditions).

Block exemption

17–12   Section 6 provides for the adoption of block exemptions, although it is doubtful whether many block exemptions will be adopted now that the exclusion order for vertical agreements will apply.

Parallel exemptions

17–13   Section 10 introduces a novel concept of parallel exemption of an agreement from the Chapter I prohibition. A Community exemption is automatically extended to cover a Chapter I prohibition if: the agreement is exempt from the relevant Community prohibition (a) by virtue of a regulation, (b) because it has been given exemption by the Commission or (c) it has benefited from the opposition procedure or (d) it would be block exempted under Community law if it affected trade between Member States. A parallel exemption does not extend to a comfort letter.

The availability of a parallel exemption for licences of intellectual property is of crucial importance. It means that a licence agreement will be exempt from the Chapter I prohibition if the agreement falls within the terms of Regulation 240/96 even if the agreement would not infringe Article 81(1) because of a lack of any effect on interstate trade.

---

[11]   See paras 16–12 to 16–37.
[12]   See para. 16–19.
[13-14] See below at para. 17–13.

Notification to DGFT

17–14 Notification to obtain guidance or exemption is provided by sections 12–16, CA.

Guidance

17–15 Section 13 allows a party to obtain guidance as to whether an agreement falls within the prohibition or is likely to be exempted. If an agreement has been notified to the DGFT for guidance, an interim immunity against a penalty is provided, the immunity lasting from the date of notification and ending when and if the DGFT notifies the parties that the agreement is prohibited.[15] If guidance is given that the agreement is not prohibited, no penalty can be imposed if it is subsequently found that the agreement is in fact prohibited.[16]

Further, once the DGFT has given guidance that the agreement is not prohibited, the DGFT will not investigate further or take any further action unless there is a material change of circumstance, the guidance was based on materially false, misleading or incomplete facts, a complaint is made by a third party or one of the parties to the agreement notifies it.[17] If materially incomplete, false or misleading information is provided by a party to the agreement, the immunity may be removed retrospectively.

Notification

17–16 Section 14 allows a party to notify the agreement for a decision on clearance, exclusion, a formal exemption. Thus the DGFT may decide that the agreement does not fall within the section 2 prohibition at all (*e.g.* because it is *de minimis*); or that the agreement is excluded (*e.g.* it is a vertical agreement) or that it satisfies the criteria for individual exemption in section 9.

If an agreement is notified, no penalty may be imposed, if the agreement is in fact prohibited, during the period beginning with the date of notification and ending when DGFT notifies the parties of the infringement. The DGFT can re-examine any decision if a material change of circumstance occurs, or the decision was based on materially incomplete, false or misleading information and in the latter case immunity may be withdrawn retrospectively if the information was provided by a party to the agreement.[18]

Applications are to be made on Form N, with an original and two copies, and an annex for confidential material. Fees payable are currently £5,000 for guidance and £13,000 for a decision.

C. Abuse of dominant position

Chapter II prohibition

17–17 Section 18 contains the Chapter II prohibition and mirrors substantially the prohibition in Article 82. Dominant position means a dominant position within the United Kingdom: Section 18(3). This is important because section 18(1) would appear to catch a dominant position merely in any market in any Member State. There is probably nothing to exclude the Chapter II prohibition if the dominant position is both within the United Kingdom and outside. Lord Simon in the House of Lords

---

[15] section 13(4).
[16] section 15(3).
[17] section 15(2).
[18] section 16.

explained that dominance could extend beyond the United Kingdom provided that the dominance included some part of the United Kingdom territory.[19]

Dominance

17–18   The definition of dominant position will be the same as applied to Article 82. It "relates to a position of economic strength enjoyed by an undertaking which enables it to prevent effective competition being maintained on the relevant market by affording it the power to behave to an appreciable extent independently of its competitors, customers and ultimately of its consumers."[20]

Abuse and effect on trade

17–19   The concepts of abuse will also be borrowed from E.C. competition law. The abuse of the dominant position in the United Kingdom may be felt outside the United Kingdom. But before the Chapter II prohibition applies, there must be effect on trade within the United Kingdom.

No exemption

17–20   There is no possibility of exemption, but recourse may be had to the concept of objective justification. However, it is possible to seek the DGFT's guidance as to whether certain conduct is prohibited by Chapter II (section 20, CA).

No transitional period

17–21   Sections 2–10 of the Competition Act 1980, which previously regulated abuse of market power, have been repealed. There is no transitional period for abuse of a dominant position. Therefore, abusive conduct falls to be considered under Chapter II from March 1, 2000 irrespective of whether the conduct was carried on prior to that date.

D.   COMPETITION PROCEDURE

Power to investigate

17–22   The provisions on investigations are contained in sections 25–31 of the Act. Under section 25, the precondition for investigations is that the DGFT has "reasonable grounds" for suspecting an infringement of the Chapter I or Chapter II prohibitions. Where such reasonable grounds are present, the DGFT has the power to compel the production of information, and the power to enter premises.

Production of information

17–23   Section 26 provides that, in order to obtain information, the DGFT must give notice in writing to the addressee. That notice may require the production of a specified document or specified information which he "considers relates to any matter relevant to the investigation". A document is defined as including information recorded in any form — this would extend for example to information held on microfiche or

---

[19]   *Hansard*, HL col. 308, November 13, 1997.
[20]   Case 27/76 *United Brands v. Commission* [1978] E.C.R. 207, para. 65.

computer. The request need not be limited to specific named documents: categories of information may be requested, for example minutes of Board meetings.

The DGFT is entitled to require information not only from parties suspected of participating in the infringement, but also from third parties whom the DGFT believes to possess relevant information, for example customers or suppliers of the undertakings under suspicion.

Entry of premises without warrant

17–24 Under section 27, any officer of the DGFT authorised in writing by the DGFT may enter any premises in connection with an investigation under section 25. "Premises" normally means business premises, but extends to domestic premises where they are used in connection with the affairs of the relevant business, or if relevant documents are to be found there; and extends also to any vehicle, for example cars.

Where the premises are occupied by an undertaking suspected of having infringed the Chapter I or Chapter II prohibition, the investigating officer may arrive unannounced. However, in respect of any other occupier, the officer must take all steps as are reasonably practicable to give two working days' written notice of the intended visit.

The investigating officer may require any person on the premises to produce any document which he considers relates to any matter relevant to the investigation, and to provide an explanation of it; and he may require any person to state where a document is to be found. Copies or extracts may be taken of documents which are produced; and where information is held on a computer which is accessible from the premises the officer may require it to be produced in a form which can be taken away (*e.g.* on disc) and in visible and legible form.

Entry of premises with warrant

17–25 Section 28 provides that the DGFT may apply to the court for a warrant to enter premises in three cases:

(a)   if he has asked for documents to be produced either by written notice under section 26, or during a visit without a warrant under section 27, and the documents have not been produced, and there are reasonable grounds for suspecting that there are such documents on the premises;

(b)   if an investigating officer has attempted to enter premises without a warrant under section 27 and has been unable to do so, and there are reasonable grounds for suspecting that the premises contain relevant documents;

(c)   the "dawn raid" situation where there are reasonable grounds for suspecting that the premises contain relevant documents, and that such documents if requested would be concealed, removed, tampered with or destroyed.

Where a warrant is obtained, the authorised officials are entitled to enter the premises using such force as is reasonably necessary and to search them for the documents which are the subject of the warrant application. Once documents or information are obtained, the authorised official may take copies or extracts, and may take any steps necessary to preserve documents (including removing them from the premises), may require an explanation of a document or information as to where it may be found and may also require computerised information to be provided as above.

Privilege

17–26   The DGFT's investigatory powers are subject to the right not to produce or disclose information or documents protected by legal professional privilege (section 30). In particular, unlike the corresponding E.C. provisions, the Act protects not only communications with external legal advisers but also in-house lawyers. Further, by virtue of section 60, the rights to privilege against self-incrimination developed in the E.C. case law will apply also in respect of investigation by the DGFT.

Enforcement

17–27   Sections 32–41 deal with enforcement of the Competition Act. In summary, the DGFT may give directions in relation to an agreement or conduct, may enforce such directions by court order, may order interim measures and may impose fines.

Directions to terminate

17–28   The primary provisions are sections 32 and 33, whereby the DGFT may give directions in writing to terminate or modify an agreement which infringes the Chapter I prohibition, or modify or bring to an end conduct which infringes Chapter II, respectively. If the directions are not complied with, the DGFT may apply under section 34 for a court order to enforce them.

Interim measures

17–29   If the DGFT has a reasonable suspicion that the Chapter I or II prohibitions have been infringed, but has not completed his investigation, he may prescribe interim measures under section 35 where he deems it necessary to act as a matter of urgency in order to prevent serious irreparable damage to a person or category of persons, or to protect the public interest. Before doing so he must, however, give written notice to the proposed addressees of the interim measures, and must give them an opportunity to make representations.

Fines

17–30   The Act contains detailed provisions on penalties in sections 36–41. In addition to any directions which may be given regarding a Chapter I or II infringement, if the DGFT considers that the infringement has been committed intentionally or negligently he may impose fines of up to 10 per cent of the infringing undertakings' total turnover in the United Kingdom. The fines in individual cases will be calculated according to the gravity and duration of the infringements, and the existence of aggravating or mitigating factors. Small agreements, and conduct with minor significance, may enjoy limited and provisional immunity from fines; as may agreements which have been notified to the DGFT.

CHAPTER 18

UNITED STATES ANTITRUST LAW

Michael L. Weiner, Partner, Skadden, Arps, Slate, Meagher & Flom LLP

INTRODUCTION

18–01   United States intellectual property laws provide incentives for innovation and its commercialisation by establishing enforceable property rights for the creators of new and useful products, more efficient processes, and original works of expression. In contrast, U.S. antitrust laws are intended to promote innovation and consumer welfare by prohibiting certain actions that may harm competition. Accordingly, although as discussed below they may at times appear to operate in an inconsistent manner, both the intellectual property laws and the antitrust laws share the common purpose of promoting innovation and enhancing consumer welfare.

This Chapter will analyse the interplay between U.S. antitrust law and intellectual property law, focusing primarily on the issues that may arise out of international intellectual property licensing agreements under U.S. antitrust law. Part A discusses the scope of U.S. jurisdiction over international licensing agreements. Part B provides a brief description of the legal framework for U.S. antitrust law. Part C summarizes the approach taken by the Antitrust Division of the U.S. Department of Justice ("DOJ") and the Federal Trade Commission ("FTC") with respect to licensing agreements. Part D examines the differences between European Union and U.S. treatment of antitrust exemptions under their respective antitrust statutes. Part E briefly discusses the circumstances under which an intellectual property owner enforcing its rights under the intellectual property laws may lose antitrust immunity. Finally, Part F provides a summary of other relevant statutory provisions, both federal and state, that may govern licensing agreements.

A. WHEN DOES UNITED STATES ANTITRUST LAW APPLY?

The Sherman Act

18–02   The Sherman Act applies to "every contract, combination . . . or conspiracy, in restraint of trade" and to every person who "monopolize[s], or attempt[s] to monopolize, or combine[s] or conspire[s] . . . to monopolize" trade or commerce "among the several states or with foreign nations."[1] In 1982, Congress amended this language found in both the Sherman Act and Section 5 of the Federal Trade Commission Act through its enactment of the Foreign Trade Antitrust Improvements Act of 1982 (the "FTAIA"). The FTAIA eliminated U.S. jurisdiction over conduct:

> "involving trade or commerce (other than import trade or import commerce) with foreign nations unless (1) such conduct has a direct, substantial and reasonably foreseeable effect (A) on [domestic or import commerce]; or (B) on export trade or export commerce . . . of a person engaged in such trade or commerce in the United States."[2]

The author wishes to acknowledge the contribution of his colleague, Linda Wong, Esq.
[1]   15 U.S.C. § 1.
[2]   15 U.S.C. § 6a(1).

Thus, it is well-established that the Sherman Act now "applies to foreign conduct that was meant to produce and did in fact produce some substantial effect in the United States."[3]

Antitrust Enforcement Guidelines for International Operations

18–03 In 1995, the DOJ and the FTC (collectively, the "agencies") released joint Antitrust Enforcement Guidelines for International Operations (the "1995 International Guidelines") clarifying their position with respect to jurisdiction over international licensing arrangements. As to the subject matter jurisdictional reach of U.S. courts, the 1995 International Guidelines state that "anticompetitive conduct that affects U.S. domestic or foreign commerce may violate the U.S. antitrust laws regardless of where such conduct occurs or the nationality of the parties involved."[4] For example, a cartel formed by foreign producers with no U.S. subsidiaries or production to raise the prices of products imported into the U.S. will be subject to Sherman Act jurisdiction because there is an intended and foreseeable effect on U.S. commerce.[5]

Assuming the objective test for jurisdictional requirements set forth in the FTAIA is satisfied, the alleged licensing arrangement involving foreign intellectual property may be subject to scrutiny under the U.S. antitrust laws.[6]

B. THE LEGAL FRAMEWORK FOR UNITED STATES ANTITRUST LAW

18–04 United States antitrust law generally applies the same principles to cases involving intellectual property rights as to cases involving any other property rights. Thus, the agencies have stated that conduct with respect to intellectual property that may have anticompetitive effects, is "neither particularly free from scrutiny under the antitrust laws, nor particularly suspect under them."[7] To understand how these general principles apply to cases involving intellectual property, it is essential to understand in general terms what sections 1 and 2 of the Sherman Act prohibit.

---

[3] *Hartford Fire Ins. Co. v. California*, 509 U.S. 764, 796 (1993) (finding subject matter jurisdiction where challenged conduct of foreign reinsurers outside the United States produced substantial effects in the United States); see also *Fleischmann Distilling Corp. v. Distillers Co.*, 395 F. Supp. 221, 226 (S.D.N.Y. 1975) ("direct and substantial adverse effect on interstate or foreign commerce" conferred subject matter jurisdiction).
[4] International Guidelines, 4 *Trade Reg. Rep.* (CCH)¶ 13,107 at § 3.11 (1995).
[5] *ibid.* Illustrative Example A. The agencies also referred to the test articulated by the United States Supreme Court in *Summit Health Ltd v. Pinhas*, 500 U.S. 322, 330–31 (1991), which looks to the "potential harm that would ensue if the conspiracy were successful, not ... whether the actual conduct in furtherance of the conspiracy had in fact the prohibited effect upon interstate or foreign commerce," in determining whether the assertion of U.S. antitrust jurisdiction is appropriate. More recently, in *United States v. Nippon Paper Indus. Co.* 109 F.3d 1 (1st Cir. 1997), the First Circuit Court of Appeals extended application of the "substantial effects" test to a criminal case, in which a foreign corporation that was alleged to have engaged in price-fixing activities solely in Japan, was held subject to the U.S. antitrust laws.
[6] *See, e.g., United States v. Pilkington plc*, 1994–2 Trade Cas. (CCH)¶ 70,842 (D. Ariz. 1994) (consent decree) (DOJ filed complaint against British flat glass manufacturer challenging various licensing agreements). Competitive restrictions in international licences may also be subject to a number of statutes, such as the National Cooperative Research and Production Act of 1993 (the "NCRPA"), that limit the scope or operation of the antitrust laws. The NCRPA provides that production and research joint ventures will not be illegal *per se*, but rather will be "judged on the basis of [their] reasonableness, taking into account all relevant factors affecting competition." 15 U.S.C. § 4302. The statute further provides that by filing notification of the venture with the agencies, joint venture parents may limit the extent of potential antitrust exposure under both federal and state laws. 15 U.S.C. § 4303(e).
[7] *See* DOJ and FTC, "*Antitrust Guidelines for the Licensing of Intellectual Property*", 4 *Trade Reg. Rep.* (CCH)¶ 13,132 at § 2.1 (1995) [hereinafter Intellectual Property Guidelines].

Sherman Act section 1: per se violations

18–05 Section 1 prohibits any agreement between two or more economically independent entities that unreasonably restrains competition.[8] Thus, agreements between horizontal competitors to fix minimum prices or to carve up markets, even if they involve intellectual property rights, are so inherently anticompetitive that they are treated as *per se* illegal,[9] and may be condemned without further analysis "as to the precise harm [they have] caused or the business excuse for [their] use."[10] In addition, the DOJ has criminally prosecuted conduct that constitutes a clear and purposeful *per se* violation of section 1, such as horizontal price fixing, bid rigging, and market or customer allocation.[11]

Sherman Act section 1: rule of reason

18–06 By contrast, conduct that is not unambiguously anti-competitive is not ordinarily subject to the *per se* rule, but instead, is analysed under the rule of reason. Thus, vertical licensing agreements which generally do not appear *prima facie* anti–competitive are analysed under the rule of reason, where the decisive question is whether the restraint is likely to have anti-competitive effects and, if so, whether the restraint is reasonably necessary to achieve procompetitive benefits that outweigh those anti-competitive effects.[12] Since the United States Supreme Court's decision in *Broadcast Music Inc. v. Columbia Broadcasting Systems,* 441 U.S. 1, 8–9 (1979), where the Court noted that characterising challenged conduct as falling within a particular category of restraint that is treated as *per se* unlawful "do[es] not always supply ready answers," courts have sparingly applied the *per se* rule.[13]

Penalties for violations

18–07 Criminal or civil sanctions may be imposed for section 1 violations. Individuals may be imprisoned for up to three years and fined the greatest of three alternatives: (i) $350,000, *see* 15 U.S.C. § 1; (ii) twice the pecuniary gain the individual derived from the offense; or (iii) twice the pecuniary loss suffered by the victims, *see* 18 U.S.C. § 3571. A corporation may also be fined the greatest of three alternatives: (i) $10 million, *see* 15 U.S.C. § 1; (ii) twice the pecuniary gain the corporation derived from the

---

[8] Section 1 of the Sherman Act states that "[e]very contract, combination in the form of trust or otherwise, or conspiracy, in restraint of trade or commerce among the several States, or with foreign nations, is declared to be illegal." 15 U.S.C. § 1.

[9] *See United States v. Univis Lens Co.* 316 U.S. 241, 252 (1942) (patent licence agreements that imposed restrictions on the licensee's resale prices constitute unreasonable restraints within the meaning of the Sherman Act); *Ethyl Gasoline Corp. v. United States* 309 U.S. 436, 458 (1940) (licensing agreements that maintain prices of articles moving in interstate commerce eliminate competition, and thus violate the Sherman Act); *see also* Intellectual Property Guidelines, § 3.4 ("among the restraints [in intellectual property licensing arrangements] that have been held per se unlawful are naked price-fixing, output restraints, and market division . . . as well as certain group boycotts . . .") and § 5.2 ("[T]he Agencies will enforce the per se rule against resale price maintenance in the intellectual property context.").

[10] *Northern Pac. Ry v. United States* 356 U.S. 1, 5 (1958).

[11] See, *e.g., California Dental Ass'n v. FTC* 526 U.S. 756 (1999) (holding that FTC inappropriately applied "quick look" analysis to advertising restrictions that were imposed on the association's members); *United States v. Hayter Oil Co.* 51 F.3d 1265 (6th Cir. 1995)(conspiracy to fix gasoline prices); *United States v. Brown* 936 F.2d 1042 (9th Cir. 1991) (conspiracy to restrain competition for billboard sites).

[12] See *Federal Trade Commission v. Indiana Fed'n of Dentists* 476 U.S. 447, 457 (1986); *NCAA v. Board of Regents of the Univ. of Okla.* 468 U.S. 85, 103 (1984).

[13] See, *e.g., Hammes v. AAMCO Transmissions Inc.* 33 F.3d 774, 782 (7th Cir. 1994) (*per se* rule would apply to customer allocation scheme but only after the court first determined whether challenged conduct constituted such a scheme); *Capital Imaging Assocs. v. Mohawk Valley Med. Assocs.* 996 F.2d 537, 543 (2d Cir. 1993) (noting that most cases fall outside the scope of *per se* application).

offense; or (iii) twice the pecuniary loss suffered by the victims, *see* 18 U.S.C. § 3571. In addition to these criminal sanctions, entities who suffer injury to their business or property as the result of an antitrust violation may recover treble damages in private litigation, brought either individually or on behalf of a class of all similarly-situated entities, pursuant to section 4 of the Clayton Act, 15 U.S.C. § 15. Finally, section 1 violations may be enjoined under section 16 of the Clayton Act, 15 U.S.C. § 25, which authorises preliminary and permanent injunctive relief in private litigation against threatened loss or damage by a violation of the antitrust laws.

Sherman Act section 2: unilateral conduct

18–08 Cases involving unilateral abuse of intellectual property rights often arise under section 2 of the Sherman Act, which does not require an agreement, but rather actual or likely monopoly power. Actual monopolisation is defined as "(1) the possession of monopoly power in the relevant market and (2) the wilful acquisition or maintenance of that power as distinguished from growth or development as a consequence of a superior product, business acumen, or historic accident."[14] Attempted monopolisation requires exclusionary conduct with "specific intent to monopolize" which creates a "dangerous probability" of monopolisation of a relevant market.[15] Under these standards, merely having monopoly power by virtue of intellectual property ownership is not problematic, but exclusionary practices with the purpose of extending the scope or duration of such power will be subject to antitrust scrutiny under the rule of reason.[16] These core antitrust principles embodied in sections 1 and 2 of the Sherman Act provide the foundation for the agencies' policy statements in the Intellectual Property Guidelines, which are discussed below.

C. THE INTELLECTUAL PROPERTY GUIDELINES

18–09 The Intellectual Property Guidelines, which were issued by the agencies in April 1995, articulate three guiding principles that underlie the agencies' approach to analysing intellectual property licensing arrangements: (1) for antitrust purposes, intellectual property is treated like any other form of property,[17] (2) the agencies will not presume that intellectual property creates market power,[18] and (3) licensing, which allows firms to combine complementary factors of production, is generally procompetitive.[19] Significantly, the antitrust laws neither proscribe a licensor's efforts to appropriate the full inherent value of the intellectual property[20] nor require the licensor to create competition.[21] The Intellectual Property Guidelines recognise that while licensing arrangements are typically viewed as procompetitive, certain restrictions in licensing agreements may raise antitrust concerns. This is particularly true where the

---

[14] *United States v. Grinnell Corp.* 384 U.S. 563, 570–71 (1966).

[15] *Spectrum Sports Inc. v. McQuillan* 506 U.S. 447, 456 (1993).

[16] See Intellectual Property Guidelines § 2.2; see also *Eastman Kodak Co. v. Image Technical Servs.* 504 U.S. 457, 479 n.29 (1992) ("[P]ower gained through some natural advantage such as patent, copyright, or business acumen can give rise to liability if a seller exploits his dominant position in one market to expand his empire into the next."); *Abbott Labs. v. Brennan* 952 F.2d 1346, 1354–55 (Fed. Cir. 1991). The U.S. concept of market power appears to be consistent with the European law concept of dominant position. See, *e.g.*, *Deutsche Grammophon v. Metro* [1971] E.C.R. 487, § 16.

[17] Intellectual Property Guidelines § 2.1.

[18] *ibid.* at § 2.2.

[19] *ibid.* at § 2.3.

[20] *ibid.* at § 2.2.

[21] *ibid.* at § 3.1. Associated with this notion is the intellectual property owner's inherent right not to deal with others. This right, however, is not absolute. See below n.52.

challenged arrangement adversely affects "the prices, quantities, qualities, or varieties of goods and services either currently or potentially available" within the relevant markets for the goods and services affected by the arrangement.[22]

Analysing licensing arrangements

18–10   In analysing licensing arrangements the agencies first examine the relationship between the licensor and the licensee and the nature of that relationship. A "vertical" relationship exists where the licensor and the licensee are in a "complementary" relationship,[23] while a "horizontal" relationship exists where the licensor and the licensee are actual or potential competitors in one or more markets.[24] The existence of a horizontal relationship does not indicate that the arrangement is anti-competitive; it merely aids the determination of whether such arrangement may have anti-competitive effects.[25] Similarly, the presence of a vertical relationship does not signal the absence of anti-competitive effects.[26]

Vertical licensing arrangements may pose anti-competitive hazards if they foreclose access by competitors to alternative technologies, increase competitors' costs of doing business, or facilitate co-ordination among horizontal competitors. Such arrangements are *per se*[27] unlawful if the owner of the property fixes resale prices in a downstream market,[28] or if the intellectual property owner conditions the licence on the customer's purchase of a second product where the seller/owner has market power in the tying (or first) product.[29] Similarly, licence agreements between horizontal competitors are *per se* illegal if they involve a naked agreement to fix prices, allocate customers or territories, or if the licence prevents the licensee from licensing, selling, distributing or using competing technologies.[30] Although exclusive licenses, such as cross-licensing arrangements between parties collectively possessing market power,[31] grantbacks,[32] and acquisitions of intellectual property rights[33] are not generally *per se* illegal, they do pose special concern to the agencies because of their potential to harm competition. Most licensing arrangements, however, do not involve naked restraints, and thus will generally be evaluated under the rule of reason.

Cross-licensing and pooling arrangements

18–11   While cross-licensing and pooling arrangements may have procompetitive effects by "integrating complementary technologies, reducing transaction costs, clearing blocking positions, and avoiding costly infringement litigation,"[34] they may also

---

[22] Intellectual Property Guidelines § 3.2. Section 3.2 delineates three types of markets: "markets for final or intermediate goods made using the intellectual property," technology markets consisting of the licensed technology and its close substitutes, and innovation markets which consist of research and development of new or improved goods or processes and its close substitutes.
[23] *ibid.* at § 3.3.
[24] *ibid.*
[25] *ibid.*
[26] *ibid.*
[27] The agencies will apply the *per se* rule only if the restraint's "nature and necessary effect are so plainly anticompetitive" and the restraint is not reasonably related to any "efficiency enhancing integration of economic activity." *ibid.* at § 3.4.
[28] *Univis Lens Co.* 316 U.S. at 252.
[29] *See United States v. Paramount Pictures, Inc.* 334 U.S. 131 (1958); *International Salt Co. v. United States* 332 U.S. 392 (1947).
[30] Intellectual Property Guidelines § 5.2.
[31] *ibid.* at § 5.5.
[32] *ibid.* at § 5.6.
[33] *ibid.* at § 5.7.
[34] *ibid.*

have anti-competitive effects, including price fixing or market allocation,[35] output restrictions, diminished rivalry, and reduced innovation.[36] This is particularly true in the infringement settlement context, where the cross-licence is typically between firms that are actual or potential competitors. Under such situations, the agencies have stated their policy to "consider the anticompetitive effect of settlement among competitors that involve cross-licensing."[37]

Grantbacks

18–12   Grantbacks — arrangements under which a licensee agrees to extend to the licensor any improvements to the licensed technology — are evaluated under the rule of reason.[38] The agencies view exclusive grantback provisions as suspect because they arguably may reduce a licensee's incentives to engage in research and development, thereby limiting rivalry in the innovation markets. By contrast, the agencies consider a non-exclusive grantback provision, which allows a licensee to license its technology to others, "less likely to have anticompetitive effects."[39] A significant factor in evaluating a grantback provision is whether the licensor has market power in a relevant technology or innovation market.[40] If so, the agencies' antitrust concerns will be heightened.

Safety zones

18–13   Finally, the Intellectual Property Guidelines set forth antitrust "safety zones," which provide that a licensing arrangement that falls within any of the enumerated zones will not likely be challenged, except in extraordinary circumstances.[41] The agencies will not challenge restraints in a licensing arrangement that affects the goods market if: (1) the restraints are not *prima facie* anti–competitive and (2) the parties to the arrangement collectively account for no more than 20 per cent of each affected relevant market.[42] The agencies also will not challenge restraints in a licensing arrangement that affects a technology or innovation market if the restraints are not *prima facie* anti–competitive and if there are four or more alternative independently controlled technologies and independently controlled entities in the affected technology market or affected innovation market, respectively.[43]

D. Antitrust exemptions: U.S. versus E.U.

18–14   Unlike their European counterpart, U.S. antitrust authorities lack the authority to promulgate "block exemptions" from the antitrust laws for certain types of

---

[35] The agencies will likely challenge cross-licensing and pooling arrangements that are mechanisms for price fixing or market division under the *per se* rule. *ibid.* at § 5.5.

[36] Pooling arrangements that require members to grant each other licences to current and future technology at minimal cost, especially when such arrangements affect a large fraction of the potential research and development in an innovation market, are generally suspect and thus subject to challenge by the agencies. *ibid.*; see also *United States v. Manufacturers Aircraft Ass'n, Inc.* 1976–1 Trade Cas. (CCH)¶ 60,810 (S.D.N.Y. 1975) (ordering dissolution of pooling arrangement and directing each defendant to grant to others a non-exclusive, non-discriminatory licence).

[37] Intellectual Property Guidelines § 5.5.

[38] *ibid.* at § 5.6. This is consistent with the U.S. Supreme Court's treatment of grantback provisions in *Transparent-Wrap Machine Corp. v. Stokes & Smith Co.* 329 U.S. 637 (1947), where the court held that a rule of reason analysis should be used to evaluate the grantback provision in an exclusive licence agreement involving a patented packaging machine.

[39] Intellectual Property Guidelines § 5.6.

[40] *ibid.*

[41] *ibid.* at § 4.3.

[42] *ibid.*

[43] *ibid.*

licensing agreements. Accordingly, and notwithstanding the similarities in the antitrust laws of the United States and the European Union, a licence agreement that qualifies for block exemption under the European Union antitrust laws may at least theoretically be found to be anti-competitive under U.S. antitrust analysis.

E. WHEN DO INTELLECTUAL PROPERTY RIGHTS LOSE THEIR ANTITRUST IMMUNITY?

18–15   Intellectual property rights are conditioned by a public purpose[44] and, therefore, any "misuse," for example by seeking to extend market power beyond the scope of the right of exclusion, may result in the loss of intellectual property protection. U.S. courts will not enforce intellectual property rights while those rights are being "misused,"[45] or if misused, until after the misuse has been purged.[46]

The "misuse" doctrine typically arises in the patent context, and is generally asserted as a defense to a patent infringement claim. In certain situations, patent misuse may rise to the level of an antitrust violation.[47] A patent is misused if it is used "to foster price-fixing arrangements or extend the patent monopoly to other products through tying arrangements."[48] Other forms of misuse occur if the licensor prohibits the licensee from dealing in goods that compete with the patented product,[49] conditions the licence of one patent on the licence of others,[50] or requires the licensee to pay royalties on unpatented items.[51] Merely seeking to enforce ones's rights, refusing to license,[52]

---

[44]   *Univis Lens Co.*, 316 U.S. at 251–52 (1942).

[45]   See *Morton Salt Co. v. G.S. Suppiger Co.* 314 U.S. 488, 492 (1942) ("It is a principle of general application that courts . . . may appropriately withhold their aid where the plaintiff is using the right asserted contrary to the public interest."); *Senza-Gel Corp. v. Sieffhart* 803 F.2d 661, 668 (Fed. Cir. 1986) ("[A] successful defense of patent misuse means [ ] that a court of equity will not lend its support to enforcement of a misuser's patent.").

[46]   See *United States Gypsum Co. v. National Gypsum Co.* 352 U.S. 457, 465 (1957) ("It is now, of course, familiar law that the courts will not aid a patent owner who has misused his patents . . . or thereafter until the effects of such misuse have been dissipated, or "purged" as the conventional saying goes."); *Morton Salt Co.* 314 U.S. at 492–94.

[47]   See Donald S. Chisum, *Chisum on Patents* § 19.04[2], at 19–298 (1998) ("Use of a patent to violate the antitrust laws will constitute misuse. However, conduct which in some respect falls short of an antitrust violation may still constitute misuse."); see also Intellectual Property Guidelines § 6 ("The Agencies may challenge the enforcement of invalid intellectual property rights as antitrust violations.").

[48]   *Boston Scientific Corp. v. Schneider (Europe) A.G.* 983 F. Supp. 245, 269 (D. Mass. 1997). The FTC alleged this type of patent misuse against Summit Technology, Inc. and VISX, Inc., contending that the companies pooled their patents into a new partnership in order to "raise, fix, stabilize and maintain" licence fees. See Complaint pp. 8, 25, *In Re Summit Tech. Inc* (FTC March 24, 1998) (No. 9286). The companies subsequently reached an agreement with the FTC to dissolve the pooling arrangement.

[49]   See *National Lockwasher Co. v. George K. Garrett Co.* 137 F.2d 255, 256 (3d Cir. 1943) ("A patentee's right does not extend to the use of the patent to purge the market of competing non-patented goods except, of course, through the process of fair competition.").

[50]   See *United States v. Loew's Inc.* 371 U.S. 38 (1962) (holding that "block booking" of copyrighted films constituted a Section 1 violation); *Hazeltine Research, Inc. v. Zenith Radio Corp.* 388 F.2d 25, 33–34 (7th Cir. 1967) (affirming finding that licensor misused its patents by attempting to force a licensee to accept a package of licences).

[51]   See *International Salt Co. v. United States* 332 U.S. 392 (1947).

[52]   Above, n.17. At least several courts, however, have eroded the well-established "no duty to deal" exception in the intellectual property context. In *Image Technical Servs. v. Eastman Kodak Co.* 125 F.3d 1195 (9th Cir. 1997), the court abandoned the traditional doctrine that a lawfully acquired patent immunizes the holder from antitrust liability for lawful conduct under the patent or copyright laws, including, at least under certain circumstances, the right to exclude others. Instead, the court applied a Section 2 duty-to-deal analysis to Kodak's refusal to license its patents, stating that "[n]either the aims of intellectual property, nor the antitrust laws justify allowing a monopolist to rely upon a pretextual business justification to mask anticompetitive conduct." *Ibid.* at 1219. Similarly, in *Intel Corp.*, the Federal Trade Commission prohibited Intel from withholding or threatening to withhold certain advance technical information from any one of its customers for reasons relating to an intellectual property dispute with that customer. 64 Fed. Reg.

and/or failing to practice one's own invention,[53] however, do not constitute patent misuse.

F. Other applicable federal and state laws

Filing notification under the Hart-Scott-Rodino Act

18–16 In addition to the Sherman and Clayton Acts, intellectual property licensing arrangements may also be subject to the Hart-Scott-Rodino Antitrust Improvements Act of 1976 (the "HSR Act"),[54] which requires that notification be provided to the agencies of stock and asset acquisitions that give the acquirer an interest of more than $15 million in the acquired company's stock or assets, where certain size thresholds of both the acquiring entity and acquired entity are met.[55] The FTC Premerger Notification Office, which is the arbiter of the notification rules under the HSR Act, view some intellectual property licences as asset acquisitions.[56] Generally, only licences that are exclusive, *i.e.*, the licensor does not retain any rights to use the licensed technology, are subject to HSR notification. However, the notification rules have been extended to *prima facie* nonexclusive licences that are *de facto* exclusive because of devices such as royalty structures that penalise additional licensing, or a side agreement that forecloses competition.[57]

CONCLUSION

18–17 The analysis of whether restraints in a licensing arrangement will harm competition and thus violate the U.S. antitrust laws is quite complex. While the intellectual property laws provide holders of intellectual property with rights to exploit the value of their intellectual property, this right is not absolute. To remain beneath the antitrust radar, a licensing arrangement must neither unreasonably restrain trade nor adversely affect competition.

20,133 (FTC April 23, 1999)(Analysis to Aid Public Comment). This duty to deal has been extended into the copyright context. In *Data Gen. Corp. v. Grumman Sys. Support Corp.* 36 F.3d 1147, 1187 (1st Cir. 1994), the First Circuit Court of Appeals stated in dictum that an author's desire to exclude others from use of its copyrighted work may in rare cases give rise to antitrust liability if the refusal to license is deemed exclusionary, and the challenged conduct affects markets beyond that of the copyrighted work. Notably, the courts were willing to impose antitrust liability in these cases because the refusal to deal was not based on a desire to protect intellectual property rights but rather premised on anti-competitive objectives.

[53] See Patent Misuse Reform Act, 35 U.S.C. § 271(d)(4) ("[N]o patent owner otherwise entitled to relief for infringement . . . shall be denied relief or deemed guilty of misuse . . . by reason of having . . . refused to license or use any rights to the patent.").

[54] 15 U.S.C. § 18A *et seq.*

[55] This "size-of-person" threshold is met where one party has at least $10 million and the other at least $100 million in annual net sales or total assets. See 15 U.S.C. § 18a(a)(2). At the time of print, the U.S. Congress was considering a Bill that would make significant changes to the filing requirements under the HSR Act. These changes would (1) increase the size of the transaction threshold from $15 to $50 million, and (2) eliminate the size-of-person test.

[56] See Stephen M. Axinn, et al., *Acquisitions Under the Hart-Scott-Rodino Antitrust Improvements Act* § 4.01[5] (1996 ed.).

[57] See *United States v. S.C. Johnson & Sons* (1995) 1 *Trade Cas.* (CCH)¶ 70,884 (N.D. Ill. 1994) (consent decree) (enjoining Bayer's licence to S.C. Johnson of its patented insecticide because the effect of the licence was to deprive consumers of a likely market entrant without any countervailing benefits).

CHAPTER 19

INTRODUCTION TO CHAPTERS ON AGREEMENTS

Jennifer Pierce, Charles Russell

The remit of this section

19–01 In the following chapters, we describe some of the transactions that inventors and promoters of technology will encounter, together with information on the possible legal structures for research, development and exploitation. We have aimed to do this in a practical way, in many cases relating the factual information to the process of structuring and negotiating the agreement. From the manner of presentation, readers might be tempted to conclude that the information is exhaustive. Whilst this section contains a considerable amount of information, it has not been possible to include everything, or to cover all possible circumstances. The aim has been to present and to enable readers to understand the broader picture. They may then wish to consult the highly specialised technical material that is available on individual topics. We have chosen to use footnotes, so that the text can be read on two levels. The broad overview is contained in the main text, whilst the more specialist material is set out in the footnotes.

Variation between sectors and jurisdictions

19–02 There is considerable variation between national laws relating to intellectual property and its exploitation. Agreements and commercial arrangements also vary between sectors of industry, depending primarily on the intellectual property rights concerned, the length of time that it takes for products to reach the marketplace, and the regulatory and other requirements of the relevant sectors. A licence relating to a pharmaceutical product may therefore differ considerably from one relating to a new type of bottle opener. The documents that are described in this section are generic, and whilst some of the factors that lead to modifications in specific sectors have been mentioned, it has not been possible to cover everything.

General matters relating to contracts

19–03 The authors have deliberately avoided including a chapter on general contractual principles, on the basis that a single chapter would be insufficient and could easily be misleading. Such information is available from a number of good sources. That said, it is worth reminding readers of some of the basic guidelines for negotiating contracts, for the benefit of any readers who are not lawyers. Above all, it is crucial that the contract contains all the essential terms in a manner that is unambiguous. It is tempting to gloss over points of contention by using vague language, but regrettably it is this sort of language that tends to become the subject of disputes at a later stage. Negotiators also need to be careful not to misrepresent the contract or any of its terms to the other party. Many contracts limit the representations on which the other party can rely to those that are set out in the contract. Such provisions are not necessarily enforceable in all circumstances, especially if the agreement does not record the basic understanding of the parties. Finally, if there is no intention to enter into a binding agreement until the formal contract is signed, it is safest to state this specifically in correspondence.

Taxation

19–04 We planned to have a chapter on taxation, written by Professor Adrian Shipwright of Pump Court Tax Chambers, which would have covered the transactions that are described in the following chapters. Regrettably, the taxation of intellectual property is in the process of a radical revision. The Finance Bill 2000 provides for the abolition of stamp duty on most intellectual property, and for a new relief in respect of expenditure on research and development. The full details of the latter provisions were not known until after this book had gone for typesetting. In addition, it is anticipated that in the near future there will be changes to the taxation of revenue derived from intellectual property. It has not therefore been possible to cover taxation in the way that we would have wished. There are some passing references in circumstances where there is an impact on the actual wording of licences and assignments, but not otherwise. Readers should, however, consider the overall tax position in respect of each transaction.

Funding agreements

19–05 We are conscious of the fact that we have covered funding but there is no chapter on funding agreements. Funding agreements are equally diverse, and it would not have been possible to cover them properly. However, many of the matters considered in the context of research and development agreements are also relevant to funding agreements, more especially the ownership of intellectual property. There is also a section on venture capital in Chapter 21.

CHAPTER 20

RESEARCH AND DEVELOPMENT AGREEMENTS

Jennifer Pierce, Charles Russell

INTRODUCTION

20–01   If a participant in a research and development[1] project wishes to retain and to exploit the intellectual property that is created during the project, a collaboration agreement can be invaluable. The agreement sets out the respective interests of the participants in the intellectual property created in the course of the work and the way in which that property is to be exploited and managed. It also deals with arrangements relating to other intellectual property that the participants will need to use if they are to commercialise the results of their work. In addition, the agreement may set out the way in which the project is to be administered. Regrettably, it is often the case that participants do not wish to spend time dealing with legal matters at the outset. Consequently, some valuable joint research work is never exploited commercially, as the participants never agree on how to do so, or they discover that they are unable to use the results of their work due to rights held by other people.

For the sake of brevity many of the references in this chapter are simply to "research", and not to "research and development".

Types of agreement

20–02   Research agreements vary substantially in both complexity and structure, depending on the circumstances. They may be used for:

(i)   contract work, where services are provided for a fee or, perhaps, for a share in the results of the project. This can range from a single project, with one research worker, to outsourcing of part of the research function;

(ii)   funded projects, such as the research funded by the European Commission and the DTI, that are described in Chapters 2 and 4;

(iii)   collaborative projects, where the participants enter into agreements to share the risks and rewards of a project or a programme, but do not create a separate legal entity, such as a company, for the purpose[2]; and

(iv)   joint ventures where a separate legal entity is established to hold the assets, and bear the risks, of the collaboration, and where there may be an exchange of technology between the joint venture and its parent companies.

In some cases, the participants will enter into more than one type of agreement. For example, funded research could be conducted through a joint venture company that uses contract researchers. There are variants on these basic arrangements such as the consortium described in Chapter 6.

---

[1] Research is, clearly, distinct from development. For the sake of simplicity they are treated together in this chapter.

[2] See Chapter 21 below.

This chapter describes agreements in respect of collaborative research. Unlike licence agreements, which are affected by an appreciable amount of law from various sources, collaboration agreements are far more practical in nature. The topics set out below are not an exhaustive list of the matters that need to be considered, or to be included in such an agreement, as they vary a great deal, depending on the circumstances and the technology involved.

If the research is to be funded externally, on the funder's terms, it is likely that those terms will cover similar subject matter to the agreement between the participants. The collaboration agreement may have to be drafted to take account of the funder's terms, and if the funding agreement is not negotiable its terms will determine those in the collaboration agreement.

PREPARATION

Confidentiality

20–03 Discussions relating to research projects almost inevitably entail the disclosure of confidential information at some stage. As described in Chapter 10, obligations of confidence can arise without a confidentiality undertaking. However, an undertaking gives greater legal certainty, and may be easier to enforce. It is therefore standard procedure in many research-based organisations to execute a confidentiality undertaking before any confidential information is disclosed.

When entering discussions with some potential sources of funding, it can be difficult, if not impossible, to obtain any written assurance of confidentiality before disclosing any information. Also, some funders use the services of external assessors, and it could be difficult to trace a breach of confidence back to them. In these circumstances, researchers should be wary of making disclosures unless they have applied for a patent in respect of the more important technical information that is to be disclosed.[3]

Description of the project

20–04 One of the most crucial elements of the agreement is the description of the project, as most clauses ultimately relate to it. The description can also be invaluable when carrying out the preliminary investigations, such as searches for intellectual property that could be infringed by exploiting the results of the project, as described below. It is, therefore, advisable to start preparing the description at the outset. In some projects the description is deliberately vague, on the basis that it will give greater flexibility to the researchers. This may defeat the object of the exercise, and may lead to disputes. As described below, it may be possible to change the project plan if it becomes necessary, so there may be no need for the plan to be vague. More impor-

---

[3] In the U.K. an invention is not patentable unless it is "new". See Patents Act 1977 ss. 1(a) and 2 and Chapter 8 above. There is a similar, but not identical, requirement for registered designs—see Registered Design Act 1949, ss. 1(2) and 1(4). See also Directive 98/71 of the European Parliament and of the Council of October 13, 1998 on the Legal Protection of Designs (the "Designs Directive") Arts 3,4,6. This is included with the U.K. references as it is to form the basis of future U.K. legislation. See Chapter 12 for further information on designs. Even if a patent has been applied for, it may be preferable to keep the subject matter confidential, as further applications can be made. Patent applications that are filed but unpublished on the priority date are not deemed to be part of the state of the art for the purpose of deciding obviousness—see Chapter 8. Also, if a patent is granted it will only be possible to claim damages backdated to the date of publication. Furthermore, confidential information ceases to be protectable once it is in the public domain—see Chapter 10 and *Att.-Gen v. Guardian Newspapers Limited (Spycatcher)* [1988] 3 All E.R. 545.

tantly, a formal mechanism for changing the project plan should alert the participants to the need to evaluate the changes to the project, in particular with regard to possible infringement of other peoples' intellectual property. If the project is funded by an external funder, the funder is likely to seek a detailed description before deciding whether to provide funding. It may also be necessary to inform the funder, or to seek the funder's permission before changing the project.

Existing rights

20–05   The primary objective of the research project is usually to produce results that can be exploited commercially, either by licensing, or through production. In either case, it is essential to avoid producing research results that cannot be used without infringing somebody else's rights. Searches[4] for such rights are usually conducted in respect of a specific product, and not a field of research. They can be costly and time-consuming. However, in many cases, it should be possible to undertake a basic search for relevant patents in at least one major jurisdiction. As with a "freedom to use" search, the results cannot be guaranteed, but they can give a good indication of the patent cover in the sector. In addition to formal searches, it is worth making detailed enquiries about the rights held by or known to each of the participants in the project. These rights may not be confined to intellectual property. They could, for example, include biological material. Pre-existing rights held by participants that are needed for the project are often known as "background".[5]

Arrangements regarding background and other rights

20–06   If enquiries reveal the existence of essential background intellectual property, or other rights held by third parties, which could be needed to exploit the results of the project, the participants will need to gain access to those rights. This could be through acquisition, by licensing, or by an option to acquire or take a licence of the rights. In the case of rights held by people who are not participating in the project, it may be a condition of access that the collaborators enter into a cross licence in respect of the rights created in the course of the research.[6] As with all acquisitions of intellectual property, it is vital for the licensor to investigate the rights to be acquired at an early stage. The requisite due diligence relating to licences is described in Chapter 23.[7] In this context it is especially important to check that none of the property and none of the participants are subject to contractual restrictions which could be triggered by use or participation in the research project. Restrictions of this type are most likely to be found in previous funding or collaboration agreements. There may be an implied licence of rights to background intellectual property, even if there is no express provision in the contract,[8] so it is advisable to state the terms on which such rights are to be provided.

---

[4] See paras 9–65 to 9–69.
[5] The term "background" was used by the European Commission and other funding bodies. Now that the Commission uses the terms "pre-existing know-how" it is possible that the terminology will change.
[6] Note that cross licences are outside the scope of Commission Regulation (E.C.) No. 240/96 of January 31, 1996 on the Application of Article 85(3) of the Treaty to certain categories of technology transfer agreements (the "Technology Transfer Block Exemption") if the licences are between competitors in relation to the products covered by the licences, and there are territorial restrictions regarding: (1) manufacture, use or putting products on the market; or (2) use of the licensed technology—see Art. 5.1(3) and 5.2(2).
[7] See paras 23–36 to 23–45 in particular.
[8] See *Ibcos Computers Ltd v. Barclays Finance Ltd* [1994] F.S.R. 275. In this case ownership was implied, but the circumstances were very different from most research and development contracts.

Competition law

20–07   E.C. competition laws prohibit agreements that may affect inter-state trade in the European Economic Area, and which have the object or effect of distorting competition within that area.[9] Agreements on the joint execution of research work or the joint development of the results of the research, up to but not including the stage of industrial application, generally do not infringe E.C. competition laws.[10] In certain circumstances, however, such as where the parties agree not to carry out other research and development in the same field, E.C. competition laws could be infringed.[11] Agreements that provide for both joint research and development work and joint exploitation of the results may infringe E.C. competition laws, because the parties jointly determine how the products developed are manufactured or the processes developed are applied or how related intellectual property rights or know-how are exploited.[12] English competition law is based on E.C. competition law, although some of the finer detail of the implementation is different.[13]

20–08   As with many agreements that infringe E.C. competition laws, it is possible to gain exemption if there are compensating benefits to the production or distribution of goods, or if it would result in the promotion of technical or economic progress, whilst allowing consumers a fair share of the resulting benefit.[14] Joint exploitation may be exempted under the "Block Exemption" for research and development agreements. However, that exploitation "must relate to products or processes for which the use of the results of the research and development is decisive".[15] Joint exploitation is not "justified where it relates to improvements which were not made within the framework of a joint research and development programme but under an agreement having some other principal objective, such as the licensing of intellectual property rights".[16] There are restrictions on the applicability of the exemption if the parties are competitors and their combined market share exceeds certain limits.[17] Under English competition law, agreements that are exempted under E.C. competition law gain automatic exemption, although they can be made subject to conditions or obligations, or the exemption may be cancelled.[18]

---

[9] See Art. 81 of the Treaty Establishing the European Community (the "E.C. Treaty"). See also Commission Regulation of December 19, 1984 on the application of Article 85(3) of the Treaty to categories of research and development agreements (418/85), the "R&D Block Exemption". The R&D Block Exemption itself refers back to a Commission notice of 1968. The R&D Block Exemption expires on December 31, 2000. A draft of a proposed replacement has been issued dated April 27, 2000 (the "Proposed Exemption"). Note that the former Art. 85 is now Art. 81 of the E.C. Treaty. See Chapter 16 for more detail.

[10] See Recital 2 of the R&D Block Exemption, which itself refers back to a Commission Notice of 1968. See also Recital 3 of the Proposed Exemption.

[11] See Recital 2 and Recital 3 immediately above.

[12] See Recital 3 of the R&D Block Exemption. There is no equivalent provision in the Proposed Exemption, which refers to joint exploitation as the "natural consequence of joint research and development".

[13] See Chapter 17 and the Competition Act 1998.

[14] See Article 81(3) of the E.C. Treaty (formerly 85(3)), and Competition Act 1998, s.9.

[15] See Recital 7 and Art. 2(d) of the R&D Block Exemption. Note also that joint exploitation of results is only covered if the results are protected by intellectual property or constitute know-how that substantially contributes to technical or economic progress. See also Recital 14 and Art. 2.4 of the Proposed Exemption. Note that Art. 2.4 does not include know-how on a literal interpretation, although it is possible that this is loose drafting.

[16] Again, see Recital 7. There is no equivalent in the Proposed Exemption, although Recital 17 refers to restrictions that are anti-competitive and which are not indispensible to attain the positive effects of R&D.

[17] See Chapter 16 and Art.3 of the R&D Block Exemption. See also Recitals 5,15,16,17 and Art. 3 of the Proposed Exemption.

[18] See Competition Act 1998, s. 10.

Regulatory matters

20–09 Regulatory issues must be considered carefully, as they may affect the project, and therefore the agreement, in important respects. For example, there are regulations covering the radiation of electro-magnetic energy,[19] which can affect researchers working with transmitters and a range of other apparatus. It may therefore be necessary to undertake the project in facilities that will contain the radiation, in order to ensure compliance. The importation of samples or equipment for use in the project may even be regulated. For example, it is possible that in other jurisdictions there is legislation on the lines of the Convention on Biological Diversity 1992, which could affect the use of genetic resources.[20] In particular, signatory states have sovereign rights over their natural resources, which includes the right to determine access to genetic resources.[21] If the result of the research is likely to be a regulated product, it is worth investigating the regulatory requirements at this stage. If the results of any experiments will be required in order to gain registration or authorisation, they could be built into the programme of research, if that is acceptable to the relevant authority.

Separate vehicle

20–10 In the case of a collaborative arrangement, it is worth considering whether to set up a separate legal entity, probably a subsidiary company, in order to undertake the research. The researchers may then be able to "ring fence" the research. This could be prompted by a desire to separate out the intellectual property that is to be used in the project, and in particular to prevent access to background intellectual property rights that are to remain separate from the project. It could also be beneficial to establish a separate vehicle if the research constitutes a diversification of the participant's business, and the conduct or exploitation of the research has a different profile from the existing business. This may be due to greater risks, or because the owner wishes to separate the businesses in the future. If research is to be undertaken in another jurisdiction, preferential terms may be available if the contract is undertaken by a legal entity established in that jurisdiction. Chapter 21 describes the various potential vehicles.

THE AGREEMENT

Matters relating to the research project

20–11 The manner in which the project is administered depends to a large extent on its complexity, and on the management style of the participants. A well-managed project may, however, obtain external funding more easily, and the participants are less likely to quarrel if their roles are well defined. Participants need to consider some or all of the following:

Project description

20–12 Further refinement of the description, including the duration of the project, and the details of any "milestones" in the course of the project, together with any "change control" provisions. "Milestones" are either crucial goals during the project, or events at particular intervals in the course of the work. Reaching the milestones

---

[19] See Wireless Telegraphy Act 1949, ss. 10 and 11, and regulations made under s.10.
[20] In particular, see Arts 3,4,15,16, and 19.
[21] See Art. 15(1), but see also Art. 15(2) relating to conditions for facilitating access to resources.

can, for example, trigger payments, reports, or evaluation of the progress of the project. In addition to milestones, the project description may also include any required standards that the participants must work to. The "change control" provisions are basically the circumstances in which the project description, the project plan, or any other provision of the contract, can be changed, and the mechanism for reaching agreement on the changes. Change control clauses provide a structure for reaching agreement, but are not effective in a deadlock situation unless a dispute resolution mechanism is included. The ultimate sanction for failure to meet milestone targets or to agree on material changes to the programme could be termination, as this could indicate that there are serious problems with the project. It is worth noting that it is a condition of the R&D Block Exemption that the objectives and the field of the research are defined.[22]

Project management

20–13 A management structure may be required, even if it is a simple one. In the most basic structure, where all the participants are on an equal footing, there could be a single co-ordinator, co-ordinating representatives, or a small committee with members from each participating organisation. External funders may insist that one of the participants takes specific responsibility for co-ordination. Single co-ordinators can obtain and distribute reports, equipment and consumables, and money from funders. They may also assist with communication between the participants in the project. Representatives of the participants, and committees, may have the ability to make decisions relating to the management of the project, and can have some responsibility for change control. If decisions are to be made by committee, it is advisable to agree on the rules that govern the committee, in a similar manner to meetings of the board of directors of a company.

Researchers

20–14 Some researchers are indispensable. It is crucial that all such researchers are contractually committed to working on the project. This could entail a fixed term contract, or a lengthy notice period if they terminate the contract, and possibly covenants restricting work on competing projects for a specified period following termination of their contract.[23] If a particular researcher is irreplaceable and his contribution is essential to the success of the project, the participants may consider taking out insurance to cover the potential damage suffered if he were unable to continue. If a researcher is not employed by one of the participants,[24] it is essential that he enters into a written agreement transferring his intellectual property rights to the participant who engaged him, as it is only title to designs (including semiconductor topographies) that is transferred automatically to the commissioner[25] of the right.[26] Such an agreement should also include a confidentiality undertaking, together with provisions for termination, and

---

[22] This is a condition of the application of the R&D Block Exemption —see Art. 2(a). See also Recital 4. This obligation has not been retained in the Proposed Exemption.
[23] See paras 10–52 to 10–59.
[24] Particular care should be taken with students as they are unlikely to be employees. Similarly, some academics may not be employees.
[25] If a design is created in pursuance of a commission, the commissioner for money or money's worth is the first owner of that design—see Copyright, Designs and Patents Act 1988, ss. 215(2),263(1), RDA 1949, s. 2(1A). See also PA 1977, ss. 7,39, CDPA 1988, s. 11, The Copyright and Rights in Database Regulations 1997 ("Database Regs") regs 14,15, The Design Rights (Semiconductor Topographies) Regulations 1989 ("Topographies Regs") reg. 5.
[26] See Chapter 14.

possibly a restrictive covenant, similar to the other researchers. It may be worth checking that none of the participants, or their researchers, have entered into a conflicting contract. In the case of academics, this could include their institution.

Facilities and equipment

20–15 It may seem obvious, but it is customary to state which research facilities will be used, and the terms on which they are to be used, especially if they are to be used by the researchers of more than one participant. Participants such as academic institutions may seek to levy a charge for the use of their facilities, or put forward the cost of the facilities as part of their contribution to the project. Also, licences and other approvals may relate to specific facilities. If it will be necessary to acquire expensive equipment to undertake the project, it is customary to provide for this in the agreement, and to specify who will own it at the end of the project. In contrast, there is often a budget for consumables.

Administration of the project

20–16 Reports In the course of most projects, reports are required at some stage. The frequency and format of the reports will vary with the circumstances, and in the case of an externally funded project will depend on the requirements of the funder. If necessary, the format of the reports can be specified in the agreement. In addition to reports, participants and their advisers may wish to visit research sites, inspect work on the project, and talk to researchers. If the parties are to share intellectual property created in the course of the project (often known as "foreground"[27]), it may be necessary for there to be an obligation to inform the other participants of the making of any patentable invention, or material know-how, as soon as reasonably practicable. These obligations, together with general reporting requirements relating to know-how,[28] could continue after the rest of the agreement is terminated, if the participants are to grant each other licences or other rights in improvements. In addition to reporting the results of the research, the agreement can impose obligations to report matters that could affect the project or the exploitation of the results.

20–17 Confidentiality The importance of keeping results confidential has already been emphasised. Failure to do so will render any invention made in the course of the project unpatentable, as the invention will no longer be new.[29] Similarly, confidential information ceases to be protectable, and therefore loses its value, if it is not kept confidential.[30] Some research and development agreements therefore incorporate the participants' protocols relating to confidentiality. These can extend beyond the customary requirement for confidentiality undertakings and restrictive covenants from employees and consultants, to restricting access to certain research facilities, and enforcing specific security measures. In the case of academic researchers, it may be necessary to provide for publication of theses and papers to be delayed, and possibly postponed indefinitely, if necessary.[31] In any event, the obligations of confidence

---

[27] Note that the European Commission is now referring to "knowledge".
[28] The extent of the know-how to be included in the report will depend on the project. It may be necessary to specify that it should enable a reasonably skilled worker to perform certain tasks.
[29] See n. 3 to 20–03 above.
[30] See paras 10–05 to 10–06 —for consideration of *Att.-Gen v. Guardian Newspapers Limited (Spycatcher)* [1988] 3 All E.R. 545 and *Mustad v. Dosen* [1963] R.P.C. 41.
[31] It will usually be necessary to recognise the importance of publication to an academic institution, and to agree on a procedure for determining what should be published—see para. 6–35 for an idea of the current benchmarks.

should survive termination of the agreement to the extent that information is still valuable and secret, and it is consistent with normal commercial exploitation.[32]

20–18 Laboratory notebooks Agreements sometimes specify in some detail the way in which laboratory notebooks are to be kept. If laboratory notebooks are kept in accordance with United States legal requirements, the participants should be in a better position to prove the date when any invention was conceived, and the subsequent steps taken to reduce the invention to practice. This could be essential in the event of a dispute in the United States if somebody else claims that they were the first to make the invention, and are therefore entitled to the grant of the patent relating to it.[33]

20–19 Sub-contractors The participants need to decide whether work on the project should be delegated to sub-contractors. In some cases it is necessary to obtain specialist skills from an external organisation, or it is more cost-effective to do so. If sub-contractors are to be employed, they are potentially less easy to control, and the involvement of an additional party increases the risk of a disclosure of confidential information. Sub-contractors may also require licences of background intellectual property in order to perform their work. If sub-contracting is permitted, it is crucial that the relevant obligations and restrictions applicable to the primary contractor are imposed on the sub-contractor.

20–20 Employment matters If employees are to work on the premises of other participants, it is necessary to consider the arrangements for their health and safety and for insurance, especially if there are any inherent risks involved in carrying out their work. It is also worth considering whether there is any possibility that any proposed arrangements which involve employees could be construed as the transfer of a business, with the consequence that any contracts of employment would be deemed to have been transferred.[34] Any restrictive covenants that are imposed on the participants can be reflected in employees' contracts, to the extent that they would be enforceable.[35] If employees are to work closely with other participants, restrictions on poaching them may assume greater importance.

Intellectual property

Background and other pre-existing rights

Acquiring and taking licences of rights

20–21 As explained in paragraph 20–05 above, the participants will need to acquire, take a licence,[36] or take an option to acquire or take a licence of all the pre-existing

---

[32] A confidentiality undertaking is covered by the R&D Block Exemption, even if the obligations continue after the expiry of the agreement — Art.5(d). Note that this does not appear in the Proposed Exemption as the "whitelist" has been deleted in favour of reliance on a combination of the black list and market share requirements.

[33] Unlike U.K. and European patent law, where the first true inventor (or, inventors, in the case of an invention made jointly) to file a patent application in respect of a patentable invention is entitled to the patent, in the U.S. a subsequent applicant who was the first to make the invention may be entitled to the grant of the patent. See paras 9–52 *et seq.* and 9–64 for a description of the relevant U.S. patent law, and for suggested practice when keeping laboratory notebooks.

[34] See Transfer of Undertakings (Protection of Employment) Regulations 1981 (as amended). General employment law is outside the scope of this publication.

[35] See paras 10–52 to 10–59.

[36] A commentary on the terms of assignments (transfers of title) is to be found in Chapter 22, and there is one on licences in Chapter 23.

intellectual property that they would infringe if they were to undertake the project or to exploit the results of the research commercially. The relative merits of outright acquisition and licensing are dealt with below. The extent to which any of the participants need to acquire a licence (as a basic minimum) in order to carry out their initial research will depend on the nature of the project, and on exceptions from infringement in the jurisdiction where the research is to be undertaken.[37] In particular, research to conceive a new product or process as opposed to conducting trials in respect of a known product, may merit exemption under English law.[38] In addition to third parties' rights, the participants need to consider the rights that they will licence between themselves. Each participant may be required to warrant in the agreement that he has disclosed all the background intellectual property of which he is aware. If property is to be acquired from other participants, it is vital to consider whether the obligation to provide the property will continue even if the participant providing the property leaves the project. It is also possible that external funders will seek some assurance that the participants have the necessary background rights to carry out the project and to exploit the results.

Reducing the cost

20–22   However the participants gain access to that intellectual property, it is likely that there will be a cost associated with this, so the parties need to be satisfied that the rights are necessary, and that the terms offered are the best that are likely to be available. For example, it could be preferable to take an option for a licence, rather than an immediate licence, if it will not be essential to have a licence to complete the project itself. An option could be substantially cheaper, and if the results of the research are uncertain, it may save the participants from taking a licence that will never be required.

Rights created in the course of the project

Ownership of intellectual property

20–23   As described in Chapter 14, the law relating to ownership of intellectual property differs between employees and consultants, between rights, and in the case of employee inventions it could depend on the circumstances.[39] In certain cases, such as those of students, research workers may be neither consultants nor employees, but a specific contract still needs to be entered into. Also, employed researchers could claim that they are acting outside the scope of their employment, and that they own the rights arising from their work.[40] Even in cases where the ownership can be established through other contracts, it is common to take a specific assignment of the rights from the creators, at some stage,[41] if only for registration purposes. Whatever the situation,

---

[37] See PA 1997, s. 60(5) and paras 8–94 et seq.

[38] See PA 1997, s. 60(5)(b) under which research undertaken for experimental purposes relating to the subject matter of the invention is exempted from infringement, but see also Monsanto v. Stauffer (No. 2) [1985] R.P.C. 515, CA.

[39] See PA 1997, s. 39(1) under which the circumstances in which an invention is made may be taken into account in determining the ownership of the invention as between employer and employee. See also Chapter 14.

[40] See PA 1997, ss. 39, 42,43, CDPA 1998 ss. 11, 215 and 263(1), Database Regs, regs 14 and15, RDA 1949, s. 2, Topographies Regs, reg. 5. Also, in the case of copyright, database right, and semiconductor topographies CDPA 1998, s. 11(2), Database Regs, reg. 14 (2), and Topographies Regs, reg. 5, provide for a possible agreement to the contrary.

[41] Note that in the United States patent applications must be filed in the names of the inventors and may be assigned at grant—see para. 9–53.

the participants may impose obligations[42] on each other relating to contracts with employees, consultants and others, and, in specific cases, may ask to see the agreements.

20–24   Rights created by consultants   With the exception of registered designs and design right, intellectual property created by contractors and other non-employees is not transferred automatically to the commissioner.[43] If all the relevant rights are to be held by the commissioner, it is necessary for their initial contracts to contain an assignment of future inventions,[44] copyright,[45] database right,[46] and designs,[47] together with the right to require a further assignment document.[48] In the case of patents and registered designs, the assignment will include the right to apply for and to be granted patents (and similar rights) and registered designs. Assignments will refer, ultimately, to a description of the project, which will determine the scope of the assignment. In each case there is also an obligation to:

(i)   disclose the relevant invention, design, copyright work, or database to the person taking the assignment as soon as it is created;

(ii)   keep the subject matter of the rights created confidential;

(iii)   assist the person taking the assignment with registration;

(iv)   sign all documents and do everything else that is necessary to ensure that the rights are assigned;

(v)   waive all moral rights that they may have in works assigned (to the extent that they exist and the person is able to do so); and

(vi)   assign rights in the physical property in all records and embodiments of the relevant inventions, works, and designs.

There could also be some kind of restrictive covenant. A further, confirmatory, assignment may be entered into when the rights have been created.

20–25   Rights created by employees   In the case of employees, whose rights generally vest in their employer by operation of law if they are made in the course of employment,[49] the same provisions may be included in their employment contract, so that the employees are aware of the situation. Under United Kingdom patent law employers usually own employee inventions if they are made in the course of the employee's normal duties, or in the course of duties specifically assigned to him.[50] It

---

[42] These could include warranties.

[43] In the case of designs, the rights are commissioned for money or money's worth—see CDPA 1988, ss. 215(2) and 263(1), RDA 1949, s. 2(1A), Topograhies Regs, reg. 5.

[44] Note that there is no specific statutory provision for assignment of future inventions, although PA 1977, s.7 refers to future assignments. It is therefore customary for agreements to contain provisions for assignment when those rights are created, and for them to be held on trust pending that assignment.

[45] See CDPA 1988, s. 91.

[46] See the Database Regs, reg. 23, which provides that database right is treated in the same way as copyright in this regard, so CDPA 1988, s. 91 applies.

[47] See CDPA 1988, s. 223, for design right, Topographies Regs, reg. 5. There is no equivalent provision in respect of registered designs, but note RDA 1949, s.3(2)—"An application for the registration of a design in which design right subsists shall not be entertained unless made by the person claiming to be the design right owner".

[48] In certain circumstances, an assignment or a licence could be implied but the result is uncertain unless there are specific terms, see paras.14–46 to 14–50.

[49] See n. 40 to para. 20–23 above.

[50] See PA 1977, s. 39(1)(a), which provides for the employer to own inventions which are made during the course of an employee's normal duties, or which are made outside the course of an employee's normal duties, but specifically assigned to him, provided that an invention might reasonably be expected to result from the carrying out of those duties. See also Chapter 14 for details.

is advisable that the employee's duties be carefully documented in his employment contract, or, in the case of an employee specifically assigned to the project, that his assignment to the project be recorded in writing. Inventions made in the course of employment, which at the time the invention is made fall outside these duties, may still belong to the employer if, due to the duties and responsibilities of the employee, the employee has a special obligation to further his employer's interests.[51] It is unwise to rely on anything other than the employee's job description, or formally assigned duties, as it is not possible to alter the statutory provisions relating to the ownership of an employee's inventions by contract.[52]

20–26 Alternative arrangements regarding ownership In joint projects, the statutory provisions regarding ownership can sometimes produce unfair results, as they do not reflect the respective contributions of the parties. This could be for numerous reasons. The research and development agreement may therefore assign the property in accordance with different provisions. For example, the agreement could provide that the participants will own the intellectual property arising from work on specific parts of the project if they have made a material contribution to that specific part. Alternatively, all the participants could be co-owners, which has substantial drawbacks, as the consent of the other co-owners is required for certain dealings in the rights (see paragraph 20–27 below). However, in view of the limited protection afforded to licensees of unregistered rights (see paragraph 20–31 below), co-ownership is worth considering. The division of rights is likely to depend on the respective contributions and interests of the participants, and on the possibility of cross-licensing the results between them (see paragraph 20–34 below). It may also depend on whether the prospective owner is likely to have sufficient competence and funds to manage and protect the property.

20–27 Co-ownership Any arrangement relating to the ownership of intellectual property, other than outright ownership, is potentially problematic. The basic alternatives are to be a co-owner or for there to be an owner and a licensee.[53] Under English law a co-owner's dealings with the property are restricted. For example, the consent of all co-owners is required for the grant of licences, and for assignments.[54] In the case of patents and trade marks, a co-owner is permitted to do anything for their own benefit that would infringe the patent or trade mark, without reference to the other co-owner, unless there is an agreement to the contrary.[55] The situation is different in the case of copyright, as co-owners of copyright require the consent of the other co-owner before they exercise the right themselves.[56] This is also the case with both registered designs and design right.[57] It is therefore crucial that co-owners of United Kingdom

---

[51] See PA 1977, s. 39(1)(b) and paras 14–07 *et seq* for details.

[52] See PA 1977, s. 42. Note also that the exercise of rights in an invention which belong to an employee according to PA 1977, s. 39 does not infringe copyright or design right in any model or document relating to the invention in which copyright and design right belong to his employer, but that s. 42(2) may not be construed as derogating from any duty of confidentiality owed to the employer, see s. 42(3).

[53] See para. 20–32, below for the drawbacks of being a licensee.

[54] See PA 1977, s. 36(3), CDPA 1988, ss. 173(2) and 258(2), RDA 1949, ss. 19(4) and 44(1), TMA 1994, s. 23. The RDA 1949, is not as clear as the other Acts, but if it is interpreted in a similar way to the CDPA 1988, this seems the most likely meaning.

[55] See PA 1977, s. 36(2) and TMA 1994, s. 23(3). But see also *Young v. Wilson* [1955] R.P.C. 351, in which the implied licence on the sale of patented products (see para. 23–11) was held to constitute a licence for which a co-owner was to account. The judgement was under the Patents Act 1949 under which co-owners were also entitled to exploit the patent without accounting to the other, unless there was an agreement to the contrary.

[56] See *Robin Ray v. Classic FM Plc*, [1998] F.S.R. 622.

[57] See RDA 1949, ss. 7(2) and 44(1). The *Robin Ray* case (above) relates to copyright but in view of the fact that the statutory provisions relating to transactions in design right are very similar it seems that consent will also be required from a co-owner of design right.

intellectual property reach a written agreement relating to the exploitation of that property.[58] It is also worth stating the nature of the co-owners' interests. Will they be joint tenants, in which case the interest passes to the other co-owner on death, or will they be tenants in common in specified shares?[59] To complicate matters, the law relating to co-ownership varies between jurisdictions.

Applications for registered rights

20–28   It is vital that steps be taken to ensure the optimum legal protection for the results of the project, so it is usually provided for in the agreement. Such provisions would normally include the filing of applications for registered property, if appropriate. One of the participants may have ultimate responsibility for filing and managing each application, and for paying registration and professional fees. The ultimate owner of the property is likely to have the greatest incentive to do this, so the task usually falls to him. The owner may, however, decide that he does not wish to make the application. In that case, it would be common for the other participants to have an option to take an assignment of the right concerned, and to make the application, provided that they granted a licence back to the assignor. There could also be a right of first refusal, granted to the other participants, in the event that the right is later to be abandoned, assigned, or is not exploited. Whoever manages the applications, assistance from the inventor(s) can be invaluable, so the agreement should provide for this, and a similar provision should be inserted in the contract with the inventor.

Potential time and expense of managing registered rights

20–29   The cost of acquiring and maintaining registered intellectual property should not be under-estimated. Drafting and prosecuting (*i.e.* registering) applications for registered rights can be crucial, but it can also be an expensive and lengthy process. It is not uncommon to have contested proceedings in the course of, or immediately following, patent prosecution, more especially an opposition to the grant of a European patent,[60] or an interference in the case of a United States application.[61] After grant, renewal fees need to be paid at regular intervals, and in increasing amounts. More importantly, the activities of licensees,[62] and the market in general, need to be monitored, and proceedings may need to be taken against infringers and licensees who are in breach of the terms of their licence. Proceedings

---

[58] In this context, note PA1977, s. 10, which provides that in the event of a dispute between joint applicants for a patent as to whether, and how, an application should proceed, a party can request the comptroller to give directions, enabling the application to proceed in the name of one or more of the parties, and/or for regulating the manner in which the application proceeds.

[59] Under PA 1977, s.36, if a patent is granted to co-owners they are entitled to an equal undivided share (*i.e.* a tenancy in common) in the patent, subject to any agreement to the contrary.

[60] See para. 9–35 for details.

[61] Interference may occur in the course of patent prosecution in the U.S. where two separate inventors claim that they are entitled to the grant of a patent on the grounds that each was the first to conceive of the invention, or that one was not the first, but the first inventor did not reduce the invention to practice with sufficient diligence. See para. 9–58 for details.

[62] To ensure that the correct revenue is collected, that the licensee does not stray outside the terms of his licence, and that the licensee complies with obligations such as marking licensed products. As explained below in connection with unregistered rights, all embodiments of patents and registered designs should be marked in order to prevent infringers from claiming that they were not aware of the existence of the rights, and had no reasonable grounds for supposing that they existed, and are not therefore liable to pay damages. In the case of patents, the word "patent", or "patented", or other words implying that a patent has been granted or applied for in respect of an aspect of the product, should be accompanied by the patent (or application) number, and in the case of a registered design the word "registered", or an abbreviation, or words implying registration should be accompanied by the registered design (or application) number. See PA 1977, s. 62(1), and RDA 1949, s. 9(1).

for infringement of registered rights can be particularly lengthy, costly, and problematic, as infringers are likely to counterclaim that the infringed property is invalid, complicating the issues,[63] and adding further risks. Those who have immediate control of the foreground intellectual property are likely to need funds, expertise, and sufficient time to devote to it. Participants may wish to share the costs of managing registered property.[64]

Management of unregistrable intellectual property rights

20–30 Copyright, database right, and design right (including rights in topographies), subsist without the formalities of registration. However, if the rights are to be enforced, it is necessary to provide evidence of subsistence and ownership. Records relating to the author (or their employer), designer (or their commissioner, or employer, or the first marketer) or maker (in the case of databases) will assist in proving qualification for the right.[65] In addition, the date of creation needs to be recorded to show that the right has not expired. The chain of title from the initial creator (who may not be the first owner) must also be documented. In practice, participants may therefore require that all original copies of works,[66] which are to be protected by unregistrable rights, are dated, marked with the name of the author, or designer, or maker (as relevant), catalogued, and stored safely, together with the above documents of title and records (where relevant). Details of the first marketing of works protected by design right should also be kept. In the case of copyright, the work should be marked with the symbol ©, and the name of the copyright owner, as well as the year of first publication, in order to claim protection for the work in countries that are signatories to the Universal Copyright Convention. Marking all works also prevents imitators from being innocent infringers,[67] so that damages[68] can be recovered from the infringer.

Property which falls outside the scope of the project

20–31 It is in the nature of research that it may produce unexpected results, or that inventions may be made unintentionally. Even if such inventions are to be the property of the inventor, and are not to be licensed to the other participants, it is worth stating this specifically. If such inventions are treated differently from the other results, the drafting of the project description assumes an even greater importance, especially if it transpires that the scope of the initial research requires modification. Alternatively, the property could be owned by the participants, and treated in a similar manner to the results.

The participants could also be prohibited from undertaking work independently in the same field as the project, or in a closely connected field, for the duration of the project, and, perhaps for a short period after that,[69] so that commercialisation of the

---

[63] See Chapter 24.
[64] Note that sharing the costs of litigation is covered by the R&D Block Exemption Art. 5(1)(e)(iii). In common with the other items on the "white list" this has not been included in the Proposed Exemption.
[65] See CDPA 1988, Part I, Chapter IX for copyright, and ss. 217–221 in respect of design right, Copyright and Rights in Databases Regulations 1997, regs 14, 18, Design Rights (Semiconductor Topographies) Regulations 1989, regs 4. See also Chapters 11,12, and 14 for an explanation of qualification and ownership.
[66] And, in the case of design right, articles made to the design — see CDPA 1988, s.213(6).
[67] See CDPA 1988, s. 97(1) and s. 233 in respect of copyright, database right, and design right (including rights relating to semiconductors).
[68] Full damages in the case of secondary infringement of design right — see CDPA 1988, s. 233(2).
[69] Under the R&D Block Exemption, such a provision is exempted during the execution of the programme, but not afterwards — see Art.4 (1)(a).The Proposed Exemption has a similar provision — see Art. 1.2.

project results is not affected. After termination of the project, the participants could continue with related research either jointly or individually. Any additional intellectual property, which is created in the course of this research, or as a result of the participants' experience of commercialising the property, may also be made subject to the intellectual property provisions of the agreement.[70]

Licences

20–32 Drawbacks of licences If participants do not own intellectual property, they can still gain access to it through licensing. In any event, all of the participants must have access to the results from the project if the R&D Block Exemption is to apply.[71] The rights conferred by the licence depend on the nature of the property and the type of licence granted. Licensees of registered rights are protected against those acquiring subsequent interests[72] provided that the licence is registered before any conflicting interest is created and registered.[73] However, registration is no protection against a licensor's failure to pay renewal fees.[74] In the case of copyright and design right,[75] the licensee's position is less secure, as a subsequent purchaser from the owner who has no notice of the licence will not be bound by it if the purchase is made in good faith and for valuable consideration.[76] A non-exclusive licensee of rights other than trade marks[77] has no right to bring infringement proceedings under English law. Furthermore, unless the participants agree on the extent to which the owner can grant licences to people other than participants, licensees have no control over the exploitation of the property.

20–33 Exploitation by licensing If the participants intend to commercialise the results of the work in different technical fields of use, it may be possible for each participant to receive an exclusive licence within their particular field.[78] Similarly, it may be possible to divide the licences, or possibly the property, on a territorial basis.[79] It is less likely, but always possible, that the property could be divided by the acts which

---

[70] Note R&D Block Exemption Art. 4(1)(g), which covers communication of experience gained in exploiting the results and non-exclusive licences relating to improvements or new applications. There is no such explicit provision in the Proposed Exemption, which does not have a "white list".
[71] This is a condition of the R&D Block Exemption—see Art.2 (b). Similarly, if the agreement is confined to joint research and development, each party must be free to exploit the results of the joint research and development, together with any pre-existing technical knowledge, which may be necessary to do so, independently—see Art. 2(c). Note also Recital 4, which refers to "results of the programme that interest it", and specifically mentions that universities may not be interested in industrial application. See also the Proposed Exemption Recital 14 and Art.2.3, which contain provisions similar to the old Arts 2(b) and 2(c), except that it is stated specifically that academic bodies and research institutes may confine their exploitation to further research.
[72] This is on the basis that the licensor is entitled to the property.
[73] See PA 1977, s.33, RDA 1949, s. 19, TMA 1994, s. 25.
[74] This will be important for licensees who are seeking some element of exclusivity, and who therefore wish to maintain the property. Renewal fees are payable at intervals in respect of each item of registered property in each jurisdiction.
[75] In the case of design right, an assignment of the right is taken to be also an assignment of the right in the registered design unless a contrary intention appears — so if the registered right is also to be assigned, entries on the register may constitute notice — see RDA 1949, s. 19(3B).
[76] See CDPA 1988, ss. 90(4), 91(3), 222(4) and 223(3).
[77] Note TMA 1994, s.30(3)
[78] This is covered by the R&D Block Exemption — Art. 4(1)(e), except where any of the participants are competitors (as defined) at the time the agreement is entered into. See also the "blacklisted" provision in Art. 6(e). Note Art. 5.1(e) of the Proposed Exemption, which is similar to 6(e), but which only applies five years after contract products are first put on the market within the common market, and is not clarified by a corresponding entry on the "white list".
[79] The R&D Block Exemption covers territorial restrictions on manufacture, and on "active" sales for a period of five years from the first E.U. marketing—see Arts 4(1)(d) and (f). See also the "blacklisted" provisions in Arts 6(f) and 6(h). The Proposed Exemption contains a similar provision to Art. 6(f)—see Art. 5.1(g).

would otherwise be prohibited by the intellectual property, such as manufacture and sale (in the case of patented products).[80]

The financial aspects of licences of the results can be particularly problematic. Royalty and other licence payments made between the participants can alter their relative financial contributions considerably. They can also be used as a means of evening out the respective contributions of the participants.[81] In addition, licensing to outsiders of foreground intellectual property could create substantial additional revenue, which the participants may wish to share,[82] subject to sharing the relevant expenses.

Academic participants usually seek financial reward for their work, in addition to project funding, and their success in obtaining it will vary. An academic participant may be able to exploit its rights in the results of the project through a company established specifically for the purpose. Otherwise, it may wish to confine its commercialisation to licensing, usually through its licensing company. If the academic participant is not permitted to grant commercial licences, it may seek a commercial return for the use of its rights by the other participants.

As Chapter 23 deals with licensing, it would be pointless to repeat the material here. Readers should, however, be aware that there is a substantial body of law relating to licensing. It is worth noting that the Technology Transfer Block Exemption, which covers licensing, may not apply to collaborative research projects, as it only applies to bipartite agreements, and does not apply to agreements that contain restrictions on competing research and development.[83]

Patent pools and cross-licensing

20–34 Patent and know-how pools, and cross-licensing arrangements have, historically, been created by technology-related businesses in certain sectors. They have created, licensed and cross-licensed substantial numbers of patents. In some cases, appreciable numbers of businesses, and businesses that have large market shares, have been parties to such arrangements. Such activities can lead to savings in research and development, and disseminate technology, but they may also assist in consolidating and abusing a dominant position, and create substantial additional barriers to entry to a relevant market, if the technology is not available to others. They may therefore have both pro and anti-competitive effects. Also, provisions of licensing arrangements between the parties are often more likely to have a detrimental effect on a particular market, because of the scale of the licensing. This area is therefore especially problematic for competition law authorities. The European Commission has slightly relaxed the provisions relating to patent pools in the most recent Technology Transfer Block Exemption, although the Exemption still does not apply to them if there are territorial restrictions.[84] Nonetheless, joint research and development, resulting in joint

---

[80] See R&D Block Exemption Art. 4(fa)-)(fc). As with the other "white list" provisions, these do not appear in the Proposed Exemption. Note, however, the "black list" provision in 5.1(g) of the Proposed Exemption. Whilst this is similar to the old 6(f), the old 6(f) was qualified by the "whitelisted" items, which is no longer the case.

[81] This is covered by the R&D Block Exemption, as is compensation for unequal exploitation of results — see Art. 5(1)(f). Again, this does not appear in the Proposed Exemption as it was on the "white list".

[82] This is covered by the R&D Block Exemption—see Art. 5(1)(g). Similarly, this is not included in the Proposed Exemption as it was on the "white list".

[83] See Art. 3(2). Note also that it does not apply to competitors that hold interests in a joint venture if the licensing relates to the activities of the joint venture, and if the parties' market share exceeds certain thresholds —Art. 5.1 (2) and 5.2(1).

[84] The old Patent Licensing Block Exemption (a precursor to the Technology Transfer Block Exemption) used not to apply to agreements between members of patent pools. They are now covered, provided that there are no territorial restrictions— Art. 5.1(1) and 5.2(2). This is of limited use. See also Chapter 16 .

ownership, cross licensing of background rights, and, almost inevitably, a licensing policy agreed between the co-owners, may infringe competition laws.

20–35 In many cases, research and development agreements will not affect trade between states in the European Economic Area, and no dominant position will be created. E.C. Competition laws will not therefore be infringed.[85] Similarly, many agreements will not affect trade or create a dominant position in the United Kingdom, and will not, therefore, infringe domestic competition law.[86] The R&D Block Exemption gives an indication of the type of agreement that may be acceptable, and of provisions that are likely to be problematic. However, the list of terms that are not covered by it, as they are likely to infringe, and may not merit exemption, concentrates on the terms for exploiting the results of the project. It is the conditions for applicability, combined with the provisions for withdrawal of Exemption, that touch on the potentially undesirable economic effects described in the paragraph above.[87] Agreements that cover joint exploitation of improvements, where the principal aim is not research and development, but is some other commercial activity, such as licensing,[88] are likely to come within this category. Similarly, agreements between participants with high market shares will not be covered by the Block Exemption, although they may be the subject of a specific exemption.[89]

Other terms of research and development agreements

Project funds

20–36 At the outset of most projects, the participants will agree on a budget, and this will be recorded in the agreement. In the case of externally funded projects, a budget may have to be submitted and agreed as part of the application process. The budget can be particularly contentious, as the parties may have differing views on the amount and nature of the expenses that are to be re-imbursed from the project funds. For example, a university may seek full re-imbursement of its overheads, in addition to the marginal cost of participating in the project.[90] Certain participants could possibly gain some or all of their commercial return from the project through the re-imbursement of expenses, if they acquire expensive items of equipment from project funds, and are permitted to keep them. The arrangements relating to expenses need to be considered in the context of the participants' respective contributions, and their likely financial return from the exploitation of the results. It is likely that one of the participants will take responsibility for financial administration. This may be a requirement of certain funders.

Valuing participants' contributions

20–37 The participants' contributions can take many forms, and it can be difficult to compare and to value them in financial terms. For example, a participant may be able to provide a particular researcher or particular facilities. It can be especially difficult to value background intellectual property, and the same is true of the respective contributions of the participants in the course of the project. However, if the

---

[85] See Arts 81(1) and 82 (formerly 85(1) and 86 of the Treaty of Rome) of the E.C. Treaty, and Chapter 16.
[86] See Competition Act 1998, ss. 2 and 18.
[87] See R&D Block Exemption Art. 10, especially 10(a) and (b), which respectively cover circumstances where third parties' ability to carry out R&D is restricted as a result of the limited research capacity available elsewhere, and where third parties' access to the market is restricted. See also Recitals 7–10. These provisions have been retained in the Proposed Exemption—Arts 7(a) and (b), and Recitals 15, 18.
[88] See Recital 7, Art. 1, and Art.2(d). There is no such reference in the Proposed Exemption.
[89] But see Art. 3 and Recital 9, and see also Recital 10. The Proposed Exemption also contains provisions to allow for short-lived increases in market share — see Art. 6(d),(e) and (f).
[90] See paras 5–10, 5–23 and 5–51.

relative contributions of the participants and the return that they receive from them, are not taken into account, the result could be inequitable. In particular, differences between the forms of intellectual property and other rights that protect the participants' results can lead to inequality. For example, a participant may perform valuable work that is only protectable as confidential information, whilst others may create registrable intellectual property.[91] The latter can potentially give far greater protection, and a corresponding financial return, provided that it is registered. Whilst this outcome is a common fact of commercial life, it may be seen as inappropriate in a collaborative project.

Restrictive covenants

20–38   In the course of a research project, participants may work so closely together that their businesses become vulnerable to unscrupulous fellow participants. This can arise as a result of the flow of sensitive information, which may not be adequately protected by a simple confidentiality undertaking, as it can be difficult to prove that it has been breached. Furthermore, good employees may be poached. In addition, a participant could be diverting resources to other, similar, projects. It is therefore common for the agreement to impose restrictions on participants engaging in competing research,[92] and poaching employees from other participants. If the participants were chosen because they will not be competing with each other in the same product market or in the same territory to any appreciable extent, participants may agree not to manufacture or sell products in other participants' territories, or technical fields of application.[93] Other restrictions on sales and distribution could also be necessary from a commercial perspective.[94]

20–39   Restraints on trade are against English public policy, and under English law are void, unless they can be justified as reasonable restrictions that are necessary to protect legitimate business interests.[95] In addition, these restrictions are subject to competition laws in the European Economic Area,[96] including those of the E.C. and the United Kingdom.[97] In each case, it is necessary to consider the nature and scope of the restriction, in order to ascertain whether it is enforceable or infringing. The R&D Block Exemption sets out restrictions[98] for which there can be no general

---

[91] Note in particular that once confidential information has been disclosed it is no longer protectable as such (although there may be a claim in respect of an unauthorised disclosure)—see paras 10–05 and 10–06.
[92] This is exempted by the R&D Block Exemption during the execution of the programme if it is in the same field as the project—Art. 4(1)(a) and (b). This is also mentioned in the Proposed Exemption—see Art. 1.2.
[93] See R&D Block Exemption Art. 4(d),(e) and (f). Note that 4(d) applies irrespective of whether the parties are competitors, but that 4(e) does not apply if two or more of the participants were competitors (within the meaning of Art. 3) at the time the agreement was entered into. 4(f) only applies to active sales for five years from first marketing in the common market. These provisions are not included in the Proposed Exemption, in common with all the other provisions that were on the "white list". Restrictions on customers that can be served, and on putting products on the market or pursuing an active sales policy in territories within the common market, after five years from the time that contract products are first put on the market in the common market, are "blacklisted"—Art. 5(e) and (g). Restrictions on passive sales are also "blacklisted"—5(f).
[94] See R&D Block Exemption Art. 4.1 (fa), (fb), (fc), and Art. 3(3a). As with all "white list" provisions, these are not included in the Proposed Exemption, but it is worth noting that the "blacklisted" item which applied when the limit in Art. 4.1 (f) had been exceeded has been retained—see n. 79 to para. 20–33 above.
[95] See paras 10–52 et seq for more information relating to restrictive covenants in general and their use for the protection of confidential information. For general information on restrictive covenants, consult a contract law text book.
[96] There may well be equivalent laws in other jurisdictions.
[97] See E.C. Treaty, Arts 81 and 82 (formerly 85 and 86 of the Treaty of Rome), and Competition Act 1988, ss. 2 and 18. See also Chapters 16 and 17.
[98] See Art. 6. See also Proposed Exemption Art. 5.

exemption from Article 81 of the E.C. Treaty. For example, a restriction that could affect research and development in fields unconnected with the research programme, is excluded from the exemption. Similarly, restrictions on research and development after the programme has finished, would fall into this category. It is still possible to apply for an individual exemption in respect of "blacklisted" items, although it is less likely to be granted. It is also possible that these restrictions would be unjustifiable restraints of trade.

Nature of liability

20–40   The participants can incur liability to the funder (and perhaps to each other) under the funding agreement, to the other participants under the agreement between them, and to the other participants and other people as a result of implied legal obligations. The most common problems during the course of the project are likely to be overspending, research that does not meet expectations, and the insolvency of one of the participants. It is also possible, although perhaps less likely, that a participant could misappropriate funds, or that people or property could be injured or damaged in the course of the research.[99] Further liability may be incurred in connection with the exploitation of the results of the project. Whether a participant wishes to exploit the results by making and distributing products, or through licensing, the liabilities arising from these activities could be increased due to the default of another participant. The main potential causes of liability during exploitation are likely to arise from substandard or dishonest results, failure to secure rights in or under intellectual property, and failure to declare or to discover the existence of rights held by other people that would be infringed by exploitation of the results.

Managing liability

20–41   Some of the risks set out above are inherent in a business that carries on research. The difference in this case is the degree of dependence on the other participants. Whilst collaborative research is often undertaken on the basis of a commercial agreement, and a separate legal entity is not created,[1] it is a form of joint venture, and may need to be managed as such. The choice of participants is therefore vital, as are the management structure, financial controls, reporting requirements, and termination provisions. Additional protection may be gained from involvement in the administration of the project. If there are project funds, keeping them in a separate designated account, being one of the required signatories for their release, and obtaining proof of the other participants' insurance, may assist with general liabilities. As regards intellectual property, it may be possible to inspect contracts with researchers, and to brief patent agents undertaking searches for rights that could be infringed. The extent to which any such measures are permitted will depend on the agreement with the participants. Some of the above suggestions are highly intrusive and need to be dealt with sensitively. In some cases it will not be possible to obtain all such information.

20–42   The obligations of the participants need to be set out clearly in the contract, which can also apportion liabilities between them. In addition, the contract may contain requirements for reporting matters that could, potentially, give rise to liability, either in the course of performing the research or in exploiting the results. Provisions

---

[99] In these circumstances, the liability is likely to arise from directing the research, or from having employees or consultants who are injured.
[1] See Chapter 21 for details of the possible legal structures.

apportioning liability can be strengthened by an indemnity in the contract, although this may meet with resistance. Claims under an indemnity are not limited to those for which liability would be incurred under normal contractual principles.[2] Furthermore, liability incurred under an indemnity may not be covered by insurance, unless this is specifically agreed with the insurer, as such liability is assumed voluntarily. Participants can also seek to exclude or to limit certain potential liability by means of an exclusion clause.[3] One of the traditional ways of providing an incentive to perform a contract is to link payments to results and compliance. This may not be a viable option in research contracts where funding is desperately needed in order to carry out the research. Another contractual means of managing liability is to provide for possible expulsion of participants who are in material breach of the contract, which lends weight to the contractual provisions.

Partial termination

20–43   General In projects where there are more than two participants, the agreement may provide for a participant to leave the project ("partial termination"). This can be as a result of a material breach of contract, insolvency or, possibly, for change of control of a participant, if the collaboration agreement provides for this. In the case of funded projects it may be necessary to obtain the consent of the funder, which can be subject to providing evidence that it will still be possible to complete the research. In cases of partial termination, the process of separation needs to be managed with care. The outgoing participant can be required to comply with restrictions on the continued "use" of intellectual property (see below), to return all embodiments of intellectual property that he will no longer be licensed to "use", together with all equipment provided to him for the project. Regulatory licences granted to the outgoing participant may be needed to continue the project. It could be difficult to transfer them, so the outgoing participant may need to assist in obtaining fresh licences. The financial terms may include the refund of grant moneys paid to the outgoing participant, and payment of compensation to the remaining participants for loss suffered due to the termination. Other terms will depend on the circumstances.

20–44   Access to intellectual property Participants who leave a project would usually[4] be expected to grant world-wide licences of all their background intellectual property that could be needed to exploit the results of the project.[5] Depending on the circumstances, the licence may be royalty–bearing. As regards the outgoing participant's foreground rights, there may be a requirement that they be assigned[6] to the remaining participants. At the very least the outgoing participant would usually be required to grant an irrevocable worldwide licence of foreground rights to the other participants, to enable them to exploit their results. For the reasons set out in paragraph 20–32, above, it is preferable for the remaining participants to take an assignment. If this is not possible, an exclusive licence is preferable, not only because of the exclusivity but also because an exclusive licensee can pursue infringers.[7] Allied to these provisions could be a requirement to comply with all reporting obligations. Also, if the

---

[2] Although it is still necessary to prove that the loss was incurred.
[3] See n. 54 to para. 23–21 for further details.
[4] This might not apply in unusual circumstances such as a material breach by the other participants.
[5] The licensees may include any participant who replaces the outgoing participant.
[6] Legal terminology for a transfer document — see Chapter 22.
[7] See n. 5 to para. 23–07.

outgoing participant has created some potentially valuable foreground rights, the remaining participants may require assistance from the inventors with prosecuting patent applications, and with litigation, if necessary. Licences of "background" and "foreground" granted to the outgoing participant may well be terminated.

Full termination and its effects

20–45 The events of termination of the entire project prior to its completion are likely to differ considerably from those that lead to partial termination, unless partial termination would be impossible in the circumstances. Whilst partial termination may relate to the actions or circumstances of an individual participant, full termination is most likely to occur if the project itself is no longer viable. Termination can be by consensus of the participants (which can never be guaranteed), or on failure to produce stated results in stipulated timescales. In view of the effects of the sanction, there may be a grace period prior to termination.

It is crucial to consider which clauses of the agreement will survive termination. In particular, the clauses relating to the ownership and future management of the foreground intellectual property, confidentiality, and licences of background intellectual property may survive termination, if the results of the research can be exploited. In the case of restrictions, survival can lead to infringement of competition laws[8] and to the invalidity of the restriction concerned, if the restriction was only exempted for the duration of the research project. It may therefore be necessary to apply for exemption, in the case of infringement of competition law, or to limit the provision, in order to ensure compliance/validity of the contract. Termination clauses can also provide for the return of property, for the apportionment of liabilities to the funder, and for compensation for financial loss.

Assignment and sub-contracting

20–46 In many projects the identity of the other participants is of prime importance. The provisions of the agreement relating to assignment[9] and sub-contracting are therefore crucial. Personal contracts are not generally assignable, and it could be claimed that a research contract fell into this category. It may, however, be possible to infer that a contract was intended to be assignable. It is therefore preferable that specific provisions be included in the contract so that the position is clear. The same goes for sub-contracting.[10] In the case of a licensee company, a change of control of that company will not be covered by a clause prohibiting assignment. The agreement may therefore provide for partial or full termination of the agreement, or rights granted under it, in the event of a change of control of a participant. Funders can include prohibitions on assignment or sub-contracting, or make such transactions subject to their consent.

Dispute resolution

20–47 As with all contracts, the potential methods of dispute resolution are litigation, arbitration and alternative dispute resolution. In this case, it is worth considering alternative forms of dispute resolution more carefully, especially in respect of disputes that are technical rather than legal, as it is possible to arrange for greater involvement of an expert in the relevant field in the dispute resolution process. Unless

---

[8] See paras 20–38 and 20–39 and Chapter 16.
[9] Legal terminology for "transfer".
[10] If sub-contracting is to be permitted, it is important to provide that the relevant restrictions in the main contract also apply to the sub-contractor.

there is specific agreement for arbitration and/or alternative dispute resolution, the only formal means of dispute resolution will be litigation. It is therefore common to provide in the agreement for other forms of dispute resolution, as it may not be possible to reach agreement when the dispute arises. The precise nature of the dispute resolution, and the composition of any panel, or the means of determining it, would usually be specified, together with any relevant details of the decision making process, including any right of appeal. If the participants are in more than one jurisdiction, the decision is likely to be more complex as the place of resolution may also need to be agreed, together with the governing law, and language of the proceedings. In addition to dispute resolution, it is common to provide for an "escalation procedure" in respect of complaints, to try to prevent disputes.[11]

Choice of law

20–48 There are appreciable differences between national laws, more especially between the jurisdictions that practise common law, and those that practise civil law. In particular, courts in civil law jurisdictions have wide scope to imply contractual terms. There can still be considerable disparity, even between jurisdictions where the law is reputedly similar.[12] The choice of law is, therefore, vital, especially in the case of projects that include participants from mainland Europe. The choice of governing law and the method and place of dispute resolution are often a matter of compromise, which can lead to extreme complications. For example, the choice of different laws for interrelated agreements could lead to some surprising results. Similar obligations might be construed differently, so provisions intended to pass on obligations from one contract to a party in another might be ineffective. Equally, if the governing law is not that of the place of the dispute resolution, there may be difficulties in interpretation. Regrettably, there is often no perfect compromise, especially if there is a mixture of laws. In many cases, English law may be acceptable, on the basis that there are fewer implied terms, so the parties know where they stand. Governing laws relating to licensing pose additional problems, which are mentioned at the end of Chapter 23.

FUNDING AGREEMENTS

20–49 Funding agreements can vary considerably, depending on the funder, and on the circumstances, including the jurisdiction and the political and financial climate in that jurisdiction. Also, the terms and conditions on which governmental and supragovernmental funding is provided may be revised from time to time, to take account of previous experience. Terms in English contracts with universities have recently been in a state of flux, as a result of changes in public sector funding.

It would be misleading to make any general statements about the terms of funding agreements. Even if an agreement has an official imprimatur, it should still be considered with care, as some people have found such agreements unacceptable. There is also a current trend for certain interested parties to make unreasonable public pronouncements as to the terms on which they are prepared to participate in research projects. The public pronouncements are, in some cases, wishful thinking. In every case it is crucial to consider as many options as possible, and to keep an open mind.

---

[11] Such a procedure would usually include referral of the complaint to senior executives of each party, with a further appeal to more senior personnel. In view of the fact that the procedure may be time-consuming, and may not always be appropriate, the procedure may be drafted so that any party can abandon it entirely and resort to more formal dispute resolution.

[12] See *Macmillan Inc. v. Bishopsgate Investment Trust plc* (No. 3) [1996] 1 W.L.R. 138,CA, which shows a major difference between English and New York laws of notice.

CHAPTER 21

COMPANIES AND OTHER BUSINESS STRUCTURES

Nicholas Thompsell, Partner, Field Fisher Waterhouse

INTRODUCTION

Company and other structures for R&D and commercialisation

21–01 This Chapter is about the corporate and unincorporated structures which may be used for the purposes of developing and commercialising intellectual property. It begins with a brief overview of the structures available. We then consider in more detail the legal rules and practice affecting the use of the most popular structure: that of the company limited by shares, and the rules about promoting the sale or subscription of a company's shares. We go on to discuss the terms on which a technology-based joint venture might be carried on and the terms on which a venture capitalist might be prepared to invest in a technology-based company. There is a short summary of the legal requirements relating to flotation. Finally there is a brief description of the main legal considerations applying to other forms of business organisation. Other forms of business organisation, where the parties do not use a separate vehicle are outside the remit of this Chapter. In particular, Research and Development Agreements are the subject of Chapter 20.

Why are business structures needed?

21–02 In some cases no business structure will be required at all. The inventor will make, develop and exploit his or her invention and that will be the end of it. Where more than one party is involved, however, the relationship between the participants must be defined, and it is often at this stage that thought is given to business structures. Depending on the circumstances and the structure chosen, the relationship could be defined in many different ways. For example:

- the participants could be operating under a collaboration agreement with one another, not amounting to a partnership;

- the participants could be partners;

- the participants could be lender and borrower;

- one participant could be the employee of another;

- the participants could be employees, directors and shareholders of a specially incorporated company.

The duties and rights of the participants will differ according to the structure employed and how their relationship is categorised. Even where there is only a single participant there are many reasons why a business structure might be needed. These may include tax, the benefit of limitation of liability[1] and the ability to separate ownership and management.

---

[1] See para. 21–08, below.

WHAT TYPES OF BUSINESS STRUCTURE ARE THERE?

21–03   There are many types of business structure available, particularly because the basic structures are often combined so that, for example, it is possible to have a partnership between two companies or to combine the incorporation of a company with a Research and Development agreement or a licensing agreement. However the most relevant forms of structure under English law can be illustrated as follows:

Forms of business structure

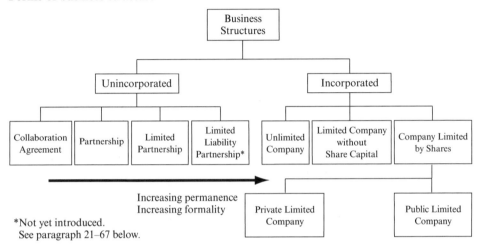

*Not yet introduced.
See paragraph 21–67 below.

Another less commonly used structure is the European Interest Grouping,[2]
    The choice of which structure to use is not always easy, and there are no hard and fast rules. Some of the factors to be taken into account, however, include the following:

- how permanent the structure is to be. In the case of a temporary arrangement, it may be inappropriate to create a separate legal entity and a collaboration agreement may be the best option. A more permanent arrangement may be better structured as a partnership or a company;

- whether the arrangement will be a personal arrangement or one where others may join or take over one person's participation. A personal relationship may lend itself to being structured as a partnership, whilst a company may be more appropriate if the membership of the business is to change;

- the tax treatment of the differing structures;

- the benefit of limited liability (see below[3]);

- whether the parties are content to follow the public filing requirements applying to companies (in particular in relation to the provision of accounts);

- whether it is important that the business should have its own identity.

In the vast majority of cases, however, where a permanent or long-term business structure is looked for, the structure chosen is a corporate structure.

---

[2] For a summary of the law and practice relating to European Interest Groupings, see M. Anderson, *European Economic Interest Groupings* (Butterworths).
[3] At para. 21–08, below.

CORPORATE VEHICLES

21–04 Under United Kingdom company law, three main types of incorporated vehicle are available:

- company limited by shares

- company limited by guarantee

- unlimited company.

Companies limited by guarantee are rarely used as a commercial or trading vehicle. Most typically they are used as vehicles for not-for-profit organisations such as charitable or academic institutions, not-for-profit research bodies, trade associations, etc. They fall, therefore, outside the scope of this Chapter.

Unlimited companies show all the characteristics of limited companies, as discussed below, except that the members of the company do not enjoy the benefit of limited liability. Limitation of liability is one of the key attractions of incorporation and unlimited companies are, therefore, relatively rare. In some cases, however, unlimited companies are chosen as a vehicle for incorporation because they enjoy one advantage over the limited liability company: they are not required to file accounts. Consequently, where privacy is more important than limitation of liability, an unlimited company may offer a suitable vehicle.

The vast majority of United Kingdom registered companies are companies limited by shares. This structure is the most popular for all types of business enterprise, including technology start-ups.

FORMATION OF A LIMITED COMPANY

21–05 The formation of a limited company is a relatively straightforward matter. A form is lodged with the Companies Registry (Form 10) including such information as the name of the company, its principal activities, the address of its registered office, details of its first directors and secretary and of its initial authorised and issued share capital. The form is lodged with the fee, and with a copy of the proposed Memorandum and Articles of Association of the company (the documents which make up its constitution and regulate the relationship between the company, its shareholders and its directors[4]).

The company can normally be formed within five working days on payment of a fee of £20 to the Registrar of Companies. In many cases ready made or "off-the-shelf" companies are purchased from company formation agents in standard format. These can be made available immediately and it is a simple matter to change the name, appoint new directors and secretary, amend the Memorandum and Articles and make such other changes as are appropriate.

MAIN FEATURES OF LIMITED LIABILITY COMPANIES

Advantages of limited liability companies

21–06 Limited liability companies offer a number of advantages over unincorporated business structures. These advantages include:

---

[4] See further at para. 21–14, below.

- separate legal personality

- limitation of liability

- separation of ownership and management

- ability to create a floating charge.

These points are considered in turn.

Separate legal personality

21–07   A limited liability company is regarded as a legal person in its own right, distinct from its members. Accordingly it can enter into contracts with its members,[5] it can own property in its own name[6] and it can sue and be sued in its own name.[7] When the shareholders or directors of a company change, this has no effect on the legal status of the company. Contracts entered into with the company remain valid. The company remains the owner of its property.

Limitation of liability

General principles

21–08   When a company is described as enjoying limited liability, this does not mean that there is any limitation on the company's own liabilities. It means rather that the members of the company enjoy limited liability. In broad terms, they can lose their investment in the company, but nothing more.[8] In the case of a company limited by shares[9] the liability of members is usually limited to amounts paid up on their shares and/or amounts which they have agreed to subscribe for their shares but which they have not yet paid. Someone who has a right of action, therefore, against the company in question could have recourse against all of the assets of the company but not against any other assets of the shareholders or the directors. The protection is often described as the "corporate veil", which may be thought of as a fictional curtain separating the company from its members.

Of course, those dealing with companies are normally aware of the principle of limited liability and bodies such as banks and commercial landlords may refuse to deal with a small company unless the shareholders in that company give a personal guarantee. The benefits of limited liability, therefore, can easily be overstated. Nevertheless the protection given by limited liability is useful. This is particularly the case for a company that is exploiting new technology if there is any likelihood of a challenge that the new technology infringes intellectual property rights of another person or may give rise to a risk of product liability.

---

[5] See, for example, *Farrar v. Farrars Ltd* (1888) 40 Ch.D. 395.

[6] By contrast a partnership, for example, cannot be registered as the holder of property. Thus a patent application would need to be held by individuals on behalf of a partnership.

[7] It is also possible for actions to be brought by or against the partners of a partnership using the name of the partnership, but this is as a result of the application of rules made by the court as a matter of convenience rather than as a matter of separate personality, see Civil Procedure Rules 1998, Schedule 1 (High Court) and Schedule 2 (County Court).

[8] Note however that if the shareholder or director was personally involved in some other way, *e.g.* through a guarantee, or as being involved in some fraudulent or negligent act, the shareholder or director could suffer personal liability through an action being brought directly against that person.

[9] In the case of a company limited by guarantee, the limitation is to the amount of the guarantee given by each member. This is usually set at a low figure such as £1.

Lifting the corporate veil

21–09   There are limited circumstances where a court will lift the corporate veil. These arise essentially where the corporate form is being used artificially to defeat a legal requirement. For example in *Re Bugle Press Limited*[10] two shareholders incorporated a company which they used to make an offer for shares of another company, including an offer for shares held by the two shareholders. The court held that the company was a mere sham and would not permit it to take advantage of a statutory compulsory acquisition procedure.

Another example can be seen in the case of *Jones v. Lipman.*[11] The defendant contracted to sell a particular property but then decided to retain it. In order to avoid a court order to comply with this agreement, he transferred the property to a company controlled by him. The court held that a court order would be made against the company even though the company had not been a party to the contract.

Although these cases may be cited as examples where the "corporate veil" is lifted, in the sense that the court has ignored the separateness of the identity of the company and its owners, none of them affects the principle of limited liability.

Separation of ownership and management

21–10   The company format allows a high degree of flexibility in structuring the relationship between the participants in the enterprise. The owners of a company are its shareholders. The ultimate management of a company is its board of directors. These can be the same people or can be entirely different. The way in which the rewards, rights and responsibilities of involvement in the company are shared is largely determined according to what rights are attached to the shares. For example, it is possible to create different classes of share with different voting rights or different rights to income. It is possible through a number of different means for participants to exercise either positive control over a company, or a power of veto over specified activities. Whilst there are few constraints on the rights that can be attached to shares, in practice the major types of share capital found can usually be divided into the following classes:

- **Ordinary shares**. These normally have a general right to vote on all issues for shareholders, a right to receive dividends (after first satisfying dividends payable to preference shareholders) and a right to share in any surplus on a winding up. Sometimes a company has more than one class of ordinary share, for example "A" Ordinary Shares and "B" Ordinary Shares. The two classes usually have similar economic rights but may have different voting rights (so that, for example, the holders of each separate class of share have separate rights to appoint a director, or to veto specific proposed actions by the company).

- **Preference shares**. These give the holder a preferential right to receive dividend and/or a share in the surplus available on a winding up in preference to ordinary shareholders. The preference dividend is normally of fixed amount, usually expressed as a percentage of the nominal value of the share and accrues in a similar way to interest, although unlike interest, the dividend can only be paid if

---

[10] [1961] Ch. 270.
[11] [1962] 1 All E.R. 442.

the company has profits out of which the payment can be made. Usually, preference shares do not have voting rights unless the dividend payable on the preference share is in arrears or a resolution is proposed which affects the rights of preference shareholding.

Preference shares offer a form of investment that may be thought of as half way between an equity shareholding (which participates fully in all the ups and downs of the fortunes of the company) and a loan to the company. The preference shareholder does not have the same security as a lender: his dividends will only be paid if there are profits available to pay the dividends and, on a winding up of the company, creditors must be satisfied before the preference shareholders can be paid. Preference shares are often issued to institutional investors who require (often as part of a larger package) a fixed rate of return on their investment in the company.

- Deferred shares. These give the holder a right to receive dividends and/or a share of surplus on a winding up only after preference and ordinary shareholders have been satisfied. Often the hurdle for their payment is set at a rate where the deferred shareholders have little realistic chance of receiving a return. Deferred shares are often used as part of tax planning. Sometimes ordinary shares are converted into deferred shares to obtain an effect similar to that of redeeming or of having the company buy back its ordinary shares, whilst avoiding the restrictions on the redemption of share capital.

In the case of private companies, and in particular joint venture companies, shares may be created with "weighted voting rights" so that the shares of a particular class, or held by a particular shareholder, would be given more votes than is warranted by their percentage of the capital of the company.

Ability to create a floating charge

21–11   One unique benefit of the company structure is that only companies can create floating charges. A floating charge is a charge giving security over the general assets and business of a company. The ability to create a floating charge widens considerably the assets that may be made available to creditors as a security, which would otherwise be limited to the assets that can be made subject to a fixed charge. Although floating charges are not as satisfactory as fixed charges[12] in many ways, the ability to create a floating charge improves the ability of companies to obtain secured debt finance because it extends significantly the range of assets over which security may be given.[13] It also gives the lender a technical advantage in being able to appoint a "receiver" with

---

[12] For example floating charges are not given priority over preferred creditors such as certain claims of employees for wages, etc., and Customs & Excise in respect of Value Added Tax.

[13] A fixed charge must be over defined assets. These assets may be future assets, as well as existing assets of the company, but the essential feature of the fixed charge is that the property that is subject to the charge will not be dealt with until the consent of the chargee is obtained. As a result it is inappropriate and impractical to take a fixed charge over a company's stock in trade. A floating charge, on the other hand, charges all the assets of a company, including its stock in trade. With a floating charge, the company can continue freely to dispose of the charged assets until enforcement of the security is needed. The terms of the charge will include terms listing the events which "crystallise" the charge. These will include such matters as winding up of the company, the appointment of a receiver, the company ceasing to carrying on business, failure to repay the loan on demand. Once the charge crystallises it effectively becomes a fixed charge over the relevant assets.

power to realise the charged assets to pay the debt. This can represent a practical way of enforcing security over a wide range of assets without calling for the liquidation of the company.

PRIVATE AND PUBLIC COMPANIES

21–12 Limited liability companies may be either private companies (the names of which will always end with "Limited" or "Ltd"[14]) or they may be public limited companies (the names of which will always end with the words "public limited company" or "plc"[15]). Public companies are designed to be used where the shares are to be widely held, whilst private companies are designed to be used where the shares are to be held by only a few shareholders and where there is to be no offer of the shares to the public.[16] A private company may not be admitted to the Official List of the London Stock Exchange.[17] Private companies are subject to lighter regulation than public companies. Various rules apply to public companies which do not apply to private companies[18] in particular:

- there are minimum capital requirements for public companies[19];

- directors of a public company are under greater restrictions in their personal dealings with the company and group companies, *e.g.* as regards loans, guarantees and other transactions with directors[20];

- if shares are issued by a public company for a consideration other than cash, an expert's (auditor's) valuation report is normally required and a special procedure

---

[14] Or the Welsh equivalent *Cyfynngedig.* See sections 25–26 of the Companies Act 1985.
[15] Or the Welsh equivalent *Cwmni Cyfynngedig Cyhoeddus.* See sections 25–26 of the Companies Act 1985.
[16] Section 81 of that Act contains an absolute prohibition on private companies offering their shares to the public, although this provision has now been partially repealed (broadly so the section no longer operates in relation to open-ended investment companies or in relation to issues where the Public Offers of Securities Regulations 1995 apply).
[17] This is provided both by law (Financial Services Act 1986, section 143(3)) and by the Rules of the Stock Exchange. In addition the London Stock Exchange will not accept private companies suitable for listing on the Alternative Investment Market.
[18] Other differences, apart from those listed in the main text, include the following:

- if a public company's net assets fall to half or less of its paid up capital, it must hold a Shareholders' Meeting to decide what action to take. (Section 142 of the Companies Act 1985);
- a public company is more restricted in giving financial assistance for the acquisition of its own shares than a private company. See Companies Act 1985, ss. 151 *et seq.* See also para. 21–25, below;
- a public company cannot purchase or redeem its shares out of capital, and the procedures for permitted redemptions or purchases of its own shares are more complicated than for a private company. See sections 159–170 of the Companies Act 1985;
- private companies only may dispense with holding Annual General Meetings (section 366A of the Companies Act 1985 (AGMs)), laying accounts before AGMs (section 386 of the Companies Act 1985) and appointing auditors annually. Public companies must carry out these procedures;
- a public company must present its accounts to shareholders and file them with the Companies Registry more quickly than a private company. See section 244 of the Companies Act 1985;
- Late filing penalties (imposed by the Companies Registry) are more expensive for public companies than private companies. See section 242A of the Companies Act 1985.

[19] A public company must have an allotted share capital of not less than £50,000 paid up at least as to one quarter of its nominal value plus the whole of any premium. See section 45(2)(b) of the Companies Act 1985.
[20] In the case of a public company the provisions are more widely drawn and breach of them is a criminal offence. See section 330 of the Companies Act 1985.

followed. This can be important where shares are to be issued in return for a transfer of intellectual property[21];

- a public company must have a minimum of two directors. (A private company need only have one[22]);

- the company secretary of a public company must possess certain qualifications[23];

- a public company must have at least two shareholders. Only a private company can be a single member company[24];

- a public company must appoint auditors. Private companies can delay appointing first auditors and do not need auditors in some limited circumstances[25];

- public companies are subject to additional restrictions on the distribution of profits which do not apply to private companies[26];

- a public company, even if not listed, is subject to the provisions of the City Code on Takeovers and Mergers.

In most cases, unless the additional cachet of being a public company is required or a listing being sought, it will be more straightforward to incorporate as a private company. Procedures exist for the conversion of a private company into a public company (and vice versa).

CONTROLLING THE VEHICLE COMPANY

The rights of shareholders

21–13 Potential shareholders will be concerned to ensure that they are given appropriate rights of control and protection before they will agree to participate in the company. Shareholders are well advised to check that they are given express protections since the position of a shareholder, in particular a minority shareholder, is given less than satisfactory protection by law. If a shareholder finds that the company is being mismanaged, or worse, by the directors there may be little that he can do about it directly. Unless the shareholder is able to control a majority of the votes of shareholders,[27] or has special rights under the company's articles of association, he will not

---

[21] See section 103 of the Companies Act 1985.

[22] If it has a different person as its secretary.

[23] See section 286 of the Companies Act 1985. The secretary must satisfy one of the following criteria:

    (a)   for at least three of the five years immediately preceding his appointment he must have held the office of Secretary of another public company; or

    (b)   he must be a solicitor, barrister or advocate admitted in the U.K.; or

    (c)   he must be a U.K. accountant or a member of the U.K. Institute of Chartered Secretaries and Administrators; or

    (d)   he must be a person who, by virtue of his holding or having held any other position or his being a member of any other body, appears to the Directors to be capable of discharging the functions of Company Secretary.

[24] See section 24 of the Companies Act 1985. With any company, however, there is nothing to stop the "second" member from holding its share as nominee for the other member.

[25] See sections 55, 384 to 385 of the Companies Act 1985.

[26] See sections 263 to 264 of the Companies Act 1985.

[27] Under section 303 of the Companies Act 1985 there is a statutory power for a director to be dismissed by ordinary resolution which applies irrespective of the provisions of the articles of association.

be able to dismiss the director concerned. Under the rule in *Foss v. Harbottle*,[28] if a wrong is done to a company, only the company itself may bring an action in respect of this wrong.

There are exceptions to the rule, however, where an individual shareholder may bring an action on behalf of the company such as:

- where the act is outside the objects[29] of the company or is illegal;

- where there is a breach of a personal right (such as where the company is appointed, under the Articles, as agent to sell the shares of the shareholder);

- where the act amounts to a fraud on a minority of shareholders and the wrong-doers are in control of the company.

Other remedies available to an aggrieved shareholder include an ability under the Insolvency Act 1986 to petition the court to wind up the company on grounds that it is "just and equitable" to do so[30] or to seek relief from the court under section 459 of the Companies Act 1985. If the shareholder can persuade the court that the company's affairs have been conducted in a way that is unfairly prejudicial to all or some members of the company,[31] the court has wide powers[32] to make an order regulating the company's future conduct, to authorise the petitioner or someone else to bring an action in the name of the company or to provide for the purchase of shares of any members of the company by other members or by the company itself. Enforcing these statutory remedies, however, is a lengthy, and usually expensive, undertaking.

In view of the difficulty of relying entirely on the general law, protection should be sought in other ways. There are at least three ways in which control can be exercised over a vehicle company: through the memorandum and articles of association of the company; through a shareholders' agreement; and through other agreements with the company.

Memorandum and articles

Memorandum

21–14   The memorandum and articles of association of a company make up its constitution and regulate the relationship between the company, its shareholders and its directors. The memorandum of association of the company follows a standard form prescribed by regulations made under the Companies Act 1985.[33] It includes the name of the company, the jurisdiction in which the company is established, if the company is a public company, a statement to that effect, a statement that the company has limited liability, a description of the company's authorised share capital and an itemisation of the objects and powers of the company.

The "objects clause" setting out the objects and powers of the company defines the activities of the company. If the company purports to undertake an act that is outside the scope of the objects clause the act is deemed *ultra vires* (*i.e.* outside the

---

[28] *Foss v. Harbottle* (1843) 2 Hare 461.
[29] See para. 21–14, below.
[30] See sections 124–125 of the Insolvency Act 1986.
[31] The members prejudiced must include the shareholder making the complaint.
[32] Under section 461 of the Companies Act 1985.
[33] The Companies (Tables A to F) Regulations 1985 (S.I. 1985 No. 805) as amended. Note it is the form of the memorandum that is prescribed, not the contents. In particular the contents of the objects clause differs substantially amongst companies.

company's powers) and can be challenged by a shareholder. However, under United Kingdom company law, once such an act is undertaken both parties to any contract are bound and the contract is enforceable.[34] In addition, a third party is not bound to check that a company's objects clause allows the company to do the act.[35] Nonetheless, company's directors would be liable to the shareholders for any losses suffered by the company as a result of the *ultra vires* act, unless the shareholders ratified the act and indemnified the directors. By drafting a narrow objects clause, it is possible to limit the scope of the company's activities.

Articles

21–15 The articles of association of the company set out the detailed provisions for such matters as the issue of shares, the transfer of shares, voting rights attaching to shares,[36] the constitution and powers of the board of directors, proceedings at board meetings and shareholder meetings and provisions on a winding up. Although there is statutory guidance available through a prescribed standard form of articles of association known as "Table A",[37] which will apply unless excluded, there is no obligation to follow this form[38] and articles can be tailored to satisfy the requirements of the company and its shareholders.

Shareholder and board meetings

21–16 Under most articles of association, the board of directors is given full authority to manage the company. Nevertheless shareholders' meetings may be needed from time to time, and will be required for the following matters:

(i) any increase of authorised share capital of the company[39], and the issue of new shares unless they are issued to existing shareholders in the same percentage as they own the existing share capital[40];

(ii) any change in the company's Articles of Association[41] or in the objects clause in its Memorandum of Association[42];

(iii) any change of name of the company[43];

(iv) certain directors' service agreements.[44]

In cases where shareholders' consent is needed, the consent may need to be given in the form of an ordinary resolution (requiring more than 50 per cent of those voting to vote in favour) or a special or extraordinary resolution (each requiring at least 75 per

---

[34] See sections 35 and 35A of the Companies Act 1985.
[35] *ibid.*, s. 35B.
[36] Although sometimes these rights are contained instead in the resolution issuing the shares.
[37] The Companies (Tables A to F) Regulations 1985 (S.I. 1985 No. 805) as amended.
[38] In the case of a company limited by shares. In the case of an unlimited company or a company limited by guarantee there is a requirement to follow the statutory form. See section 8 of the Companies Act 1985.
[39] See section 80 of the Companies Act 1985. Sometimes this authority is given through the Articles of Association instead. Note also the provisions of section 80A of the Companies Act 1985.
[40] See section 89 of the Companies Act 1985. This provision may be excluded in the case of private companies through the company's Articles of Association.
[41] See section 9 of the Companies Act 1985.
[42] See section 4 of the Companies Act 1985.
[43] See section 28 of the Companies Act 1985.
[44] See section 319 of the Companies Act 1985. The section applies broadly where the contract is not terminable by the company within five years.

cent of those voting to vote in favour).[45] Matters requiring a special resolution or extraordinary resolution include:

- amendment to the company's objects;

- amendment to the company's articles;

- reduction of share capital;

- purchase by the company of its own shares and redemptions of own shares[46];

- financial assistance by the company in relation to a purchase of its own shares.[47]

21-17   Apart from the few issues reserved to shareholders by law or by express provisions of the articles of association, the conduct of the business of the company is in the hands of its directors. How far the decisions are to be taken by means of a formal resolution at a board meeting, and how far decisions are made informally, or even by individual directors, varies from company to company. There are, however, some decisions which must be made by the board acting as a whole. These include resolutions:

(i)   to issue new shares which have been authorised by the shareholders;

(ii)   to refuse to register the transfer of existing shares not being fully paid up shares and to refuse to register the transfer of shares on which the company has a lien (*i.e.* a right to withhold the shares or the rights attaching to them from the holder of such shares);

(iii)   to appoint anyone to be an agent of the company;

(iv)   to appoint new directors to fill a casual vacancy or as an addition to the existing directors provided the new number does not exceed the maximum number of directors fixed by the articles of association;

(v)   to approve annual accounts and the directors' report;

(vi)   to resolve what, if any, dividend should be paid each year by the company. The shareholders can only accept or reject the amount so decided;

(vii)   to appoint all company officers (Chairman, Managing Director, and Company Secretary);

(viii) to authorise the use of the company seal;

(ix)   to approve any change in the registered office of the company.

In other cases a third party such as a bank or a landlord may insist on seeing a formal board resolution before dealing with the company.

---

[45] See section 378 of the Companies Act 1985. A special resolution normally requires 21 days' notice whilst an extraordinary resolution and an ordinary resolution normally requires 14 days' notice, although in some circumstances it is possible for the meetings to be held on short notice or for the resolutions to be passed as written resolutions.

[46] See para. 21–24, below.

[47] See para. 21–25, below.

Voting rights and voting power

21–18 Unless special voting rights are given to shares of a particular class (and in the absence of any shareholders' agreement as discussed below) the extent of control that can be exercised through holding a particular percentage of a company's ordinary shares, can be summarised as follows:

Percentage of ordinary shares held	Comments
Above 5%	Power to prevent a shareholders' meeting being held on short notice
Above 10%	Power to avoid/block compulsory purchase of shares where other shareholders accept an offer for their shares[48]
Above 25%	Power to block a special resolution or extra-ordinary resolution of shareholders.
50%	Ability to block an ordinary resolution
Above 50%	Legal control. Ability to pass an ordinary resolution, and therefore to appoint or remove directors.
75%	Ability to pass a special resolution or extraordinary resolution.
95% and above	Power to pass a shareholders' resolution on short notice[49]

Shareholders' agreement

21–19 A shareholders' agreement is an agreement between shareholders setting out terms relating to how they will use their rights as shareholders. Often the terms of a shareholders' agreement deal with similar issues to those contained in the articles of association. For example the agreement may contain restrictions on the transfer of shares, or an agreement to procure that the company will not take certain actions (such as disposing of property) without the consent of all or a certain number of shareholders. It may also include obligations on shareholders to provide funding to the company and set out information rights for shareholders, provisions for the adoption of and adherence to a business plan and a proposed dividend policy.

From the viewpoint of an individual shareholder, putting restrictions in a shareholders' agreement rather than in Articles of Association has the advantage of privacy (Articles of Association are available to the public at the Companies Registry) and the advantage that the agreement cannot (usually) be changed without each party's consent, whilst Articles of Association can be changed by special resolution,[50] which may not require the consent of each party.

---

[48] See sections 428–430 of the Companies Act 1985.

[49] Although in addition, those wishing to pass a resolution on short notice must constitute a majority in number of the members of the company.

[50] That is by a 75 per cent majority. Note also that changes affecting the rights of a particular class of shareholder normally require sanction by an extraordinary resolution of members of that class: See section 123 of the Companies Act 1985.

On the other hand putting the restrictions in the Articles of Association has the benefit that, in many cases, breach of the provision becomes impossible. For example if shares were issued or transferred in breach of the Articles, the issue or transfer would be invalid. If this were merely a breach of a shareholders' agreement, the transfer would be valid and the dissenting shareholder would be left only with a remedy in damages.[51]

Often the company involved is also made a party. Making the company a party has advantages in that obligations can then be enforced directly against the company. However there are limits on what constraints may be put on the company.[52]

A consideration of the terms commonly found in a shareholders' agreement in the context of an investment in a technology based company appears at paragraph 21–44, below. Paragraph 21–50 includes a consideration of the terms often inserted in a shareholders' agreement where the shareholders are financial investors in a technology company.

Control through other agreements

21–20   Where a company is incorporated to deal with intellectual property rights, there will often be another agreement which has an important bearing on how the company's business is conducted. This may be a collaboration agreement or an intellectual property licence. The terms of such an agreement could include restrictions on the activities of the company, a right to terminate the agreement or licence if there is a change of control of the company and/or options to acquire shares in the company.

CONSTRAINTS OF COMPANY LAW

Flexibility, but with limits

21–21   Although there is considerable flexibility in structuring a company to meet particular circumstances, this is not to say that there is complete flexibility. Certain constraints are imposed by company law. Many of these constraints arise from the application of a principle known as maintenance of capital. Because the creditors of a limited company can look only to the assets of that company, the law has developed a number of statutory rules and prohibitions to protect the capital of the company. Other prohibitions arise to enforce fair dealings between shareholders and between the company and its directors. Some of the more important rules are listed below.

Rules applying to the issue of shares

21–22   Shares may be issued by a company either in return for cash or in return for a non-cash asset. Shares may be issued at a subscription price that is equal to or

---

[51] Always assuming he could show he had suffered any damage.

[52] *Russell v. Northern Bank Development Company* [1992] B.C.L.C. 1016. In this case the company and its shareholders entered into an Agreement which purported to take precedence over the company's Articles of Association. Clause 3 of the Agreement said that no further share capital would be issued without written consent from the parties to the Agreement. When the company sought to issue further share capital some years later a shareholder sought to use the Agreement to stop the company doing so. The House of Lords held that the company's agreement not to issue further shares was unenforceable because the company could not agree to limit the exercise of its statutory powers in this way. However, it was held that an agreement between the shareholders as to how they would vote on any matter relating to issuing further share capital is valid.

above[53] the share's nominal or "par" value. They may not be issued at any lower price. For example a £1 share can be issued for £1 or for £10, but not for 50p.

Where shares are issued in return for a non-cash asset (such as in return for a transfer of intellectual property to the company), the directors of the company must satisfy themselves that the value being ascribed to that asset is fair, in order to satisfy themselves that the shares are not being issued at a subscription price that is below the net asset value and that amounts are being credited correctly to the share capital and share premium accounts.[54]

Shares may be issued as partly paid shares. This means that the shareholder in question has committed to pay the full subscription price but has not yet done so. The articles of the company, and the terms on which the shares are issued, will govern when the outstanding payment is due and will usually contain terms allowing the company to require forfeiture of the shares if payment is not made. If the company is wound up before the shares are fully paid, the shareholder in question can be required to pay the amount outstanding.

Rules applying to distributions and dividends

21–23 Companies may pay dividends or otherwise make "distributions" to members only where the company has "profits available for distribution". What is meant by profits available for distribution is defined at length in the relevant provisions of the Companies Act,[55] and these provisions are supplemented by Generally Applicable Accounting Standards.[56]

In the case of start-up technology companies, there may well be substantial expenditure in the early years. A choice will need to be made as to whether to charge this expenditure to the profit and loss account of the company or to "capitalise" the expenditure, reflecting it as an asset in the books of the company. If the expenditure is charged to the profit and loss account, then these early losses may take some time to correct and meanwhile it may not be possible to pay a dividend.

If during the period before the company starts to realise net profits, the founder seeks to obtain an income out of the company, this must be through a means other than dividend. Possibilities include the payment of licence fees by the company in respect of intellectual property licensed to it or the payment of a salary. Another possibility is that the founder should fund the company partly through loan, rather than share capital. Any surplus then accruing to the company can be used either to pay interest on the loan or to repay the capital of the loan even when the company is not yet in profit. These choices need to be considered at the outset as part of the structuring of the business plan of the company.

Rules applying to the redemption or purchase of own shares

21–24 A company can redeem its shares if the relevant shares were originally issued as redeemable shares. Redeemable shares may be issued as long as the company also

---

[53] Where the issue price is greater than the par value, the par value is credited in the company's accounting records to a "share capital account" and any amount paid in excess of the par value is credited to a "share premium account". See section 130 of the Companies Act 1985.

[54] Where the company in question is a public company (a plc), there is in addition a requirement for the company to obtain a valuation report from an independent valuer who is qualified to act as an auditor of the company. See section 103 of the Companies Act 1985.

[55] See generally Part VII of the Companies Act 1985.

[56] The Accounting Standards Board Limited has been appointed to prescribe these standards.

has in issue at least one share that is not a redeemable share. In addition, in certain circumstances a company may purchase its shares whether or not they were issued as redeemable shares. The effects of purchasing own shares and redeeming shares are in fact very similar. The principal difference between the two procedures is that with redemption of shares, the terms of the redemption, and circumstances in which the redemption may take place are set when the share is issued. With the purchase of own shares, the terms of the purchase are set at the time of the purchase.

The circumstances in which a company can purchase or redeem its own shares are strictly limited. The issue of redeemable shares and/or the purchase of own shares must be authorised by the company's Articles of Association.[57] Usually the purchase must be made out of distributable profits[58] and/or out of the proceeds of a fresh issue of shares. A private company, however, if specifically authorised by its articles may purchase or redeem its shares out of assets representing its existing share capital, although only to the extent that the purchase cannot be effected out of distributable profits or the proceeds of any fresh share issue made for the purpose of funding the purchase or redemption. A formal procedure must be followed to authorise the payment out of capital. This involves:

- the passing of a special resolution approving the proposal;

- a statutory declaration being made by the directors of the company stating that the company will, following the purchase, be able to pay its debts and continue as a going concern for at least a year;

- a report by the company's auditors backing up the directors' declaration;

- the publication of a notice in the *London Gazette* and also in a national paper[59];

- statutory rights for members or creditors to object.

Prohibition on financial assistance

21–25 A further provision designed to ensure the maintenance of capital,[60] makes it a criminal offence for a company to give "financial assistance" in connection with the acquisition of its own shares, or in order to discharge a liability assumed for the purpose of acquiring such shares. This prohibition is intended to prevent creditors of a company being disadvantaged through the company reducing its assets to help with the funding of a transaction in its own shares. However the prohibition, and particularly the definition of what constitutes "financial assistance", is drafted in broad terms and represents something of a trap for the unwary. For example, on the sale of the shares in a small owner-managed company it might be agreed that the company should transfer certain assets of the company (for example a piece of intellectual property that was not vital to the business) to the seller of the business prior to or immediately following the sale. If this property was being transferred at less than its market value, and the undervalue was material, the offence could be committed.[60a]

There is a *"whitewash"* procedure available for private companies which allows financial assistance to be given where the net assets of the company will not be dimin-

---

[57] See sections 159 and 162 of the Companies Act 1985.
[58] Within the meaning of section 263 of the Companies Act 1985.
[59] (Or individual notification to shareholders and creditors).
[60] See sections 151–158 of the Companies Act 1985.
[60a] Such a transfer might also constitute an unlawful distribution (see *Aveling Barford Ltd v. Perion* [1989] B.C.L.C. 626) and a "transaction at an undervalue" for the purposes of section 238 of the Insolvency Act 1986 so that the transfer could potentially be set aside by the court in certain circumstances.

ished by the assistance. The procedure involves the passing of a special resolution approving the proposal, a statutory declaration made by the directors of the company stating that the company will, following the rendering of the financial assistance, be able to pay its debts and continue as a going concern for at least a year and a report by the company's auditors supporting the directors' statutory declaration.

Continuing requirements

21–26 Once a company has been registered, the company has a duty to keep the company's registration with the Companies Registry up-to-date and must observe certain formalities. This involves:

Annual shareholders' meeting

21–27 In each year (and within fifteen months since the last such meeting) a company must hold an Annual General Meeting (AGM) of shareholders. In the case of a new company, the meeting must be within eighteen months from the date of registration.[61]

Annual return

21–28 Once in every calendar year a company must deliver a form called an Annual Return to the Registrar of Companies.[62] This form shows the share capital and the names of the Directors and Company Secretary and Shareholders.

Filing of accounts

21–29 The balance sheet, profit and loss account and directors' report on the company must be approved by the board of directors, accepted at a shareholders meeting and delivered to the Registrar of Companies within ten months (seven months for a public company) of the end of the company's financial year. There are financial penalties for delay, and the company can be threatened with winding up if there is a serious delay.

Elective resolutions

21–30 The Companies Act 1989 introduced[63] the concept of an elective resolution with a view to reducing the procedures for small private companies. It allows a private company to pass elective resolutions (unanimous resolution passed on 21 days' notice and capable of being revoked by Ordinary Resolution) for the following:

(i)   dispensing with the need to appoint auditors each year;

(ii)   dispensing with the need to hold AGMs and lay accounts before a general meeting; and

(iii)   granting the directors authority to allot up to the authorised share capital of the company for an unlimited period of time.

---

[61] At the AGM of the shareholders, the following matters are normally resolved by a simple majority:

(a)   acceptance of the Accounts and of the proposed dividend (if any);

(b)   re-election of directors (if re-election is required under the Articles of Association); and

(c)   appointment or re-appointment of auditors.

[62] See section 363 of the Companies Act 1985.
[63] By inserting a new section 379A into the Companies Act 1985.

Records required by the Companies Act 1985

21–31 The Company Secretary is required to keep all the official records which between them comprise the company's "statutory books".[64]

RESPONSIBILITIES OF DIRECTORS

Duty of good faith

21–32 Directors have a duty of good faith to their company. Any director, even a director in a small private company, cannot afford to ignore this duty.

The duty of good faith has been found to contain various elements including:

• a duty not to make secret profits[65];

• a duty to exercise powers for a proper purpose (and, for example, to issue shares only to raise money for the company — not to bolster the position of a particular shareholder)[66];

• a duty not to fetter discretion as a director[67]; and

• a duty to avoid conflicts of interests[68]

Usually, however, there is a practical way of dealing with a potential conflict of interests. If a director wants to enter into a contract with his company, or have another company with which he is associated enter into a contract, usually the Articles of Association of his company will allow him to declare the nature of any interest he may have in the transaction at a board meeting. Depending on the articles, the director may or may not then be allowed to vote on the question, but in either case, he will not be liable for a breach of duty.

Duty of skill and care

21–33 A director also owes a legal duty of skill and care. This duty was first formulated in 1925 and was found to have three essential elements[69]:

• a director is not an expert and should not be presumed to have skills he does not in fact possess;

---

[64] These comprise, the Register of Members, Register of Debenture holders (if any), Register of Applications and Allotments of Shares, Register of Share Transfers, Register of Directors and Secretaries, Register of Interests of Directors (and their families) in the shares and debentures of the company, and certain related companies, *e.g.*, parent company; Register of Charges or security interests granted by the company, Minutes of proceedings of General Meetings of Shareholders and Directors' Meetings .

[65] The point is well illustrated from the case of *Regal (Hastings) Limited v. Gulliver* [1967] 2 A.C. 134; (1942) 1 All E.R. 278. The directors of the company owning the Regal Cinema in Hastings identified new cinemas for their company to acquire. The company did not have the money to undertake this venture and so the directors bought these cinemas themselves. The venture was a success and they resold the cinemas at a profit. An action was bought against them and it was found that they had breached their duty of good faith to their company by "profiting from their office". They were made to hand over their profits to the company. See also *Cook v. Deeks* (1916) 1 A.C. 554; *Boardman v. Phipps* (1967) 2 A.C. 46.

[66] *Hogg v. Cramphorn Limited* [1967] Ch. 254; *Punt v. Symons and Co. Limited* (1903) 2 Ch. 506.

[67] *Clark v. Workman* [1920] 1 Ir. R 107.

[68] *Guinness v. Saunders* [1990] 2 A.C. 663.

[69] *Re City Equitable Fire Insurance Co.* [1925] Ch. 407.

- a director should display, but need only display, in relation to his duties the reasonable care of an ordinary man;

- a director is entitled, in the absence of suspicious circumstances, to delegate.

This formulation should be viewed cautiously in the light of modern standards, as the expectations of directors have risen. The formulation probably still largely holds good for a non-executive director but for executive directors, normally employed for their expertise, there is probably a presumption that they have and should display the necessary skills to perform their role. A finance director will be expected to be able to read a set of accounts, for example. Of course, if a director does in fact have particular qualifications or experience, he is obliged to apply it when carrying out his duties.

Liability of directors and protections

21–34   Breach of these duties can render a director liable to civil action. Remedies against the director could include the director being forced to turn over to the company profits he has made in breach of duty, and contracts entered into by him or his associates with the company can be rendered void by the company. If a director fails in his duties to the company he can have unlimited liability for any resultant loss, whether or not he has made a personal gain.

The Companies Act prevents the company from exempting a director from liability for negligence, default, breach of duty or breach of trust.[70] However certain types of breach can, in some circumstances, be ratified by shareholders. For example shareholders can ratify an allotment of shares for an improper purpose, a failure to disclose an interest in a contract, the obtaining of a secret profit not available to the company, and a non-fraudulent failure to exercise due skill and care. However the company cannot ratify any infringement of the individual rights of shareholders, any act that is outside the powers of the company or any fraudulent or dishonest breach of duty.

Statutory duties

21–35   Overlaid on these common law legal duties is a multitude of particular statutory duties, which apply to directors. Directors are responsible:

- for the maintenance of the statutory books and accounting records of the company;

- for the company's compliance with innumerable laws such as health and safety, environmental protection, compliance with statutory licensing requirements.

In addition, the Companies Acts include various provisions which are designed to protect a company from particular abuses that have arisen in practice. For example:

- directors' remuneration may not be paid on an after tax basis[71];

- certain payments to directors for loss of office cannot be made without shareholders' approval[72];

---

[70] See section 310 of the Companies Act 1985. However the company can indemnify the director against the costs of a successful defence of an action and there is now the ability for the company to pay for director's insurance.

[71] See section 311 of the Companies Act 1985.

[72] See section 312 of the Companies Act 1985.

- except in certain circumstances, loans in excess of £2,500 cannot be made by a company to any of its directors[73];

- directors' service contracts must not exceed five years unless members approve[74];

- a director must disclose interests in shares and debentures.[75]

One provision that may be of particular importance where an investor is transferring intellectual property rights to a company, is that a director may not enter into an arrangement to acquire from or transfer to the company a "non-cash asset" except in certain limited circumstances or with the approval of shareholders.[76]

Directors of listed or quoted companies also need to be concerned about the provisions (contained in the Part IV of the Criminal Justice Act 1993) regarding insider dealing.

Duties on insolvency

21–36   In general directors owe their duties to shareholders.[77] However when a company is insolvent or is approaching insolvency then the focus of directors' duties changes. The director begins to owe duties to creditors and in practice it is in this sort of circumstance that a director is most likely to run into difficulties. Directors can be made personally liable to compensate the company or creditors if they act wrongfully in connection with an insolvency. They can also be disqualified from acting as a director of any other company.

Directors can be made personally liable to compensate the company or creditors if they:

- misappropriate the company's funds or otherwise breach their duty to the company[78];

- are guilty of "fraudulent" or "wrongful" trading;

- break rules on re-using the names of insolvent companies[79];

- involve themselves in the management of a company while disqualified from acting as a director.[80]

Fraudulent trading means carrying on business with the intention of defrauding creditors.[81] It is a criminal offence, punishable by imprisonment or a fine under the Companies Act 1985. It also gives rise to civil liability under the Insolvency Act 1986. A director can be called upon to contribute to the company's assets.

Wrongful trading[82] gives rise to personal liability but is not a criminal offence. It

---

[73] See section 330 of the Companies Act 1985.
[74] See section 319 of the Companies Act 1985.
[75] See section 324 of the Companies Act 1985.
[76] See section 320 of the Companies Act 1985. The exception applies if the value of the non-cash asset is less than £1,000 or the lower of £50,000 or 10 per cent of the company's net assets. Failure to obtain the necessary consent allows the transaction to be rendered void by the company or for the company to claim damages for loss.
[77] There is a statutory provision which requires directors also to consider the interests of employees, but it is unclear what the practical effect of this is. See section 305 of the Companies Act 1985.
[78] See section 212 of the Insolvency Act 1986.
[79] See sections 216–217 of the Insolvency Act 1986.
[80] See section 15 of the Company Directors Disqualification Act 1986.
[81] See section 213 of the Insolvency Act 1986.
[82] See section 214 of the Insolvency Act 1986.

applies if the company goes into insolvent liquidation and, at some time before the commencement of the winding-up, a director knew or ought to have concluded that there was no reasonable prospect of the company avoiding insolvent liquidation and the court is satisfied that the director did not take every step he ought to have taken to minimise the potential loss to creditors.

A director can be disqualified from involvement in the management of a company for a period of up to 15 years.[83] It is a criminal offence to act as a director while disqualified.[84]

ATTRACTING INVESTMENT

Securities legislation

21–37 There is relative freedom to solicit purchasers for an idea, or investment in a business.[85] However, the ability to offer shares in a company and to advertise this investment opportunity, is constrained to a greater degree. Shares and debentures,[86] even in a small private company, constitute "investments" for the purposes of securities legislation in the United Kingdom and elsewhere. The relevant legislation needs to be considered before issuing any invitation to anyone to acquire (or dispose) of shares.

Is a prospectus needed?

21–38 Where it is proposed to offer shares (or debentures) of a company to the public the requirements of Public Offers of Securities Regulations 1995[87] ("POS Regulations") apply. These regulations require a formal prospectus to be produced and filed with the Companies Registry. The regulations set out detailed requirements for the contents of the prospectus. Compliance with these provisions can be onerous. However, this requirement arises only where the promotion is to the public in the United Kingdom. If shares are being offered in a foreign jurisdiction, the laws of that jurisdiction must also be considered. There are many exceptions to the requirement. These exceptions (which can be operated cumulatively) include exceptions where:

- fewer than 50 people are approached;

- the offer is to certain securities professionals;

- the offer is to existing members, or employees or their families;

- the offer is in the context of a takeover offer.

---

[83] See generally the Company Directors Disqualification Act 1986.

[84] See section 13 of the Company Directors Disqualification Act 1986.

[85] This is as long as the investment does not constitute an "investment" as defined by Schedule 1 to the Financial Services Act 1986. The definition of investment is lengthy and complex. The term would include shares (or loan stock or similar securities) of a company (including an overseas company) but would normally exclude putting capital into or giving a loan to a general partnership. Participation as an investor in a limited partnership, however, would usually comprise investment for the purposes of the Financial Services Act 1986 as a participation in a "collective investment scheme" so that the legislation described in para. 21–40, below would apply, as would separate restrictions on the promotion of shares or units under section 76, Financial Services Act 1986. The definition of collective investment scheme (in section 75 of the Financial Services Act 1986) is complex and can apply to other investment arrangements not amounting to ordinary companies or commercial partnerships. In addition to the special restrictions on promotion of collective investment schemes, acting as an operator of a collective investment scheme requires authorisation under the same Act.

[86] *i.e.* tradable debt securities of the company.

[87] S.I. 1995 No. 1537.

Where the exceptions do not apply and the company is marketing its shares to the public, a prospectus is required. The prospectus must be lodged at the Companies Registry and must contain information stipulated by the POS Regulations. The publication of a prospectus will give rise to certain responsibilities upon the company, its directors, any proposed directors and certain other persons. As well as complying with the detailed requirements for the contents of the prospectus, the prospectus is required:

- to be true, and not misleading;

- to give sufficient information to enable investors to make an informed investment decision; and

- not to omit anything likely to affect an investor's investment decision.

Responsibility for the prospectus

21–39 Under the POS Regulations, the persons responsible for the prospectus include the company, the directors and any person who accepts (and is stated in the prospectus as accepting) responsibility for the prospectus or for any part of it. The publication of a prospectus which is inaccurate or misleading can result in the persons responsible for the prospectus being subject to several types of civil liability.[88] In serious cases, misleading statements in a prospectus can give rise to criminal liability for fraud or under section 47 of the Financial Services Act 1986.[89] A process of "verifying" the accuracy of the prospectus and all related information prior to the admission is used to reduce these risks of liability.[90]

Investment advertisements

21–40 Even where a prospectus is not required, some other document may be produced which would constitute an investment advertisement. It is generally an offence to issue an investment advertisement unless the issuer is an authorised person (*i.e.* a person authorised to undertake investment business under the Financial Services Act 1986) or the advertisement is approved by an authorised person. "Investment advertisement" is widely defined and would catch almost any document designed to promote a sale of shares, including for example, a circular letter to prospective investors and a business plan, if it is to be used in such a way. There are, however, a number of exceptions to the requirements of section 57, and two of these exceptions are particularly useful. It is not an offence to issue an unapproved investment advertisement:

- to people whom are reasonably believed to fall into certain defined categories of investor[91];

---

[88] For example, liability may arise (a) under the POS Regulations to compensate investors who suffer loss by reason of false or misleading information in a prospectus, or as a result of the omission of any matter required to be included in it; (b) under the general law of contract for breach of a specific contractual term embodied in the prospectus, or for misrepresentation; (c) in tort for a negligent mis-statement or omission (whether in the prospectus or otherwise) on which an investor relies and suffers loss as a result; or (d) in respect of a fraudulent misrepresentation.

[89] See further at para. 21–43, below.

[90] The process involves the preparation of "verification notes" which take the form of a detailed list of questions to which answers must be given by the directors and/or by other responsible persons able to verify the relevant facts and provide supporting material.

[91] These categories include:

- an "authorised person" or an "exempted person" under the Financial Services Act 1986;

- business journalists;

- where the investment advertisement is published with a view to bringing about the sale of 75 per cent or more of the share capital of the body corporate in question.[92] The exemption applies if the proposed transaction will be between parties each of whom is a body corporate, a partnership, single individual or a group of connected individuals. This exemption does not assist with capital raising but should be applicable in the majority of company sales. However, care needs to be taken as there are some circumstances in company sales where the exemption would not apply.[93]

Unsolicited calls

21–41 The Financial Services Act also includes provisions[94] restricting the solicitation of investment agreements by means of unsolicited calls.[95] This is not a criminal offence but any investment agreement that is entered into as a result of the unsolicited call is not enforceable against the person on whom the call was made, and that person is entitled to recover any money or property paid or transferred by him under the agreement together with compensation for any loss.

There are a number of exemptions. The restriction does not apply where the caller is not acting by way of business, in certain cases regarding corporate disposals or acquisitions or management buy-outs and for calls between business partners or close relatives.[96] Also the court has discretion not to apply this provision if it is satisfied that the person called was not materially influenced by anything said or done during or as a result of the call. It may also not apply the provision where the agreement was made following discussions over a period such that the agreement could be fairly regarded as a consequence of those discussions rather than of the call where the person was aware of the nature of the agreement and any risks involved in entering into it.

- U.K. or foreign governments, U.K. local authorities and any international organisation of which the U.K., or another member state of the European Union is a member.

- a body corporate, either:
  - (a) being a body corporate with more than 20 members and share capital or net assets of not less than £500,000 (or a subsidiary of such a body corporate); or
  - (b) a body corporate with fewer members but which has, or has a holding company or subsidiary which has, share capital or net assets of not less than £5,000,000; or
  - (c) an unincorporated association with net assets of not less than £5,000,000.

- a director, officer or employee of a person falling within any of the above categories and who is involved in investment business.

- the trustee of a trust with gross assets of £10,000,000.

[92] See the Financial Services Act 1986 (Investment Advertisements) (Exemptions) (No. 2) Order 1995 (S.I. 1995 No. 1536).
[93] Such circumstances include:

- where the company has a complex capital structure. The shares sold must represent (together with shares already held by the purchasers) 75 per cent or more of the voting rights exercisable in all circumstances at any general meeting of the company.

- where the sellers or buyers do not fall within the defined categories of parties. The exemption does not apply where there is more than one seller or buyer and the group selling or buying does not fall within the definition of a "group of connected individuals".

[94] See section 56 of the Financial Services Act 1986.
[95] An unsolicited call means a personal visit or oral communication made without express invitation.
[96] See generally the Common Unsolicited Calls Regulations published by the Securities and Investments Board.

Nevertheless, where potential investors are cold-called, the provision creates uncertainty about the unenforceability of the agreement.

Proposed changes to the law

21–42 The law concerning the promotion of investments will be substantially amended when the Financial Services and Markets Act 2000 comes into force. This Act brings in a new regime. Under this new regime, there will be a combined restriction on all types of financial promotion including advertising, and unsolicited calls.

Under the new proposals it will be a criminal offence for any person to communicate an invitation or inducement to enter into defined categories of "investment activity" for business purposes unless that person is authorised to undertake investment business or an authorised person has approved the communication. Any contract resulting from such an unlawful communication will be unenforceable, and there will be a right to recover losses. What constitutes a "communication" is not defined in the Act and it will be left to secondary legislation to spell out circumstances in which the offence does and does not apply.

The Government has recognised that the current law regarding investment advertisements and cold calling makes it extremely difficult for start-up businesses to raise capital privately and so operates to restrict the growth of an enterprise culture. It has announced that it is proposing two exemptions to this new offence which are intended to make it easier for companies to raise equity privately. One would apply where the proposed investor is a "high net worth individual" (to be defined by reference to income and net assets not including main residence). The other would apply where a person is certified by an authorised person (*i.e.* someone authorised under the Financial Services Act to conduct investment business) as being a sophisticated investor able to judge the risks involved in a particular investment.

Liability for misleading investors

21–43 Whether or not the requirement for approval of an investment advertisement applies, care must be taken to avoid falling foul of further provisions of the Financial Services Act 1986. The Act provides that it is an offence knowingly or recklessly to make a statement, promise or forecast that is misleading, false or deceptive or to dishonestly conceal facts for the purpose of inducing someone to enter into an investment agreement (or to refrain from doing so).[97] The Act also provides that it is an offence to do any act, or engage in any course of conduct, which creates a false or misleading impression as to the market in or value of any securities if it is done for the purpose of creating that impression and of thereby inducing the acquisition (or disposal) of such securities.[98] These are regarded as serious offences. The maximum penalty for breach is up to seven years' imprisonment and/or an unlimited fine.

---

[97] See section 47(1) of the Financial Services Act 1986. These provisions are due to be repealed by provisions contained in the Financial Services and Markets Act 2000. However, the new Act contains provisions having a similar effect.
[98] See section 47(2) of the Financial Services Act 1986. These provisions are due to be repealed by provisions contained in the Financial Services and Markets Act 2000. However, the Act contains provisions having a similar effect.

CONSIDERATIONS FOR A TYPICAL IP BASED JOINT VENTURE

21–44 A typical intellectual property joint venture might involve two or more people, companies or institutions coming together to undertake collaborative research. Alternatively, one party might provide complementary assets or skills such as financial backing or business or marketing know-how. Joint ventures are often formed by the parties incorporating a company in which they will each be shareholders. This form of business structure is referred to as an "incorporated joint venture". However, as discussed in paragraphs 21–02 and 21–03 above, other business structures are also available and are frequently used. The documentation for a typical incorporated joint venture would involve a shareholders' agreement, specially tailored Articles of Association and possibly other documents such as intellectual property licences from one or both parties in favour of the joint venture company, employment contracts, or agreements for one or both parties to provide services to the joint venture company or to allow it to occupy property. There may also be separate loan agreements.

The following issues apply in the case of an incorporated joint venture, although similar issues will also arise in the case of other forms of joint venture.

Financing and share capital

21–45 Perhaps the single most important matter to be determined is the number of shares to be issued to each party and the rights to be attached to each type of share. As we have seen[99] the percentage of share capital held, and the rights attaching to those shares will have an important effect in determining the rights of the parties to control the venture. The number and types of shares to be issued will reflect the respective contributions of the parties. Shares may be issued for cash or in return for the contribution of intellectual property or of some other asset. One party may be expected to contribute more time and effort to the affairs of the company and this may be reflected in his shareholding.

It would be unusual for a joint venture company to be funded wholly by share capital and the documentation of the joint venture should reflect a clear understanding of how the venture is to be financed, based on the projected financial needs of the company. An agreement will be needed to cover the initial financing of the joint venture. Possibilities include a mixture of equity capital (shares) and loans from the shareholders. Alternatively it might be envisaged that one party or the other might guarantee bank funding or lease finance. If shareholders are to make loans to the company, the documentation will need to include the terms as to interest and repayment. Equally the agreement should deal with the obligations of the parties to provide further funds to the joint venture if the venture should overrun on costs. The agreement will often establish a procedure for agreeing annual budgets. It should also deal with dividend policy and state in particular whether profits are to be distributed or retained.

Control

21–46 If the parties are equal shareholders, then usually the documentation will provide for a "deadlock" with each party retaining an equal shareholding and equal rights to board representation, so that either party can veto any proposal at board or shareholder level. If the parties are not equal shareholders in the joint venture, the

---

[99] See para. 21–10 and paras 21–13 *et seq.* above generally.

majority holder will wish to retain control. The minority joint venture partner will wish to protect his position. The usual compromise is to give both sides rights to appoint directors. In addition there might be limited veto powers so that either party could prevent certain matters occurring. These powers need to be carefully thought out. A typical list of matters which might require the consent of the minority is as follows:

- capital expenditure in excess of agreed budget;
- share issues;
- changes in Memorandum or Articles of Association;
- sales/purchases/mortgages of land;
- entering into or terminating any long term agreement or an agreement involving more than a fixed sum;
- unusual or onerous contracts;
- change of business;
- acquisition by the joint venture of shares;
- unusual disposals of joint venture assets;
- loans by joint venture;
- factoring of the joint venture's book debts;
- agreements of the joint venture company with one or other of the parties;
- senior staff remuneration;
- employee share schemes;
- change of name;
- abnormal borrowing;
- business acquisitions or disposal;
- repurchase of joint venture's own shares;
- early redemption of joint venture's loans;
- changes of year end, auditors or registered office;
- accounting write downs;
- changes in accounting policies or reporting practices.

What happens in the event of a deadlock varies from agreement to agreement. Some-times no provision is made at all for deadlock, relying on the parties' respective self–interest to sort out any dispute. Often, where the joint venture is between two companies, there is a formal mechanism for disputes to be referred up the hierarchy within the organisations involved so that the most intractable disputes are eventually referred up to director level. Sometimes the agreement will provide for a deadlock (or more often, a deadlock on certain provisions) to be resolved by reference to a third party. Sometimes an unresolved dispute will trigger termination mechanisms.[1]

---

[1] See para. 21–49, below.

Any veto rights, and any special provisions for a shareholder to appoint directors should normally be incorporated into the Articles so that breach of the rights is not merely a breach of contract but would be outside the powers of the directors.

Board of directors

21–47 The joint venture agreement (and/or the joint venture Articles) should contain provisions dealing with such practicalities as:

- how the board is nominated;

- how the chairman is appointed. Often there are provisions for the chairmanship to alternate;

- whether the chairman will have a casting vote[2];

- whether directors will be allowed to appoint alternates?[3-4]

- policy on the remuneration and expenses of directors;

- venue, quorum and frequency of board meetings.

These matters may seem like trivial details in the early stages of a joint venture, but they can assume major importance, if at a later date relations between the parties become difficult.

Competition of partners with the joint venture

21–48 Competition with the joint venture is often controlled. Typical provisions might include undertakings:

- not to use know-how or confidential information acquired through participation in the joint venture during the existence of the joint venture and for a period following termination of the joint venture;

- not to solicit customers of or suppliers to the joint venture during the existence of the joint venture and for a period following termination of the joint venture;

- not to solicit joint venture staff;

- not to compete with the business of the joint venture during the existence of the joint venture;

- not to compete with the business of the joint venture for a period following termination of the joint venture.

This last restriction is often more controversial than the previous restrictions, particularly where the participant in the joint venture is an individual who may have no way of earning a living outside the business area in which the business operates. The courts are generally unwilling to enforce restrictions which operate in restraint of trade unless the restriction (having regard to the scope, geographical area and length of the restriction) is reasonable in the circumstances. Generally speaking, however, the courts will be readier to accept a restriction as reasonable if it is made in the

---

[2]   That is a second vote in the case of a deadlock at board level.
[3-4] That is, another person who can take the place of the director and attend and vote at board meetings when the appointed director is absent.

context of a joint venture or a partnership than if it is made in the context of an employment relationship.

In the case of large joint ventures, such restrictions, and indeed the very existence of the joint venture, could be capable of having a significant anti-competitive effect. As a result national and European Union competition law may be relevant and should be considered at the outset. Chapters 16 and 17 deal with these considerations in more detail.

Exit provisions and share transfers

21–49   It will be necessary to agree on the intended duration of the joint venture. Is it intended that the joint venture should continue indefinitely or is it for a fixed term? Provisions are often included setting out how the joint venture will end. This might be through the purchase of shares by one of the parties, through sale to a third party or by winding up the company and distributing its assets to its shareholders. Often provisions are included where one party has the right to buy out the other party's interest or to force the other party to buy its interest, as a means of bringing the joint venture to an end in certain circumstances or to force a winding–up of the company. These provisions might be activated, for example, on:

(i)   breach of the joint venture agreement;

(ii)   breach of any associated agreement;

(iii)   change of control of a joint venture partner;

(iv)   failure to provide funding; or

(v)   in the event of an insoluble dispute.

If one party is to buy out the other, the documentation will need to set out how the price is to be fixed. This might be on the basis of a valuation by reference to a formula, such as a multiple of historic earnings or turnover or by reference to the company's net assets. Alternatively the price might be fixed by an assessment by an expert valuer of the fair market price (usually stated to be on the assumption of a willing buyer and a willing seller). This could be stated either as a pro rata valuation of the whole company, or a valuation of the block of shares held (which may imply a discount to reflect a minority holding or a premium if the block effectively gives control).

Linked with the question of exit is the question of share transfers. Usually the transfer of shares is restricted in the case of a joint venture. Each party does not normally want to be faced with a new joint venture partner which it does not yet know. However the nature and extent of the restriction may vary from case to case. Sometimes there is an absolute restriction. Sometimes transfers are allowed but the other shareholders are given pre-emption rights to allow them to prevent a transfer by buying the shares themselves. Often there is an exception for intra-group transfers or for transfers within a family or to a family trust to allow for tax planning. Where such transfers are allowed there is almost invariably a requirement for the transferee to sign a deed of adherence agreeing to be bound by the joint venture agreement.

TYPICAL TERMS OF VENTURE CAPITAL BACKING

21–50 Chapter 3 explains how to go about obtaining venture capital backing and indicates what venture capital investors are looking for when considering making an investment. This section looks in a little more detail at the contents of a typical venture capital investment agreement.[5]

Where a venture capital house is considering backing a project, it will present its own standard terms for investment. To someone who has not seen these sorts of terms before, they can look somewhat forebidding and some terms may look at first to be downright unreasonable. The venture capital investors do, however, have legitimate interests to protect. Often they are putting substantial amounts of money into a venture and need to know that their investment matches the picture they have been given. They will also be concerned to tie the management into the structure through positive incentives and through legal restrictions, and to ensure that they have sufficient control to prevent the company from doing things that are against their interests. The investment agreement will often be presented as being "non-negotiable". In practice, if the investment opportunity is an attractive one, venture capitalists will usually accept some degree of reasonable negotiation, provided that their main objectives are satisfied. Typical terms of the investment agreement might include the following:

Conditions of subscription

21–51 The investment agreement will usually contain conditions which must be satisfied before the investment will be made. The conditions are designed to give comfort to the investors that the relevant corporate formalities have been observed and that any step that is crucial to the investment has been taken. These conditions might include:

(i)   adoption of new Articles of Association containing protections for the venture capitalist (see below);

(ii)   the transfer of intellectual property to the company;

(iii) adoption of a business plan approved by the venture capitalist;

(iv)   the signature of new service agreements by key members of management (see below); and

(v)   the passing of shareholders' resolutions permitting the transaction.

Representations and warranties

21–52 Another main way in which the venture capital investor protects itself is by obtaining warranties about the company and its business. The warranties are a series of statements made about the company and its business. If the warranties are untrue, the investor has a right to sue for damages. The investor will require these warranties to be given by both the investee company and its founders, so that the founders will have personal liability.

What is appropriate to be warranted will vary from case to case, but the venture capitalist's general principle will be to see warranties to cover every major area of the company's activities. The list of warranties can be forebidding and may run to many pages. In other cases the investor may produce a shorter list of warranties, but this may include sweeping warranties, such as a warranty that all information that has been

---

[5] For a fuller explanation see D. Cooke, *Venture Capital Law and Practice* (Sweet and Maxwell).

given to the investor is true and that there is no information which might reasonably influence the decision of the investor, which the investor has not been told about.

Items to be specifically covered by warranties might include:

- the accuracy of any financial information given, including warranties that assumptions underlying projections in the business plan have been made reasonably;

- the accuracy of any factual information given on which the investor has relied;

- the contracts that the company has entered into;

- the ownership and quality of intellectual property, and whether other people's intellectual property rights may be infringed by carrying out the business plan;

- the ownership of real estate;

- the identity, terms of employment and pensions arrangements of staff;

- the tax position of the company;

- the licences and consents that the company has and needs to carry on its business;

- whether the company is involved in or is likely to become involved in any litigation.

The usual format is that the warranties are given as sweeping statements of fact such as "*The Company is not engaged in any litigation.*" If any such statement is untrue (as is often the case) then the warranty is qualified by a separate letter called the disclosure letter written by the warrantors to the investors[6] setting out in detail where the warranties are untrue. The investors then have notice of the problems and cannot take action for the breach of warranty.

Although venture capital investors will rarely be persuaded to move away from the principle of requiring warranties, it is usually possible to negotiate the terms of the warranties to some extent and to provide some protections for the warrantors. Protections commonly sought (and given) include:

- qualifying some of the warranties with words such as "*as far as the warrantors are aware*" so that they are not given as absolute statements of fact;

- a financial limit of liability (often set at the amount to be invested by the venture capitalist) or possibly, in the case of individual warrantors, as a small multiple of their annual income from the company (such as three times salary);

- a time limit so that a claim for breach of warranty needs to be made within a certain period, usually somewhere between one and three years;

- a *de minimis* level so that individual claims below a certain level fall out of account;

- an overall *de minimis* (often referred to as the "basket" or the "threshold") so that claims cannot be made until the aggregate of claims exceeds a particular amount.

---

[6] Sometimes the letter is from the warrantors' lawyers to the investors' lawyers in an attempt to benefit from legal professional privilege so that the letter might be protected from a requirement for disclosure in a court action brought by a third party. This could be valuable if, for example, the disclosure letter contained a frank exposition of the strengths and weaknesses of the company's position in relation to some piece of litigation.

Investment in the company

21–53   Different venture capital investors have differing requirements. Some look only to growth. For others it is important to obtain a continuing income return by way of interest or preference share dividends. Also, the financing needs of the company may be best served through a combination of different types of funding. The venture capital firm's investment may, therefore, be provided either entirely as ordinary share capital or through a mixture of ordinary shares, preference shares and/or as loan capital which may include so-called "mezzanine lending". Mezzanine lending is a term used to describe debt which stands half way between ordinary debt and equity. It is subordinated to ordinary debt, so that on an insolvency, it will be repaid only after the ordinary debt is repaid. This subordination is an advantage to the company since it allows the company to raise further borrowings from other lenders who will be willing to lend in the knowledge that they will be first in the queue. In return for accepting this subordination, the mezzanine lender is usually given a higher rate of interest than might be expected for ordinary debt and will usually receive an "equity kicker", that is warrants (options) to subscribe for ordinary shares on favourable terms.

Ratchet mechanisms

21–54   A fairly common feature of venture capital structures is the inclusion of a ratchet mechanism. The effect of the mechanism is to vary the percentage of equity returns given to management according to the performance of the company. The purpose is to incentivise the management by providing it with enhanced returns where the management achieves performance targets. The ratchet can also operate as a way of settling a dispute between the investor and the founders as to the value of the company, in cases where the assessment of value depends on the view to be taken of the company's future prospects.

There are two main types of measurement criteria. The first is to measure the company's financial performance. This can be by reference to profits or turnover but is more usually through the calculation of an internal rate of return (IRR). The IRR is essentially a discount rate applied to discount the cashflow received by the investors by way of revenue (interest and dividend) and capital return (on disposal of their shares). Targets can be set by applying a chosen rate of IRR and performance can then be measured against these targets. The second type of measurement is known as an "Exit ratchet". Exit is the term used by venture capitalists to describe the means by which they will realise their investment. An exit ratchet operates by reference to the value achieved by the company on exit, such as the sale price on a trade sale or the price achieved on a flotation. An exit ratchet has the advantages of (relative) simplicity and of incentivising the management to achieve an exit.

Ratchets can work either positively by enhancing the management's share of equity in the case of good performance or negatively by reducing the share in the case of weak performance. There are a number of mechanisms that can be used to achieve the ratchet. These include:

- issuing (either to the management or to the investors) deferred shares which convert into ordinary shares at a rate determined by the ratchet targets;

- by granting either the management or the investors options to subscribe for shares at a fixed price;

- by providing for share transfers at set values.

Ratchet mechanisms are inherently complex and professional advice is needed on their effect, the type of mechanism to be used and the tax implications of the arrangements.

Financial assistance

21–55   If the venture capital firm is looking to put its investment in the company by way of loan capital, it will want to ensure that its loan is secured on the assets of the company. If the transaction is being undertaken in conjunction with the acquisition of shares by the venture capitalist in the company, this could breach the law against financial assistance discussed above. The venture capitalist will require the directors of the company to join in a "whitewash" procedure which will permit the financial assistance.[7]

Investor protection clauses

21–56   The venture capital investor will look to protect its investment through the Investment Agreement and the Articles of Association. It will usually require board representation, or at least the right to board representation, coupled with the right to see all board papers. It will also insist on including terms to prevent the company from undertaking certain acts without the written consent of the investor, or in some cases, without the consent of the investor's nominated director. The activities restricted are likely to include:

(i)   undertaking any business outside the company's main business;

(ii)   disposing of (including any exclusive licensing and the creation of any security over) property and in particular the key intellectual property of the company;

(iii)   taking on new employees or increasing the remuneration of existing employees or directors;

(iv)   forming a subsidiary;

(v)   entering into any partnership or joint venture;

(vi)   alterations to share capital such as the consolidation or subdivision of shares;

(vii)   a new issue of shares;

(viii) any repurchase or redemption of shares.

Transfer of shares

21–57   The venture capitalist will wish to ensure that the ownership of shares does not change, and in particular that the management remains locked into the company. This will usually be effected by restrictions on the transfer of shares not held by the venture capital fund. The precise terms of these restrictions will vary from case to case and could include an absolute restriction on transferring shares, a requirement to offer the shares first to other shareholders, or a restriction on selling to someone other than an existing shareholder. Often these restrictions can be softened in negotiations, at least to the extent of allowing transfers to family members or family trusts so as to allow tax planning.

In relation to its own shareholding, the venture capital fund may accept restrictions

---

[7] See above at para. 21–25.

on transfer, but will usually insist on the ability to transfer between its own funds and to transfer to investors in the venture capital fund.

It is fairly common to deal separately with takeovers and to include so called "tag along" and "drag along" provisions. "Tag along" provisions state that where some shareholders find a buyer for their shares they must procure that the buyer also makes an offer to buy the remaining shares on the same terms. "Drag along" provisions state that where some shareholders (usually a majority or a super majority of shareholders) find a buyer for their shares they may force the other shareholders to sell the remaining shares on the same terms.[8]

Service agreements

21–58 Another way in which the venture capitalist will wish to lock management into the company is through the terms of service agreements. The venture capitalist will usually insist on new service agreements being entered into which may include:

- obligations to give the company a full-time commitment;

- lengthy notice periods for termination coupled perhaps with "garden leave" provisions;

- an acknowledgment that all intellectual property produced in the course of employment will belong to the company;

- restrictive covenants applying during the term of the employment and for a period afterwards preventing competition with the company, use of the company's confidential information and/or poaching of customers, suppliers or employees.

Exit

21–59 The Investment Agreement will often include some indication of the intended time scale for exit, and how exit is to be achieved, although these terms may constitute more of a statement of intent than anything that is meant to be legally binding. The most common means of exit are:

- trade sale;

- buyout by management;

- redemption/repurchase of shares by the company;

- flotation.

Venture capitalists are experts in finding and helping to implement an exit, and this is a stage at which they can definitely add value. Where venture capitalists will be unhelpful, however, is in their attitude to warranties. On any sale or flotation the purchasers or sponsors are likely to require warranties to be given by the selling shareholders. Venture capitalists will almost always refuse to join in giving warranties.

---

[8] Note that there are statutory provisions having a similar effect in Part XIIIA of the Companies Act 1985. This applies where a person makes an offer to acquire all the shares of any class of the company, and receives acceptances of that offer in respect of nine-tenths in value of the shares of that class. It provides procedures under which the offeror is given a right to purchase the remaining shares, and a duty to purchase such shares if required.

Their arguments for doing so may be good or bad, but this is a point on which they rarely bend.[9]

In some circumstances on a trade sale, a venture capital investor will require a "No embarrassment" provision to be inserted in the sale documentation. This is a clause that says if the buyer sells the company or its business on within a period (often set at two to three years) and realises a profit above a certain threshold, the venture capital investor will receive extra consideration so that it will effectively share in this profit. This sort of clause is used in particular where an exit involves a sale to a purchaser connected with the company's existing management. In such cases the clause protects the investor against selling at an undervalue as a result of the management failing to keep it fully informed of the company's potential for growth.

STOCK EXCHANGE LISTING

21–60 For some entrepreneurs the ultimate dream is to take their own company to the stock market. This also is often contemplated as a possible exit route by venture capitalists backing a technology business at its early stage. The advantages of a flotation include the possibility of raising new capital for growth of the company, the possibility of realising the existing investment, an ability to make future acquisitions by issuing listed shares as consideration and the possibility of increasing the profile of the company.

It should be appreciated, however, that these advantages are not enjoyed automatically by every stock exchange entrant. The ability to raise money and to trade shares on an exchange will only come about if the company is able to attract sufficient investors to create a healthy market in the shares. Smaller companies, particularly on the smaller markets, can find that there is little liquidity in their shares.

Going public has its price. First, there is the cost of the exercise itself, which can be substantial. Secondly, by going public, the company takes on responsibilities to its investors and to the Exchange. All the markets have continuing requirements for reporting and rules concerning corporate governance which can prove onerous.

The markets

21–61 A number of different stock exchanges are available for these purposes. Suitable stock exchanges commonly accessed by United Kingdom technology companies include, in the United Kingdom, The London Stock Exchange, either through its main market,[10] its junior market, the Alternative Investment Market (AIM), or Techmark, its new market for technology companies, EASDAQ[11] in Europe, and NASDAQ in the United States.[12] In addition, capital can be raised through OFEX. Technically, OFEX is a dealing facility rather than a market, although, it does provide a way of raising capital. Other new markets which are seeking to attract technology

---

[9] Except perhaps that they may be prepared to allow a limited part of their share of the sale proceeds to be kept for a short period in a retention account to be used to cover warranty claims, and they may be prepared to allow their share of the sale proceeds to be affected post completion by a price adjustment by reference to net assets. These devices can come close to having the investment house participate in warranties without them breaching their self-imposed rule that they will not act as warrantors.

[10] The major part of the task of regulating the main market in London has been transferred to the Financial Services Authority as the U.K. Listing Authority. To obtain a full listing in London it is also necessary to obtain admission to the Official List maintained by the Financial Services Authority and for the securities to be admitted to trading by the London Stock Exchange.

[11] The European Association of Securities Dealers Automated Quotations.

[12] The National Association of Securities Dealers Automated Quotations.

investments overseas include the German Neuer Markt and the French Second Marché.

Formal requirements

21–62   Each market has its own formal requirements for listing or quotation, although in practice the requirements for listing may be more onerous than those theoretically applicable. For example rules applicable for listings on the London Stock Exchange allow a company to come to that market with an initial market capitalisation of only £750,000 but in practice it would be difficult to find a sponsor or institutional investors who would support a new entrant joining the market at that level. Professional advice is required to know whether a company is suitable for listing and which of these markets is likely to prove the most suitable.

Preparation for listing

21–63   Preparing for a flotation requires planning and preparation. An early step is to ensure that the accounts and the accounting procedures are completely up to date and will bear a high level of scrutiny. A formal business plan should be prepared and discussed with the company's proposed sponsor. The major shareholders will need to be persuaded to support the flotation proposal.

   Once the company is ready for a flotation, the advisers need to be chosen. The relationship with the advisers is likely to continue for many years and advisers should be chosen for their compatibility with management and their sympathy with the aims of the company as much as on the level of their fees.

Documentation for listing

21–64   Further preparation for the flotation comes about when reporting accountants undertake substantial financial due diligence, often producing a "long form report", or perhaps a "short form report" detailing the major financial, management and market issues applicable to the company. At the same time a working capital report will probably be needed to show that the company has (or will have following any capital raising) sufficient working capital for its contemplated needs. This will involve preparing cash flow projections forwards for 18 or more months in advance. Relying heavily on the long form report, the company and its advisers proceed to draft (and verify) the formal document which describes the company for the purposes of the flotation. This document may be called listing particulars or an admission document or a prospectus, depending on which market is involved.

   The precise mechanics of the flotation will depend on the circumstances and how any new money is to be raised.

PARTNERSHIP

21–65   Partnership is a common form of business structure, although its use is not particularly common for technology based businesses, where it is often felt that the business should be incorporated so that technology rights can be owned by the business itself. Under the Partnership Act 1890 a partnership arises automatically whenever two or more people carry on business together with a view to profit. Partnerships, therefore, can arise informally without the parties entering into a formal Partnership Deed or Partnership Agreement. Where no Partnership Agreement is entered into, the relationship of the participants is governed entirely by the Partnership Act 1890. More usually, however, participants enter into a formal Partnership Deed which will modify the provisions of this Act and which will set out:

- the scope of the business to be pursued;
- participants' funding obligations in relation to the business;
- how partnership assets are to be held;
- how profits (and losses) are to be shared between the partners;
- how decisions are to be made within the partnership;
- limitations on the authority of partners to bind one another;
- how new partners may be introduced;
- what happens when partners retire;
- arrangements for the dissolution of the partnership.

It can sometimes be argued that a partnership arises in the course of collaborative research and/or development. The distinction is important in that if the relationship of partners has arisen, under the Partnership Act 1890 certain terms are deemed to apply.[13] These implied terms may not be the terms desired or expected by the participants. For example, each "partner" will be entitled to pledge the credit of the other partners.[14] It would be usual to state in a Collaboration Agreement that the parties agree that no relationship of partnership (or of agency) arises. However, such a clause is not conclusive and would not stop a court from finding that a partnership existed if the facts bore this out. Two important indicators are whether there is yet anything that can be thought of as a "business"[15] and whether there is an agreement to share in profits and (especially) losses.[16]

Partnership may be suitable as a business structure if:

- the parties are looking for a structure that is transparent for the purposes of tax, that is to say where the business entity itself does not have its own tax liability (as would be the case with a company). With a partnership, each partner is responsible for tax on that partner's own share of the profits and the vehicle itself is not separately taxed;
- limitation of liability is not important;
- the promoters of the business wish to avoid the formalities involved in incorporation, and in particular the requirement to disclose accounts.

Partnership can also be useful as part of tax planning. By making a spouse a partner in a venture, it is possible to transfer part of the (taxable) profits to the spouse and so allowing use of the spouse's tax allowances or lower tax rates. Whilst the spouse does genuinely need to participate in the partnership, there is no legal requirement for the profits to be shared proportionately to the effort involved or contribution made.

---

[13] See Partnership Act 1890, in particular section, 24.

[14] Partnership Act 1890, sections 5, 6 and 7.

[15] See *Keith Spicer Limited v. Mansell* [1970] All E.R. 462.

[16] More guidance is given in section 2 of the Partnership Act 1890. See also *Fenston v. Johnstone (HM Inspector of Taxes)* [1940] 23 T.C. 29).

Nevertheless, partnership is not a particularly common way of dealing with the exploitation of intellectual property, and it is not the intention of this work to provide a full account of the topic. See *Lindley & Banks on Partnership* for further information.[17]

LIMITED PARTNERSHIP

21–66 Limited partnership can be thought of as a hybrid form of business structure falling halfway between a partnership and a limited company. Unlike a company, a limited partnership under English law has no legal personality of its own.[18] However, unlike a partnership, participants in a limited partnership (other than a general partner) can enjoy limited liability.[19] The interests in the partnership (usually referred to as "participations" or sometimes "shares") of limited partners are usually tradable (sometimes with restrictions), whereas an interest in a general partnership is usually regarded as being personal to the partner involved and is not tradable.

Partnership law as set out in the Partnership Act 1890 applies also to limited partnerships. However, additional provisions applying only to limited partnerships, are set out in the Limited Partnerships Act 1907. This Act provides that a limited partnership arises where a partnership is registered as a limited partnership under that Act. There must be a Partnership Agreement, and the Agreement must provide for two categories of partner. First there must be a general partner (or partners) who will be responsible for the day-to-day management of the partnership, and who will have unlimited liability for the obligations of the partnership. Secondly, there must be limited partners who will enjoy limited liability. That is to say that their responsibility for the debts and obligations of the limited partnership will be limited to the amount of partnership capital that they provide to the partnership. The price paid for this advantage, however, is that they may not participate in the management of the partnership, and may not withdraw their partnership capital until the limited partnership is terminated. In view of this latter point, it is common for limited partners to fund the partnership through a mixture of capital and loan. Although the capital cannot be repaid, the loan element can be.

Unlike a general partnership, which requires no formalities for its creation, a limited partnership is created only when the limited partnership agreement (satisfying requirements of the Limited Partnership Act 1907) is registered at the Companies Registry.

Limited partnership is a common form of structure for a venture capital fund. It is not common for limited partnerships to be used as a vehicle for holding, and exploiting, intellectual property. However, where the participants include a passive investor or group of investors who have no desire to be involved in management, then a limited partnership vehicle may be a suitable option. Limited partnerships do offer certain tax advantages, in that, like ordinary partnerships, they are regarded as "tax transparent".

---

[17] Roderick C. L'Anson. (Sweet & Maxwell).
[18] The position is different however in some other jurisdictions where a limited partnership may be regarded as having a separate legal personality. In particular Scottish partnerships (including limited partnerships) are regarded as having a separate legal personality — see section 4(2) of the Partnership Act 1890.
[19] See section 4 of the Limited Partnerships Act 1907.

LIMITED LIABILITY PARTNERSHIPS

21–67 The Limited Liability Partnerships Act 2000 will, when it comes into force, establish a new type of business structure to be known as a "Limited Liability Partnership" ("LLP"). The new structure is intended to combine the organisational flexibility and tax status of a partnership with limited liability for its members. Unlike the older form of Limited Partnership, there is no requirement for there to be a general partner with unlimited liability, and there is no restriction on any partners participating in management.

This new structure is being introduced as a result of pressure from professional firms (chiefly accountants) which traditionally practise as partnerships but which have felt an increasing need for limited liability in the light of a general increase in litigation being brought for professional negligence and in the size of claims. However, this new vehicle will be available to all types of businesses.

To form an LLP, there must be two people who subscribe to a document called an "Incorporation Document". The Incorporation Document must be delivered to the Registrar of Companies and must contain various items of information such as the name and place of business of the LLP. The Incorporation document is lodged on a public register maintained by the Registrar of Companies. The LLP comes into existence by means of the issue of a certificate by the Registrar.

Like a company, an LLP will be regarded as a legal person in its own right. Unlike a company, where the doctrine of *ultra vires* applies, an LLP will have unlimited capacity, so that it is able, legally, to do anything that a natural person could do. Members of (partners in) an LLP will enjoy limited liability in a similar way to shareholders in a limited company. They will become members of an LLP either by signing the Incorporation Document or by agreement with the existing members.

The relationship between the members of an LLP will be governed by any agreement between them (subject to the (scanty) provisions of the Incorporation Document). There is no obligation to have any agreement, and no requirement to publish it but in practice it is expected that LLPs would have a partnership agreement in a very similar format to that for a general partnership.

As with a partnership, each member of the LLP is an agent of the LLP and may represent and act on behalf of the LLP in all its business, unless that member is given no authority to act on behalf for the LLP and the person dealing with the member is aware of this (or did not know that the member was in fact a member of the LLP).

It remains to be seen whether this business format proves popular for businesses other than professional businesses. The format appears to have sufficient flexibility to adapt itself to a wide range of uses. It offers a different tax treatment to the limited company format but with similar advantages as regards limited liability, separate existence and continuous succession.

CHAPTER 22

ASSIGNMENTS

Jennifer Pierce, Charles Russell

The assignment process

22–01 Assignments of intellectual property must be in writing in order to be effective.[1] Assignment documents themselves are relatively simple, and there are many suitable precedents available. The assignment process (for registered rights) is equally simple. Regrettably, both the document and any subsequent formalities are based on assumptions of ownership of intellectual property, which may ultimately prove to be false. It is therefore crucial for both assignors[2] and assignees[3] to investigate the title to the rights that are to be assigned. Laws relating to title are dealt with in Chapter 14. The practical means of investigating the assignor's title, and the suitability of the rights themselves are both included in Chapter 23. If the assignee fails to investigate the property and to obtain sufficient comfort, he may fail to satisfy a subsequent purchaser or licensee who has more stringent requirements. In particular, a person taking an assignment from the inventor, author, designer or maker[4] may be more easily convinced of the authenticity of the initial documents of title than a subsequent assignee.

Present and future assignments

22–02 Assignments can cover property that is already in existence, or property that has yet to be created. In the case of copyright and design right, the United Kingdom legislation[5] provides specifically for assignment of present and future copyright and design right. Under such a future assignment, the property is automatically assigned when it comes into existence, provided that at that time the assignee would be entitled as against all other persons to require the right to be transferred to him.[6] There are no equivalent provisions in respect of patents or registered designs. There is, however, provision for a patent to be granted to an assignee by virtue of a future assignment.[7] An application for a registered design must be made by the person claiming to be the design right owner.[8] The assignor is usually required to hold any rights that are not assigned on trust for the assignee, pending signature of an assignment document. As a practical matter, the assignee may not know of the creation of any future property

---

[1] There are statutory requirements that assignments be in writing — see Patents Act 1977, s. 30(6), 31(6), Copyright, Designs and Patents Act 1988, ss. 90(3), 222(3), Trade Marks Act 1994, s. 24(3). In view of the requirement of registration, there is, effectively, an equivalent provision for registered designs — see Registered Designs Act 1949, s. 19.

[2] People who dispose of rights by means of a transfer (assignment).

[3] People who acquire rights through a transfer (assignment).

[4] The person from whom the title is ultimately derived, even if they are not the first owner — see PA 1977, ss. 7, 39, CDPA 1988, ss. 11, 215, the Copyright and Rights in Databases Regulations 1997 ("Database Regs") regs 14 and 15, the Design Right (Semiconductor Topographies) Regulations 1989 ("Topographies Regs") reg. 5, and RDA 1949, s. 2.

[5] Assignment of future rights is not permitted in all jurisdictions.

[6] So if, for any reason, such as the existence of prior rights, the assignee could not claim specific performance, the assignment will not take place automatically. See CDPA 1988, ss. 91 and 223. Note that CDPA 1988, s. 91 applies to database right — see Database Regs reg. 23, and that CDPA 1988, s. 223 applies to semiconductor topographies — see Topographies Regs reg. 3.

[7] Note PA 1977, s. 7, which refers to "an enforceable term of any agreement entered into with the inventor before the making of the invention".

[8] See RDA 1949, s. 3(2).

unless the assignor is required to inform him of all of the details of the relevant property. The time limits for providing information and documents will need to be stated specifically, and will vary according to the particular circumstances.

Individual rights

Patents and patent applications

22–03 The description of patent property in the assignment document will vary depending on the stage that the property has reached in the prosecution process. If no application has been made, the rights to be assigned will be the invention and any further material disclosed by the assignor to the assignee for inclusion in a patent application, together with the right to apply for and to be granted a patent in respect of the invention.[9] In addition to the usual further assurance provision that is described below, the assignee may need assistance from the inventor in the patenting process. For example, the assignee may require the inventor to comment on the patent specification, or on the results of prior art discovered in searches. If the application has already been filed, the application should also be assigned, together with the right to claim priority from it, and all further applications, or patents, such as re-issues, extensions,[10] divisions, continuations, and corresponding applications. The territorial scope of the assignment should be specified.

22–04 In the United Kingdom, there is statutory provision for rights in patents and applications to be assigned, as well as entire patents.[11] Partial assignment leads to joint ownership, under the terms of the agreement and of the relevant legislation. Unless otherwise agreed, each co-owner requires the consent of the other to grant a licence, or to assign or mortgage their share.[12] There is also statutory provision for an assignment of the right to bring proceedings for previous infringement, but the right needs to be exercised specifically.[13] In addition to the rights in any patent, or application, the assignor may be required to assign copyright and any design right in the specification, as those rights can be exercised separately.[14] If the assignment is of an application that has not been published, a confidentiality undertaking may be included in the assignment. Assignments relating to patents and applications must be signed by all parties in order to be effective.[15]

---

[9] Note that in the United States it is a requirement that the application is in the name of the inventor, irrespective of the assignement. Note also that the right to grant is subject to compliance with requirements for patentability. Also, whilst PA 1977 s.7 refers merely to the grant of a patent it is wise to refer to the right to make the application in case there are different requirements in another jurisdiction.

[10] The assignor could also be required to assign the right to apply for a supplementary protection certificate using the patent as the basic patent. This will, however, be of limited use, as the certificate is applied for in respect of a particular product, which is the subject of an authorisation. Any provision in an assignment relating to a supplementary protection certificate will need to recognise this. See Council Regulation (E.E.C.) No. 1768/92 of June 18, 1992 concerning the creation of a supplementary protection certificate for medicinal products (as amended), and Regulation (E.C.) No. 1610/96 of the European Parliament and of the Council of July 23, 1996 concerning the creation of a supplementary protection certificate for plant protection products.

[11] See PA 1977, s. 30(2).

[12] In the United Kingdom except for Scotland this is PA 1977, ss. 30(2), 36, and in Scotland it is PA 1977, ss. 31(3), 36.

[13] See PA 1977, ss. 30(7) and 31(7).

[14] In *Catnic Ltd v. Hill Smith* [1978] F.S.R. 405, Whitford J. expressed the view (obiter) that the copyright in the drawings for the patent specification could not be enforced separately from the patent, but it is suggested that courts may prefer to follow *Werner Motors Ltd v. A.W. Gamage Ltd* (1904) 21 R.P. C. 621, in which it was held that a design could validly be the subject of both a patent and a registered design.

[15] See PA 1977, s. 30(6), but note that under PA 1977, s.31(6) different provisions apply in Scotland.

22–05 Registration of assignments of patents and applications is vital to preserve the rights of the assignee. If the assignment is not registered it can be defeated by a subsequent assignment if the person claiming under the later document did not know of the earlier document at the time of the later transaction.[16] Furthermore, if the assignment is not registered within six months from its date, the assignee cannot claim damages or an account of profit in respect of the period before the assignment is registered.[17] There is, however, no obligation to register before commencing proceedings for infringement.

Copyright

22–06 In the United Kingdom[18] an assignment of copyright may be total or partial.[19,20] A partial assignment can be limited to one or more of the things[21] that the assignor has the exclusive right to do, or may be limited to part of the period for which copyright subsists.[22] It is therefore prudent to describe both the nature of the work[23] and the extent of the rights that are to be assigned. The assignment of the right to bring proceedings and to claim damages in respect of past infringements needs to be mentioned specifically. In addition to the United Kingdom copyright, the copyright in other jurisdictions can be included. Even if the assignment is to relate to all the rights comprised in copyright for the duration of those rights throughout the world, it is prudent to state this specifically. Whilst there are no registration formalities for an assignment in the United Kingdom, it is advisable to obtain the originals[24] of all the other documents in the chain of title, together with the necessary details and proof of the author and date of first publication in order to demonstrate ownership in future. It is still possible to register copyright in some other jurisdictions, most notably in the United States.

22–07 Assignments of copyright are subject to the licences that have already been granted, except for assignments to purchasers in good faith for valuable consideration, without notice of the licence (actual or constructive).[25] It is therefore crucial for the assignor that the assignment contains details of all subsisting licences. Otherwise the assignee may take the copyright free of the licences, leaving the licensees with a claim against the licensor for breach of contract. It is preferable that the assignee also receives copies of the licences prior to assignment, as he will then have notice of any unusual provisions.

---

[16] See PA 1977, s. 33. See also para. 23–37.

[17] An exception is made if the court or the comptroller is satisfied that it was not practicable to register the assignment before the end of the six month period, and that it was registered as soon as practicable thereafter — see PA 1977, s. 68.

[18] Note that in Germany copyright may not be assigned, except on death, due to recognition of the moral rights of the author. In such cases, an exclusive licence may be granted instead. See, for example, U. Suthersanen, *Design Law in Europe*, para. 14–039 for further details.

[19] CDPA 1988, s. 90(2).

[20] In the case of copyright and assignments made prior to August 1, 1989 it is necessary to consider the terms of the previous legislation. As much of the copyright that is assigned or licensed in connection with technology is relatively new, the provisions of the old legislation have not been set out here.

[21] Note that this does not refer to "acts", as a partial assignment is not limited to the "prohibited acts". A partial assignment may, for example, refer to hardback rights — see para. 23–56 for an explanation of prohibited acts.

[22] In the event of litigation, it is the proprietor of the particular right that is infringed who may take action. See CDPA 1988, s. 173(1).

[23] In cases where there have been a series of drafts or sketches, it may be necessary to take an assignment of all works on which the final work is based.

[24] Or certified copies in cases where there is a partial assignment.

[25] CDPA 1988, s. 90(4).

22–08   Moral rights, which subsist in copyright works,[26] cannot be assigned,[27] but they may be waived under English law, provided that the waiver is in writing and signed by the person waiving the right.[28] If a copyright work is to be assigned, the assignee may require that the author waives his moral rights under English law, and under other laws, to the extent that he is able.

Design right

22–09   Assignment of design right is very similar to an assignment of copyright, although there are some important differences. First, moral rights do not subsist in designs in which copyright does not also subsist. Also, design right was created under United Kingdom legislation, and is not the subject of specific international treaties, like other intellectual property,[29] so similar rights may not subsist elsewhere. The other important difference is the link with the title to the corresponding registered design. Where a design is registered and the proprietor of the registered design is also the design right owner, an assignment of the registered design is taken to be an assignment of the design right, unless there appears to be a contrary intention.[30] There is also a parallel provision in respect of assignments of design right, under which the corresponding registered design is automatically assigned.[31] The provisions are not as efficient as they first appear, as it is possible that there could be a number of separate design rights. The assignment may be construed as transferring all such rights, but if it failed to do so, the assignee should have an implied licence under the remaining rights, provided that they are held by the assignor. In any event, cross licences might be necessary if that were not the case. Copyright can also subsist in design drawings and should therefore be included in the assignment, if relevant.

Registered designs

22–10   As described above, the title to a registered design is linked to that of the corresponding design right. In addition, it is not possible to register an interest in a registered design unless the person to be registered is also entitled to a corresponding interest in the design right.[32] Moreover, if there is a corresponding design right, it is only the proprietor of that right who is eligible to apply for registration.[33] The assignment should therefore specify the registered design, the corresponding design right, if it is also held by the assignor, and any related design right, if possible. In any event, if no registration has been made, the right to apply for and to be granted registration can be assigned.[34] As is the case with patents, it is possible to claim priority from the initial application for the purpose of applying for rights in other territories, so the right to claim priority from the first application can also be included, although it is currently of limited value, as the United Kingdom has yet to accede to the Hague Agreement.[35]

---

[26] With certain notable exceptions, such as computer programs — see CDPA 1988, ss.79(2), 81(2) in respect of the right to be identified as author and the right to object to derogatory treatment of a work.

[27] See CDPA 1988, s. 94.

[28] See CDPA 1988, s. 87. Note that the waiver may be conditional, unconditional or subject to revocation. Note also that the requirement that the waiver be in writing does not exclude the possibility of estoppel.

[29] The position is different in respect of topography rights.

[30] See CDPA 1988, s. 224.

[31] See RDA 1949, s. 19(3B).

[32] See RDA 1949, s. 19(3A).

[33] See RDA 1949, s. 3(2)

[34] See RDA 1949, s. 2. This may be implied from the wording;.

[35] U.K. nationals, who are not able to claim domicile, habitual residence, or a real and effective industrial or commercial establishment in a contracting state, therefore have no right to file an international application — see Art. 3 of the Geneva Act of the Hague Agreement.

22–11 Unlike patents, copyright and design right, there is no specific provision for an assignment of a part of a registered design. However, it is clear that co-ownership is envisaged.[36] As registrations relate to specific articles, it is possible for registrations in respect of different articles to be in separate ownership. Registrations in other jurisdictions are clearly separate items of property. It is also possible, either through the circumstances of creation or through choice, that the design right may be held separately from the corresponding registered design. It will be evident from the information above that this arrangement has substantial drawbacks, but it is possible for it to exist. As with the other rights described above, the right to bring an action and to claim damages in respect of any previous infringements needs to be included in the assignment. It is also possible that in addition to design right, copyright will subsist in the representation of the design that is filed, so this should also be assigned specifically.

22–12 Registration of assignments, and any share in a registered design, appears to be mandatory.[37] Only the registered proprietor has the rights given by registration,[38] and if a document has not been registered it can only be produced as evidence of title in court if the court so directs, or in proceedings to rectify the register.[39] The registered proprietor has power to assign, subject to other peoples' rights that are entered on the register, with the proviso that any equities may be enforced in the same manner as for other personal property.[40] Furthermore, it is only the registered proprietor who can sue for infringement.[41] In any event, failure to register could lead to somebody else registering conflicting rights, and to the assignment being subject to them.[42]

Trade marks

22–13 In the United Kingdom, trade marks and applications can be assigned independently of the goodwill in the business in which they are used.[43] However, as explained in Chapter 13, a proprietor of a registered trade mark may need to demonstrate a reputation in connection with the use of the mark in order to register it, or to claim successfully in respect of an infringement. Furthermore, in some other jurisdictions it may not be possible, or it may be difficult, to assign a trade mark without including the goodwill associated with it in the assignment. It is therefore prudent to assign the associated goodwill together with the mark, if possible. However, assignments of goodwill and any associated intellectual property[44] together with equipment may, for the purposes of employment legislation, amount to a transfer of an "undertaking". In these circumstances there could be a deemed transfer of

---

[36] RDA 1949, s. 2(2) refers to a design being held "jointly with the original proprietor", and s. 19(1) refers to a "share" in a registered design.

[37] See RDA 1949, s. 19(1) "Where a person becomes entitled by assignment . . . he shall apply to the registrar . . ."

[38] See RDA 1949, s. 7.

[39] Note, however, the wording of RDA 1949, s. 19(1), which refers to registration "where a person becomes entitled by assignment...". Note also RDA 1949, s. 19(5).

[40] See RDA 1949, s. 19(4).

[41] See RDA 1949, s. 7.

[42] See RDA 1949, s. 19(4).

[43] Note, that in the case of a Community trade mark, a transfer of the whole of an undertaking includes the transfer of the Community trade mark unless there is an agreement to the contrary, or the circumstances clearly state otherwise. See Council Regulation of December 20, 1993 on the Community trade mark (40/94) (the "CTM Regs.") — Art. 17(2). Note also that the licensing chapter does not contain information about Community trade marks.

[44] The representation of the mark could also be a copyright work, or in the case of shape marks, a design document recording design right.

employees, leading, potentially, to substantial liability.[45] In addition to the assignment of the mark, and in view of the assignment of goodwill, the assignment can contain a restrictive covenant prohibiting the assignor from using the mark. If the assignment is to include the right to take action in respect of past infringements, this should be stated specifically.

22–14   In the course of prosecuting applications for trade marks, it is often necessary to provide evidence of the previous use of the mark.[46] So if a trade mark application is to be assigned, it may be necessary to obtain such evidence at the time of the assignment, or to provide for assistance from the assignor after assignment. It is possible to make further applications in other jurisdictions, claiming priority from the original United Kingdom application. If the application/mark is assigned at this stage, it is common to assign the right to claim priority together with the mark. It is possible to assign part of a trade mark, relating to some of the goods or services for which it is registered, or in relation to use in a particular manner or in a particular locality.[47] Use by the other proprietor could, however, lead to the mark becoming generic, or becoming liable to mislead,[48] which would constitute grounds for revocation.[49] In the case of such an assignment it would be necessary to control the use of the mark in a similar manner to a licence. In respect of Community trade marks there are requirements relating to the nationality of the applicant or proprietor.[50]

22–15   In common with other registrable rights, there are distinct advantages to registration. Until an assignment is registered it is not effective against a person acquiring a conflicting interest in or under a registered trade mark in ignorance of it.[51] Furthermore, if an assignment is not registered within six months of its date, the assignee is not entitled to damages or an account of profits in respect of infringements occurring after the date of the assignment but before the date of registration.[52] This does not apply if the court is satisfied that it was not practicable for the registration to be made within that period, and that the application was made as soon as practicable afterwards. In the case of a Community trade mark, an assignee may not enforce the mark unless the transfer has been registered.[53]

Confidential information (or know-how) and passing off

22–16   Substantial value is ascribed to confidential information in the context of some transactions, and it is often classed together with intellectual property.

---

[45] See the Transfer of Undertakings (Protection of Employment) Regulations 1981 (as amended). Regrettably, employment law is outside the scope of this work.
[46] This is common in order to demonstrate that a mark, which does not appear to have sufficient inherent distinctiveness for registration, has acquired distinctiveness through use — see Trade Marks Act 1994, ss. 1(1) and 3(1)(b) and (c).
[47] See TMA 1994, s. 24(2).
[48] Note that on an assignment of a Community trade mark, if it is clear from the transfer documents that because of the transfer the mark is likely to mislead the public in respect of the nature, quality or geographical origin of the goods or services in respect of which it is registered, the Office may not register the transfer unless the registrant agrees to limit the registration so that it is not likely to mislead. See CTM Regs Art. 17(4)
[49] See TMA 1994, s. 46.
[50] See CTM Regs Art. 5. Note also that under Art. 2 of the Protocol Relating to the Madrid Agreement Concerning the International Registration of Marks there are also requirements relating to the nationality of the applicant or proprietor.
[51] See TMA 1994, s. 25(3)(a).
[52] See TMA 1994, s. 25(4).
[53] See CTM Regs Art. 17(6).

Regrettably, it does not have the same attributes as intellectual property; in fact it does not have the basic attribute of property.[54] "Title" to confidential information cannot be transferred. It is, however, possible to transmit the information, and for the person transmitting it to undertake not to disclose or to use it in future, and to erase all other records of the information in their possession.[55] The transferor can also provide additional assistance with patenting. If there has been a wrongful disclosure of the information prior to the transfer, the assignee could possibly encounter difficulties if he wished to take over any right of action in respect of this, as it is a personal action. Material that is classed as "know-how" is often the subject of copyright and design right, which can be assigned as described above. The right to take action for passing off does not exist independently of the goodwill of the business to which it relates, so it is not an individual right and cannot be assigned separately.

Further assurance

22–17   Further documents may need to be executed to perfect the assignee's title. They are most likely to be needed in the context of registered or registrable rights, especially if the rights extend to a number of territories, as documents may be required for each jurisdiction, to comply with differing legal obligations. It is common practice to sign an "umbrella" agreement to assign that covers the rights in all jurisdictions. This is usually followed by individual assignments in each territory.[56] In any event, it is common to register a simple form of assignment, which gives the basic details of the assignment, and does not disclose sensitive commercial details, unless it is a requirement in the relevant jurisdiction. In addition to these formalities, in the case of future patents and registered designs, a further assignment is often required when the invention or design has been made.[57] Also, if there is some impediment to the operation of the statutory provisions for future assignment, a further assignment of those rights may be needed.[58] In addition to the standard "further assurance" clause, a power of attorney can be included, which permits the assignee to sign documents on behalf of the assignor. This may not be effective in all jurisdictions.

Embodiments, other necessary property and licences

22–18   Embodiments of intellectual property can be expensive and time-consuming to produce. They may even embody additional know-how, which is not recorded in any other form, and which needs to be kept confidential. In the case of copyright and designs, the original drawings and models could be important evidence for the purpose of proving title, in the case of litigation. In some cases, the property may relate to biological resources, without which the property itself is of little value. In all such cases, it is important that the acquisition of the ancillary material be included in the

---

[54] See, however, *Buchanan v. Alba Diagnostics Ltd* [2000] R.P.C. 367, a Scottish case in which Lord Hamilton referred to the distinction between "assignable as an item of property and know-how in the form of skill, experience and learning".

[55] But see *Buchanan v. Alba Diaggnostics Ltd op. cit.*, in which Lord Hamilton stated that the enforceability of a restrictive covenant relating to know-how in the form of skill, experience and learning would be questionable.

[56] If the umbrella assignment is effective to transfer the property, it should be registered, as the subsequent assignment document is not then the true assignment, and the registration formalities are not completed by registering it — see *Coflexip Stena Offshore Limited's Patent* [1997] R.P. C. 179. Registration of the umbrella assignment entails public disclosure of the document. The only alternative in respect of U.K. rights is to have an umbrella agreement to assign followed by an individual assignment.

[57] See para. 22–02, above.

[58] See also para. 22–02, above.

assignment, together with the intangible property. The ancillary material can, however, be transferred by delivery in order to minimise stamp duty.[59] In the case of rights that relate to products that are the subject of regulatory approvals, the assignee could ask for an assignment of the approval as well, but this may not be possible. However, it may be possible for the assignor to provide the data, and any other information on which the original application was based, together with a copy of the previous application, and correspondence with the relevant authority.[60]

Consideration, and the form of the document

22–19   It is common to provide for at least a nominal consideration of £1. This enables the assignee to obtain specific performance of the agreement, and classifies the assignee as a purchaser for value. As a purchaser for value the assignee will, for example, take copyright and design right free of any licences of which the assignee does not have notice.[61] If the assignment contains a power of attorney the assignment will need to be by deed.

The payment structure can change the nature of the assignment. The differences between an exclusive licence and an assignment may be limited. If the consideration for the assignment is payable over a relatively long period of time, and the assignor has the ability to terminate the contract for non-payment, there could be little difference between the two. In these circumstances, the assignment could be treated as a licence. In particular, the Technology Transfer Block Exemption[62] refers to the possibility that "exclusive licences restrictive of competition" might be presented as assignments.[63] It categorises assignments with licences where "the risk associated with exploitation remains with the assignor". In particular, it mentions agreements where the consideration is dependent on the extent of the assignee's turnover or the quantity of products made[64] using the assigned technology.[65]

Warranties and liability

22–20   Assignees are likely to seek warranties that the property to be assigned subsists, that there is no reason why it should cease to do so prior to its expiration, and that the assignor is entitled to transfer the property. As a basic minimum, the assignment may be with "full title guarantee", or with "limited title guarantee",[66] unless there are special circumstances. The assignee could also seek comfort in respect of infringement of other rights by the exercise of the rights to be assigned. Some of the difficulties presented by such warranties are explained in Chapter 23.[67] There will be differences in this case, as the duty of a licensor to prosecute applications and to maintain granted property will not apply in these circumstances. In addition to the warranties described in Chapter 23, assignees could seek comfort in respect of any

---

[59] Stamp duty is still payable in respect of many assignments of assets other than intellectual property.
[60] In the case of a future application for a supplementary protection certificate, a copy of the marketing approval will be required, although this may be obtained by the granting body from the relevant authority — see *Biogen Inc. v. Smithkline Beecham Biologicals S.A.* [1997] R.P.C. 833.
[61] See para. 22–07 above.
[62] Commission Regulation (E.C.) No. 240/96 of January 31, 1996 on the application of Article 85(3) of the Treaty to certain categories of technology transfer agreements.
[63] See Chapters 16 and 17 for an explanation of E.C. and U.K. competition laws.
[64] There is an equivalent provision relating to the use of a process.
[65] See Recital 9 and Art. 6.
[66] See The Law of Property (Miscellaneous Provisions) Act 1994 for the details of the covenants implied by these expressions.
[67] These warranties may be heavily qualified — see paras 23–76 to 23–80.

licences or other rights, such as mortgages, to which the rights to be assigned may be subject.

The warranties may be accompanied by the customary exclusion clause, which limits the assignor's liability, and further defines the warranties. There can be financial and temporal limits to the claims under the warranties, and certain liabilities may be specifically excluded. For example, the assignor can expressly exclude liability in respect of the future prosecution of applications for registered rights, provided that the assignor has disclosed all relevant facts that the assignor is aware of.

Stamp duty

22–21   Stamp duty is no longer payable in respect of assignments of patents, registered designs, copyrights, design rights, trade marks, plant breeders' rights and corresponding and similar rights in other territories. This exception applies to assignments made after March 27, 2000.[68] Database right and confidential information are notable omissions from this list. Similarly, transfers of goodwill, which is usually assigned with a trade mark, are subject to stamp duty. Also, in respect of foreign rights, the extent to which they must correspond to or be similar to the United Kingdom rights listed above, in order to qualify for the exception, is uncertain. For example, there is no form of second tier protection for inventions (such as the utility model) in the United Kingdom. It is not clear whether this is a similar right to a patent. As foreign rights could be subject to a separate assignment, which could be kept outside the jurisdiction, this problem is unlikely to arise in many cases.

Governing law and dispute resolution

22–22   In view of the international nature of intellectual property, and the disparity between the provisions of different laws relating to the transfer of property, the governing law may have a substantial bearing on the terms of the contract. However, contractual provisions cannot modify the effects of national laws relating to subsistence of the property, or formalities for registration. Similarly, they cannot affect the rights of third parties, which may arise under national laws. Assignments are also subject to specific laws relating to assignments, and to any relevant competition laws that apply in the territories where the rights subsist. The choice of forum and method of dispute resolution are unlikely to be effective to prevent the application of foreign laws in such circumstances. As explained in Chapters 20 and 23, additional difficulties may be encountered if the forum for dispute resolution and the governing law are from different jurisdictions.

---

[68] See Finance Act 2000 s. 129 and Schedule 34.

LICENSING

Jennifer Pierce, Charles Russell

ABOUT THIS CHAPTER

23–01 In the course of reading this chapter it will become apparent that many different national laws, and branches of those laws, may affect a transaction. Regrettably, it is not possible to give details of all relevant laws in a chapter of this sort. The aim has therefore been to describe a licensing method, to cover aspects of relevant laws that may be particularly useful in practice, and to make the reader aware of some other laws that may unexpectedly affect a transaction. Further details of some of the specialist areas of law that are mentioned can be found elsewhere in this book. Amongst the omissions are general contract law, as it is readily available from other sources, and matters that are specific to software licensing, as there are many other books on the subject. All references are to the laws of England, unless the contrary is stated. Laws from other jurisdictions are mentioned occasionally in order to give an idea of the considerable jurisdictional differences in laws. The chapter is divided into three sections: the decision to licence: preparation: drawing up the licence.

THE DECISION TO LICENCE

Licensing is not necessarily the best option

23–02 Licensing is just one method of exploiting intellectual property and is by no means the best one in all circumstances. At its most basic level, a licence is a contract not to sue for infringement of the licensed rights. The salient feature of a licence is therefore a dependence on legal rights to generate income. In some instances, the owner of the rights would be better off manufacturing and distributing products, or having them made under contract and acting as a distributor. In others an outright sale of the rights could be the best option. In certain cases it may be appropriate to combine methods of exploitation. The decision to licence will often depend on a combination of legal and commercial factors. Some of the factors set out in the next section of this chapter are likely to influence the choice of exploitation method.

Commercial considerations — licensor

23–03 From the licensor's perspective, licensing can be extremely profitable. It forms the basis of substantial businesses, and can be used to diversify established ones. Such diversification can be particularly successful where a business's intellectual property and related rights are not being used in a way that fulfils their economic potential. The most common examples are where the rights owner is unable to meet the market demand for products made under the rights, and where a licensor is only manufacturing products in one potential field of use for an invention. In these circumstances, the licensee can exploit an untapped or under-exploited market for a developed product, and the overheads associated with the creation and maintenance of the intellectual property are spread over a second source of income.

Licensing can also be extremely unprofitable. It can appear attractive, if the initial costs are compared to those of establishing production facilities and a distribution chain. However, the potential income stream for a licensor in respect of a product is often a fraction of that earned by a manufacturer, and the long-term expenses can be considerable. Leaving aside the obvious costs of creating, maintaining and protecting the underlying property, licensors are often surprised by the costs of finding licensees, entering into agreements, and managing those agreements. Regrettably, the expense of enforcing the underlying property will often be greater for licensors who are not perceived as having sufficient financial strength to litigate. Potential licensees can be reluctant to pay royalties if they consider that the risk of being pursued for infringement is slim.

Commercial considerations — licensee

23–04   Similarly, licensing can be commercially attractive to the licensee. It can be the only means of acquiring access to vital technology. In a number of cases the licensee is already infringing the rights to be licensed, and may, therefore, have made a substantial investment in the technology. Sometimes licensing is used in a similar manner to franchising, in that the licensee is seeking to acquire proven technology for a new business or product line, and does not want the initial expense or risk of research and development. There can, however, be substantial drawbacks. These often relate to the suitability or state of development of the technology or to licence payments that are not commensurate with the commercial value of the rights acquired. The licensee who approaches licensing like franchising may be disappointed, as the degree of assistance given by many licensors is minimal.

Preparation — and further factors influencing the decision to licence

23–05   It is advisable for both licensors and licensees to undertake some due diligence[1] before entering into a licence, to ascertain the feasibility of licensing, and to assist in structuring and drafting the licence. Whilst this ought sensibly to precede the detailed licence negotiations, it often takes place contemporaneously.

In the majority of bipartite agreements, the concerns of each party will differ considerably, so the interests of licensors and licensees are treated separately in this section. Some items are relevant to both parties, and they are not repeated, so it is advisable to read both parts of the section in order to see the full picture.

Licensor

The rights to be licensed

23–06   Are the rights strong enough?   The first and most important factor to consider is the strength of the rights that are to be licensed. As a licence is essentially a contract not to pursue the licensee for infringement of rights in return for royalties, the licensor's ability to command payment often depends on his ability to sue. That, in turn, will depend on the scope of the monopoly provided by the rights, as well as the validity and enforceability of those rights. In many cases the strength of the rights will be uncertain, as the validity of patents and other rights for which there is a requirement of inventiveness or originality depends on the extent of the prior art, which will always remain

---

[1] The term "due diligence" is a generic term used to refer to the investigations that precede transactions relating to assets.

an unknown quantity[2] to some extent. In addition to the strength of the available rights, it is also worth considering the length of time until anticipated expiration. In any event the value of the licence is likely to depend on whether the rights to be licensed constitute a sufficient "threat" to somebody, whether it be real or imaginary.

23–07 Exclusive licensees have a different perspective The person who feels "threatened" need not necessarily be the licensee in question. If the licence is exclusive in some respect, the whole or part of the licensor's monopoly will be licensed. The licensee will therefore be in a similar position to the licensor as regards other potential users of the rights. In the United Kingdom, an exclusive licensee of any right under a patent, copyright, database right,[3] or design right,[4] can pursue infringers of those rights in much the same way as the owner of the right.[5] In these circumstances, the licensee may be more prepared to accept shortcomings in the rights in view of the potential advantage of the monopoly that is offered. However, if the rights are perceived to be weak, the licensee is more likely to seek to limit its obligation to pay royalties in the event that the property ceases to exist.

23–08 Possible advantages of registered rights In the context of inventions, the most suitable rights to licence are often registered ones. In the United Kingdom it can be difficult to rely on the legal protection afforded to confidential information in the long term, as information ceases to be protectable if it is not kept confidential.[6] Information can, theoretically, be kept confidential indefinitely, so protection could outlast the term of a patent. However, in a licensing context the likelihood of disclosure can increase considerably. Also, the protection given to confidential information, together with all other unregistered rights, is not a true monopoly and does not extend to works created independently.[7] Independent creation is common if there is further development work, as in the industrial context the rights may be specific to certain embodiments of the technology. In any event the licensor of such unregistered rights is more likely to need to continue development as confidential information is disclosed to outsiders or designs are updated. As far as unregistered "trade marks" are concerned, under

---

[2] In the U.K. an invention is not patentable unless it is "new" — see Patents Act 1977, ss. 1(a), 2 and Chapter 8 above. There is a similar, but not identical, requirement for registered designs — see Registered Designs Act 1949, ss. 1(2), 1(4). See also Directive 98/71 of the European Parliament and of the Council of October 13, 1998 on the Legal Protection of Designs (the "Design Directive"), Arts 3,4,6. This is included with the U.K. references on the basis that it is to form the basis of future U.K. legislation. For further information on designs see Chapter 12.
[3] The Copyright and Rights in Databases Regulations 1997 (S.I. 1997 No. 3032) (the "Database Regs"), create a new database right.
[4] This includes designs that are semiconductor topographies. See the Design Right (Semiconductor Topographies) Regulations 1989 (S.I. 1989 No. 1100) — ("Topographies Regs."), that modify certain of the provisions of the CDPA 1988, insofar as they relate to design right in semiconductor topographies.
[5] PA 1977 ss.30(7), 67, 130(1), Copyright, Designs and Patents Act 1988, ss. 92(1), 101, 225(1), 234, Database Regs, reg. 23. There is no equivalent provision for registered designs, although an exclusive licensee of a registered design may be able to be registered as a proprietor — see (1) *Isaac Oren* and (2) *Tiny Love v. Red Box Toy Factory Limited* [1999] F.S.R. 785. There are separate, and differing, definitions of "exclusive licence", depending on the right concerned. The situation may vary between jurisdictions. For example, under U.S. patent law only a licensee of substantially all the rights under a patent has the right to bring proceedings. There is a different regime for trade marks; a non-exclusive licensee of a trade mark may bring proceedings in certain circumstances, and a licence may provide for an exclusive licensee to have the same rights and remedies as an assignee — see Trade Marks Act 1994, ss. 29–31. Regrettably, it is not possible to deal with Community trade marks and Madrid Protocol filings in a chapter of this sort.
[6] See paras 10–05 and 10–06 for consideration of *Att.-Gen v. Guardian Newspapers Limited (Spycatcher)* [1988] 3 All E.R. 545 and *Mustad v. Dosen* [1963] R.P. C. 41.
[7] In the U.K., none of the prohibited acts relating to these rights affect the exercise of rights that are created independently. See CDPA 1988, Part I, Chapter II and ss. 226–228, Database Regs, reg. 16, Topographies Regs, reg. 8 for infringing acts.

English law nobody has a monopoly in a name or logo as such. They do not subsist independently of the goodwill of the business in which they are used,[8] so if they are to be licensed independently registration offers significant protection to the licensor.[9]

23–09 Practical example   Using the example of the bicycle on page 105, the design would be disclosed on the sale of products, so it could not be protected indefinitely as confidential information. The value of being the first in the market could, however, be recognised in accordance with the "springboard" doctrine if there were a breach of confidence before the product appeared on the market.[10] Copyright in the design drawings would not protect against copying in three-dimensional form,[11] and design right in the three dimensional version would only protect against substantial copying of the design.[12] Broadly speaking, neither copyright nor design right would protect the underlying concept as distinct from an embodiment of it.[13] A registered design would be preferable to design right, as it would protect against independent creation of the same design. Nevertheless, it would still only protect a particular design, as opposed to the general idea.[14] The best means of protection in these circumstances would be a patent to cover the basic concept, coupled, perhaps, with a registered design to cover features of the design that have "eye appeal".[15]

23–10 The route to market   In some industries there is a complex production and sales chain, involving a variety of specialist manufacturers and sellers of components. The price at which products change hands in that chain can reflect the costs of labour and materials, or the ultimate usefulness of the product at that stage. It may not reflect the value of the technology that underlies any of the components. If this is the case and the licensor is rewarded by royalties based on the selling price of a cheap product at an early stage in the sales chain, the licensor could lose substantial potential revenue. A careful examination of the sales chain in each likely territory, and between territories, should enable the licensor to structure the licence to take account of this. The licensor will aim to reduce the rights that are licensed to each link in the sales chain[16] and make the most of each right in each territory. The licensor may, however, have to modify the optimal strategy to take account of competition laws and laws relating to the free movement of goods.[17]

23–11 Is the protection broad enough in practice?   The licensor usually seeks broad protection. It should ideally cover the initial concept, improved versions, and, if possible and appropriate, enable licences to be granted in separate technical fields of

---

[8] Under TMA 1994, s.2(2), it is not possible to take action for infringement of an unregistered trade mark as such. The section states that this does not affect the law relating to passing off, so, effectively, the only available remedy is in passing off, and for that it is necessary to prove, amongst other things, that there is goodwill associated with the mark.

[9] See *Harrods Limited v. Harrodian School Limited* [1996] R.P.C. 697, and also see *Harrods Ltd v. Harrods (Buenos Aires) Ltd and Another* [1997] F.S.R. 420, and [1999] F.S.R. 187, CA. The latter relates to a good example of the potential pitfalls.

[10] See paragraphs 10–20 and 10–21 for an explanation of the "springboard" doctrine.

[11] CDPA 1988, s. 51 and Sched.1, s.19.

[12] CDPA 1988, ss. 226–228.

[13] But see paras 11–09 and 11–10 for the "idea/expression" dichotomy in copyright and note Database Regs, reg. 6, Topographies Regs, reg. 8

[14] It could, however, last for five years longer than a patent. PA 1977, s. 25, R.D.A. 1949, s. 8

[15] The classical example of the advantages of unregistered rights is the mythical "Coca-Cola formula" which is reputed to have been protected successfully as confidential information. If, indeed, it ever remained truly secret, it would have been less likely to do so if it had been the subject of a major licensing programme.

[16] See paras 23–56 to 23–59, which describe ways of reducing the rights granted.

[17] See paras 23–58 and 23–59 and Chapters 16–18.

use.[18] In view of the problems described in the paragraph above, it is advisable to consider the commercial value in the distribution chain of the acts that the rights would potentially prohibit. Under English law, it is not possible to take action for infringement[19] of intellectual property when all acts that would potentially infringe are the subject of an express or implied licence. For example, in respect of a patented article,[20] a licence to use, and to re-sell, at least in the United Kingdom, is usually implied on the first sale by the right owner or its licensee, unless the licence granted by sale is expressly limited.[21] Such limitation can be achieved by means of a label licence, as described below.[22] It is worth noting that repairs to products do not generally constitute infringement unless they involve remaking the product. However, the supply of consumables may not be treated as repair.[23] The extent of the implied licence to re-sell in the United Kingdom will need to be considered in the context of the relevant laws on the free movement of goods and competition, as explained below.

23–12 Practical example for paras 23–10 and 23–11 Consider the licensing problems that might be posed by an invention relating to a computer chip. Suppose that it is best manufactured by companies in a separate part of the electronics market from the manufacturers of the hardware with which it will operate. The price of each chip needs to be low, in view of the large numbers that each user will require. The price of chips is already too high, and the chip manufacturer will have to invest in expensive research and development to bring the price down. It is therefore unlikely that the chip manufacturer will pay a substantial royalty. Furthermore, the venture is particularly speculative as the hardware manufacturer has to modify the hardware to use the chip and neither the technical result nor a sufficient market share for the finished product is guaranteed. In these circumstances the licensor may consider developing the necessary hardware itself, creating further rights in the process. Those rights might then command a higher royalty.[24]

23–13 Extent of territorial protection The extent of protection afforded by intellectual property needs to be considered territory by territory, as each right in each territory is a separate piece of property. Each could, theoretically, be licensed separately. The licensor may consider creating registered rights in at least one territory in which each licensee is likely to perform at least one act, which would otherwise be prohibited by the rights, in the course of bringing each of its products to market. As there are usually a number of territories where products could potentially be manufactured, it is common for owners of rights to choose territories where the products are likely to be sold. If they are too numerous for it to be economical to acquire rights in each, it may

---

[18] See para. 23–59, below.
[19] See PA 1977, s. 60, CDPA 1988, Part I, Chapter II and ss. 226–228, Database Regs, reg. 16, Topographies Regs, reg. 8, RDA 1949, s.7, Design Directive, Arts 12,13, TMA 1994, ss. 10–12 for infringing acts.
[20] The extent of the implied licence may depend on the intellectual property right and on the trade customs in the particular industry by which it is exercised.
[21] See *Betts v. Willmott* (1871) L.R. 6 Ch. App. 239, *Solar Thomson Engineering Co. Ltd and Another v. Barton* [1977] R.P.C. 537, *Dellareed v. Delkim Developments* [1988] F.S.R.329, and *Canon Kabushi Kaisha v. Green Cartridge Co. (H.K.) Ltd* [1997] F.S.R. 827, and *United Wire v. Screen Repair Services Ltd* [2000] F.S.R. 204 and House of Lords judgment of July 2000. In *Canon* it was observed that repair which does not constitute manufacture does not infringe. In United Wire it was held that the scale of the "repair" amounted to manufacture and therefore infringed. See also, *British Leyland Motor Corp. v. Armstrong Patents Co.* [1986] R.P.C. 279 and the *Canon* case above — on the possible implied right to repair as a matter of public policy, due to the exercise of monopoly power in an aftermarket. See also para. 23–58.
[22] See para. 23–58.
[23] *See Canon Kabushi Kaisha v. Green Cartridge Co., op. cit.*
[24] The alternative is label licensing (see para. 23–58), which could be problematic in this case, as it may be difficult to trace the chips after they have left the chip manufacturer. The licensor may therefore know that the terms of the label licence are not being complied with, but may find it hard to find the culprits.

be worthwhile considering territories where there are permanent and/or especially important manufacturing facilities.[25]

23–14   Restrictions on sales of licensed products between territories can be desirable commercially, but may be contrary to laws on the free movement of goods[26] and to competition laws.[27] In particular, restrictions relating to the export or import of goods put on the market with the consent of the right owner within the European Economic Area are contrary to such laws.[28] That consent needs to be considered in conjunction with the extent of any implied consent (*i.e.* licence) to re-sell, which is referred to above.[29] The most recent English case in respect of consent to sales of products and the subsequent export of those products relates to trade marks[30] and is a High Court decision.[31] In that case, the sale of products under licence outside the EEA was held to constitute consent to marketing those products in the EEA. The decision potentially conflicts with the most recent E.C. authorities on the free movement of goods,[32] which establish the principle that no member state can impose international exhaustion.[33] The above English decision has been referred to the European Courts, so there should soon be an E.C. decision on the subject.

23–15   Additional rights   Ancillary rights may be useful to both licensor and licensee. It can be difficult to force a licensee to accept rights that he does not wish to use, unless there is a technical reason for doing so. In any event this could lead to infringement of E.C.[34] and English competition law (see below).[35] In practice, the licensee often seeks technical assistance, or insists on testing a reliable prototype and/or seeing test results, before he will enter into the licence contract. Indeed, there

---

[25] There may be drawbacks to relying on the protection given by process patents in the territories where the product will be sold but not manufactured, as the protection may not be as broad. In the case of imports of patented products to the U.K., where the right concerned is a patented process, it is necessary to be able to prove that products made by the process have been obtained directly by means of the process — see PA 1977, s.60(1)(c).

[26] See Articles 28 and 30 of the Treaty Establishing the European Community (the "E.C. Treaty") (formerly 30, and 36 of the Treaty of Rome) — and see paras 16–103 *et seq.* for details. See also CDPA 1988, ss. 18, 228, TMA 1994, s. 12, Database Regs, reg. 12(5), Topographies Regs, reg. 8(2) — see also Design Directive Art.15 and Directive 98/44 of the European Parliament and of the Council of July 6, 1998 on the legal protection of biotechnological inventions (the "Biotech Directive") Art. 10, which will also lead to legislation in the U.K.

[27] See Arts 81 and 82 E.C. Treaty, paras 23–23 to 23–25 and 23–59 below for brief summary, and Chapters 16 to 18 for more detail.

[28] See n. 26 above.

[29] See paras 23–11 and 23–12.

[30] The decision contains a statement that the same principles may not apply to rights other than trade marks.

[31] *Zino Davidoff SA v. A & G Imports Limited* [1999] R.P.C. 631.

[32] There is no E.C. authority on exhaustion of rights under intellectual property laws, as distinct from laws relating to the free movement of goods and competition laws.

[33] *Silhouette International Schmeid GmbH KG v. Hartlauer Handelsgesellschaft GmbH* [1998] F.S.R. 474 (Advocate General); [1999] F.S.R. 729, and *Sebago Inc. and Ancienne Maison Dubois et Fils SA v. GB-UNIC SA* [1999] 2 C.M.L.R. 1317. Note Advocate General Jacobs' statement in the second of those cases at p. 482 that "It should be assumed for present purposes that Silhouette did not consent to its products being re-sold within the EEA. That is so even though the national court expresses some doubt as to whether the restrictions upon re-sale were passed on to the purchaser". This should be contrasted with the cases on label licensing in the notes to para. 23–58 below.

[34] See the Block Exemption, Articles 1.3 and 1.4 (see references to "secret and substantial" know-how and "necessary patents"), 4.2(a), 5.1(4) and 10. See also Recitals 4,5, and 6. Compulsory use of a trade mark is a notable exception. This is permitted under Art. 2.1(11). See paras 23–23 to 23–25 below and paras 16–40 *et seq.*

[35] The Competition Act 1998 is modelled on E.C. competition law and references in this chapter to competition law should be read as references to English law where there may be an effect on competition within the U.K., unless otherwise stated, although the legislative references will differ. Note also that CA 1998, s. 10 provides for E.C. law exemptions to apply to English law prohibitions unless varied or cancelled by the Director General of Fair Trading. See also Chapter 17.

are appreciable numbers of licensees who would prefer to be supplied with technology on a turnkey basis. This may work to the advantage of a licensor who is particularly concerned with the quality of licensed products and their packaging. It can also assist the licensor in complying with any express warranty that the licensed technology will work.[36] However, in view of the liability that could potentially be incurred as a result of extending the transaction beyond intellectual property licensing, the possible benefits to the licensor need to be considered carefully.[37] Similarly, trade marks can be useful additional rights, as they can last indefinitely, but they, too, could lead to the licensor incurring additional liability.[38]

23–16   Are the rights valid?   If a licensor spends a great deal of time and money investigating the validity of his registered rights, the exercise could be counter-productive. If the licensor investigates the rights, and concludes that they are probably invalid, he may be in a difficult position. Under English law, if he executes a licence in these circumstances, and the property is subsequently found to be invalid, the licensee may take action to rescind the licence, and possibly claim damages, on the ground of the licensor's misrepresentations.[39] Also, if the licensor were to be forced to give an express warranty of validity to the best of his knowledge, the licensor would be in breach of that warranty. Furthermore, in subsequent court proceedings concerning the validity of the rights, written material created during such investigations might have to be disclosed, if it were not privileged, and could be very damaging.[40] Nonetheless, the licensor may look for obvious flaws in the property, which he might be expected to be aware of, such as records of prior disclosures or publications.[41]

23–17   There are types of property, which by their very nature can present difficulties to the licensor due to inherent problems of validity or subsistence. For example, United Kingdom design right (unregistered) does not subsist in certain features of designs, including "features of shape or configuration of an article which enable the article to be connected to, or placed in, around or against, another article so that either article may perform its function".[42] This exception can cause considerable confusion in practice, and in some instances it could potentially reduce the scope of the right to the point where it is of little commercial value. It may, however, be less of a problem with entire objects or features of or parts of entire objects, when assembled to make a larger article, as opposed to individual parts.[43] In addition, design right is not the subject of international treaties in the same way as other intellectual property rights, so there may be no reciprocal protection for these rights in other countries where a licensor might hope to exploit them.[44]

---

[36] The licensee may require such a warranty and in some cases it might be implied from the circumstances. In civil law jurisdictions it may be implied unless specifically excluded — see N. Byrne, *Licensing Technology*, (2nd ed.) p. 171.

[37] See paras 23–32 and 23–86, below.

[38] See para. 23–32 below.

[39] See *Lawes v. Purser* (1856) 6 El. & BL. 930, *McDougall Brothers v. Partington* (1890) R.P. C. 216, and *Ashworth v. Law* (1890) R.P. C. 231. These cases refer to possible repudiation on the grounds of fraud, and do not refer to any specific representations, but this is probably due to the age of the cases. It is most likely that these days the licensee would seek to rescind the contract and/or claim damages, based on misrepresentations and possibly breach of contract. It is not clear whether the act of entering into the licence, which must be on the basis that the licensor has property to licence, would, itself, be a representation that the licensor believed that he might have a valid patent to licence.

[40] See para. 24–66 — Disclosure.

[41] See n. 2 to para. 23–06 above and paras 8–49 *et seq.*

[42] CDPA 1988, s.213 (3)(b) and para. 12–11.

[43] *Ocular Sciences Ltd v. Aspect Vision Care Ltd & Ors.* [1997] Part II at 421, and *Baby Dan AS v. Brevi SRL* [1999] F.S.R. 377.

[44] There is greater reciprocal protection for topography rights.

23–18   Are the rights enforceable in practice?   Many people concentrate on the validity and scope of rights and forget to assess their enforceability, which is of equal importance. In particular, a valid patent may be of little use if, in practice, it will be almost impossible to obtain the necessary evidence to prove that it has been infringed. For example, a patent that covers an improvement to a manufacturing process may produce the same products as the pre-existing technology. It may not, therefore, be possible to detect infringement without inspecting the production line. In England it can be difficult to carry out this type of inspection by legitimate means,[45] unless there is some evidence of infringement. Whilst it may be possible to obtain an order for pre-action disclosure, or inspection, it is necessary to have sufficient evidence to demonstrate that there are likely to be proceedings.[46]

23–19   Does the licensor own the rights?   A warranty of the right to grant the licence could be inferred from the circumstances of the grant of the licence.[47] In addition, the prospective licensee can seek a warranty of title to the rights to be licensed.[48] Ascertaining the true ownership of intellectual property rights can be problematic. In the United Kingdom, title to rights other than trade marks is derived from the inventor,[49] author, designer, maker, or marketer (in the case of design right), even if he is not the first owner.[50] Documents of title are, therefore, based on assumptions regarding the identity and status of the inventor(s), author(s), designer(s), makers(s), or marketer(s) that may be incorrect.[51] In practice the investigation of title is likely to be relatively simple, as it is often the licensor who created the right, or employed the person who did so. If this is not the case, it could be difficult for the licensor to look beyond the documents available. It is also necessary to consider rights to ancillary matter to be provided to the licensee, such as cell lines and plants.[52]

23–20   Ways of dealing with problems   If the rights are found to be weak, the licensor has other options to proceeding with the current licensing strategy. He can attempt to strengthen the rights, perhaps by filing applications for registered property, or for more registered property, if that is still possible. Alternatively, he might consider

---

[45] See para. 24–116 — Search orders.

[46] See *Civil Procedure 2000* (The White Book) at 25.1.26, 25.1.27, 25.1.28, 25.1.29 initially.

[47] Under English law there are no general implied warranties applicable to licences of intellectual property, unless they can be inferred in a particular case — see *Mills v. Carson* (1893)10 R.P. C. 9. Note the cases where a freedom to use warranty was implied in n. 55 to 23–80 below.

[48] In England the Sale of Goods Act 1979 (SGA 1979) and the Supply of Goods and Services Act 1982 (SGSA 1982), which include provisions for implied warranties of title, are intended to cover the sale and hire of goods and the provision of services. Such warranties are separate from those described in n. 47 to 23–19. It does not seem likely that the SGA and the SGSA apply to intellectual property, as opposed to embodiments of that property and services relating to it — see *St. Albans City & D.C. v. International Computers Limited* [1997] F.S.R. 251, where it was held that software that was not provided on disc was not goods. However a warranty was implied that the software was to be reasonably fit for the purpose for which it was sold. The SGA 1979, the SGSA 1982, and the Unfair Contract Terms Act 1977 will apply to prototypes and other goods and services provided by the licensor. Readers should consult a work on contract law for further information on this subject. Implied terms are more likely to be found in civil law jurisdictions and the licensor may incur substantial liability if they are breached — see N. Byrne, *Licensing Technology* (2nd ed.) p. 166.

[49] See PA 1977, ss. 7 and 39.

[50] See CDPA 1988, ss. 11, 215, and 220, Database Regs, regs 14,15, Topographies Regs, reg. 5, and RDA 1949, s. 2

[51] See PA 1977, ss. 7–12,37,38,72,74 (note s.37(9)) CDPA 1988, ss. 11,215, 220, Database Regs, regs 14,15, Topographies Regs, reg. 5, RDA 1949, s.2. See Chapter 14 for an explanation of title to intellectual property.

[52] Laws relating to plants are outside the scope of this book. However the reader may find the following a useful starting point — Plant Varieties and Seeds Act 1964, Agricultural (Misc. Prov) Act 1968, Council Regulation 2100/94 of July 27, 1994 (as amended).

It is possible that in other jurisdictions there is legislation on the lines of provisions of the Convention on Biological Diversity 1992, in particular Arts 3,4,15,16,19, which could affect access to genetic resources.

modifying the intended method of exploitation. For example, as know-how loses much of its value if it is disclosed, it may still be reasonable to grant a licence, provided that a significant proportion of the payments are made, or fall due, on signature, or in the early part of the licence term. Otherwise, the licensor will have to rely on contractual remedies over a long period when the licensee may gain no further commercial advantage from the rights. Contractual remedies are less extensive and effective than remedies for infringement of intellectual property.[53] The licensor could, alternatively, decide on a different method of exploitation. He may also postpone licensing if better protection is more likely to be obtained after further development work has been carried out.

Other peoples' rights

23–21 The risks Intellectual property held by other people, which would be infringed by use of the licensed technology, can prevent exploitation of that technology. Licences do not generally contain an implied warranty by the licensor of non-infringement of such rights, although there is nothing to prevent a court from implying one. Licensors are therefore inclined to exclude such warranties.[54] The licensee may, however, require an express warranty that the licensor is not aware of any potential infringement. In the case of unregistered rights, some licensees will go further and seek an unqualified warranty[55] of non-infringement. Under English law, infringement of copyright and design right depends, broadly, on substantial copying of protected material, so the licensee may claim that the licensor should know whether he is infringing.[56] The licensor may be able to manage this risk by using rights that he has created himself, or that he has acquired with a warranty of non-infringement from a good source. It is possible to infringe registered rights unknowingly, so it is rare for unqualified warranties to be given in view of the unknown, and therefore unquantifiable, risks of infringement.

23–22 Searching for other peoples' registered rights Searching for patents owned by other people, that can affect commercialisation of rights, and assessing their effects, is difficult and can be very costly and time-consuming.[57] A licensor who is manufacturing a product, may already have conducted his own searches in the territories where the rights are to be licensed, in order to avoid liability in respect of his own manufacture and sale. Similarly, if he has used the technology openly in those territories over a long period without receiving claims for infringement, the risk can be reduced. In other cases the licensor will need to consider whether to search himself, and if so, to

---

[53] See Chapter 24 for remedies available in actions for infringement.

[54] Exclusion clauses are governed by the Unfair Contract Terms Act 1977, which may render them invalid. Any contract, so far as it relates to the creation or transfer or termination of a right or interest in any intellectual property and "technical or commercial information", is exempted from s. 2 (exclusion of liability arising from negligence), ss. 3 and 4 (liability under certain contracts with consumers). The exemption probably applies to licences , although it could be claimed that a licence was a right "under" such property, not "a right or interest in"it. Note that the exemption only covers a contract "so far as it relates"to the creation, transfer or termination of those rights, so it is of limited use. See Schedule 1. UCTA 1977. See also *The Salvage Association v. CAP Financial Services Limited* [1995] F.S.R. 654 and *St Albans City and District Council v International Computers Ltd* [1995] F.S.R. 686. However, the defendants in these cases did more than merely license intellectual property. The validity of exclusion clauses in general is outside the scope of this work. Readers should consult a work on contract law for further information.

[55] In this context it is a warranty which is not qualified by reference to the knowledge of the warrantor. See also paras 23–76 to 23–80, below.

[56] But see para. 11–38 — inference of copying from similarity.

[57] This is on the basis that the searcher does not know the identity of the inventor, or the proprietor, or any other details of the patents. See paras 9–65 to 9–69.

what extent. Unless the licensee is bound to make a specific product it can be difficult for a licensor to know what to search for. As with other investigations, these searches can create problems as a potential licensee may insist on seeing the results, and if there are adverse findings they may have to be disclosed in the course of future legal proceedings,[58] unless they remain privileged. Conversely, if the licensor is obliged to give warranties, such a search can be an invaluable means of reducing the licensor's risk.

Legal restrictions on licensing

23–23 Competition and similar laws Intellectual property rights, especially registered rights, can be used to establish a strong position in the marketplace, which may be abused, and may be used to distort competition. In many jurisdictions there is legislation to prevent rights owners from taking unjustified advantage of their rights.[59] Some of these laws are classified as competition laws, and some as intellectual property laws; the distinction can be artificial.[60] The licensing of intellectual property that subsists in a particular territory is likely to be subject to the competition laws of that territory, and of the territories in which it has anti-competitive effects. Contravention of these laws may lead to fines, to the whole or part of the licence being void, or to the underlying property becoming unenforceable during the period of the breach, and in some jurisdictions may include imprisonment.[61] It is generally possible to negotiate commercial terms which comply with competition laws, or which are acceptable to the relevant authorities. However, there are instances where there is a strong reluctance to modify the commercial terms to comply with competition laws. The licensor should therefore consider the potential effects of these laws at an early stage, owing to the severity of the potential sanctions for infringement.

23–24 Contravention of competition and similar laws Contravention of competition and other, similar, laws is most likely to occur when a right owner seeks to do something that will have effects beyond the scope of the rights granted. For example, provisions such as those that are on the E.C. Technology Transfer Block Exemption[62] "black"[63] and "grey"[64] lists are of a type that may infringe, whilst those on the "white list"[65] are exempted from the effects of the relevant prohibition. There are similar guidelines in other jurisdictions.[66] Note that agreements that are exempted under E.C. law are also exempted under English competition law, although there are powers to impose conditions or obligations on such exemptions, or to cancel them.[67] Also, the grant of exclusive patent licences that command substantial payments may be subject

---

[58] See paras 24–47 *et seq.*—Discovery. Furthermore, if the search discloses prior art in relation to a patent, and this has not been disclosed to the U.S. Patent and Trade Mark Office, disclosure may have to be made to that office if the patent is to remain valid — see para. 9–59 for an explanation of re-examination in the U.S.
[59] For U.K., E.C. and U.S. laws in general see Chapters 16–18.
[60] For example, PA 1977, ss. 48–54 (as amended), which relate to compulsory licensing, and powers exercisable by the Competition Commission, and Art. 82 E.C. Treaty (formerly Art. 86 of the Treaty of Rome) together with CA s.18 which may be used to bring about compulsory licensing. Similarly, in the U.S. both antitrust and patent misuse laws affect such arrangements.
[61] See para. 18–07.
[62] Commission Regulation (E.C.) No. 240/96 of January 31, 1996 on the application of Article 81 (3) (formerly 85(3)) of the Treaty to certain categories of technology transfer agreements.
[63] See Art.3 for list of provisions that are outside the exemption.
[64] See Art. 4 for provisions that are subject to the opposition procedure under which provisions may be notified and exempted unless the Commission notifies the contrary — see para. 16–44.
[65] See Arts 1 and 2 for details of the exempted provisions.
[66] See also Chapter 18 — U.S. Department of Justice and Federal Trade Commission Guidelines. Unfortunately it has not been possible to refer to specific guidelines in this chapter.
[67] See the CA 1998, s. 10.

to merger control legislation,[68] as is the case in the United States.[69] If a potentially problematic provision is commercially important, it may be best to sound out the relevant authorities in good time before signature, and, if possible and appropriate, obtain the necessary clearances.[70]

23–25   Is a right owner obliged to grant licences under competition law?   Under European Community competition law, a licensor is entitled to refuse to licence its rights.[71] This does not constitute an abuse of a dominant position under Article 82 (formerly 86). Similarly, there can be no contravention of Article 81 (formerly 85), which applies to agreements or arrangements which may affect trade between member states, and which restrict or distort competition between states in the European Economic Area, if the refusal is the result of a truly unilateral decision.[72] This should also be the case under the English Competition Act 1998,[73] as that Act provides for consistency of interpretation with E.C. competition laws as far as possible, having regard to any relevant differences. The same is true in the United States.[74] However, under European Community and English laws, there is a fine dividing line between a refusal to licence intellectual property rights and using those rights to assist the licensor in doing something which would infringe Article 82.[75] Again, there are similar laws in the United States.[76]

23–26   Is a right owner obliged to grant licences under patent law?   In most cases English patent law does not interfere with the licensor's prerogative not to grant a licence of a United Kingdom patent. United Kingdom patent law does, however, provide for compulsory licensing of patents in certain limited circumstances.[77] Very broadly speaking,[78] proprietors from World Trade Organisation countries[79] may be obliged[80] to grant licences where:

(i)   demand in the United Kingdom[81] for a product that is the subject of a patent is not being met on reasonable terms;

(ii)   due to the refusal by the patent owner to grant a licence or licences on reasonable terms, (a) the exploitation in the United Kingdom of any other patented invention which involves an important technical advance of considerable economic significance, in relation to the patented invention, is hin-

---

[68] This is not currently the case under English or E.C. law, although the acquisition of a patent or a patent licence may assist in creating or consolidating a dominant position, which could then be abused. See para. 23–25, below and see para. 16–79 and Chapter 17.

[69] See para. 18–16.

[70] See Chapters 16 and 17 but note the potential drawbacks of notification– see para. 16–93.

[71] See *Volvo v. Eric Veng (U.K.) Ltd* [1989] 4 C.M.L.R. 122, relating to a registered design.

[72] See, however, para. 16–04, as the result may be different if there are additional parties involved.

[73] See Competition Act, 1998, s. 60.

[74] See para. 18–08.

[75] See Chapter 16 part C for *RTE v. Commission (Magill)* [1995] E.C.R. I-743. See also *Phillips v. Ingman* [1999] F.S.R. 112.

[76] See Chapter 18, section B.

[77] PA 1977, ss. 48 to 54 (as amended).

[78] It is imperative to consider the detailed amendments made to the PA 1977, by S.I.1999 No.1899, the Patents and Trade Marks (World Trade Organisation) Regulations 1999.

[79] Separate provisions apply to other proprietors.

[80] Note that these provisions do not apply to inventions in the field of semiconductor technology.

[81] These provisions have been amended in the wake of *Re Compulsory Patent Licences: E.C.C. v. U.K. and Italy* [1993] R.P.C. 283. It is apparent from that case that there is a fine line between national provisions that have been considered to infringe E.C. laws relating to the free movement of goods, and those that have not been viewed as doing so.

dered; or (b) the establishment or development of commercial or industrial activities in the United Kingdom is unfairly prejudiced;

(iii) due to conditions imposed on the grant of patent licences, or to conditions relating to the disposal or use of the patented product (or on the use of a patented process), dealings with materials that are not protected by the patent, or the establishment or development of commercial or industrial activities in the United Kingdom are unfairly prejudiced.

Patentees may also be subject to compulsory licensing as a result of a report of the Competition Commission.[82] The majority of these provisions are not applicable for three years from the date of grant of a patent.[83] In practice the compulsory licensing provisions are used rarely. In addition to the current compulsory licensing regime, there is statutory provision for Crown use.[84] There are also compulsory licences of registered designs,[85] and schemes for licensing other rights, except for rights in semiconductor topographies.[86]

23–27 Dual Use Technology In many jurisdictions, including England, there are restrictions on exporting technology that can be used for military as well as civilian purposes. In the United Kingdom, all patent and registered design applications may be screened on behalf of the Ministry of Defence to ascertain whether they should be kept secret in view of potential military applications.[87] Technology that is the subject of a patent or registered design, which has been examined in this way, and which is not subject to secrecy order, should, in theory, be less likely to contravene these export restrictions. However, in view of the potentially disastrous commercial consequences of a secrecy order, great care is usually taken to avoid detection, so it would be unwise to rely on the MOD inspection. Other jurisdictions have similar export restrictions, which can be particularly stringent, and may include provisions relating to the re-export of goods originating from their country. Detailed consideration of this topic is outside the scope of this work, but there are some references to relevant material in the footnotes.[88]

Impact of different national laws

23–28 One of the factors that increases the expense of licensing is the extent to which foreign legal advice is required. The subsistence of licensed property in other jurisdictions is subject to the intellectual property laws of those jurisdictions. Similarly, licensees in other jurisdictions are subject to the laws of those jurisdictions, including those that relate to licensing. Some of those laws are substantially different from English law. Developing countries may consider that the effect of obligations to protect intellectual property under TRIPS[89] will be further costs that they can ill-afford. They may also fear that foreign licensors will not give them access to the latest

[82] See PA 1977, s. 51.
[83] PA 1977, s. 51 is an exception.
[84] See PA 1977, ss. 55–59 and paras 8–109 *et seq.*
[85] See RDA 1949, s. 10.
[86] CDPA 1988, Part I, Chapter VII, ss. 237,238, Sched. 1, para. 19, Database Regs, reg. 24, Schedule 2, Topographies Regs, reg. 9. The Biotech Directive will lead to additional compulsory licensing and cross licensing relating to plant varieties — see Art. 12.
[87] PA 1977, s. 22, RDA 1949, s. 5.
[88] Import, Export and Customs Powers (Defence) Act 1939, E.C. Council Regulations on the Export of Dual-Use Goods (implemented in the U.K. by the Dual-Use Items (Export Control) Regulations 2000 (S.I. 2000 No. 2620), Export of Goods (Control) Order 1994 (S.I. 1994 No. 1191) as amended, Chemical Weapons Act 1996, Official Secrets Act, *White Paper on Strategic Export Controls* 1998.
[89] Agreement on Trade-related Aspects of Intellectual Property Rights.

inventions and that they will receive outdated technology in return for inflated licence payments. In response to this, certain nations have imposed technology licensing regulations, which limit licensors' ability to licence on normal commercial terms. This can, in practice, modify the intended effect of TRIPS.

23–29   Some local laws go as far as prescribing the form of the licence agreement. Others simply provide for the agreement to be subject to local laws. The effect of such provisions should not be underestimated. They may substantially reduce the potential income from licensing, to the extent that it is an unprofitable method of exploitation. They can also, unwittingly, promote the licensing practices that they are supposed to prevent. It may not be possible to give a licensee the latest technology on the prescribed terms, if those terms are better than the terms offered to licensees from other territories. For example, under certain national terms and conditions, licence payments made by local entities are limited to three per cent of net sales, the licence may be of limited duration, and after termination the licensee can continue using the technology.[90] Other, more general, foreign laws such as exchange control and withholding taxes may also have a substantial impact on licensors.

23–30   Possibility of negotiating terms   It is advisable to consider the possible impact of local laws at an early stage while there is still time to approach the trade mission of the licensee's country to negotiate terms. There can be no guarantee of success, but it is worth trying. The authorities in question may be aware of the potentially adverse effects of their regulations and might therefore be prepared to be flexible in suitable cases. In addition to these negotiations, it may also be necessary to obtain formal approval or clearance of the contract after signature. If guidance has already been obtained from the relevant authorities, this could be little more than a formality. Failure to obtain approval may, for example, lead to the agreement being unenforceable, or to fines, so they should be taken seriously, especially if know-how is to be transferred at the outset.

Regulatory requirements

23–31   Regulations relating to the products that will be made under the licence can affect the licence terms. Regulation of this type is not uncommon; for example, sticking plaster is classed as a medical device.[91] Regulations such as those relating to medical devices and medicines contain requirements to be met prior to marketing, including product approvals. In some cases this could hinder or delay a product launch, and in the case of pharmaceutical products it is likely to do so. Consequently, it may be necessary to structure the agreement to accommodate the lack of royalty income prior to product approval, and to provide incentives to obtain that approval. Furthermore, the licensor may wish to co-ordinate the regulatory applications being made in respect of different territories. He may also wish to gain access to the licensee's test results and other regulatory material in case it is necessary to assist another licensee to obtain approval. Regulations can also affect any prototypes or samples that are to be supplied to the licensee. For example, a prototype may work on a radio frequency that the licensee is not permitted to use in the normal way.[92]

---

[90] This example is from the Malaysian regulations. It would, however, be unfair to single out Malaysia.
[91] See Council Directive 93/42 of June 14, 1993 concerning medical devices, implemented in England by the Medical Devices Regulations 1994 (S.I. 1994 No. 3017).
[92] See Wireless Telegraphy Act 1949, ss. 10,11, and regulations made under s. 10.

Liability

23–32 The liability of the licensor will depend on the licensed property, the ancillary assistance and products provided, and the territories where this is taking place. It is thought that under English law the liability of a licensor of intellectual property, other than trademarks,[93] is confined to the liability that he assumes expressly in the contract, on the basis that such liability would not be implied. However, there is nothing to prevent a court from implying such liability in a contract, if it could be inferred in the circumstances. Similarly, it might be possible for a court to find a licensor liable in tort, depending on the facts. For example, there is a fine dividing line between licensing technical information, which may be treated as intellectual property for these purposes, and providing consultancy services, in respect of which there is an established duty of care. In view of the lack of case law in this area, and the resultant uncertainty, it is common to exclude liability specifically. Certain provisions of the legislation that governs unfair contract terms do not apply to such exclusions insofar as they relate to the creation, transfer or termination of rights in intellectual property.[94] Exclusion clauses relating to such contracts are therefore likely to be more effective.[95]

In contrast, if the licensor provides trade marks, know-how,[96] consultancy services, prototypes or samples, or otherwise assists in or directs the licensee's production process, the licensor risks incurring liability in contract and tort together with liability under various statutes.[97] The nature of the liability will depend, in each case, on the nature of the property, goods and services provided. The licensor can also attempt to limit this liability, although such limitations may not be as effective as those that relate to the licensing of intellectual property.[98] In other jurisdictions the situation can be very different.[99] Licensors need to weigh the commercial advantage of providing ancillary rights, goods and services against the potential liabilities that they could incur if they do so.

Confidentiality undertakings

23–33 If confidential information is to be disclosed in the course of licence negotiations it is wise to enter into a written confidentiality undertaking at the outset. If no undertaking is signed, the basic rights and obligations of the parties will be uncertain. An undertaking gives greater legal certainty, and it is likely that it will be easier to enforce it. As discussed in Chapter 10, an obligation of confidence is the principal means of protecting confidential information. It is especially important if technical information is to be disclosed prior to patent filing. If an invention is disclosed in

---

[93] Under the Consumer Protection Act 1987, an "own brander" may incur liability in respect of the branded goods, on the basis that they are holding themselves out as the manufacturer — see s. 2(2)(b). The licensor may attempt to avoid liability by requiring the licensee to state that the product has been made under licence by the licensee. A person who presents themselves as the manufacturer may also be treated as the producer of a product for the purpose of the General Product Safety Regulations 1994.

[94] See n. 54 to para. 23–21 above.

[95] See para. 23–86 below.

[96] Know-how may be classed as "technical information", and therefore as intellectual property, for the purpose of the U.C.T.A. exemptions, so it may be less of a problem. Whilst it is included in the list of intellectual property in the exemption, it is not strictly intellectual property.

[97] This includes liability under terms implied by SGA, SGSA and other legislation relating to the sale of products. In the case of tortious liability, it is most likely to be in negligence, although liability for other torts, such as nuisance, cannot be ruled out.

If the licensor is assisting with or directing production, other liabilities may be incurred, but the nature of these liabilities is beyond the scope of this book.

[98] See n. 54 to para. 23–21 above.

[99] For example, in the United States the licensor of a trade mark may be held jointly and severally liable with the licensee for liability caused by the licensee's defective products to which the trade mark is affixed.

breach of confidence, the invention does not cease to be new, and therefore patentable, if a patent application is filed within six months of the date of the disclosure.[1] However, if a significant number of disclosures are made in confidence, the invention could, nevertheless, cease to be confidential.[2] There are similar, but not identical, provisions relating to registered designs.[3]

Finding out about the licensee

23–34 If the licensee is to be required to make long term payments, or to make a substantial investment in the technology to be licensed, the licensor often investigates the licensee. Such an investigation is likely to include a company search, with particular emphasis on the report and accounts. The extent of any further investigation will depend on the perceived risk to the licensor. If the licensee wants to postpone or reduce payments during the initial period of the licence, the licensor may wish to inspect the licensee's development and marketing plans to see if they are realistic. In these circumstances, the licensee is effectively acquiring part of the life of the licensor's rights in return for assurances of future business performance. The licensor is therefore in an analagous position to an investor, and may also wish to see the licensee's financial and business plans, although the licensor could find it difficult to obtain them.

Approaching the licensee

23–35 Licensing is often an alternative to litigation. Some licensors are therefore tempted to begin their negotiations with a letter threatening legal proceedings. The licensor should, however, be aware that this might lead to the potential licensee taking retaliatory action. Any person aggrieved by unjustified threats of infringement proceedings in relation to a patent, design right, registered design or trade mark may take action against the person making those threats, provided that the infringer is doing something other than manufacturing[4] or importing. In some instances, the threats will be justified, and will not therefore be actionable, but in others the licensor will run an unacceptably high risk of proceedings in which damages could be claimed, and where the validity of patents and registered designs might be put in issue. It is, however, permitted to draw attention to the existence of intellectual property rights without risking threats proceedings.[5]

Also, if the licensee is to be permitted to "use", or continue "using" the technology during the licence negotiations, the licensor will need to establish the basis for this "use" at the outset. This will avoid any implication that the licensee will be entitled to continue to do so if negotiations are broken off.[6]

---

[1] PA 1977, s. 2.
[2] See *Re Dalrymple's Appln.* [1957] R.P.C. 449 — where over 1,079 disclosures were made.
[3] RDA 1949, s. 6 — see also Design Directive Art.6.
[4] Or using a process, in the case of a patent.
[5] PA 1977, ss. 70, 74, CDPA 1988, s. 253, RDA 1949 s. 26, TMA 1994 s. 21. There is no equivalent provision for copyright, but a declaration of non-infringement may be available — see *Leco Instruments (U.K.) Ltd v. Land Pyrometers Ltd* [1982] R.P.C. 133.
[6] In *Tweedale v. Howard and Bullough Ltd* (1896) 13 R.P.C. 522 it was held that the circumstances constituted a licence, although the parties acted as if there were an agreement in place after negotiations had ceased.

Licensee

Investigating the licensor's rights

23–36 General The licensee's investigation of the rights to be licensed will often be more thorough than the licensor's. The extent of the licensee's inquiries is likely to depend on the proposed payment for the licence, and the licensee's investment in further research and development, and in manufacturing plant. In each case, the licensee will wish to assess the ownership, strength, scope and enforceability of the rights.[7] The licensee will balance the cost of a licence against the risks of being sued for infringement if he does not take one. It is not common for licensees to make inquiries of licensors on the scale of the due diligence questionnaires that are used in corporate finance transactions and company acquisitions. However, in view of some of the potential problems that can occur, the licensee needs to ask sufficient questions to satisfy himself that he has adequately protected his investment in the technology.

23–37 Scope of investigation of registered rights The extent of the due diligence required to protect a licensee of United Kingdom registered rights[8] from acquiring those rights subject to undisclosed interests is uncertain.[9] In the United Kingdom, the effect of registration of patents and registered designs is set out in the relevant statutes.[10] If a transaction relating to a patent[11] is registered,[12] the registered right will take precedence over prior unregistered rights, provided that the person registering did not have prior notice of those rights.[13] In the case of registered designs, the registered owner can deal with the right as he wishes, subject to any rights of which notice is entered on the register.[14] In addition, patents are personal property[15] and registered designs are referred to as being personal property.[16] As such, a person with actual notice of a valid right takes the property subject to that right. It is also possible that a purchaser or licensee of most[17] intellectual property will be bound by rights if he has constructive notice of them. In this context constructive notice is likely to mean notice

---

[7] See paras 23–06 to 23–19, above.

[8] In the U.K., copyright is not registrable, but it is registrable in certain other jurisdictions, most notably in the U.S. (where registration is optional, but may serve as a predicate for certain legal claims).

[9] There is a distinction between third party rights that affect the title to the property to be licensed, and rights held by third parties that are not to be licensed, but which may affect the licensee's exploitation of the licensed rights. This para relates to the former.

[10] PA 1977, s. 33 and RDA 1949 s. 19.

[11] See PA 1977, s. 33 for list of registrable transactions which include licences, assignments and mortgages. In the case of applications, rights are notified to the comptroller.

[12] It is sufficient for these purposes that an application for registration has been submitted to the Patent Office — see PA 1977, s. 33(4). There is therefore a period during which a transaction may be treated as registered, but it is not possible to obtain details through a search. Shortly before going to press, the Patent Office had a Chartermark target of four weeks for the registration of assignments.

[13] PA 1977, s. 33(1). The position with trade marks is similar — see TMA 1994, s. 25(3).

[14] RDA 1949 s. 19(4).

[15] PA 1977, s. 30(1). Trade marks are also personal property — see TMA 1994, s. 22.

[16] RDA 1949, s. 19(4)

[17] The situation is less clear in the case of patents. See *New Ixion Tyre and Cycle Company Ltd v. Spilsbury and Others* (1898) R.P.C. 567 at 572, *per* Chitty, L.J. "This is not constructive notice but actual notice which is required where there is a case turning on the statutory right to register. It has been decided under the Statute of Anne. It was actual notice." This would not necessarily be the case today, in view of the position with copyright (see below), although actual notice has been required in previous cases relating to other personal property, in which courts have prevented action that would prejudice equitable rights — see *Swiss Bank v. Lloyds Bank* [1979] 1Ch 581 and *Law Debenture Corpn. v. Ural Caspian Ltd* [1993] 1 W.L.R. 138.
    Note that CDPA 1988, ss. 90(4) and 222(4) refer to constructive notice in the context of purchasers of copyright and design right.

of something which the purchaser should reasonably have known or had cause to suspect.[18]

23–38 Investigation in practice In practice, the meaning of constructive notice in the context of registered intellectual property, and the implications for a careful licensee, are less clear. For example, would a licensee be expected to enquire whether the licensor has entered into any other agreements that might affect the exploitation of the technology? Hopefully, it would not be necessary to do so. If, however, the licensor disclosed that another licensee had been granted rights that might conflict with the rights to be granted, the licensee would probably be deemed to have notice of them. Clearly, it is not in the interests of a licensee to ask too many questions. If he would not be expected to make such enquiries, he may be better off not doing so. The process can be time–consuming and the licensee would be bound by any rights that he might discover. However, a balance needs to be struck in view of the potential risks.

23–39 Inspecting the registers The licensee should, ideally, obtain copies of all entries on the relevant registers. If the copies are obtained from the licensor, this will save search fees at the outset, and in the case of registered designs the licensor will have information that cannot be obtained by searching.[19] In the United Kingdom it is a criminal offence to falsify a copy document of title to a United Kingdom registered intellectual property right,[20] so it is possible to use the licensor's United Kingdom copy entries, at least at the outset. Inspection of the register still gives an added degree of comfort, as it protects against the licensor's fraud or failure to produce a complete record. There is, however, a delay in recording transactions, and those that have been filed in respect of patents, but which have yet to be entered, will still be binding.[21] The licensee should therefore consider undertaking a further search immediately before signature. If a licensee's budget will stretch that far, searches should ideally be undertaken in all jurisdictions where the rights are held. If this is too expensive, a compromise could be to search in the jurisdictions where the licensee's greatest source of income is likely to be. It is unwise to rely solely on the register without knowledge of the local law.[22]

23–40 What a licensee of registered rights looks for The licensee will consider the same matters as an infringer who is assessing the merits of his case. He will consider whether the statutory requirements for validity[23] of the property appear to be met, whether the licensor owns the rights, or has the right to grant the licence,[24] and whether the licensee's proposed dealings in products would infringe the licensor's

---

[18] See also *Macmillan Inc. v. Bishopsgate Investment Trust plc (No.3)* [1996] 1 W.L.R. 387, CA. This case also illustrates the potential jurisdictional differences in laws relating to notice.
[19] In the U.K., it is only possible to search for granted rights, and not for applications. Furthermore, the material provided to searchers is limited to a representation or specimen of the design, certain supporting evidence (but not all), and it is necessary to search specifically for the articles to which the design is applied. There are more stringent rules relating to certain types of registration — see RDA 1949, ss. 22, 23.
[20] PA 1977, s. 109, RDA 1949 s. 34, TMA 1994 s. 94.
[21] PA 1977, s. 33(4). The position is the same for trade marks — see TMA 1994 s. 25(3). There is no equivalent provision for designs. Shortly before going to press, enquiries of the Patent Office revealed a Chartermark target of four weeks for assignments.
[22] For example, certain transactions, such as charges, may not be registerable in some jurisdictions.
[23] See Chapters 8,12, and 13 and PA 1977, ss. 1–4,72,73,74,77, RDA 1949, s.1,11, Design Directive Arts 3,4,5,6,7,8,9,11, TMA 1994 ss. 3,5,46,47.
[24] This may be a licensee, with a right to grant sub-licences. In the case of a sub-licence, the head licence needs to be examined to see the basis on which it is terminable, the scope of the rights granted, and any other relevant conditions.

rights.[25] The licensee will only want to take a licence if he thinks that there is a possibility of being sued for infringement. In many cases there are potential flaws in the property or there is some doubt as to whether the licensee's products will infringe. In these cases the licensee's willingness to enter into a licence may depend on the payments proposed by the licensor. If the royalty is low, it could be cheaper to take a licence than to litigate, or to risk litigation. The licensee should also consider whether the register discloses any evidence of other people's rights that the licensee could infringe. There could, for example, be other relevant patents cited in searches made in the course of patent prosecution.[26] Unfortunately, this material is no substitute for a "freedom to use" search.

23–41 "Freedom to use" searches[27] It can be difficult for the licensee to know whether his proposed products will infringe rights other than those of the licensor, unless a search has been conducted or the licensor or another licensee has manufactured and sold the same products in the same territories without difficulty. Even then there is no certainty. In some instances the likelihood of infringement will be remote, especially in small marketplaces where all manufacturers and producers have a reasonable idea of their competitors' activities. There are, however, many cases where the licensee is best advised to commission a formal search for other people's registered rights, particularly if he is to invest heavily in new plant. Searches can be expensive, and time-consuming and it is never possible to guarantee the results. The extent of the search will depend on the time and money available for searching and the likely benefit to the licensee. In certain jurisdictions it is necessary to commission a search in order to avoid the payment of greatly increased damages in the event of infringement.

23–42 Problems with title — registered rights Registries are generally unable to check the completeness and authenticity of the documents that are filed. Entries in the United Kingdom register only provide *prima facie* evidence[28] of the matters recorded, and can be rectified. The person who is truly entitled to the grant of a patent or application can apply for it to be put in his name, revoked or declared invalid, and in the case of an application may apply for it to be amended or refused. If a patent application is refused or a patent revoked or declared invalid, the applicant can also apply for leave to file his own application in its place. None of the above applications may be made to the Patent Office more than two years from the grant of the patent, unless the right owner knew that he was not entitled to the patent.[29] It is also possible for a person to register a transaction relating to a patent[30] knowing that a prior, and conflicting, transaction has not been registered, in which case the person with the prior right to register can successfully claim rights to the property.[31] A person who loses his licence as a result of a one of the above applications may, in certain circumstances, apply for the transfer of his licence, or for a non-exclusive licence from the new

---

[25] See n. 19 to para. 23–11, above.

[26] PA 1977, s.17.

[27] "Freedom to use" searches are different from the searches referred to in para. 23–37, above. See paras 9–65 to 9–69.

[28] PA 1977, s. 32(9) as amended by the Patents, Designs and Marks Act 1986, s. 1, Sched. 1, para. 4, RDA 1949, s. 17(8) (as similarly amended), TMA 1994, s. 72. *Prima facie* evidence is evidence that constitutes sufficient proof of a matter unless and until more reliable evidence is provided.

[29] PA 1977, ss. 8–12,37,38,72,74. It is possible that a Court may be prepared to grant relief after this period, but the exact position is uncertain, as the Patents Act has some unclear wording.

[30] Similar provisions apply in respect of trade marks — see TMA 1994, s. 25(3).

[31] See PA 1977, ss. 33.

owner.[32] Otherwise, the licensee's only potential recourse is to sue the licensor for breach of contract. The register of designs can also be rectified.[33]

23–43 Investigating unregistered rights The licensee should require details of all material unregistered rights to be licensed, to check that they subsist. When checking that a work qualifies for copyright protection,[34] and its ownership, in the United Kingdom,[35] the licensee will require details of the subject matter, and the author or place of first publication.[36] In the case of United Kingdom design right, the licensee will require details of the design and the commissioner or, if there is no commissioner, the employer of the designer. If there is no commissioner or employer, he will want details of the designer himself.[37] Designs that would not otherwise qualify, can still qualify for protection by reference to the place of first marketing,[38] in which case the person who undertakes that marketing would be the first owner. In the case of database right, the licensee will seek details of the maker.[39] Additionally, the licensee could ask to see all documents that trace the chain of title back to the first owner.[40] The licensor may, however, regard this as an imposition. Whilst the author, designer or maker may not be the first owner, and not, therefore, entitled to dispose of the property, the chain of title will depend on him and the contracts that he entered into. As with registered rights, the extent of the requisite due diligence in relation to unregistered rights is uncertain.

23–44 Problems with the licensor's title — unregistered rights Similar problems to those affecting registered rights arise with unregistered rights, although the lack of registration results in an added degree of uncertainty. There is no reliable way of giving notice of ownership and other interests in copyright, database right, or design right, nor are there independent means of verifying documents of title to those rights in the United Kingdom.[41] Furthermore, a licence relates to intellectual property, but the licensee only has a contract with the licensor, and cannot rely on anything else. If the licensor were to transfer his rights, without telling the purchaser about the licence, and the purchaser had no notice (actual or constructive) of the licence, the purchaser would not have to honour the licence.[42] In these circumstances the licensee would need a new licence from the purchaser and could only take action against the old licensor for breach of contract. The licensee should therefore assure himself that the licensor is sufficiently reputable, and could decide to include a provision to withhold some payments until the end of the period of the licence to reduce the risk.

23–45 Typical problems with all rights Defects in title are not uncommon, and are becoming more prevalent as researchers seek funds from more than one source. In

---

[32] PA 1977, ss. 11,38.
[33] RDA 1949, s. 20. Note also RDA 1949, s. 19(4), which refers to equities being enforced in the same manner as with other personal property.
[34] See paras 11–30 *et seq.* above and CDPA 1988, s. 1(3), Part I, Chapters IX,X, Database Regs, reg. 18. Note that in the case of copyright created prior to August 1, 1989, when the CDPA 1988 entered into force, the transitional provisions and the previous legislation need to be considered.
[35] Subsistence of foreign rights will depend on the laws of the relevant jurisdiction — but see CDPA 1988, s. 160, under which copyright originating from a country that fails to give adequate protection to British works may not be recognised.
[36] In the case of copyright but not database right — see Database Regs. reg. 18.
[37] See CDPA 1988, ss. 213(5),214,215, 217–221, Topographies Regs, regs. 4,5 and paras 14–27 to 14–30.
[38] See CDPA 1988, ss. 215 and 220, Topographies Regs, regs. 4,5, and para. 14–28 above.
[39] See Database Regs, regs 14,15,18 and para. 14–40.
[40] See Chapter 14 for further information on title.
[41] Note that in the U.S. copyright is registrable.
[42] CDPA 1988, ss. 90,91,222,223. ss. 90, 91 CDPA 1988, apply to database right — see Database Regs, reg. 23.

addition, funding contracts can be complex, non-negotiable, and there can be a deadline for signature. Consequently, the small print may be ignored, and conflicts between the contractual provisions relating to ownership and licensing of intellectual property may be overlooked. Another common source of problems is the failure of joint owners to agree on licensing. In the United Kingdom licences under intellectual property cannot generally be granted without the consent of all joint owners.[43] Furthermore, it can be difficult to determine who made the inventive contribution for the purpose of determining inventorship or authorship and, therefore, ownership.[44] To complicate matters further, some employees who are inventors, claim ownership of inventions on the basis that they were not made in the course of their employment,[45] and are not therefore owned by their employer. Inventors may also try to claim compensation from their employer.[46] In the case of consultants it is necessary for them to assign or otherwise grant rights, or else the extent of the licensor's rights (other than registered designs and design right[47]) will be uncertain.[48]

23–46 Warranties The licensee's response to problems revealed by due diligence could be to ask for warranties, backed up by an indemnity, even if the licensor is unlikely to be able to meet any claims. Regrettably, this may not solve the licensee's problems. If there is a known risk of infringing somebody else's rights by using the technology, the financial consequences could be such that warranties are of little practical use. If the problems relate to the "validity" of the rights the response may depend on whether the licence is exclusive or non-exclusive. A non-exclusive licensee can gain from the defect in the rights; there is no need for a licence of an invalid right, and as the licensor never offered any exclusivity, he will lose nothing other than the payments that he will already have made. An exclusive licensee will lose his monopoly, and, possibly, a significant initial payment for the grant of the licence. Licensees can take out insurance, to cover their risks relating to intellectual property, although the scope and cost of such insurance may not meet their requirements.

Alternatives to taking a licence

23–47 If it seems unlikely that the licensee will infringe the licensor's rights, he should re-consider whether it is necessary to take a licence. In the United Kingdom, the alternatives are, broadly, to do nothing, to agree that there would be no infringement, to obtain a declaration of non-infringement of a patent from a court,[49] or to bring an action for registered rights to be revoked.[50] In the short term, it is often cheaper and easier to take a licence. The other options are similar to, or could lead to, the infringement proceedings that the licensee wishes to avoid. The long-term costs of taking a licence can, however, be considerable. The licensee must balance the cost of the proposed licence against that of the other options.[51] He must also consider the

---

[43] PA 1977, s. 36(3), CDPA 1988, ss. 173(2),258(2), TMA 1994, s. 23(4). RDA 1949, s. 19(4) is less clear on the subject, but is seems likely that it would be interpreted in a similar way to the CDPA 1988. Note that there is no equivalent to CDPA 1988, s. 173(2) in respect of database right, although this may be implied. Laws in other jurisdictions may be different.
[44] See, for example the situation in *Henry Brothers (Magherafelt) Ltd v. Ministry of Defence and Northern Ireland Office* [1999] R.P.C. 442.
[45] See PA 1977, s. 39 and paras 14–07 to 14–15.
[46] See PA 1977, ss. 40, and 41 and paras 8–112 to 8–116.
[47] See CDPA 1988, ss. 215, 263 and RDA 1949, s.2.
[48] See paras 14–08, 14–09, 14–38, 14–40, 14–41 and 14–42 to 14–50.
[49] See PA 1977, s. 71 and cases where a declaration has been granted relying on the court's inherent jurisdiction — see *Wyko Group Plc v. Cooper Roller Bearings Co. Ltd* [1996] F.S.R. 126.
[50] PA 1977, s. 72, RDA 1949, s. 11, TMA 1994, ss. 46, 47.
[51] See sections on damages and costs in Chapter 24.

possibility that if he does not take a licence the licensor might grant an exclusive licence to somebody else, preventing the licensor from granting further licences.

23–48 Is it the right time to take a licence? The licensee does not, strictly speaking, need to take a licence of an application for a United Kingdom patent before that right is granted. However, after grant, the owner of the right can claim damages for infringement, backdated to the date of publication of the application.[52] Similarly, if the licensee is still conducting research, its activities may not yet require a licence, even if the research is undertaken with an ultimate commercial purpose in mind.[53] From a commercial perspective, it could also be premature to take a licence if the licensee is not familiar with the technology or if the technology is in the early stages of development.[54] This may not be a problem unless the licensee will be required to make a substantial payment on signature of the licence, or to enter into other irrevocable financial commitments. In such circumstances the licensee has the alternatives of an evaluation agreement or an option for a licence. An option has the advantage of keeping open the possibility of taking a licence at a later date.

Assessing the licensor and the technology

23–49 If the licensee intends to rely on the licensor for anything other than his promise not to sue for infringement of registered rights, the licensee should consider making inquiries about the licensor. These are likely to include the licensor's financial standing, and his ability to comply with his other contractual commitments. This is particularly important if the licence is exclusive, if it includes important unregistered rights, or if the licensee will rely on the licensor's technical expertise or warranties. Investigations usually start with a company search, but they can be more extensive, if necessary. In the context of international licensing, where it is obligatory to register licences of intellectual property in some jurisdictions, other licences of the same property may be available for inspection. This can be an invaluable source of information for the licensee. The licensee could also seek an independent assessment of the technology.

DRAWING UP THE LICENCE

Scope of the licence — the property to be licensed

General — licensor

23–50 It goes without saying that the licensor needs to take a licence of sufficient property to enable the licensee to manufacture specific products.[55] The range of such products is usually the subject of negotiation. If the licensor confines the licence to rights that are needed by the licensee, he can put the remaining rights to other uses, or may require further payment for them. In some instances, the licensor may exclude certain property from the licence, in order to avoid unnecessary liability. The licensor can also retain rights as a means of managing his financial exposure to the licensee in

---

[52] See PA 1977, s. 69.
[53] See PA 1977, s. 60(5)(b) and *Monsanto v. Stauffer (No. 2)* [1985] R.P.C. 515 CA.
[54] In some industries, such as pharmaceuticals, it is common to conclude licences at an early stage.
[55] In the case of rights relating to processes, there may be no manufacture of products, but the broad principles are the same. In some cases the means of charging royalty, and certain other elements of the licence may need to be adjusted.

the early phases of the relationship. Having decided on the scope of the licence, the licensor will need to decide whether the licence will be non-exclusive, sole, or exclusive.[56] The licensor could decide on a combination. For example, there could be an initial period of exclusivity, followed by a non-exclusive period. If the licence is to be exclusive, the licence may infringe competition laws, and the licensor may therefore wish to comply with the Block Exemption or its equivalents in other territories.[57]

Describing the property — licensor

23–51 Whilst the licensor may intend to grant a licence in respect of specific products, there are drawbacks to expressing it that way in the licence agreement. It is possible that the licensor will be understood to have granted rights under property that he does not own. If somebody else owns other rights that would be infringed by manufacturing the product, the licence could be construed as a licence of those other rights. As mentioned above, even the most comprehensive of searches may not find all the rights that could be infringed, and if such rights exist, the liability of the licensor could be considerable.[58] Some manufacturers view this as an ordinary business risk, but the profits that a manufacturer can earn from each product are often substantially greater than those of a licensor. However, if products have been manufactured and sold in the territories to be covered by the licence, without notice of infringement, the risk of infringement may be substantially reduced.

23–52 The licence will have to provide for future developments, such as events in the course of patent prosecution.[59] In the case of patents, it may be possible to limit the licence to certain patent claims.[60] If the licensor intends to apply for a supplementary protection certificate in respect of a licensed patent, this can be included, and the licensee may be required to provide all necessary assistance in obtaining the certificate.[61] If know-how is documented carefully, this should not only avoid disputes over its extent, but also assist the licensor to take advantage of the Block Exemption.[62] The licensor can add a licence of a trade mark and make its use compulsory.[63] This can

---

[56] A non-exclusive licence permits the licensee to use the technology, but the licensor is free to licence others and to "use" the technology himself. A sole licence restricts the licensor to one licensee but the licensor may continue "using" the technology. An exclusive licence gives the licensee the right to use the technology to the exclusion of all others, including the licensor. Whilst the above descriptions are those which are generally understood, it may be worthwhile defining what is meant in the licence, as there is some authority to the contrary. There are statutory definitions of "exclusive licence" — see n. 5 to para. 23–07 — but they do not necessarily apply in other contexts such as this. Furthermore, the general understanding of the meaning of these terms may be different in other territories. For example, an "exclusive" licence may have the same effect as a sole licence unless the exclusivity is specifically defined.
[57] See paras 23–23 and 23–24, above and Chapters 16–18.
[58] See paras 9–65 to 9–69.
[59] For example, the rights may include a recent PCT application under which the licensor can elect to continue prosecution in any of the PCT territories. It is highly likely that the filing programme will be narrowed down in future, so the licence will need to specify the territories in which it will proceed — see paras 9–40 et seq.
[60] This type of division may produce an effect that is similar to a field of use restriction, but without the restrictive wording, although that does not affect the competition law analysis.
[61] Otherwise the licensee is not obliged to assist, and it may be necessary for the authority awarding the certificate to obtain a copy of the marketing approval from the authority that granted the approval — see *Biogen Inc. v. Smithkline Beecham Biologicals S.A.* [1997] R.P.C. 833.
[62] The Block Exemption covers know-how agreements if "the parties have identified in any appropriate form the initial know-how and any subsequent improvements to it . . .". This may also help to demonstrate that the know-how is "substantial", which is another requirement of the Block Exemption — see Art. 1.3
[63] See the Block Exemption, Art. 1.1(7). Note that the Block Exemption does not apply to trade mark licences unless the licensed trade marks are ancillary — see Art. 5.1(4). Also, Art. 1.1(7) states that trade mark licences are covered by the Block Exemption provided that the licensee is not prevented from identifying itself as the manufacturer.

sometimes be a useful means of prolonging the licensor's income stream. A trade mark need never expire, and the licensee might decide to take a licence of the mark after the other rights have expired. All other rights are of a limited duration, except for rights in confidential information, which, effectively, only expire when the information ceases to be confidential. At that point the Block Exemption[64] no longer applies to a continuing know-how licence, although it does apply to the payment of royalties for the remaining term of the licence.[65]

Improvements — licensor

23–53   Improvements to the licensed technology can also be included by express agreement. Improvements are inherently speculative, and the precise commercial and legal effects of the terms relating to them are likely to be uncertain.[66] The licensor has to strike a balance between giving away too many rights, and providing the licensee with technology that is rapidly outdated. At either of these extremes the licensor could lose revenue. It may be best to postpone the decision until the improvements are made, unless the licensee is likely to make useful improvements, which could be licensed on a reciprocal basis to the licensor.[67] If the licensor includes improvements, he can seek to limit the ambit of the provision. For example, improvements could be confined to technology that comes within the scope of the claims of licensed patents and/or that could not be used independently of the licensed rights.[68] The agreement should state when disclosures of improvements will be made, how they may be used, and the duration of the obligations.[69] The Block Exemption provides guidance as to the terms that are likely to be acceptable under E.C. competition law.[70]

Describing the property — licensee

23–54   The licensee's objective will be to ensure that he has access to all the property that he needs. If property is not within the scope of the licence wording, the licensee may infringe the licensor's retained rights. If the licensee needs to extend the licence at a later stage, there is no guarantee that the rights will be available. Even if they are available, it could be costly to acquire them. In particular, the descriptions of the licensed rights must stand the test of time. For example, the definition of patents should provide for events in the course of prosecution of the patents, such as extensions, and supplementary protection certificates, if relevant.[71] The licensee can also

---

[64] Block Exemption, Art. 1.

[65] See Article 2.1(7)(a).

[66] Note also that the term "improvement" has no precise legal definition. As with many matters of construction, the meaning will depend to a great extent on the individual context. The term may include inventions that are the subject of future patent applications, although it is unlikely to include improvements that are so radical as to make them "quite distinct" — see *Buchanan v. Alba Diagnostics Ltd* [2000] R.P.C. 367 (–note that this is a Scottish case). See also *Linotype and Machinery Ltd v. Hopkins* [1908] R.P.C., CA; [1910] R.P.C. 109, HL, where "improvement" was understood as including an improvement to the technology that was the subject of the rights. Greater precision and certainty may be achieved through definitions.

[67] If the licensor requires the licensee to surrender improvements, applicable competition laws may be infringed. The surrender of the licensee's improvements is included on the "black list" in the Block Exemption — see Art. 3(6). Reciprocal licences are, however, on the "white list" — Art. 2.1(4). Note that in the case of severable improvements (which are not defined) the licence back is outside the Block Exemption if it is exclusive — see Art. 2.1(4).

[68] The licence needs careful wording in this respect, as the nature of the improvements covered by it may depend on fine points of interpretation.

[69] See *National Broach & Machine Co. v. Churchill Gear Machines Ltd* [1967] R.P.C. 99, HL and *Regina Glass Fibre Ltd v. Werner Schuller* [1972] R.P.C. 229, CA for examples of potential pitfalls.

[70] See Arts 2.1(4), 3(6),3(7), 8(3), Recital 14 and paras 16–30, 16–46 and 16–47.

[71] Laws in some territories outside the U.K. may provide for patent extension or restoration. Strictly speaking there is no such provision in the U.K., but there are Supplementary Protection Certificates, which have equivalent effect. See also paras 8–117 to 8–122 and 9–56.

seek to include improvements to the licensed technology, although this will often prompt the licensor to seek a reciprocal licence. In the case of registered designs, the licensee should require a licence of the corresponding design right so that his licence can be registered, and will therefore be enforceable against subsequent assignees of the registered design.[72]

Physical property

23–55 The licensor may also provide physical property in the form of prototypes, samples, or other embodiments of intellectual property, in order to assist the licensee. These items are usually treated like any other goods that are sold or hired, although it is necessary to take into account their experimental nature when considering any warranties and exclusion clauses. In addition, the licensor may make available materials or matter such as cell lines, which can be the most valuable element of the technology to be provided. Such material will not necessarily be an embodiment of intellectual property, and it will not be information as such. If this is the case, the agreement relating to it will be for the hire of the material. Whilst certain legal duties can be implied in hire arrangements,[73] it advisable to set out the obligations of both parties in an agreement, as the implied terms could well not be suitable. In particular, if living matter provided to the licensee produces further matter, the further matter is likely to belong to the licensee, unless specific provision is made in the agreement.[74]

Dividing the property — licensor

23–56 Dividing the property — by prohibited acts As mentioned above, the licensor needs to consider the likely manufacturing and distribution network for licensed products. From this he can deduce the rights that will be needed by each person making, assembling, selling or hiring the future licensed products. The licensor may take a royalty in respect of each of the acts which is prohibited by the intellectual property, unless that act is the subject of a licence.[75] The prohibited acts depend on the nature of the intellectual property concerned and on the jurisdiction where the intellectual property subsists. In the United Kingdom:

(i)   the acts prohibited by a patent are, broadly, manufacture, sale, use, importation, keeping products, using patented processes to make them,[76] and supplying the means relating to an essential element of the invention for putting the invention into effect;[77]

(ii)   the main acts prohibited by copyright relating to technology are copying, adapting and issuing copies to the public, importing, possessing, selling on or hiring infringing copies, or providing means for making infringing copies;[78]

---

[72] See RDA 1949, s. 19(3A).

[73] This subject is beyond the scope of this work. Readers may wish to consult a contract textbook that covers the sale and hire of goods and bailment.

[74] See *Tucker v. Farm and General Investment Trust Ltd* [1966] 2 Q.B. 421 — the issue of hired livestock are the property of the hirer of the livestock unless the contract provides otherwise.

[75] See n. 21 to 23–11 and 23–11 generally.

[76] Licensors of process patents may adopt a different licensing strategy. They may, for example, decide to hire out apparatus which is needed to carry out the process — see PA 1977, s. 60(2). The fees for hire may relate to the throughput of the apparatus.

[77] There are other prohibited acts — see PA 1977, s. 60.

[78] There are other prohibited acts — see CDPA 1988, Part I, Chapter II.

(iii)   in respect of design right, the prohibited acts are, broadly, making a design
         document recording the design to enable an article to be made, making an
         article to the design, importing it, possessing it, or selling or hiring it;[79]

(iv)   in the case of registered designs the acts are, broadly, making or importing for
         sale or hire, or for use in a business, selling, hiring, and providing means for
         manufacture;[80] and

(v)   in relation to database right, the relevant acts are basically extraction and
         re-utilisation.[81]

In many of the cases set out above, there is a requirement that the activity be commercial. See the chapters dealing with the relevant rights for further details.

23–57 Dividing the property — licence wording   In cases where the licence covers all
the prohibited acts in a manufacturing context, it is common to refer simply to the
"manufacture, use and sale" of the product. It is possible that the prohibited acts in
other jurisdictions will differ, so it can be difficult to be more precise. However, if the
licensor does not wish to confine the licence to certain prohibited acts, it may be simpler and more precise to refer to all acts that would otherwise be prohibited. If the
licence is to be limited, the draughtsperson can specifically exclude prohibited acts
that are outside the scope of the licence, in order to avoid ambiguity.[82] It may also be
possible to restrict the licensee further than by reference to prohibited acts. For example, manufacturing could be confined to a specific site or sites, although the European
Commission might view this as potentially infringing competition law.[83] If any manufacturing is to be undertaken by a third party, it is advisable to provide for this specifically in the licence, as manufacture by an agent could be construed as sub-licensing.[84]

23–58 Dividing property — label licensing   In the United Kingdom, a licence to use
and to re-sell a patented[85] product within the United Kingdom is implied[86] on the
authorised sale of that product. If an implied licence is to be limited, the seller must
ensure that this is drawn to the attention of the purchaser at the time of the sale.[87] The

---

[79] Again, there are other prohibited acts. See CDPA 1988, ss. 226,227,228, Topographies Regs, reg. 8.
[80] See RDA 1949, s. 7. See also Design Regs, reg.12 for prohibited acts, which also include export, and
stocking products.
[81] See Database Regs, reg. 16.
[82] This is particularly important in the case of know-how, which is not, strictly speaking, an intellectual
property right, and dealings in it are not regulated by statute, so the licence is the only means of doing so.
[83] Note also that restrictions on the quantity of products that the licensee may produce may be contrary to
Art. 81(1) (formerly 85(1)), but that the Block Exemption applies where the licence is to be granted so that
a customer may have a second source of supply, or where the licensee may only manufacture the quantities
he requires for his own products — see Arts 1(8), 2.1(13), and 3(5). The wording after the semi colon in
Art. 2.1(12) has been used to support an argument that a single site licence infringes Art. 81.
[84] See *Howard and Bullough, Ltd v. Tweedales and Smalley* (1895) 12 R.P.C. 519, and *Allen & Hanbury's
(Salbutamol) Patent* [1987] R.P.C. 327, but see also *contra Henry Brothers (Magherafelt) Ltd*, above. An
independent contractor could, effectively, be a sub-licensee. If the licensor wishes to prohibit sub-licensing,
it is important to avoid any wording which suggests that an independent contractor may perform any of
the acts to be licensed.
[85] Note that the extent of any implied licence may depend on the intellectual property concerned, and on
the market in which the right is exercised.
[86] See para. 23–11.
[87] See *The Incandescent Gas Light Company v. Cantelo* (1895) 12 R.P.C. 262, *National Phonograph Co of
Australia Ltd v. Menck* (1911) R.P.C. 229, and *Roussel UCLAF S.A. v. Hockley International Ltd.* [1996]
R.P. C. 441. Contrast with the *Silhouette* case relating to consent for the purpose of E.C. laws on the free
movement of goods — see n. 33 to 23–14 — but note the judgement that is awaited in respect of *Zino
Davidoff SA v. A & G Imports Limited* [1999] R.P.C. 631 — see para. 23–14.

adequacy of the notice will depend on the circumstances.[88] In some instances, notice is given by labeling products, or by informing the purchaser at the time of sale. These notices need not set out the restrictions in full,[89] but the buyer must be made aware of the nature of the limitation,[90] and where to find the relevant details. Extreme care needs to be taken to ensure that these licences do not contravene competition and similar laws. For example, the implied licence might be restricted to use and sale in conjunction with another product. However, such a restriction would constitute tying, which could be contrary to E.C. and English competition laws.[91] Label licensing lends itself to attempts to artificially extend the scope of a patent monopoly by contract, but such restrictions may not be justifiable.[92]

23–59   Dividing property — field of use and territory   It is possible to reduce the scope of the licensed property, by confining the licensee's activities to a field of use. This can be expressed as a technical field of use, or a product market. For example, it may be possible to grant separate licences for human and veterinary use.[93] Other field of use restrictions may infringe competition law.[94] It is also possible to divide the property by reference to territories. Within the European Economic Area, certain territorial restrictions on the first sale may be tolerated. The degree of tolerance depends on the property concerned, and whether active or passive sales are restricted.[95] The use of intellectual property rights to prohibit the importation of goods that were lawfully put on the market in the EEA[96] by or with the consent of the licensor is generally prohibited.[97] However, it may be possible to prevent patented goods, which were lawfully put on the market outside the EEA, from being imported into the area.[98] As mentioned above, the licensor could have to limit the extent of the implied licence given on the first sale[99] under United Kingdom patent law.

---

[88] In relation to trade marks, they have included the nature of the goods, the circumstances under which they were put on the market, the terms of any contract for sale and the applicable law of that contract — see *Zino Davidoff SA v. A & G Imports Limited* above. However, it is clear from that case that trade marks may be viewed differently from other intellectual property — but see para. 23–14, above — this case has been referred to the European Courts.

[89] See *Columbia Gramophone Co. Ltd v. Murray* (1922) R.P.C. 239.

[90] An innocent purchaser will not be bound by the restriction. See *Hazeltine Corporation v. Lissen Ltd* (1939) R.P.C. 62.

[91] See para. 23–60 below and para. 16–28 above.

[92] See para. 16–29 — and *Windsurfing International Inc. v. E.C. Commission* [1986] 3 C.M.L.R. 489.

[93] The Block Exemption includes licensing by technical fields of application and licensing to "one or more product markets" on the "white list" — see Art. 2.1 (8), whereas market sharing by competitors is included on the "black list" — see Art. 3(4). Note also the exception made in the "white list" in respect of a second source of supply — Art. 2.1(13).

[94] See, for example, Art. 3(4) of the Block Exemption which relates to customer allocation.

[95] For example, the Block Exemption applies to agreements where the licensee is subject to a ban on (i) passive sales into territories allotted to other licensees for a period of no more than five years from the date of first marketing within the EEA, provided, in the case of patents, that the product is protected by patents in the respective territories, and (ii) active sales into territories allotted to other licensees provided, in the case of patents, that there is protection in the respective territories, and in the case of know-how for a period of ten years. See Art. 1.2– 1.4 of the Block Exemption and para. 16–43 for further details. Blanket non-compete clauses and some attempts to stifle parallel trade are "blacklisted" in the Block Exemption — Arts 3(2) and 3(3). See also Chapter 16.

[96] It is possible that in the EFTA countries of the EEA, "the principle of free movement of goods as laid down by Articles 11 to 13 applies only to goods originating in the EEA" — see *Mag Instrument Inc. v. California Trading Company Norway*, Ulsteen Case E-2/97, advisory opinion of December 3, 1997 — referred to by Advocate General Jacobs in Silhouette — see n. 33 to para. 23–14 above.

[97] See E.C. Treaty, Articles 28 and 30 (formerly 30 and 36 of the Treaty of Rome) and paras 16–103 *et seq.*

[98] See *Silhouette* in n. 33 to para. 23–14 and n. 87 to para. 23–58 above, which related to trade marks, but which may have wider application. However, see also *Zino Davidoff* in n. 87 to para. 23–58 above, and comments in para. 23–14, above. Note, however, that it may only apply to trade marks, and that it has been referred to the European Courts.

[99] The implied licence extends to the importation of products that have been sold by the patentee or his agent. The licence does not, however, extend to products made by licensees under foreign patents — see

Tying

23–60   Some licensors, especially patentees, require the licensee to acquire further intellectual property, goods or services, in addition to the primary licensed rights. This can be for sound technical reasons. However, the licensor could also use this as a means of extending the scope of the monopoly granted by the licensed rights, more particularly through label licensing. This may be contrary to competition and/or intellectual property laws. The E.C. Block Exemption makes it clear that obligations to:

(i) accept quality specifications, further licences, goods or services, which are not necessary to achieve technically satisfactory production; or

(ii) conform to quality standards that are not applied to the licensor and all licensees;

may be contrary to Article 81(1).[1] Tying of goods or services is, however, exempted if the goods or services are technically necessary or if they are needed to ensure that products conform to quality specifications which are applicable to the licensor and all licensees.[2] Tying can also constitute an abuse of a dominant position, under Article 82.[3]

Quality control

23–61   Quality control in general   Quality control provisions serve several purposes in licences. They can help to create and to preserve the reputation of the licensed products, and, perhaps, of the licensor. Also, if defective products made by the licensee could give rise to the licensor incurring liability, quality control may help to avoid such claims. The licensor is most likely to incur liability where his involvement is such that he could be deemed to be holding himself out as the producer of the licensed products,[4] or if the licensor has directed production. If the licensor has merely licensed intellectual property, and given no additional advice or assistance, he may wish to avoid such involvement in the production process. Assistance with quality control could lead to the licensor assuming additional liabilities. If the licensor is to inspect products, the basis of such inspection should be made clear, preferably in the contract. As with tying, if the agreement is to be within the Block Exemption, minimum quality specifications imposed should be necessary for the technically proper exploitation of the technology, or should apply to the licensor and all other licensees.[5]

Société Anonyme des Manufactures de Glaces v. Tilghman's Patent Sand Blast Co. (1884) 25 Ch.D 1; *Sterling Drug Inc. v. C.H. Beck Ltd* [1973] R.P.C. 915; *The Wellcome Foundation Ltd v. Discpharm Ltd* [1993] F.S.R. 433, but see also *Betts v. Willmott* (1871) L.R. 6 Ch. App.239. However, in view of the *Zino Davidoff* case mentioned in n. 87 to para. 23–58 above, it seems that the law in this area may change. Until the nature of the required consent is settled by the European Courts, it is advisable that any restrictions be as explicit as possible.

[1] Block Exemption, Art. 4.2 (a). These provisions are subject to the opposition procedure, so they need to be considered in their economic context. See para. 16–48.

[2] Article 2.1(5).

[3] See, for example, *Hilti AG v. E.C. Commission* [1992] 4 C.M.L.R. 16. See also paras 16–79 *et seq.* for more detail.

[4] E.C. Council Directive 85/374, implemented in England by the Consumer Protection Act 1987, applies to producers and to "any person who, by putting his name on the product or using a trade mark or other distinguishing mark in relation to the product has held himself out to be the producer of the product" — s.2(2). Amongst the defences available to a licensor could be the fact that the licensor had never supplied the product — s.4(1)(b). Licensors may therefore consider carefully the potential consequences of providing samples to licensees for evaluation. The licensor may attempt to avoid liability by requiring the licensee to state that the products are made by the licensee under licence.

[5] See Art. 2.1(5).

23–62   Quality control and trade marks   Quality control provisions also play a vital part in maintaining the validity of any licensed trade mark. In the United Kingdom, there is no formal requirement to impose such conditions. However, the grounds for revocation of a United Kingdom trade mark include use of the mark, in relation to the goods or services for which it is registered, by the owner (or with his consent) in a way that may be liable to mislead the public.[6] This applies in particular to the nature, quality, or geographical origin of the goods or services. Consequently, it is common practice to include and to enforce quality control provisions in licences if a trade mark is included. Similarly, a trade mark can be revoked if, through "the acts or inactivity of the proprietor, it has become the common name in the trade for a product or service for which it is registered".[7] Many licensors therefore stipulate the way in which the mark should be used.[8] As a trade mark is an ancillary right for the purposes of the Block Exemption, additional restrictions that relate to the trade mark, and not to patents or know-how, are outside the scope of the Exemption.[9] Certain other common provisions in trade mark licences are mentioned below.[10]

Confidentiality

23–63   The only means of protecting confidential information is to maintain its confidential status. One of the most important terms of licences of such information is, therefore, the licensee's obligation of confidence. Care needs to be taken with the wording of the undertaking. It can be difficult to impose an absolute obligation to keep all information confidential. Some licensees are only prepared to undertake to use reasonable endeavours to keep information confidential. An absolute obligation is infinitely preferable for the licensor. It is difficult enough to prove that confidential information has been disclosed due to the licensee's default, and even harder to show that this was the result of a failure to take reasonable measures to protect the information. If the terms of the undertaking apply to information that is in the public domain, or which the licensee can obtain legitimately elsewhere, the undertaking may be an unreasonable restraint of trade. It is therefore common to exempt from the obligation information that the licensee should be able to use legitimately without restriction, or which the licensee will be unable to protect. It is crucial that the obligation of confidence continues after the termination of the agreement, if it is likely that the information will still be protectable.[11]

Financial terms

General

23–64   The financial terms are likely to be determined by a number of factors. They will usually include the nature of the technology, and of the intellectual property, the

---

[6] TMA 1994, s.46(1)(d).

[7] TMA 1994, s.46(1)(c).

[8] Trade mark licence precedent books such as Cook and Horton, *Practical Intellectual Property Precedents* (Sweet & Maxwell), contain sample clauses.

[9] See Arts 5.1(4) and 10(15).

[10] If the licence is non-exclusive, the licensee is entitled to pursue infringers if the proprietor of the mark refuses to do so, or fails to do so within two months of being called upon — see TMA 1994, ss. 30(2) and 30(3). Some licensors exclude these rights. Also, the proprietor may be required to hold proceeds of a monetary remedy on behalf of licensees — see TMA 1994, s. 30(6). Again, the licence may stipulate that the licensee is not entitled to this.

[11] Such restrictions are covered by the Block Exemption — see Art. 2.1(1).

stage of development of the technology, the profile and creditworthiness of the licensee, and the licensor's costs. These factors are likely to influence the amount, type and timing of payments, and any related provisions for termination of the agreement or of rights under it. Like any other commercial agreement, the terms will depend on the relative bargaining positions of the parties. However, a licence is unlike some other commercial agreements in that the licensor's prosperity will depend to a large extent on that of the licensee. The licensor's approach may be similar to that of a venture capitalist, in that the licensor may consider the licensee's development plans and be more accommodating in the early phases of the licence, when the licensee is establishing his business. Licensing can, of course, be combined with other types of financial arrangement, such as equity investment, and loans.

Fixed payments

23–65   As a basic minimum, licensors usually aim to cover the costs of maintaining the property. Minimum payments can be expressed as fixed amounts, payable on signature and/or at specific intervals during the term of the licence. They may also be in the form of an additional payment that is made at the end of each royalty period in which the licensee's payments fall below a certain level. If minimum payments are made in advance, they may sometimes be treated as credits against future royalty payments. Some licensors expect minimum payments to be made from "earned" royalties. This might be preferable to a requirement for a "minimum throughput", as the licensor is usually more interested in royalty than in the scale of manufacture. If the licensee is permitted to grant sub-licences, a percentage of the initial fee (if any) may be payable to the licensor, in addition to any royalties payable on the sub-licensee's production. Many licensors also seek the flexibility to terminate the licence, or any exclusivity granted, if the licensee is not meeting the licensor's financial expectations.[12] The licensor can then grant further licences. This is especially important in the case of exclusive licences, where the licensor may have no other source of income from the property.

Royalties

23–66   Running royalties can be set on the basis of a specified payment in respect of products sold, or as a percentage of the selling price of products.[13] In the latter case, efficient charging depends on choosing the right article(s) on which to base the payment. The bare minimum is likely to be the product, or part for a product, that infringes the licensed technology, calculated at the time of the first sale. The licensor may seek to calculate royalties by reference to other articles[14] if the licensed product is part of a larger article, which would have a lesser value if sold independently. Similarly, royalties could be based on finished products, if the price of those products would substantially exceed the price of "raw" products. Licensors often forget that they are able to charge royalties for prohibited acts other than sale, more particularly manufacture and hire. On termination of the licence, a licensor may charge royalty on products that are manufactured but not sold, or postpone royalty payments in respect of those products until they are sold. If no royalty has been paid on stock, the licence may prohibit further sales and possibly provide for the licensor to purchase existing stock.

---

[12] See para. 23–70 for competition law implications.
[13] Different provisions may have to be made in respect of throughput under process patents, but they will depend to a large extent on the nature of the process.
[14] See para. 23–70 for competition law implications.

23–67 Deductions from royalties It is common to pay royalty on the full price at which products are offered for sale, net of certain deductions which can substantially reduce the amount of royalty payable. The precise nature of the deductions is a matter for negotiation, but will often include taxes, insurance and the cost of carriage. Discounts and commission are more contentious. They can be used as a means of cross-subsidising sales of other products, but, equally, it may be essential for the licensee to discount products in a falling market. In some markets, discounts for bulk purchases are customary. A compromise could be to prohibit cross-subsidy in the licence, to monitor the market carefully and to insist that all deductions are documented on the invoices. It is important to draw the distinction between permitting deductions for the purpose of calculating royalty, and determining the licensee's discount structure. The former is permissible, but the latter is a form of price fixing and may breach competition law.[15] If it is not intended that deductions be made, it is preferable to state this specifically.

23–68 Problems with royalties Particular difficulties can arise where the licensed product is sold together with other products ("bundled"). In such cases it may be difficult to ascribe the right value to the licensed product, more especially if the licensee sells at a discount. This problem can be exacerbated when the deductions are taken into account. Some licensors require that products are sold independently if problems arise as a result of bundling. Even if licensed products are sold independently, it is still possible to evade royalty by selling them at a non-arm's length price. The licence may provide that royalty is calculated on the arm's length selling price, or if there is no arm's length price, a fair value may be estimated, or a formula chosen, based on manufacturing and other costs plus a margin. In some cases a minimum royalty per product can be used to act as a further anti-avoidance measure. An alternative to basing royalty on the price of products is to levy a "turnover tax". This is payable by reference to the gross profits of the licensee, less deductions. It is used rarely due to its imprecise nature.[16]

Timing of payments

23–69 There can good reasons for choosing both to bring forward payments and to postpone them. If the licensed rights are weak, it might be preferable for a licensor to take a larger downpayment and reduced long term royalties. If the licensee has insufficient funds, it may be necessary to postpone at least some payments until after the licensed rights have expired.[17] This weakens the licensor's position. His remedies against a licensee that fails to pay are substantially reduced. After expiry of the licensed property, the licensor can no longer obtain an injunction to prevent the licensee from using the technology;[18] he can only recover the unpaid royalties as a debt. Licensees, on the other hand, are likely to want to postpone payments until the technology and market opportunity are proven. In the case of registered rights, there is no requirement in the United Kingdom to take a licence until the right is granted.

---

[15] See Art. 81(1)(a) (formerly 85(1)(a)) and Block Exemption, Art.3(1), Competition Act, s.2, paras 16–17 and 16–37 and Chapters 16 to 18 in general.

[16] Again, there are competition law implications — see para. 23–70.

[17] Payment of royalties after the expiry of patents could be in restraint of trade, but see *Tool Metal Manufacturing Company Ltd v. Tungsten Electric Company Ltd* (1955) R.P.C. 209 and *Bristol Repetition Ltd v. Fomento (Sterling Area) Ltd* [1961] R.P.C. 222, where such provisions were enforceable. See also para. 23–70.

[18] See Chapter 24 section E.

The owners of patents, registered designs and trade marks can, however, claim royalties in respect of periods prior to the date of grant.[19]

Competition law aspects of payment provisions

23-70 A payment structure can infringe competition laws if it attempts to extend the monopoly of the rights granted by requiring payment for acts that would not infringe the licensed property. Under E.C. and English law, it may be permissible to charge royalty on an article that contains non-patented components if it would be difficult to establish the royalty in respect of the individual components, or if there is no separate market for them.[20] The extent of the flexibility permitted to licensors in this regard is uncertain.[21] The Block Exemption does, however, cover provision for the payment of royalties after the expiry of the licensed patents, if the aim is to "facilitate payment".[22]

Royalty provisions can be used to achieve anti-competitive objectives. A requirement for a licensee to make certain minimum payments under the licence is generally deemed to be acceptable.[23] However, if the amount of royalty payable would prevent the licensee from producing or distributing competing products, the agreement may be ineligible for exemption.[24] If the licensee has the power to determine the prices at which licensed products are sold, he also has the power to evade royalty by selling products at artificially low prices, financed by cross-subsidies. Some licensors consider that the legitimate protection described above is not sufficient, and are tempted to solve the problem by agreeing the prices at which licensed products will be sold. Price fixing is generally prohibited under E.C. and English competition laws.[25] Those laws also prohibit excessive and discriminatory pricing.[26]

Administrative provisions relating to payments

23-71 Failure to include licence provisions relating to currency exchange, indexation and auditing, can be costly for licensors. The licensor can specify the method of calculating the exchange rate, and the time when the payments are to be converted. Otherwise, the licensee could take advantage of exchange rate fluctuations. In certain territories outside the European Economic Area, there are exchange control regulations. This can lead to a loss of revenue and further administration, so the licensor may seek additional payments, some form of security, or an alternative source of payment, to compensate. Also, the value of fixed payments can diminish considerably over the life of the licence, unless they are increased in line with inflation, using a suitable pricing index. Similarly, the licensor is likely to charge interest

---

[19] See PA 1977, s.69, under which royalties may be claimed back to the date of publication. See also RDA 1949, s.3(5), and TMA 1994, s.40(3), under which registered designs and trade marks are registered with effect from the date when the application was made, or, in the case of registered designs, treated as having been made.

[20] See *IMA AG v. Windsurfing International Inc* — Commission case (83/400) [1984] 1 C.M.L.R .1, and *Windsurfing International v. E.C. Commission* [1986] 3 C.M.L.R. 489.

[21] For example, it is unclear how the charging of royalty by means of turnover tax would be viewed.

[22] See Block Exemption Art. 2.1(7)(b). It also covers know-how that enters the public domain during the period of the agreement — see Art. 2.1(7) (a).

[23] See Block Exemption, Art. 2.1(9).

[24] See Block Exemption, Recital 21 — setting of royalty rates to achieve "blacklisted" objectives in Art. 3 renders agreement ineligible for exemption. Blacklisted provisions include restrictions on competition with the licensor or with others in respect of R&D, production, use or distribution of competing products..

[25] Art. 81(1) (formerly 85(1)) and Competition Act, s.2. See also Block Exemption, Art. 3(1) and paras 16–17 and 16–37.

[26] See para. 16–84.

on overdue payments.[27] Auditing provisions are of crucial importance to licensors, as licensees often fail to calculate royalty correctly and the shortfall can be considerable. The licensor will usually seek unrestricted access to the licensee's accounts and other records,[28] preferably on the licensee's premises and the right to ask questions of the licensee's employees.[29]

Taxation

23–72 Licences can form part of a complex tax arrangement, but it is unlikely that this will be reflected in the licence terms. The taxes that are likely to be provided for in the licence are withholding tax, and VAT. Withholding and similar taxes may be payable directly by the licensor in respect of the licensee's payments, and, as such, can reduce the licence income considerably. It is therefore common for licensors to seek to increase the licensee's payments so that the amount actually retained by the licensor is not reduced by the tax. VAT may also be payable on licence payments, so the licensor must provide a VAT invoice where appropriate, or, where appropriate, make arrangements with Customs and Excise for the licensee to self certify the royalty payable, and therefore the VAT that is due. Royalties charged to business licensees outside the United Kingdom are currently outside the scope of United Kingdom VAT.

Other obligations of licensee

Further incentives for licensees

23–73 Licences may contain additional incentives for licensees. Financial obligations can be insufficient, for example where the licensee must undertake research and development, or must spend time "tooling up", before he is able to commence production and pay royalties. Some licensors therefore require the licensee to keep to a development plan or to equip production facilities sufficient to satisfy customer demand for licensed products.[30] The licensee's obligations could therefore include completion of the phases of a development, production, or marketing plan. Even if a licensor does not wish to include specific provisions, he may require a general covenant that the licensee will use his best[31] or reasonable endeavours to exploit the technology.[32] Such obligations are, however, of limited use, as they could be interpreted in a number of ways, depending on the circumstances. Some licensors also seek some form of

---

[27] *Redges v. Mulliner* (1893) 10 R.P.C. 21 provides authority for the proposition that interest is payable on patent royalties irrespective of whether there is an express term in the licence. It is, however, safer to provide specifically for the payment of interest.

[28] See *Fomento (Sterling Area) Ltd v. Selsdon Fountain Pen Company Ltd* [1958] R.P.C. 8 — the auditing clause should be drafted so that it is clear that the audit is not confined to books of account, but permits inspection of all other information that is necessary or appropriate to determine the royalty payable.

[29] The licensee may object to such provisions on the grounds that other confidential information might be obtained by the licensor. This problem can be overcome by engaging an independent auditor. An alternative is for the licensee to obtain a certificate from their auditor, but the auditor is unlikely to have the same incentive to detect underpayments. It is common to require the licensee to pay the audit fees if material underpayments are disclosed in the course of an audit.

[30] This may be an important provision in cases where there are territorial restraints on licensees, as under Arts 7(2) and 7(3) of the Block Exemption the exemption may be withdrawn, if the licensee fails to meet certain customer demand. The licensor may wish to terminate the agreement for breach of such a provision.

[31] An obligation to use "best endeavours" does not usually oblige the licensee to take measures that would be ruinous — see *Terrell v. Mabie Todd & Co. Ltd* (1952) R.P.C. 234 for interpretation in the context of a patent licence.

[32] See Block Exemption, Art. 2.1(9) — production of a minimum quantity is exempted, as is an obligation to use best endeavours to manufacture and market the licensed product — see Art. 2.1(17), but see also Art. 7.(4), which provides for withdrawal if the parties were competing manufacturers when the licence was granted, and the provision has the effect of preventing the licensee from using competing technologies.

restrictive covenant. Certain limited covenants are permitted under the Block Exemption,[33] but general restrictive covenants prohibiting competition, may infringe E.C. and United Kingdom competition law.[34] Under English law they may also be void as an unreasonable restraint of trade.[35]

Marking

23-74 In the United Kingdom, it is not possible to claim damages for infringement of a patent, copyright, design right, or registered design if, at the time of the infringement, the infringer was unaware and had no reasonable grounds for supposing that the right existed.[36] It is therefore common for licensors to oblige licensees to mark all products that would infringe with notice of the relevant right, or that an application is pending. In the case of patents and registered designs, inclusion of the registration number in the notice is preferable to a mere reference to "patented" or "registered" (respectively), which cannot be guaranteed to be effective.[37] If it is impossible to mark the product itself, the associated packaging should still be marked. It is also worthwhile marking packaging with a notice of the rights that would be infringed by copying the packaging itself.[38] Care needs to be taken when marking. In the United Kingdom it is a criminal offence to falsely represent that a right is registered, or an application made.[39] Marking constitutes such a representation. In the case of patents, there is a grace period in the event of refusal or withdrawal of a patent application or expiry or revocation of a patent. In the case of registered designs the prohibition is confined to subsequent marking, and not to sales of pre-marked products.[40]

Further obligations of licensor

Most favoured licensee

23-75 Licence payments can place a substantial burden on a licensee, which is usually tolerated on the basis of the competitive edge provided by the licensed technology. If, however, the licensor makes the technology available to others on more favourable terms, the licensee could well be severely disadvantaged, and this may infringe competition law.[41] Some licensees require the licensor to offer such preferential terms to them. This can present considerable problems for a licensor. Unless the other licences are in standard form, it can be difficult to assess which is preferable. If the licensee wishes to adopt other terms and conditions, he should be obliged to adopt them in their entirety, so the licensee cannot choose only the most favourable provisions. The

---

[33] Covenants not to use the licensor's technology to construct facilities for third parties — Art. 2.1(12), and a provision allowing the licensor to terminate the exclusivity granted and stop licensing improvements if the licensee enters into competition with the licensor (or with other undertakings connected with the licensor) in the Common Market in respect of R&D, production, use or distribution of competing products — see Art. 2.1(18)

[34] See Block Exemption Art. 3(2).

[35] Aspects of this topic are covered in outline in Chapter 10, but otherwise are outside the scope of this publication. Readers should consult a work on contract law.

[36] PA 1977, s. 62(1), CDPA 1988, ss. 97(1), 233, RDA 1949, s.9. See also *Lancer Boss Ltd v. Henley Forklift Co. Ltd* [1973] F.S.R. 14.

[37] See PA 1977, s.62(1) and RDA 1949, s.9(1).

[38] Even an instruction leaflet may be protected as copyright. See para. 11–41 and *Elanco Products Limited v. Mandops (Agrochemical Specialists) Limited* (1980) R.P.C. 213.

[39] See PA 1977, ss.110,111, RDA 1949, s.35, TMA 1994, s.95.

[40] The statutory provision for trade marks is different again, as the offence is only committed if a person knows, or has reason to believe that the representation is false — TMA 1994, s.95.

[41] E.C. Treaty, Art.81(1) (formerly 85(1) of the Treaty of Rome) specifically mentions applying "in relation to customers in the trade unequal conditions in respect of equivalent transactions, placing them thereby at a competitive disadvantage" as an example of a prohibited activity.

situation may be complicated further by the existence of implied licences, or by licences granted to settle past infringements, which the licensor may need to exclude from the arrangement. In practice, some licences prohibit disclosure of terms to other licensees, making it difficult to operate the most favoured licensee clause.

Warranties

23–76 It is common to agree on the licensor's warranties, and for the licensor to exclude all other warranties and representations and certain general liabilities. In addition to the express warranties set out in the agreement, it is possible that certain warranties may be implied, so the exclusion clause is extremely important.[42] The combination of the warranties and exclusion clauses therefore apportions risks between the licensor and licensee.[43] Licensees may seek a variety of warranties, depending on the circumstances.[44] They are often concerned that:

(a)   the licensed property subsists, and will continue to do so;

(b)   the licensor owns the rights and/or has the right to grant the licence, and will continue to do so;

(c)   the use of the licensed technology will not lead to an infringement of other peoples' rights;

(d)   the licensed technology will work.

The risks to each party will depend on the nature of the rights to be licensed, the market place, and the available knowledge about that market place. It will also depend on individual circumstances. If, for example, a patent has already been the subject of litigation, in which the validity of the patent was upheld, some of the risks could be considerably reduced.

23–77 Warranties of validity   Registered property clearly subsists if it is on the register. Whether it will be revoked or declared invalid is almost impossible to judge in many cases, unless it has been found valid during litigation that cannot be appealed. The requirement that a patentable invention, or registerable design, must be new means that without the sort of rigorous assessment that takes place in the course of litigation it is difficult to know whether such a right is valid. An absolute warranty of validity of a patent or registered design may therefore constitute a substantial risk to the warrantor, which he will usually avoid. Warranties such as this, which relate to unquantifiable risks, can be qualified by reference to the warrantor's knowledge and belief.[45] It is more reasonable to expect a licensor to warrant that he does not know of any reason why a registered right would be held invalid, and that he has disclosed details of existing or pending litigation. Even if the warranty is given subject to qualification, it can still present difficulties to the licensor. If the licensor makes any disclosures, they may be used later to challenge the validity of the patent.

---

[42] See nn. 47 and 48 to para. 23–19.
[43] This is subject to the exclusion clause being valid. Exclusion clauses may be held to be invalid — see paras 23–21 and 23–32.
[44] In the case of "turnkey" contracts, the licensee may require warranties in respect of the performance of the technology, independent of the warranties relating to intellectual property.
[45] Another means of reducing this risk is to disclose all information known to the licensor, which could lead to a breach of warranty, and exclude the risks disclosed from the ambit of the warranty. It is not possible to do this if no facts are known to the licensor.

23–78   Whilst it is difficult to give an absolute warranty of the validity of a registered right, it can be easier to give and to obtain warranties in narrower terms. For example, the licensor may warrant that the property will not be permitted to lapse through failure to pay renewal fees,[46] or to comply with other procedural requirements. In the case of applications, there may be a warranty that the licensor will prosecute the applications diligently. When giving a warranty to prosecute patent applications, the licensor should specify the territories where he intends to pursue applications, if the election has yet to be made. Otherwise, it could be implied that he has agreed to prosecute applications in all possible territories, within the scope of the licence, which could be costly. Some licensees also require a warranty that the licensor will not apply for or consent to the grant of any compulsory licences. The licensee of a patent that is to be the subject of a supplementary protection certificate in respect of a medicinal or plant protection product[47] could require a warranty that the licensor will apply for the certificate.[48]

23–79   Warranties of subsistence   Warranties of the subsistence of unregistered rights are not as problematic as warranties of validity of registered rights. However, care must still be taken, as the property must fulfill the basic statutory requirements in order to subsist, and the legal position may be different in other jurisdictions.[49] Certain rights and subject matter will pose special problems. In particular, there may be an appreciable risk that design right does not subsist in elements of designs due to the exceptions from the right.[50] The licensor and licensee should be equally able to evaluate whether the work or design is of a type that is protectable, so a warranty in this regard could be inappropriate. However, it is only the licensor who is likely to know, for example, whether a work or design has been copied from a pre-existing design and does not, therefore, meet the requirement of originality.[51] The licensee may seek a warranty of compliance with such requirements, although an absolute warranty could present risks for a licensor who had no personal knowledge of these matters.

23–80   Warranties of ownership and freedom to use   A warranty that the licensor has the right to grant the licence could be implied,[52] irrespective of any express warranties. It is important to distinguish this warranty from a warranty that the licensee would not infringe somebody else's rights if he were to manufacture a particular product. Products that could be the subject of registered rights present particular problems, as it is possible to infringe registered property unknowingly.[53] As was explained above, searching for other peoples' registered property can be relatively costly and time

---

[46] The obligation to pay renewal fees may not be inferred — see *Re Railway and Electric Appliances Co.* (1888) 38 Ch.D 597, in which a covenant to pay renewal fees was not implied. In that case there was an assignment, with a continuing consideration that was calculated on the basis of continuing sales.
[47] See Council Regulation (EEC) 1768/92 of June 18, 1992 Concerning the Creation of a Supplementary Protection Certificate for Medicinal Products, Council Regulation (EEC) 1610/96 concerning the creation of a Supplementary Protection Certificate for Plant Protection Products and paras 8–117 *et seq.* for a general description of supplementary protection certificates.
[48] The licensor may require the licensee to assist the licensor and to give the licensor all necessary documents and information to enable the licensor to make the application, although this is no longer essential — see n. 61 to 23–52.
[49] For example the interpretation of the requirement that certain copyright works be original — see CDPA 1988, s.1(1)(a) — may result in works that would be the subject of copyright in England being denied protection elsewhere — see para. 11–08.
[50] See CDPA 1988, s.213(3).
[51] This requirement relates to literary, dramatic, musical and artistic works — CDPA 1988, s.1(1)(a), and to designs — CDPA 1988, s.213(1) and 213(4).
[52] See para. 23–19.
[53] Note that a product may be protected by a variety of patents, and that ownership of one patent does not mean that the proprietor is entitled to work the other patents.

consuming, and the result is never guaranteed.[54] Many licensors are therefore unwilling to give freedom to use warranties,[55] unless they are manufacturing the licensed products themselves, as the risk will then be reduced. In contrast, infringement of unregistered property occurs, essentially, as a result of imitation of a work or design. That imitation can, theoretically, be subconscious,[56] but in most cases an infringer should be able to tell that he is infringing, so a warranty is more appropriate, although, again, it presents risks for a licensor without personal knowledge of the circumstances in which the work was created.

Other provisions

Invalidity

23–81   If the property is declared invalid, the licensee will not receive a return for any continuing royalty payments. The licence may therefore provide for royalty payments to cease if all the licensed property is declared invalid, but if the decision is reversed on appeal back royalties are usually payable.[57] If the property is only partially valid, it can be difficult to judge whether there should be a corresponding reduction in royalty, especially at the time when the licence is concluded. It is difficult to legislate for such matters in advance. Also, any royalty apportionment might affect the amount of damages that could be awarded to a licensor in an infringement action. Furthermore provisions for reducing royalties present difficulties to a licensor who needs to reduce the scope of the licensed rights voluntarily, in order to make them less vulnerable to challenge.[58] It is likely that at some point the licensee will insist on reducing the royalty. Indeed, it may be necessary for the licensor to do so if a reduction in the licensed property would lead to the licensor charging royalty in respect of products that would not infringe the licensed property.[59] The reduced rate can be negotiated as part of the licence, and a formula for calculation included in the licence, or, more likely, a means of dispute resolution may be included to deal with failure to agree terms at a later stage.[60]

Infringement by others — licensor

23–82   Licensors may require the licensee to inform them if they become aware of infringement. However, due to the cost of litigation and the risk that the infringer will successfully counterclaim that registered property is invalid, licensors can be reluctant to accept an absolute obligation to take action against infringers. In the United

---

[54] See para. 23–41 above and paras 9–65 to 9–69. Note that it may be difficult to search if the precise nature of the product has not been determined.

[55] Note that a warranty of freedom to use may be implied from the wording and context. See *Frayling Furniture Limited v. Premier Upholstery Limited* [1998] Patents Court — under a licence of "the rights" it was construed that there was an implied term that the licensed right had not been the subject of unlawful copying. See also *Microbeads AC and Anor v. Vinhurst Road Markings Ltd* [1975] 1 All E.R. 529, CA in which a covenant for quiet possession was implied.

[56] See para. 11–38 — inference of copying from similarity.

[57] Payments may be made to an interest-bearing suspense account, whilst the matter is decided.

[58] See paras 8–85 to 8–89.

[59] See para. 23–70.

[60] Note that in *Licensing Technology*, (2nd ed.), Noel Byrne cites the case of *St Regis Paper Company v. Royal Industries* (1977) 194 U.S.P.Q. 52 (U.S. Court of Appeals, Ninth Circuit), where the licensed patent was held invalid, the agreement terminated, and whilst the licensor was awarded compensation for continued use of inessential know-how, the court declined to hold that royalties would continue to be payable.

Kingdom, exclusive licensees, and licensees of right, are entitled to pursue the infringer themselves.[61] The licensor could, therefore, lose the licensed property as a result of a licensee's mismanagement of an infringement action. Some licensors seek to avoid this by requiring licensees not to litigate[62] or to obtain the licensor's permission before commencing litigation. If the licensor consents to litigation, he may require the licensee to use advisers acceptable to the licensor, to consult the licensor before taking certain major steps in the litigation, and to fund all or part of the expenses of litigation. The licence can also state the proportion of any damages payable to the licensor. An alternative to permitting the licensee to litigate may be to change the royalty structure, if the licensee can show that his competitiveness is being affected by the infringement.

Settlement — licensor

23–83 Many disputes, and potential disputes, relating to intellectual property are settled by the grant of a licence. Payments under the licence are likely to include amounts in respect of royalties that would have been paid if the licence had been granted at the outset. If litigation has already commenced, other matters, such as the payment of legal costs are likely to be included in a settlement agreement. Settlement of disputes can be more difficult for an exclusive licensor, as he will not then have the option to grant a licence to the infringer. The licensee may, however, be persuaded to grant a royalty-bearing sub-licence, with limited scope, if the infringer is not in competition with the licensee. If the licensee is to be permitted to settle infringements the licensor is likely to require the licensee to obtain consent before doing so. It is also possible for the licensee to permit the licensor to grant a further licence as part of a settlement, on the basis that the licensee will share the royalties with the licensor. The alternative to settlement is to obtain a permanent injunction, which can be extremely costly. A licensee could well be more flexible with respect to further licensing if he were obliged to pay all or some of the costs of litigation.

Infringement by others — licensee

23–84 Exclusive and non-exclusive licensees are likely to take a different attitude to infringement. A non-exclusive licensee, who only takes a licence to avoid infringement proceedings, may be content with a reduction or cessation of royalty payments if the licensor does not pursue infringers. Exclusive licensees, on the other hand, usually plan and invest on the basis of exclusivity. They may seek to take advantage of their ability to litigate, or to maintain their exclusivity by obliging the licensor to do so, although the latter right needs to be specifically provided for in the licence. Infringement proceedings can be costly, especially if the licensee is required to pay royalty whilst litigating. A licensee who wishes to litigate can therefore seek to reduce the costs by sharing them with the licensor, or by offsetting expenses paid against royalty payable. He may also seek a proportion of any payment on settlement, an account of profits, or an award of damages.

---

[61] See n. 5 to para. 23–07, above for references relating to exclusive licences. Note that in the case of licences of right under patents the licensee may request the licensor to take action against infringers, and if the licensor fails to do so within two months of the request, the licensee may take action itself, unless the terms are settled by agreement and expressly provide to the contrary — PA 1977, s.46(4). Note also that in *Mills v. Carson* (1893) R.P.C. 9, a covenant for quiet enjoyment was held to have been breached when third parties were able to use the patented invention.

[62] Note that under TMA 1994, s.30, a licensee may take infringement proceedings if his licensor refuses or fails to do so within two months of being called upon, unless the licence provides otherwise.

No challenge clauses

23–85   From the licensee's perspective, licensing is an alternative to obtaining a declaration of non-infringement, or proceedings for revocation.[63] The licensor may therefore seek an undertaking from the licensee that the licensee will not:

(a)   oppose the grant of any of the property to be licensed; or

(b)   bring proceedings for revocation of any of that property; or

(c)   claim that the property is invalid in the course of any proceedings.

This would produce a similar result to a settlement agreement. Licensors should, however, be aware that a clause of this sort could be contrary to competition laws.[64] Such problems can be overcome by avoiding a blanket ban and simply providing for termination of the licence in such circumstances,[65] although it is unclear whether this would also apply to ancillary rights.[66] If there is no express undertaking not to "challenge" the licensed property, such obligations may be implied for the duration of the licence.[67]

Liability — licensor

23–86   As described above, the licensor could incur substantial liability, and may therefore try to exclude or to limit it.[68] The Unfair Contract Terms Act 1977 does not prevent exclusion of liability for negligence (or certain exclusion and indemnity clauses) in a bare licence of intellectual property.[69] Liability in respect of other aspects the arrangement between the licensor and licensee is outside the scope of the exemption.[70] It is therefore common for the exclusion clause relating to intellectual property to be separate from the limitation clause relating to other liabilities. Such provisions can be accompanied by an indemnity in respect of liability incurred by the licensor due to the licensee's exercise of the licenced rights. Note, however, that an indemnity may have a similar effect to an exclusion clause, so it needs to be drafted with care. The licensor can also seek to prevent liability from arising in the first place. Liability in negligence arises if the licensor owes the licensee a duty of care. Unless the licensee is relying on the advice of the licensor, the licence may specifically refute any implication of a duty of care, and explain why.[71] In many cases, the licensee has greater skill and knowledge than the licensor, and it is unfair for the licensee to claim that he is relying on the licensor's technical expertise.

---

[63]   See PA 1977, ss.71,72,74, RDA 1949, s.11, TMA 1994, ss.46,47.
[64]   Under the Block Exemption, such provisions are subject to the opposition procedure — Art. 4.2(b). These provisions do not contravene Art. 81(1) (formerly 85(1)) if the licence is free — see *Bayer AG v. Sullhofer* [1993] F.S.R. 414.
[65]   See Block Exemption, Art. 2.1(15). See also Art. 2.1(16).
[66]   Art. 2.1(15) only applies to patents and know-how.
[67]   *Lawes v. Purser* (1856) 6 El. & Bl.930, *McDougall Brothers v. Partington* (1890) R.P.C. 216, *Ashworth v. Law* (1890) 7 R.P.C. 231. In the latter case circumstances of fraud and mistake are mentioned as exceptions, as is "ambiguity on the patent", although it is unclear what it means. If the licensee relies on common law rights, there is no contractual term or arrangement that could breach competition law, but express contractual provisions for termination are likely to be easier to enforce.
[68]   As stated in n. 54, to para. 23–21, exclusion clauses may be held to be unenforceable.
[69]   See n. 54 to para. 23–21.
[70]   See Unfair Contract Terms Act 1977, Schedule 1 s.1 and ss. 2 to 4 in particular. Note that the exclusion clauses are those in standard terms where one party deals as a consumer. Regrettably, it is not possible to give details of exclusion clauses in general in a work of this sort.
[71]   It is also important to exclude any pre-contractual or collateral representations, using a carefully drafted "entire agreement" clause.

Insurance

23–87 The potential liability arising from the licensed activity and the provision of ancillary goods and services, may comprise infringement of other peoples' intellectual property, product liability, employer's liability, and possibly environmental liability as well as general contractual liability and liability in tort. In addition, there will be the normal risks of the licensee's business. The position of the licensor will depend on his involvement in this activity, but can be modified by the licence agreement, to the extent that it is enforceable,[72] and provided that the licensee has sufficient net worth to meet claims. The party that is to assume the risk can seek, or be obliged under the terms of the licence to seek, insurance. When insuring risks relating specifically to intellectual property, it may be necessary to use a specialist insurer. The licence can stipulate the nature and amount of the insurance, and the period for which it is to be effective. The party to the agreement who is not paying the insurance may be included on the policy as a named insured, so that they can be paid directly by the insurer. If liability is to be assumed contractually, the insurer needs to be informed of this beforehand, as there is a duty to disclose to the insurer all information that could affect his decision to insure.

Registration and government approvals

23–88 The licensee's rights may become subject to the rights of third parties if the licence is not registered.[73] In the case of United Kingdom registered designs it seems that registration by a person acquiring a licence is mandatory.[74] An exclusive licensee of a United Kingdom patent, and licensees of United Kingdom trade marks, cannot claim[75] damages, or an account of profits, for periods when the licence is unregistered unless it is during the initial six-month registration period, or during any period of grace permitted by a court.[76] Licences often contain sensitive commercial information, so there is a tendency to register a short form licence containing as few details as possible, and to provide for this in the main agreement.

In addition, government and exchange control approvals can be required in other territories, and there could be penalties for failure to comply. It is therefore common to provide that the licence will not take effect until approval is obtained, and for termination if approval is not granted within a stated period. Registration is usually a duty of the licensee, as it is most likely to be a requirement of the licensee's government.[77] However, if it is possible that the approval may be subject to amendment of the terms of the licence, the licensor will need to be in direct contact with the relevant authority.

---

[72] It is not just UCTA that will govern enforceability. For example, some legislation contains criminal penalties for contravention, which cannot be dealt with by contractual exclusion clauses.

[73] See PA 1977, s.33, TMA 1994, s.25(3).

[74] RDA 1949, s.19(1). Failure to register will mean that the document may not be produced in court unless the court so directs, or in proceedings for rectification of the register — see RDA 1949, s.19(5).

[75] In the case of trade marks the licensee may not bring proceedings unless the licence is registered — see TMA 1994, s.25(3)(b).

[76] This is in circumstances where it was not practicable to make the application before the end of the six month period and an application was made as soon as practicable thereafter. See PA 1977, s.68, TMA 1994, s.25(4). In the case of patents, both the court and the Comptroller may make the decision.

[77] There is no such requirement for registration in the United Kingdom. Requirements for governmental approval are most likely to be found outside the European Union and the USA. Governmental approvals should not be confused with the requirement to register the licence to protect the licensee's rights.

Term

23–89   An appreciable number of licences are entered into to settle past infringements, so the commencement of the term needs to be stated as well as the termination
date, or means of determining it. Even if the licence is for the duration of the licensed
rights, it is worth mentioning this specifically, as a patent licence may be terminated
on reasonable notice if there is no express termination provision.[78] In addition to the
maximum term, the licence can provide for termination on notice. The commercial
viability of such an arrangement will depend on the payment structure. If the licence
is multi-jurisdictional, the licence payments, and the general provisions of the agreement, could terminate when the last item of property expires. This would entail the
payment of post-term royalties in certain jurisdictions,[79] but could prevent avoidance
of payments.[80] Alternatively, the licences may expire in each individual territory when
the rights in that territory expire, although the licence agreement itself would continue
until the expiration of the last item of property. The duration of each clause of the
licence agreement needs to be considered separately. Some, such as the provision of
consultancy, could terminate before the rest of the agreement.

Termination

23–90   In addition to the term, most licences include termination provisions, which
can be found in commercial precedent books. These may need to be modified to
accommodate termination of specific rights in circumstances where the rest of the
licence continues. For example, the licence can terminate in respect of one or more
fields of use, or territories, or exclusivity may be terminated entirely, or in respect of
specific rights. Such provisions are commonly linked to performance targets. Other
modifications to standard clauses include termination if the licensee challenges the
licensed property,[81] and termination for breach of insurance provisions. In view of the
close business relationship that can exist between licensor and licensee, the licensor
may wish to terminate licensed rights if the licensee competes with the licensor. The
Block Exemption permits termination of exclusivity, and licences to improvements, if
the licensee competes in the common market in respect of research and development,
production, use or distribution of competing products.[82] However, the Block Exemption does not otherwise apply to such restrictions.[83]

Consequences of termination

23–91   The provisions of each clause need to be considered, to see whether it should
survive termination, and, if so, how long it should survive for. Some provisions such
as obligations of confidentiality, and a prohibition on use of the licensed rights after

---

[78] See *Coppin v. Lloyd* (1898) 15 R.P.C. 373, *Martin- Baker Aircraft Co. Ltd v. Canadian Flight Equipment
Ltd* (1955) R.P.C. 236. In the latter case there was provision for termination for breach and insolvency, but
no other provision for termination. The fact that an agreement provides for termination by one party and
not the other does not mean that it is terminable on notice — see *Guyot v. Thomson* (1894) 11 R.P.C. 541.
The position may be different in respect of copyright — see *The Modern Law of Copyright and Designs* by
Laddie, Prescott and Vitoria (2nd ed.), Vol. 1, para. 14.7.
[79] See para. 23–70 and footnotes.
[80] If the licensee sells to a third party in a territory where the rights have expired, and the licensed products
are then sold in a patent territory, it may be administratively difficult to track the subsequent infringing
sale.
[81] See para. 23–85.
[82] Art. 2.1(18).
[83] See Art. 3(2).

termination can continue until those rights expire.[84] In the case of a partial termination, or the expiration of part of the licensed rights, it may be necessary to adjust certain terms of the licence such as royalty payments. In addition to the existing obligations, there may be specific duties on termination, such as the return of documents recording the know-how, and samples of products. If the licence was registered, arrangements need to be made for the registration to be cancelled. The licensee could well have stocks of the product, which could be sold to the licensor, or, if the licensee is permitted to sell them, royalty will need to be calculated and paid, either on termination or on the eventual sale. If the licensor sells remaining stock, this may lead to the licensor incurring product liability on the resale of such products, so it may be preferable for the licensee to make the sales.

Sub-licences

23–92 Licensees under United Kingdom[85] patents and trade marks have no right to grant sub-licences unless this is provided for in the contract.[86] In respect of other rights, it is safest to include an express provision in the licence, as there is no statutory right to sub-licence in the United Kingdom,[87] nor is there a requirement for consent. As a licensee, it is unwise to rely on using an agent, unless there is a specific right to do so, as the agent might be classified as a sub-licensee.[88] If the licensor is to permit sub-licensing, he may require the licensee to impose terms that are similar to those in the head licence. An exception is usually made in the case of payments, in order to avoid infringement of competition law by price fixing. The licensor can also make the grant of the sub-licence subject to consent, which, in turn, can be subject to any further conditions that the licensor may need to impose. In particular, the sub-licensee may be required to permit the head licensor to carry out an audit. It has been held that the sub-licence terminates concurrently with the head licence unless a contrary intention can be inferred.[89] Sub-licensees need to seek comfort from the head licensor in respect of the potential determination of the head licence.

Assignment

23–93 Similarly, it is not possible to assign, or otherwise transfer rights under a patent licence,[90] unless this is provided for in the licence.[91] If, however, the licensor knowingly accepts royalties from a person who claims to be an assignee, the licensor may not subsequently deny the validity of the assignment.[92] In the case of rights other than patents a prohibition against assignment should ideally be stated specifically.[93] A

---

[84] This is covered by the Block Exemption — Art.2.1(3) — provided that the prohibition relates to patents that are in force, and to know-how that remains secret.
[85] Note that there may be a right to grant sub-licences in other jurisdictions.
[86] See PA 1977, s.30(4), TMA 1994, s.28(4). A prohibition on sub-licensing is permitted under Art. 2.1(2) of the Block Exemption.
[87] Note that the position may be different in other jurisdictions.
[88] See *Allen & Hanbury's (Salbutamol) Patent* [1987] R.P.C. 327, CA, but see also *Henry Brothers (Magherafelt) Ltd v. Ministry of Defence and Northern Ireland Office* [1999] R.P.C. 442 in which it was held that the distinction was unreal.
[89] *Austin Baldwin & Co. Ltd v. Greenwood Batley Ltd* (1925) R.P.C. 454. It is still advisable to state specifically that the termination provisions of the sub-licence must mirror those in the head licence.
[90] PA 1977, s.30(4).
[91] See Art.2.1(2) of the Block Exemption, under which a prohibition on assignment is exempted.
[92] *Lawson v. Donald Macpherson & Co. Ltd* (1897) R.P.C. 696.
[93] This is especially important, as it is possible to assign the benefit, but not the burden of a contract. See, for example, G.H. Treitel, *The Law of Contract* (10th ed.), p. 648.

licence could be classed as a personal right, which cannot be transferred. This has been held to be the case with copyright,[94] but it is unwise to rely on this.

If a licensee is acquired by a competitor of the licensor, the licensor may wish to prevent the acquirer from using the licensed technology to compete with the licensor. The most effective way of dealing with this could be to prohibit assignment, and provide for termination if there is a change of control of the licensee. This may help to avoid wide restrictions on the licensee competing with the licensor, which may infringe competition laws.[95]

Dispute resolution

23–94   The basic choice is between one or more of litigation, arbitration, and some form of alternative dispute resolution.[96] This could be further refined by choosing specific courts, or types of arbitration or dispute resolution. If the contract contains no provision for dispute resolution, and it is not possible to reach agreement if a dispute arises, the parties will have no choice but to litigate, as both arbitration and alternative dispute resolution are voluntary. See paragraph 20–47 for further details of some of the matters that need to be covered in the agreement. The topic is usually considered in conjunction with the governing law clause, as it may be more efficient to resolve disputes in the territory of the governing law. It is technically possible to choose the law of one party together with dispute resolution in the territory of the other, which is less satisfactory. It is also possible to choose the law and jurisdiction of the claimant or defendant, but this would be difficult if there were a counterclaim. Generally, a mixture of different laws and jurisdictions can cause additional difficulties. Much will depend on the nature of the contract. The party who is most likely to take action will seek the most efficient means of doing so, and a party who wishes to avoid liability may wish to choose the most lengthy and costly procedure. In many cases, however, the likelihood of a claim by either party is the same.

Governing law

23–95   There can be considerable differences between national laws, even those that are reputedly similar,[97] so the choice of the law governing the contract itself is important. In civil law jurisdictions, in particular, there is considerable flexibility to imply terms that are not specifically included in the agreement. In addition, the validity, subsistence and title to the licensed property will be governed by the law of each jurisdiction where rights are claimed. Infringement of licensed rights, and of other peoples' rights is generally subject to the law of the territory of the right infringed, although it may be possible to obtain a remedy in another jurisdiction in certain limited circumstances.[98] Exercise of the licensed rights will generally be subject to the competition law of the territory where the rights subsist, or where competition (or free movement of goods) is affected. Outside the European Union, the re-patriation of revenue may be subject to local exchange control laws. Parties may be subject to taxation in territories where revenue is generated. Finally, the form of the agreement may be subject to the licensing laws of the licensee's jurisdiction.

---

[94] *Dorling v. Honnor (Design copyright)*[1963] R.P.C. 205.
[95] See Art. 3(2) of the Block Exemption.
[96] Chapter 24 describes each of these forms of dispute resolution.
[97] See the case referred to in n. 18 to para. 23–37 for differences between English law and New York law, for example.
[98] See paras 24–25 and 24–28 for details.

CHAPTER 24

REMEDIES AND HOW TO GET THEM

Nigel Jones, Partner, Linklaters & Alliance

INTRODUCTION

24–01 Earlier Chapters have considered the legal rights that parties to research and development ("R&D") and other technology-related agreements may have when things go wrong. This Chapter focuses on how to go about enforcing those rights, and the remedies that may be available. It is not limited to a discussion of court procedures: litigation is clearly one option, but it is not the only mechanism for resolving disputes. Arbitration and various forms of Alternative Dispute Resolution (ADR) are becoming increasingly popular. The advantages and disadvantages of these alternatives are considered in Section A. Then there are strategic issues to be considered. How quickly, for example, should an aggrieved party take action? What does the agreement say about dispute resolution? How much needs to be done before "pulling the trigger"? How much will it all cost? These issues are reviewed in Section B.

A related question is where the dispute resolution process should take place. If it is to be litigation, which countries' courts have jurisdiction? And over which issues? Section C discusses these issues. Section D outlines the procedures that must be followed in English litigation involving rights. It considers the position in the "post-Woolf" era — under the new Civil Procedure Rules,[1] and discusses the extent to which the new rules have made, or are likely to make, a significant difference. Finally, Section E reviews the remedies that may be available, both in the short term (such as an interlocutory injunction, a search order (formerly known as an Anton Piller order) or a freezing order (formerly called a Mareva injunction)), and after a full trial (including injunctions, and damages or an account of profits).

Throughout, the focus is on the intellectual property rights which are most likely to be important in the context of R&D arrangements, namely patents and confidential information. The procedures relating to enforcement of rights in trade marks, designs and copyright are not considered in detail, although many of the principles discussed will apply equally to disputes involving these rights.[2] The discussion also focuses on disputes which fall to be resolved in the United Kingdom. However it does have an international perspective, recognising the increasingly global nature of the business world, and, in consequence, the cross-border nature of the disputes which arise. This aspect is given particular consideration in the discussion of strategy (Section B) and jurisdictional issues (Section C). The Chapter is not however intended to provide advice on matters of law or procedure in countries other than England and Wales. Much of what is said applies to Scotland as well, but here the procedures vary and specialist local advice should be sought.[3] Other sources should be consulted if such advice is required.

---

[1] The Civil Procedure Rules ("CPR") came into effect on April 26, 1999 and replace the Rules of the Supreme Court 1965 ("RSC") and County Court Rules 1981 ("CCR"). They implement the changes recommended by Lord Woolf in his report on the English civil litigation system, *"Access to Justice"*.

[2] This is particularly the case for disputes concerning registered designs, which, with patents, constitute Patents Court business — see Section D below.

[3] For simplicity, "England" and "the U.K." are used throughout.

A. Choosing dispute resolution mechanisms

Specifying these mechanisms in agreements

24–02 Those responsible for negotiating an R&D agreement, a licence agreement or indeed any other form of agreement, are generally reluctant to deal with dispute resolution issues as a priority. They enter the negotiations optimistic that everything will progress without difficulty, and see such issues as being an unnecessary distraction from negotiations on what they regard as the key commercial terms. Anyone who has had to try to sort out a dispute arising out of a contract negotiated on this basis will know all too well the problems that result; and will understand why most experienced negotiators and their legal advisors consider the termination and dispute resolution mechanism provisions to be among the most important in any contract.

In the context of R&D and other technology transfer agreements, such provisions are critical. If the agreement does not make clear what the parties' rights are when things go wrong, and how disputes are to be resolved, the parties are likely to end up spending as much, if not more, time and money in trying to sort out such problems than they have done in trying to get the project off the ground in the first place. It is also important to consider who is to have responsibility for protecting the underlying rights against infringement by third parties, who is to run any litigation relating to such infringement, and, equally important, who is to pay for it. Similarly, the agreement needs to deal with the possibility that one of the parties may be sued by a third party in respect of the activities it carries out under the terms of the agreement.

The options

24–03 Clearly, the question of what is likely to be the most appropriate procedure for resolving disputes can only be addressed with knowledge of the alternative mechanisms available, and the principal features and distinctions that differentiate them. The main options are as follows:

- Litigation through the courts
- Arbitration
- Alternative Dispute Resolution, or "ADR" (covering a range of procedures including executive dispute resolution, mediation, mini trials, and adjudication)

The following sections explain the key distinctions between these procedures. Anyone drafting a dispute mechanism in an R&D, technology transfer or licence agreement should review these matters (and/or discuss them with their client) to ensure that the procedure most appropriate to the particular circumstances of the agreement is documented.

Litigation

24–04 If the contract makes no provision for a particular form of dispute resolution mechanism, and the parties cannot agree on an alternative once the dispute has arisen, litigation will generally be the only choice. The procedures which a claimant[4] will need to follow if the litigation is in the English courts are considered in Section D. The main

---

[4] Formerly "plaintiff".

features of litigation which should be taken into account when considering whether it is the appropriate dispute resolution mechanism are the following[5]:

- **Publicity**: Litigation is public; most other forms of dispute resolution are not. The Claim Form[6] which a claimant must file in order to commence legal proceedings, and the Particulars of Claim[7] filed with it, are public documents. Any member of the public may inspect, and obtain copies of them. Further, all hearings (including the trial) and appeals are held in public, except in very rare circumstances.[8] So if you want to publicise the dispute, litigation is probably the best option. In contrast, if you want to keep it confidential, litigation though the courts will not be appropriate.

- **Cost**: The charges levied by the court for use of its facilities are minimal. However, the procedural requirements, even under the new CPR, will normally require significant input from lawyers. As a result, litigation is one of the most expensive mechanisms for resolving disputes.

- **Expertise**: In England, the courts which hear intellectual property disputes are staffed by judges with significant experience of handling intellectual property cases, including dealing with their technical aspects. That is not, however, always the case with other technical disputes, or in other countries. There is therefore a considerable risk in bringing litigation that the judge who will decide the outcome of the case will not have experience in dealing with the legal or technical issues specific to the dispute. This clearly increases the risk of an unsatisfactory outcome.

- **Procedure**: The rules which govern civil litigation are generally more comprehensive than any of the rules governing the alternative mechanisms addressed in this chapter. That is certainly true in the United Kingdom.[9] The rules cater for most circumstances that are likely to arise during the dispute resolution procedure, and there are specific sections dealing with intellectual property disputes.[10] One particular issue here is the courts' power to require disclosure of documents — previously known in England, and still properly known in the United States, as discovery.[11] There are many who consider this to be a major advantage of litigation over other forms of dispute resolution (though the distinction with arbitration is often limited, as arbitrators can have very similar powers). Others take

---

[5] "Litigation" here means, strictly, litigation in the English courts. However most of the features identified will be equally characteristic of litigation elsewhere — though see Section C for a discussion of some of the main differences in litigation procedure which the author has experienced in other countries.

[6] Formerly "Writ" under the RSC.

[7] Formerly "statement of Claim", *ibid.*

[8] In cases involving intellectual property disputes, the parties often want the judge to exclude members of the public when matters involving information which is commercially sensitive are being discussed. The judges are sympathetic to such requests, but also place great importance on English legal proceedings being open to the public. It is therefore on only very rare occasions, and for very short periods, that any hearing in an intellectual property dispute will be closed to the public.

[9] Both under the "old" rules (RSC) and under the new CPR.

[10] In particular, the Patents Court Practice Direction — see Section D.

[11] The provision to the other party of documents which are relevant to matters in issue in the dispute and which help or harm each side's case.

the opposite stand, identifying the disclosure process as being the main cause of the high cost and slow speed of litigation. As will be discussed in detail in Section D, there have been significant changes in the rules on disclosure in England in recent years (some introduced before the CPR came into force). These changes have significantly reduced the scope for the disclosure process to bog down litigation, but left it open for the courts to order fuller disclosure in appropriate cases. They may also be a reason why other forms of dispute resolution have not become as popular in the United Kingdom as in the United States, discovery in the United States having been cited as one of the main reasons for users' dissatisfaction with the litigation process, and for their opting for ADR instead.

- **Delay**: The traditional view is that litigation is slow. This was recognised by Lord Woolf in conducting his review of the English civil litigation system.[12] One of the key themes of his proposed reforms (now implemented) was that the timescale of litigation should be shorter and more certain. The new rules (considered in more detail below) seek to achieve this objective, and experience to-date suggests that they are being successful. In any event, measures had already been taken by the judges in the English Patents Court to streamline patent and other intellectual property litigation, resulting in cases being heard within about 12 months of starting, and trials lasting days rather than weeks. The position in the United Kingdom is much better than it used to be. In many other countries, however, delay still remains a significant problem for those seeking to resolve their disputes through the courts.[13]

- **Third parties**: In litigation, the parties have rights to sue multiple defendants or join third parties. The court also has powers to join third parties into the proceedings.[14] In certain circumstances, this can be a major advantage of court proceedings over all alternative mechanisms.

- **Appeals**: A party which is unhappy with the outcome of a decision of a first instance court can in most circumstances, and with leave of the court, appeal. There are two levels in the United Kingdom: first to the Court of Appeal and then to the House of Lords.[15] Most other jurisdictions have similar arrangements.

Arbitration

24–05   As an alternative to litigation, the parties can agree to submit the dispute to arbitration. In essence, arbitration is consensual. In England, it is governed by the

---

[12] Lord Woolf, *Access to Justice* (interim report) (London, HMSO, July 1995) and Lord Woolf, *Access to Justice* [final report] (London, HMSO, July 1996).

[13] See the discussion on international strategy in Section E below for a comparison of the speed with which courts in different countries deal with intellectual property disputes.

[14] Under the CPR, the court has a wide discretion to order that a person be added (or removed) as a party to a claim where it is "desirable" to do so, provided that the limitation period has not expired. The two bases on which the court must consider the "desirability" test are: (a) to enable it to resolve all the matters in dispute in the proceedings; or (b) to enable it to resolve a matter between an existing party and a proposed new party, which is connected with an issue in the claim (Rule 19.1(2)).

[15] See Section D for a discussion of the right to appeal and the procedures involved.

Arbitration Act 1996.[16] That Act applies provided there is an agreement in written form binding the parties to resolve the dispute by arbitration. This can be a clause included in the original agreement between the parties, or a new agreement entered into after the dispute has arisen. Needless to say, it will be much easier to agree the provisions of an arbitration agreement (or clause) before a dispute arises than later on.

One of the key distinctions between arbitration and litigation in relation to intellectual property disputes is that arbitrators do not have the power to revoke intellectual property rights. This is one of the main reasons — perhaps the main reason — why arbitration has not traditionally been a favoured route for resolving intellectual property disputes. A defendant in a patent infringement action will normally want the patent revoked if he succeeds in defending the allegation of infringement, to eliminate the risk of its being asserted against him again in the future.[17] Arbitration cannot achieve that objective; litigation can. The powers of arbitrators to award effective interim remedies are also more limited than those of the court; and again, this is often a major factor in intellectual property disputes. Arbitrators do have greater powers now than under previous English legislation — including powers to rule on their own jurisdiction, to deal with procedural and evidential matters (including giving directions in relation to property which is the subject of the proceedings, directing the examination of witnesses and giving directions for the preservation of evidence), to appoint experts, legal advisers or assessors of their own, and to order security for costs. However their powers are less extensive than those of the court, and in many instances, can be excluded by the parties.On the other hand, arbitration has a significant advantage in relation to international disputes in that it provides an opportunity for comparatively easy enforcement of awards. Arbitral awards tend to be easier to enforce internationally under the New York Convention[18] than decisions of national courts.[19]

Although less formal, the procedures involved in an arbitration can resemble court proceedings in many respects; although its consensual nature means that the parties to an arbitration agreement can to some degree determine the constitution of the Tribunal, the procedure to be followed and the powers of the Arbitrator.[20]

24–06 There are many other factors which should be taken into account in deciding whether arbitration is the appropriate mechanism for resolving any particular dispute. These have been the subject of many earlier commentaries.[21] Here, briefly, is a review of some of them, following the same order as above in relation to litigation:

---

[16] The Act came into force on January 31, 1997, and applies to all arbitrations commencing after that date. The court procedures were amended by the introduction into the RSC of a new Order 73. That Order is now reproduced largely unchanged in Practice Direction Part 49G, CPR.

[17] This may become less of a concern in the U.K. if the courts adopt the approach of Laddie J. in *Coflexip SA v. Stolt Comex Seaway (No. 2)* [1999] F.S.R. 473, in which he awarded an injunction of limited scope, not one which simply prohibited infringement of the patent in general terms — see para. 24–75.

[18] The Convention on the Recognition and Enforcement of Foreign Arbitral Awards adopted by the United Nations Conference on International Commercial Arbitration on June 10, 1958.

[19] Though this is less of an issue in Europe as the Brussels and Lugano Conventions (see Section C below) provide for mutual recognition of judgements of Contracting States.

[20] For arbitrations covered by the Arbitration Act 1996, this freedom is limited by the Act's mandatory provisions which cannot be contracted out of. These include the provisions on staying legal proceedings (Section 9), the court's power to extend the limitation period in arbitration agreements (Section 12), the court's powers to secure the attendance of witnesses (Section 43), the provisions on enforcement of awards (Section 66), and those dealing with challenges to the awards (Section 67–68).

[21] See for example the Final Report of the Commission on International Arbitration of the International Chamber of Commerce on "Intellectual Property Disputes and Arbitration" (Document No 420/362; prepared for a meeting on October 28, 1997 by Julian D M Lew, Chairman of the Working Party); and Karet and Duncan, "Arbitration in Intellectual Property Disputes" in The CIPA Journal, April 1998, at 260–269.

- **Publicity**: Arbitrations are, generally speaking, private. The proceedings and their outcome will, in principle, be confidential. This is subject to two caveats: leaks sometimes occur, and, if recourse is made to the court during the arbitration proceedings, matters may well then become public. Nevertheless, if confidentiality is desired, arbitration is a better option than litigation.

- **Cost**: Unlike judges, arbitrators charge the parties for their services. Their rates will depend on seniority, experience, and so on. By their nature, arbitrators tend to be senior, experienced professionals, and therefore tend to be expensive. In addition, the parties normally have to pay to hire the rooms in which the arbitration will take place, and in certain arbitrations (notably ICC[22] arbitrations) there is also the cost of the arbitration body concerned. The cost of professional advisors will vary significantly depending on the procedures agreed by the parties, the extent to which they are complied with in a timely manner, the extent of disclosure, the length of the arbitral hearing, and so on. In principle, arbitration proceedings should be cheaper than litigation overall. However, that will not always be the case, particularly in complex technical intellectual property disputes, and in the light of the new, streamlined, procedures in the United Kingdom courts.

- **Expertise**: As has already been said, the judges which hear intellectual property disputes in England are generally well versed in the legal and technical issues. Nevertheless, if the dispute concerns highly complex technical issues, the parties may prefer to have a more technically qualified individual assessing the merits of the case. In this respect, arbitration provides greater flexibility than litigation, as the parties can seek to agree a more suitably qualified tribunal. If the dispute involves complex legal issues as well, it is likely to be necessary to have a lawyer on the arbitral panel. In such circumstances, a panel of three arbitrators may be most appropriate, although this will necessarily increase cost.

- **Procedure**: The differences in relation to procedural aspects are more limited than some of the other forms of dispute resolution referred to here, although there is inevitably greater flexibility in an arbitration.

- **Delay**: As has been said, the traditional view is that court proceedings are long and protracted. That is changing. But arbitrations may also be becoming faster: there are provisions in the 1996 Arbitration Act to encourage arbitrators (and the parties) to deal with arbitrations more quickly.[23] However, the lack of sanctions available to an arbitrator, and the consequent need to invoke

---

[22] International Chamber of Commerce.

[23] For example, the opening words of the Act (Section 1(a)) are:

"The provisions of this Part are founded on the following principles, and shall be construed accordingly —

(a)   the object of arbitration is to obtain the fair resolution of disputes by an impartial tribunal *without unnecessary delay or expense . . .*" (emphasis added).

the court's jurisdiction in the event of failure to comply, can lengthen arbitration disputes.[24]

- **Third Parties**: Neither the parties to an arbitration nor the arbitrator himself have the power to join third parties into the arbitration. Only if the relevant third party agrees to be bound by the arbitration agreement can the dispute with it be adjudicated in the same proceedings. In certain circumstances, this may be a significant disadvantage over litigation.

- **Appeals**: The rights of appeal against an arbitrator's decision are more limited than those in court proceedings. The award can be challenged only on the grounds of "serious irregularity"[25] and, unless the parties otherwise agree, appealed only on a point of law.[26] An appeal must be brought within 28 days of the date of the award.[27]

24–07 Arbitration clauses If, having considered all of these issues, it is decided that an arbitration clause should be included in the relevant agreement, there are a number of important issues which need to be addressed. These are all covered in standard clauses recommended by the arbitration bodies. That recommended by WIPO[28] reads as follows:

"Any dispute, controversy or claim arising under, out of or relating to this contract and any subsequent amendments of this contract, including, without limitation, its formation, its validity, binding effect, interpretation, performance, breach or termination, as well as non-contractual claims, shall be referred to and finally determined by arbitration in accordance with the WIPO Arbitration Rules. The arbitral tribunal shall consist of [three arbitrators] [a sole arbitrator]. The place of arbitration shall be . . . . The language to be used in the arbitral pro-

---

[24] Arbitrators have greater powers under the 1996 Act than under previous Arbitration Acts. They are set out in section 41 of the Act, and include the power to: dismiss a claim if there has been "inordinate and inexcusable delay on the part of the claimant and the delay (a) gives rise, or is likely to give rise, to a substantial risk that it is not possible to have a fair resolution of the issues in that claim, or (b) has caused, or is likely to cause, serious prejudice to the respondent . . ." (s. 41(3)); continue a hearing without a party being present (s. 41(4)); and make peremptory orders (s. 41(6)) which, if not complied with, can then be enforced by a court on application to it by one of the parties (in the circumstances set out in s. 42 of the Act). There are also practical steps the arbitrator can take, such as fixing an early hearing date at the initial meeting. Further, parties to an arbitration may not want to delay without good cause for fear that it might incur the displeasure of the arbitrator.

[25] Section 68, of the Arbitration Act 1996.

[26] Section 69, *ibid*. This is not one of the mandatory provisions of the Arbitration Act. The parties can therefore contract out of it. Further, an agreement between the parties to dispense with the requirement for the Tribunal to give reasons for its award is deemed to be an agreement to exclude the court's jurisdiction in relation to appeals (section 69 (1)).

[27] Sections 70(3) and 54, *ibid*.

[28] The World Intellectual Property Organisation. WIPO established its own Arbitration Centre in October, 1994. Both the WIPO Arbitration Rules and the WIPO Mediation Rules contain provisions designed to accommodate the specific characteristics of intellectual property disputes. There are detailed provisions on confidentiality, including a special mechanism for the appointment of a confidentiality advisor in arbitrations involving trade secrets. There are also specific provisions on scientific and technical evidence, experiments and site visits. For trans-national dispute resolutions, the Centre offers a single procedure rather than several court actions in different countries; the Rules are designed for use in any legal system. The Centre also offers a procedure for the grant of immediate interim relief involving a panel of stand-by arbitrators who can be appointed within 24 hours of request to hear the matter and make an award almost immediately.

ceedings shall be . . . The dispute, controversy or claim shall be decided in accordance with the law of . . .".

There is no particular magic in this form of words. A binding agreement to arbitrate will come into existence, at least as a matter of English Law, without this level of detail.[29] However, delay and disagreement over procedural issues will be minimised when a dispute arises if wording along these lines is used, and the following issues (at least) are dealt with in the arbitration clause:

- Procedural rules to be applied

- Place of arbitration

- Governing law

- Number of arbitrators

- Mechanism for appointing arbitrator(s)

- Language of the arbitration

24–08 Other mechanisms There are a wide range of mechanisms available for resolving disputes without recourse to litigation or arbitration. These include expert determination and various forms of alternative dispute resolution, or "ADR". These have tended to be less widely used in the intellectual property arena than in others. The growth in ADR in the United States (where it originated) has not been mirrored (at least not yet) in the United Kingdom. A detailed review of all the options would therefore be inappropriate.[30] A few words about these alternatives may nevertheless be of assistance to the reader.

24–09 Expert determination This differs from court proceedings or arbitration in that an expert, unlike a judge or an arbitrator, is permitted (and required) to use his professional expertise to resolve the matter before him. It is therefore often used in contracts which provide for determination of technical issues. It may well therefore be appropriate for example to include in agreements relating to the supply of computer equipment or software, where the question of whether the materials perform in accordance with the specification is likely to be a key issue in any dispute which arises. This mechanism should be quick, cheap, procedurally straightforward and private. However an expert cannot seek assistance on legal issues from the court, and enforcement is more difficult.

24–10 ADR This is a somewhat vague term, covering a range of dispute resolution mechanisms. It is usually understood by practitioners to cover mechanisms which involve a neutral and independent third party, where the objective is to seek a settlement not a determination of legal rights, and which is non-binding and confidential.[31]

---

[29] Particularly under the 1996 Arbitration Act, which provides that the requirement for an *"agreement in writing"* (to render an agreement subject to the Act) is satisfied if the agreement is in writing, even if not signed by the parties, or if evidenced by an exchange of communications in writing, or if the agreement is evidenced in writing (which can include an agreement made otherwise than in writing if it is recorded by one of the parties, or by a third party with the authority of the parties to the agreement) — Section 5 of the Act.

[30] For a detailed review of available ADR mechanisms, and their benefits and potential disadvantages, see for example Bevan, *Alternative Dispute Resolution*, (Sweet & Maxwell, 1992).

[31] This is the sense in which the term is used here. It seems to have taken on an even broader meaning among certain practitioners in the U. S., encompassing everything other than full-blown litigation, including arbitration. Given this inconsistency in usage, anyone who is asked whether they are interested in resolving a dispute by "ADR" should first seek clarification as to what the other party means by the term.

Various bodies have been established in the United Kingdom to assist parties in ADR.[32] Also, and perhaps more importantly for the future of ADR in the United Kingdom, its use is strongly encouraged in the new Civil Procedure Rules. One of the court's new responsibilities is to be more pro-active in case management. This includes helping the parties to settle the whole or part of the case,[33] to encourage the parties to use ADR if appropriate, and to facilitate its use.[34]

24–11 Purpose of ADR As has been said, ADR does not seek to determine the legal rights of the parties, but to encourage settlement. A high proportion of civil cases settle before trial anyway. The objective of ADR is to focus the parties on the key issues, and encourage them to discuss them at an earlier stage and thereby save legal costs.

ADR is non-binding. This is intended to assist in disclosure and frank discussion. It requires the co-operation of the parties. Without that, it cannot succeed. It must also be confidential, and all discussions held "without prejudice" to the party's legal rights.

It is not necessarily an *alternative* to litigation or arbitration. It can be used as an *additional* tool in resolving disputes, running simultaneously with litigation or arbitration, or as a precursor to either of them.

24–12 The neutral party The success or failure of the process may in many cases depend on the characteristics of the individual appointed (for example by CEDR) to assist the parties to settle their dispute. The individual must be seen to be fair and trustworthy, be in a position to understand the parties' concerns and the legal and technical issues involved, and have good negotiating skills.

24–13 Future of ADR ADR has certainly increased in popularity in recent years. It has not, however, seen the significant growth in the United Kingdom which many had predicted. It remains to be seen whether the new Civil Procedure Rules will have any impact on the use of ADR in technology-related disputes.

Review

24–14 The preceding paragraphs have identified the main differences between various mechanisms for resolving technology-related disputes, concentrating in particular on the impact of these differences on disputes involving intellectual property rights. In many cases, litigation will be the only realistic option, notwithstanding its inherent disadvantages. In other circumstances the less formal alternatives may be better. Each agreement or dispute must be considered on its facts. Strategic issues may also be important. These are considered in the following section.

B. Strategy

General

24–15 There are numerous questions that need to be addressed before embarking on litigation or other dispute–resolution procedures. The answers will determine what rights (if any) that party has, where and how it can enforce them, what remedies are

---

[32] For example, the Centre for Dispute Resolution ("CEDR"), established in November 1990.
[33] CPR, r. 1.4(2)(f).
[34] CPR, r. 1.4(2)(e).

available to it, how quickly it needs to act to avoid losing the right to any such remedies and the procedure it should follow in order to obtain those remedies.

Commercial aims

24–16   Probably the single most important question to ask at the outset is: "*What are my commercial objectives?*" In turn, this raises further questions. What remedies do you want? Are you content to seek financial compensation, or do you also want to stop the activity complained of? If the latter, do you want to stop it as a matter of urgency? And over what geographical scope? Do you think the other party may destroy important evidence if you give him notice of your intention to take action against him; or might he move his assets out of the relevant jurisdiction to avoid your being able to enforce any claim against him?

Other questions to address

24–17   In parallel with these strategic questions, a number of other questions need to be answered before embarking on any legal process: What has the other party done wrong? Does his act give you a legal right against him? In the case of a contract, does the act constitute a breach of contractual obligations? If so, how important was that provision? Does the agreement specify the consequences of such an act? For example, does it give you the right to terminate, and if so, how is that right to be exercised? Do you want to terminate, or would you prefer the agreement to continue in force? When did you find out about that act? What is your budget? Do you want to try to keep on good terms with the other party, or is an ongoing relationship irrelevant or impossible? What does the agreement say? It may sound obvious, but the first place to look when considering how to resolve a contractual dispute is clearly the agreement itself. What does the contract say about dispute resolution? Are there any time limits? Is there a specific dispute resolution mechanism, requiring, for example, mediation or arbitration rather than court action? Does it say where disputes must be resolved, and under which law? Are there any limits on the remedies available?

Cost questions

24–18   An important new strategic consideration in United Kingdom disputes is the potential cost consequences of rushing prematurely into litigation. The new Civil Procedure Rules require the parties to any dispute to explore the possibility of settlement before launching into litigation.[35] If it is clear that a negotiated settlement is unrealistic, there may be little point in trying. For example, if you have a patent which covers an extremely valuable product (say, in the pharmaceuticals field), and you believe a major competitor is selling a product which infringes, there may be little to gain by without prejudice discussions. It may be clear to you that they would not be interested in taking a licence (even if you are prepared to consider granting one as a compromise to avoid the cost and inconvenience of litigation), and would rather take their chances in a court dispute, in which they will inevitably deny infringement, and challenge validity. However, a decision not at least to explore the possibility of some form of settlement should be taken in the knowledge that it may ultimately be taken into account by the court in assessing costs. On the other hand, laying your cards on the table at an early stage also has disadvantages, giving your opponent more time to prepare his case.

---

[35] CPR r. 1.1(2)(a), (b), and (c).

On balance, if the commercial stakes are high enough, the potential cost impact of not exploring settlement may not be significant, and may militate in favour of prompt action, without first engaging in settlement discussions. The strategy you will want to adopt in relation to any particular dispute will vary with the circumstances of each particular case.

There are no hard and fast rules. Each case must be considered on its merits, and in the light of the commercial objectives of the parties. An important factor is the likely cost of the various available approaches. No organisation has unlimited funds, and any sensible commercial entity will want to keep legal costs to a minimum. A key issue, however, will be "cost effectiveness" rather than absolute cost. Investing up front in a full preparation of your case before putting it to the other side may well result in significant costs savings down the line. Again, each case has to be considered on its merits.

International disputes

Choice of forum

24–19 If the dispute involves more than one country, other strategic and practical issues will need to be considered at the outset.[36] One of the most important of these will be choice of forum.

The rules which determine which courts have jurisdiction over any particular dispute will be discussed in Section C. The following paragraphs consider instead the strategic issues that should be taken into account, and assume that jurisdictional issues will not constrain implementation of the chosen strategy. The following factors will influence the choice of forum:

- **Available remedies**: Various forms of interim relief may be available. These include an interim injunction, search order and freezing order. You will need to consider which countries grant such remedies, the likelihood of them actually being granted in the circumstances of the case, at what cost, and subject to what conditions.[37] You should also consider how likely it is that the first instance court's decision will be reversed on appeal.

- **Scope of relief**: The geographical scope of the relief that courts in different countries will grant may also be important. This issue is considered in more detail in Section C below. As that discussion will reveal, care must be taken in deciding where to sue first, as starting litigation in the "wrong" country may prejudice your chances of success.

- **Speed**: The "efficiency" of each jurisdiction's legal system will also be a factor. In Europe, relief is generally available faster in the United Kingdom, Germany and the Netherlands (at least interim relief) than, for example, Spain and Italy.

A further factor is the relevant national courts' policy on stays. Until recently, the general practice of courts in the United Kingdom and Holland was not to

---

[36] This discussion focuses on disputes which are likely to result in parallel *litigation* in numerous countries. The issues may be less relevant for other forms of dispute resolution mechanism.
[37] For example, a cross-undertaking and/or bond to cover any damages suffered by the defendant may be called for.

stay national patent disputes pending the outcome of counterpart opposition proceedings at the European Patent Office (EPO); and the French courts tended to take the opposite stance. However, recent English and French court decisions suggest that these "traditional" approaches may be changing. So local advisers should be consulted if this is an important issue in any particular case.

The speed with which a case is dealt with will also depend on the local courts' policy in relation to expert witnesses: does the local court insist on appointing an expert, and if so, what is their role? The litigation procedure is likely to be slower where courts seek the views of technical (and in some cases also patent law) experts: for example in many continental countries (particularly Southern Europe — Spain and Italy).

- **Precedent value of decision**: The precedent value of the decisions of a particular country's courts may also be a factor. The decisions of some tribunals carry more weight that others, and this may militate in favour of a strategy involving starting proceedings in that court first, obtaining a favourable decision, and then seeking to use that to your advantage elsewhere. Similarly, suing first in a jurisdiction where local law is unfavourable, or is uncertain, may create adverse precedent, whereas suing first in a favourable forum may have the opposite effect. Much will depend on the particular judge that hears a case in a particular country. However, the patents judges in Europe now meet on a regular basis and seem committed to reducing the level of inconsistency in the approaches they take and the conclusions they reach. Attempts are also being made to achieve greater consistency between the decisions of national courts and those of the Opposition Division and Appeal Boards of the European Patent Office.[38] EPO decisions (particularly of the Appeal Boards) are generally regarded as influential, though not binding, on European (and other) courts; though there are notable exceptions.[39]

- **Procedural complexity**: The different litigation procedures adopted in different countries may also be a factor. For example, countries in which discovery (now "disclosure") and oral evidence (with cross examination) are the norm (including the United States and United Kingdom[40]) are often considered to give a fairer result but obviously at a cost. To quote from a recent article by a French author on this point:

"The system of discovery and cross-examination as a means of establishing proof are seen by lawyers of the common law system as the guarantee of the remarkable quality of the English system of justice; but is held by continental lawyers in the same reverence as for the finest of the best English motor cars: respect tinged with fear as to the price. From their side of the Channel, our English friends regard our rapid and energetic French proceedings for seizure in

---

[38] For a recent review of the different approaches taken across Europe, the difficulties this presents for litigants, and the proposals for reform, see P. Colletti, "No relief in sight: difficulties in obtaining judgements in Europe using EPO issued patents", [1999] J.P.T.O.S. 351.

[39] For example, in *Biogen v. Medeva* [1995] R.P.C. 68, a case involving Biogen's Hepatitis B patents, the U.K. Court of Appeal revoked the patent shortly after the EPO's Technical Board of Appeal had upheld it ([1995] E.P.O.R. 1). The Court of Appeal decision was later confirmed by the House of Lords ([1997] R.P.C. 1).

[40] Though see comments elsewhere in this Chapter on the more restrictive approach now being taken by the English courts on discovery and the scope for lengthy cross-examination at trial.

infringement matters as they would our plates of frogs' legs: more with fear than envy."[41]

- **Cost**: The relative cost of litigating in particular jurisdictions will also be a factor. Litigation in the United States is generally considered to be more expensive than anywhere else. In Europe, the United Kingdom has in the past been viewed as being more expensive than elsewhere. However this is not true of all cases,[42] and should become a less important factor as more streamlined procedures take effect in the United Kingdom. Further, as has already been said, it is cost *effectiveness*, not cost *per se*, which is important; and many organisations consider the United Kingdom procedures, with the right to cross-examine witnesses and to gain access to the other party's documents, to be worth every penny.

- **Importance of location**: The value of the local market in each country will also need to be taken into account. In the case of an intellectual property dispute where an injunction is sought, all other factors being equal, you will want to focus your efforts (and resources) on your most important markets, or alternatively on the defendant's manufacturing location. Obtaining an injunction to prevent manufacture will probably stop worldwide sales, subject to the defendant's ability to stock large volumes of product elsewhere and, possibly, its ability to shift manufacture (although in some fields, notably pharmaceuticals, the latter is unlikely to be possible in the short term for regulatory reasons).

Conclusion

24–20 No matter how strong (or weak) your case may be as a matter of law, your strategy in dealing with the dispute is critical. Whether you are looking to inflict a quick "mortal blow" on your opponent or to delay an eventual court decision for as long as possible (for example, to allow time for negotiations or introduce a new product which you hope will render the outcome of the dispute irrelevant) you must consider the various points raised above at an early stage. An investment in obtaining proper advice on these issues up front will pay significant dividends down the line.

C. Jurisdiction

Introduction

24–21 If two English companies disagree on whether an agreement for the provision of services in England has been complied with, it will not take long to decide what the governing law of the dispute is, nor where it should be adjudicated. However where there is an international element to the dispute, as is increasingly the case, the position will not be as straightforward. If, for example, a German company is selling a product throughout Europe, and a United States company believes that the product infringes intellectual property rights it has in the United Kingdom, Italy, France and Germany,

---

[41] Pierre Véron of Lamy, Véron, Ribeye & Associes, Lyon, France.
[42] For example, the author is aware of one case where German lawyers (who can charge on the value of the litigation rather than on an hourly rate) charged a very significant fee for drafting an initial brief, the sole purpose of which was to get the other party to the negotiating table. The charge for equivalent work in the U.K. would have been much less.

where can it sue, and what powers do the Courts have in relation to acts outside their territory?

These issues are addressed in a number of European Conventions, and have been the subject of numerous cases in a number of key European jurisdictions in recent years. Unfortunately, the decisions have not been consistent, and (in some important respects) it is difficult to draw any clear guidance from them.

The Current position

Basic rules — Brussels and Lugano Conventions

24-22 The law which governs so-called "forum shopping" in Europe is set out in the Brussels[43] and Lugano[44] Conventions. The basic rule is that a defendant must be sued in its country of domicile.[45] Exceptions to this rule (which must be construed narrowly) provide that a defendant can instead be sued in the country in which the infringing acts take place[46] or, in the case of a co-defendant, in the country in which any of the other co-defendants is domiciled.[47] In matters relating to a contract, there is a further exception, namely that the defendant may be sued in the courts of the place of performance of the obligation in question.[48] Defendants can also be sued elsewhere if they voluntarily submit to the jurisdiction or, subject to the specific exceptions discussed below, where a valid agreement binds their submission to another jurisdiction.[49]

24-23 Validity of IP rights As regards questions of the *validity* of a patent (or other *registered* intellectual property right), *exclusive* jurisdiction is given to the courts of the country of registration.[50] This is because the grant of a patent is an exercise of national sovereignty, and other States should not interfere in that exercise. The Convention also provides that where a court of one country is "seized" of a claim (*i.e.* where the relevant proceedings have been started before that court) which is *"principally concerned"* with a matter over which the courts of another country have exclusive jurisdiction (by virtue of Article 16), it must decline jurisdiction.[51]

24-24 Conflicts of jurisdiction There are also rules in the Convention to determine what should happen where the same *cause* of action or a related cause of action is being heard simultaneously in more than one jurisdiction. The aim is to avoid the risk of conflicting judgments. Where proceedings involving the same cause of action and between the same parties are brought in the course of different contracting states, any court other than the court first seized *must* decline jurisdiction in favour of the other.[52] Where

---

[43] The Convention on Jurisdiction and the Enforcement of Judgments in Civil and Commercial Matters, implemented in the U.K. by the Civil Jurisdiction and Judgments Act 1982.

[44] The Lugano Convention on Jurisdiction and the Enforcement of Judgments 1988, implemented in the U.K. by the Civil Jurisdiction and Judgments Act 1991. For present purposes, the Lugano Convention is substantially identical to the Brussels Convention.

[45] Article 1. This is because one of the stated purposes of the Conventions is to protect the interests of Defendants and, generally, a defendant will find it easier to defend its rights in its home territory.

[46] Article 5(3).

[47] Article 6(1).

[48] Article 5(1).

[49] Article 17. The general principle is that an agreement by which a party submits to a particular jurisdiction is binding if at least one party is domiciled in a contracting state, and if no party is so domiciled, the chosen court must still be given a right of first refusal. However, Article 17 is expressly subordinate to Article 16 (see below), so an agreement of the parties that the validity of, say, an English patent should be determined by a German court would not be effective.

[50] Article 16(4).

[51] Article 19.

[52] Article 21.

related actions are brought in different courts, any court other than the court first seized may stay the proceedings.[53] In addition, a court other than the court first seized may, on the application of one of the parties, decline jurisdiction if the law of that court permits the consolidation of related actions and the court first seized has jurisdiction over both actions.

Actions are "related" where they are "*so closely connected that it is expedient to hear and determine them together to avoid the risk of irreconcilable judgments resulting from separate proceedings*".

24–25 Patent disputes   These provisions are particularly relevant when disputes relating to parallel patents are pending in more than one country. It is quite common for a claimant to run several sets of patent infringement proceedings simultaneously: for example in major markets such as England and Germany. The two sets of proceedings might involve exactly the same parties and even the same infringement issues. The courts will need to consider whether the two actions involve "*the same cause of action*", such that the second court would be obliged to decline jurisdiction. The conventional answer is "no", on the basis that each national patent is territorial and gives rise to a distinct cause of action. However, it is open to the court to adopt a broader interpretation of Article 21, and decide that they are the same. In England, the court took a narrow view of the term "*same cause of action*" in a trade mark infringement and passing off action. It refused to yield jurisdiction to the Irish court, saying that the trademarks were territorial in nature.[54]

Even if the actions do not involve the *same* cause of action, it is pretty clear that the actions are "related". So the court second seized of the case has a discretion whether or not to stay. Given that the purpose of Article 22 is to avoid irreconcilable judgments and to avoid duplication of effort, it will often be preferable for the second court to stay proceedings. However, each case will depend on its particular circumstances.

Where the relevant proceedings are for provisional measures (such as a preliminary injunction in a patent infringement action) there is a further exception to the basic rules. Such measures can be granted by courts in countries other than those which would normally have exclusive jurisdiction.[55]

Application of rules in recent patent cases

24–26   The reason for the recent increase in attention given to these issues in intellectual property disputes is that, in the early 1990s, the Dutch court decided that it had jurisdiction to grant injunctions in patent infringement cases covering acts outside Holland.[56.] This followed an earlier trade mark infringement case, in which the Dutch Court initiated this trend.[57] Litigants have tried to obtain similar relief in other jurisdictions, but with limited success. In particular, the United Kingdom courts signalled fairly clearly that they did not consider it appropriate to grant such

---

[53] Article 22.

[54] *L A Gear Inc v. Gerald Whelan & Sons Limited* [1991] F.S.R. 670.

[55] Article 24.

[56] See, for example, *Philips v. Hemogram*, The Hague District Court, December 30, 1991, 1992 BIE, No. 80, 323, 1992 IER, No. 17, 76; *Vredo v. Samson*, The Hague Court of Appeals, June 4, 1992; *Pipe Liners v. Wavin*, The Hague District Court, December 28, 1990, 1991 KG, No. 80, 1991 IER, No. 6, 19, aff'd, The Hague Court of Appeals, Jan 16 1992, 1993 BIE, No. 9, 44, 1992 KG No. 85, 1992, No. 10 53; *Applied Research Systems v. Organon*, The Hague Court of Appeals, February 3, 1994, 1995 IER 8, 1995, GRUR Int 253; *Chiron Co v. Akzo Pharma-Organon Technika-UBI*, The Hague District Court, July 22, 1994, 1994 I.E.R., No. 24 150.

[57] *Interlas v. Lincoln*, HR November 24, 1989, [1992] N.J. 404,1597.

relief.[58] And more recently, the Dutch courts have taken a more restrictive approach.[59]

A number of these issues arose in two recent cases which came before the English Court of Appeal.[60] The court decided that the questions were of sufficient importance to warrant a reference to the European Court of Justice.[61] As and when these are answered, further clarification may be provided. In the meantime, to the extent that the decisions reveal common themes as to the approach an English court will take they are these:

- A claimant can bring an action in the United Kingdom for infringement of United Kingdom IP rights by a foreigner.[62]

- Defendants of different domiciles can be sued together in, say, the United Kingdom, if one defendant is domiciled here.[63]

- The English courts have jurisdiction to hear an action against an English defendant for infringement of foreign unregistered IP rights[64] on the basis of Article 2, and all his foreign co-defendants on the basis of Article 6(1), even though the courts in the other states might also have jurisdiction on the basis that the non-U.K. defendants are domiciled there.

- The fundamental question in an action for the infringement of a registered IP right is whether the defendant(s) is/are infringing a valid claim. So an English court will seldom, if ever, have jurisdiction to hear an action for infringement of a foreign registered right.

- There is not necessarily a sufficient connection between parallel national

---

[58] *E.S. Coin Controls Limited v. Suzo International (U.K.) Ltd* [1997] F.S.R. 660, where Laddie J. came to the conclusion that where both validity and infringement were in issue, as would normally be the case, the issue of validity was a principal issue in the proceedings, that both infringement and validity had therefore to be determined together; and as a result, under article 16(4), both issues had to be determined by the court in the country in which the patent was registered. In those circumstances, the English court could not decide the issue of infringement of a non-English patent. At the time, this was in stark contrast to the position being adopted by the Dutch Courts. See also *Pearce v. Ove Arup* [1997] FSR 641, upheld on appeal ([1999] 1 All ER 769); *Fort Dodge v. Akzo Nobel* [1998] F.S.R. 222, CA; *Boston Scientific v. Cordis* [1998] F.S.R. 222 (CA).

[59] *EGP v. Boston Scientific*, The Hague Court of Appeals, April 23, 1998, [1999] F.S.R. 352.

[60] *Fort Dodge v. Akzo Nobel; Boston Scientific v. Cordis*. (See above.)

[61] The following is a summary of the major questions referred to the ECJ:
   1. Where validity and infringement are both in issue: do the courts of the state where a patent is registered have exclusive jurisdiction in relation to validity and infringement under Articles 16(4) and 19 of the Brussels Convention? Should the courts of a state where a patent is not registered decline jurisdiction (under Article 19)?
   2. Does the EPC take precedence over the Brussels Convention so that the EPC rule (that a European patent confers the same rights as a national patent and infringement shall be dealt with by national law) applies?
   3. If so, does this mean that (notwithstanding the Brussels Convention) infringement proceedings may only be brought in the territory where the patent is registered and that no cross-border relief is available?
   4. Can a court grant provisional relief (under Article 24) where: it does not otherwise have jurisdiction; and there are no proceedings for final relief pending or imminent before a court with jurisdiction as to the substance of the matter?
   5. Where two national patents stem from the same European patent, is this sufficient connection for jurisdiction against a non-domiciled defendant where the litigation is pending under Article 6(1), where either: (a) the domiciled person is alleged to infringe both patents; or (b) the domiciled person is alleged to infringe "home" patent and the non-domiciled person to infringe the other patent?
   6. Are the answers to the above questions different if the proceedings are Dutch Kort Geding proceedings?

[62] Article 5(3): an action for compensation for IP infringement is a "matter relating to a tort".

[63] Article 6(1).

[64] To which Article 16(4) does not apply.

European IP cases under related rights to require them to be heard together under Article 6.

24–27   Position in Holland   As regards Holland, the main points in the light of the *EGP v. Boston Scientific*[65] decision are these:

- Where there are multiple defendants, they can be sued in Holland if there is a sufficient connection between the actions. But a remote connection will no longer suffice to give the Dutch courts jurisdiction over non-Dutch defendants alleged to have infringed non-Dutch patents. The Dutch courts will have jurisdiction over non-Dutch patents alleged to be infringed by a Dutch defendant. They will also have jurisdiction over non-Dutch defendants alleged to have infringed non-Dutch patents, but only if the defendants are part of the same group and supply identical products to the national markets, and the company with the key role in the infringement, responsible for the "common plan" or management, is Dutch — the so-called "spider in the web" approach. In other circumstances, the claimant will have to sue the non-Dutch defendant in that defendant's home jurisdiction.[66]

- Where the validity of the non-Dutch patent is put in issue in proceedings on the merits, the Dutch court should "take a cautious approach" with respect to the infringement claim. It should normally suspend (or "stay") the infringement action in respect of the foreign patent until the foreign court has given judgment on validity.[67]

- In contrast to the position in the early 1990s, where jurisdiction is based entirely on Article 24 (interim measures), the court should not grant relief by way of injunction covering countries outside Holland.[68]

The future

24–28   In addition to clarification from the European Court of Justice when it answers the questions recently referred to it, there are moves afoot to address these difficulties via new legislation. The proposals fall into two main classes. First, attempts to create a single European patent, with a new enforcement regime which would avoid the problems outlined above. Second, proposals to amend the Brussels Convention, and possibly also to bring in a new fully international Convention covering these issues. Although both have been the subject of extensive discussion (and indeed often heated debate), both proposals are at a relatively early stage, and many uncertainties remain. It would therefore be inappropriate in a work of this nature to discuss the proposals in detail. Suffice it to say that anyone reading this text more than a year or two after its publication should investigate the progress made on these initiatives before relying on what has been said above. If significant advances have been made since the date on which this text was written, the situation may by then be very different.

Conclusion

24–29   Many non-European companies already view Europe as a single market. When they encounter patent problems here, their initial (understandable) reaction is

---

[65] See above.
[66] Paragraphs 16 to 18 of the Hague Court of Appeal's decision.
[67] *ibid.*, paragraph 28.
[68] *ibid.*, paragraph 32.

to try to find a single forum in which the problem can be resolved. That search is seldom successful. Instead it leads them into the complex web of legislation and case law which this section has sought to summarise and, to some extent, unravel. There is hope that, one day, the position will be more straightforward. The European Court of Justice may answer some of the main questions which are the cause of the uncertainties; or legislative reform may achieve the same end. Neither, however, is likely to give its answers quickly.

D. PROCEDURE

Introduction

24–30 This section explains the procedures that have to be followed in litigation in the English courts. The procedures are described by reference mainly to the context of intellectual property disputes. However, they will also be generally applicable to contractual disputes. It is assumed for the purposes of this section (recapping on the issues that have been addressed in earlier sections) that the party wishing to have the dispute resolved has decided that litigation (rather than arbitration or ADR) is the appropriate mechanism, that, strategically, now is the time to bring proceedings, and that it has satisfied itself that England is the (or one of the) appropriate jurisdiction(s) in which to bring the proceedings.

It provides a general guide to the procedures, and an overview of what to expect when enforcing rights in England, or faced with defending litigation here. It does not deal with every detail or exception.[69] It is also limited to the procedures in the English courts. Other jurisdictions including Scotland have significantly different legal systems, and the procedures therefore vary from those outlined here.

This summary is based on the new Civil Procedure Rules, introduced on April 26, 1999. The general approach is similar to that under the previous regime with which many readers may be familiar but there are *important* differences. Some are fairly minor, such as the nomenclature[70]; others are more significant, such as the introduction of case management conferences and other measures giving the court powers to be more pro-active in managing the case and encouraging settlement.

These differences seek to achieve the objectives of the Woolf Reforms: namely to make the civil justice system quick, fairer and cheaper.

It is early days, but indications so far are that the judges are using their new powers proactively, and potential users of the English system need to be aware that, once proceedings are commenced, the judges are likely to push the proceedings forward at what many may regard as an uncomfortably quick pace.

Relevant tribunals

24–31 The tribunals which have jurisdiction to hear disputes about United Kingdom patents are the United Kingdom Patent Office, the EPO, the Patents Court (in the High Court) and the Patents County Court. The United Kingdom and European Patent Offices do not have jurisdiction to deal with contractual disputes; the High Court and the Patents County Court do. Further, disputes relating to infringement are

---

[69] Given the very recent introduction of the CPR, few authoritative commentaries are yet available. One such text is Blackstone's Guide to the Civil Procedure Rules, Blackstone Press Limited, 1999; though even it is already out of date, as it does not include reference to the numerous cases which have been decided under the new rules and have given guidance on their interpretation and impact.

[70] See for example the references to "Claim Form" rather than "Writ" and "Particulars of Claim" rather than "Statement of Claim" in nn. 6 and 7 above.

normally brought before one of the courts, as are most validity disputes. The following discussion therefore does not discuss further the procedures of either of the patent offices. It is the court procedures that will be discussed below.

Patents Court

24-32 The Patents Court is part of the Chancery Division of the High Court. The matters over which it has jurisdiction ("Patents Court business") include:

(a) any claim under the Patents Acts 1949 to 1961 and 1977;

(b) any claim under the Registered Designs Acts 1949 to 1961;

(c) any claim under the Defence Contracts Act 1958; and

(d) all proceedings for the determination of a question or the making of a declaration relating to a patent (or an application for a patent) under the inherent jurisdiction of the High Court.[71]

One judge presides over the court. There is never a jury.

Patents County Court

24-33 The Patents County Court (or "PCC") was created in 1990. While County Courts in the United Kingdom generally hear smaller cases, and have limited jurisdiction, there are as yet no formal limits on the jurisdiction of the PCC in patent matters.

One judge presides over the PCC.[71a] There are now also several deputy judges, appointed from among senior patent barristers.

The PCC has the same jurisdiction as the High Court over matters relating to patents (and designs). Its stated aims were to simplify the procedure and reduce the cost of patent litigation, especially to enable individuals and small and medium-sized businesses to bring or defend claims without becoming financially crippled.

The PCC has been operating unevenly since 1990. Generally it has worked well enough in small cases: it has succeeded in giving an airing to numerous simple cases where previously the patentee might not otherwise have litigated, because of the likely cost of enforcing his rights. On the other hand, the PCC has been much criticised in its handling of larger or more complex cases, some aspects of which have taken longer and cost much more than they would have done in the High Court. However, the PCC was not established for the large or complex case. The problem is that the "small man" does not only have small cases: he needs to be given access to justice even for his complex cases. It is not easy to achieve that at low cost, particularly if his opponent puts up a determined fight.

The following discussion focuses more on the High Court than the Patents County Court, reflecting the greater use presently being made of the former. In any event, the procedures are now much more uniform following the introduction of the CPR.

---

[71] Patents Court Practice Direction ("PCPD"), r. 2. Note that the Patents County Court also has jurisdiction over these matters. Further, the PCPD applies not only to Patents Court business but also to proceedings under the Copyright, Designs and Patents Act 1988, the Trade Marks Acts 1938 and 1994, and the Olympic Symbol etc. Protection Act 1995 and Olympic Association Right (Infringement Proceedings) Regulations 1995 (PCPD, Rule 1).

[71a] At the date of writing, there is in fact no such Judge, the first PCC Judge having retired and no new appointment having been made.

High Court procedures

Overview

24-34 Most IP cases and major contractual disputes take place in the High Court. This section discusses the procedures adopted in the High Court down to and including trial. It deals specifically with patent actions, identifying how these differ from other actions. Cases involving the protection of intellectual property sometimes require urgent relief to stop an infringement before the IP owner suffers irreparable harm. Urgent applications for relief are dealt with in Section E, while this section deals with the course of a normal action to full trial.

The effect of the new "CPR"

24-35 A major difference between the new and old procedures for resolving IP disputes is that the court, rather than the parties themselves, is now primarily responsible for the control and management of the litigation. Hence the court may now set timetables and give directions on its own initiative to ensure that a case proceeds in an expeditious manner in accordance with the new objectives. Other changes are that the obligation to produce documents during the disclosure stage is now slightly less onerous; and the obligation on experts to comply with their duty to the court has been set out in more detail. Information technology will be used more under the new procedures. And the court will now actively be encouraging settlement in order to meet the objectives of the new rules.

New rules on patent litigation

24-36 A special Patents Court Practice Direction has been issued which, together with the Civil Procedure Rules, governs matters of procedure in patent cases. The Practice Direction provides specific guidance on running patent cases.

The basic steps in a typical infringement action are described below.

A claimant before suing, or a defendant before defending, needs to give very careful consideration to the chances of success in the action if it should reach trial. In practice only a very small proportion of cases go all the way (only about 10 per cent of cases started in the High Court in 1999), but expenses can mount up at an early stage.

Commencing the action

24-37 Letter of demand Prior to commencement of normal legal proceedings it is usual (but not compulsory) to write a letter of demand, calling upon the infringer to stop his infringing acts and to pay damages for past infringements. Under the new rules, it will be more important to be seen to be acting in ways that will promote dispute resolution.[72] Judges are likely to be less sympathetic to a party who appears to be litigating for the sake of it, or seems to be unwilling to co-operate with his opponent, or appears unwilling even to consider settling. For these reasons it will normally be a good idea to send a fair and reasonable letter before commencing proceedings. However, the comments made elsewhere in this work concerning unjustified threats should be borne in mind at this stage.[73]

---

[72] Under rule 44.3(5) the court is required, when considering awarding costs, to look at the conduct of the parties before and during the proceedings, whether it was reasonable for a party to raise, pursue or contest a particular issue, the manner in which a party has dealt with the case or an issue and whether the claimant has exaggerated his claim. A Defendant may succeed in litigation but, having failed to comply with a pre-action protocol, be ordered to pay the Claimant's costs. (See also the discussion of this point in Section B, Strategy, above.)

[73] Paragraphs 8–103 to 8–108, above.

24–38 Claim form Formal proceedings are commenced by issuing a Claim Form (previously known as a "writ"). This is served on the infringer with detailed Particulars of Infringement.

The Claim Form sets out brief details of the infringement complained of and the relief sought. The relief sought may include damages or an account of profits, an injunction, delivery up of the infringing goods, and costs.

24–39 Particulars of claim The Particulars of Claim describe the acts of infringement that are alleged to have been carried out by the defendant. In a patent case it must state which of the claims in the specification of the patent are alleged to be infringed and give at least one instance of each type of infringement alleged.

24–40 Acknowledgment of service and defence The defendant must "acknowledge service" of the Particulars of Infringement within 14 days. If (having taken advice from his lawyers) he wishes to defend, he must file a Defence. Failure to do so may result in Judgment being entered in default. Rarely will a defendant give up at this stage. The Defence should not merely deny the truth of the claimant's case, but should state reasons for any denials. Where validity is not challenged, the Defence will generally be due within 28 days of service of the Claim Form and Particulars of Infringement.

24–41 Alleging invalidity in a patent action Where the defendant wishes to raise objections to the validity of the patent he does so by way of a Counterclaim for revocation, which will be included in the Defence. The Defence and Counterclaim must be served with Particulars of Objections setting out each of the grounds on which the validity of the patent is challenged. For example, if the grounds for objecting include lack of novelty or inventive step, the particulars must give full details of every prior publication or prior use relied upon. When validity is put in issue the defendant has 42 days from service of the Claim Form and Particulars of Infringement in which to serve these documents.

If the claimant wants to reply to any issues in the Defence and Counterclaim which require more than a mere denial, he may serve a Reply. He must serve his own Defence to any allegation made against him in the counter-claim.

24–42 Statement of truth All statements of case must contain a "statement of truth": that is, a statement signed by a person holding a senior position in the company or by its lawyers that the facts stated in the statement of case are believed by that party to be true. If any facts set out in a statement of case are subsequently shown to be false, the party (or signatory) may be in contempt of court, and thereby liable for fines or even imprisonment. If a party fails to include a statement of truth, his opponent may apply to the court for an order that the offending statement of case be struck out if it is not corrected promptly.

24–43 Summary judgment There is a procedure for either party to seek judgment at an early stage if he believes that his opponent's claim or defence (as the case may be) has no real prospect of success. He does this by making an application for summary judgment. In the past applications for summary judgment rarely succeeded in patent infringement actions, as defendants were generally able to raise enough of an argument of non-infringement or invalidity to prevent defeat at this stage. This may change under the new rules.

24–44 Extensions of time Any party who needs more time to serve a statement of case or other pleading has to apply to the court for an extension of time. It is no longer possible merely to obtain the other party's consent to a time extension. The only exception to this rule is that an extension of up to 28 days may be agreed for service of the Defence.

Case management

24–45 Within 14 days of the defendant filing a Defence or an Acknowledgement of Service form (whichever is the earlier), the claimant must apply to the court for directions on the next steps in the action.[74] These will be given at a Case Management Conference, held by a Master, or in the case of a patent action by a judge, which the parties will attend.

At the Case Management Conference the Master or judge reviews what has happened in the litigation so far and determines what are the key issues in terms of fact and law. He may discuss in general terms the costs incurred to date and the possibility of settlement; and must now enquire whether the parties have considered the option of alternative dispute resolution. The judge will set directions for the completion of the next steps in the litigation.

Admissions

24–46 The court encourages the parties to make admissions at an early stage of the action, for example in the Defence or Reply. Failing this, within 21 days of the due time for service of a Reply either party may serve on the other a Notice to Admit facts that are specified in the notice. This process is common in patent actions, though less so in other cases. A Notice to Admit is a useful way of trying to limit the matters that are in issue and that need to be determined by the court. The more that is admitted by each side, the simpler the action becomes, in particular through limiting the scope of disclosure of documents and the amount of evidence. A Notice to Admit may also include a request to identify points not in dispute and a request that a patentee identify which of his claims he intends to assert as having independent validity.

Admissions typically would cover matters such as the fact that the pleaded prior art was indeed published before the priority date of the patent; or the fact that the alleged infringing product has some of the characteristics that are required by the patent.

The party on whom a Notice to Admit is served has 21 days in which to respond with a Notice of Admissions. Failure to admit facts requested means that, if the fact is later proved, the person on whom the Notice to Admit was served must normally bear the cost of proving that fact: for example, the cost of having to call a witness especially to deal with the point. Therefore careful consideration should be given before refusing to make reasonable admissions.

---

[74] CPR, Part 49E, r. 2.5(2). This has the somewhat illogical consequence that a Case Management Conference can take place before the defendant has even served his Defence, which will clearly make it difficult for the court to assess the issues in the case and determine what the appropriate directions should be. Early indications from the Patents Court judges are that they will be reluctant to hold case management conferences before the Defence has been served.

Disclosure

24–47   The next step is normally the disclosure by each party to the other of relevant documents[75] that are under their control. Relevant documents are:

(a)   documents upon which a party will itself rely;

(b)   documents which could adversely affect a party's own case or the other party's case; and

(c)   documents which could support another party's case.

Parties are obliged to make a "reasonable search" for documents in categories (b) and (c). What constitutes a "reasonable search" depends on factors such as the number of documents involved, the nature and complexity of the proceedings, the ease and expense of retrieval and the likely significance of the documents to be retrieved.

24–48   Duty to disclose   Although the rules of disclosure are slightly more relaxed than the old "discovery" rules, disclosure is still an onerous and continuing obligation which is owed to the court. The parties' solicitors owe a duty, as officers of the court, to ensure that proper and full disclosure is given by their clients, and must take positive steps to ensure that their clients know the importance of preserving all relevant documents once the action has begun.

24–49   Privilege   A party is not obliged to allow inspection of "privileged" documents. These are documents in relation to which the party is entitled to claim privilege. The main kinds relevant to IP actions are:

- confidential communications passing between a party and his legal adviser if written for the purpose of getting or giving legal advice or assistance, whether or not in connection with the litigation ("legal advice privilege");

- confidential communications made by a legal adviser or by the client with third parties (for example, experts) with a view to getting information for the legal adviser and which have come into existence during the course of or in contemplation of the litigation ("litigation privilege");

- confidential communications relating to specified matters, including the protection of an invention, a design or technical information, passing between a person and his patent agent or made for the purpose of giving or obtaining information so as to be able to instruct a patent agent; — ("Patent agents' privilege")

- "without prejudice" communications, being documents relating to negotiations genuinely aimed at resolving the dispute between the parties.

24–50   Method of disclosure   Disclosure takes place by the parties exchanging lists of the documents to be disclosed. The list must include a "disclosure statement". This sets out the extent of the search carried out and certifies that the party itself understands the duty of disclosure and that it has carried out that duty. The documents

---

[75] Note that the term "documents" here means more than merely paper documents. It includes audio and video tapes, computer discs and other means of tangible recording of information. (CPR r. 31.4 defines the word document as, "anything in which information of any description is recorded".)

listed (apart from those that are privileged) are made available for inspection, usually at the parties' solicitors' offices. The parties can request copies of any or all of the non-privileged documents.

24–51 Implied undertaking There is always an implied undertaking to the court that each party will use the other side's documents only for the purposes of the litigation. In addition, special arrangements may be agreed or ordered to protect confidential information (whether technical or commercial) contained in a party's disclosure documents. Breach of the implied undertaking may amount to contempt of court.

24–52 Disclosure in patent actions The disclosure stage has the potential to push up costs. This is so especially if the objections to validity of the patent include assertions that the invention is obvious. Although the test for inventiveness is ultimately an objective one, the claimant will normally have to dig into his records of how the invention was made, covering the research and testing that was done prior to filing the patent application. The trawl can result in many documents for the claimant to list; and then these documents will all be looked at by the defendant, searching for evidence to assist his case.

The cost of the disclosure stage of patent litigation has been reduced considerably over recent years. Much of this reduction was achieved by a change in the rules relating specifically to patent actions which happened in 1995.[76] Since 1995, when a defendant asserts that the patent in suit is not valid because it is obvious, the claimant can limit the search for documents relating to how the invention was made to those documents which came into existence in the period beginning two years before the earliest claimed priority date and ending two years after that date. (These cut-off dates may be varied by the judge if the circumstances of the case demand it.)

The 1995 rule change also applied to the disclosure of documents relating to an objection of lack of novelty, but the effect was more limited. Novelty objections have always had to be particularised, that is, by identifying particular pieces of prior art. Thus, even if the patentee knows of another piece of prior art which, it could be argued, would anticipate his invention, he does not have to disclose this as it will not be relevant to the pleaded novelty attack (although it may well be relevant to any obviousness attack that is pleaded). This illustrates the point that the statements of case form the basic definition of the scope of the dispute between the parties.

24–53 Product or process descriptions in patent actions Documents relating to infringement are exempt, following the 1995 rule change, if the party against whom the allegation of infringement is made has already given the other side full particulars of the relevant product or process.

If a party thinks that another party has given inadequate disclosure he may apply to the court for a specific disclosure order. The court will not grant such an order if it considers that the cost and effort involved in giving disclosure will be disproportionate to the likely benefit.

The new rules and modern judicial practice have streamlined proceedings and potentially reduced their expense, whilst still making it difficult for the parties to hide the truth.

---

[76] S.I. 95 No. 2206 amended RSC, rule 104(ii), and introduced the new rules on disclosure.

24–54   Experts in patent cases   Instructing a suitable independent expert is usually extremely important in patent litigation. A good expert will help both the party who has instructed him or her, and ultimately the court, to assess the claims of the patent in suit and whether they are infringed and are valid. Where the invention is said to be old or obvious, both parties will require an expert who knows about the relevant art at the time of publication of the prior art (for assessing novelty) and at the priority date of the patent in suit (for assessing obviousness or insufficiency).

24–55   Finding an expert   Suitable experts are often difficult to find. Sometimes there are very few people who were working in the field at the relevant time; and often those that were, although very knowledgeable, may not make good witnesses to give oral evidence in court. The search for an expert should begin as soon as possible. When acting for the claimant, the expert(s) should in most cases be located before issuing the Claim Form.

24–56   Permission of court   Under the new rules, no party may call an expert or put in evidence an expert report without the court's permission. Permission must be sought in relation to a particular named expert (or a specified field at least) and, if permission is granted, it will relate only to that named expert or that field. The expert's evidence will normally have to be in the form of a written report, upon which the expert will be cross-examined at trial. Experts' reports should be restricted to covering that which is reasonably required to resolve the proceedings justly.

24–57   Contents of report   The new rules lay down certain matters that must be contained in an expert report, for example, the instructions he was given and details of all materials relied upon in making the report. It is important to note that instructions to expert witnesses will no longer automatically be treated as privileged, and the disclosure of them to the other side can be ordered if there are good reasons for doing so. This is a very significant change and one which practitioners need to be careful to take into account in their communications with experts.

A party may put written questions, on one occasion only, to another party's expert. Any answers will form part of the expert's report.

The court may order that the parties' experts meet (without prejudice) in order to identify, discuss and if possible narrow the issues in dispute. It has yet to be seen whether this possibility will be greatly used; it may simply add to the expense without eliciting much useful information.

The court also has power to order that, where two or more parties wish to submit expert evidence on a particular issue, evidence on that issue shall be given by one expert only. It is unlikely that the court will use this power very often in patent litigation.

24–58   Experts in non-patent cases   Experts are used far less often in non-patent disputes, though where the subject matter is highly technical, they will still be needed (*e.g.* in complex IT disputes).

24–59   Experiments in patent actions   Parties may also want an expert to help them consider whether experiments should be carried out, and to help design them. Although the client's technical staff may well have ideas on this, expert evidence will nearly always be needed in support, and any expert will want to have a say in what experiments he is going to have to discuss and support.

Experiments are typically used to support an assertion that the alleged infringing product or process does or does not have particular characteristics as claimed in the patent; or to demonstrate what a piece of prior art reveals. Where a party wishes to rely upon an experiment, he must serve a notice stating the facts he intends to establish and giving full particulars of the experiment proposed to establish them.

The recipient of a notice of experiments then has 21 days in which to respond by stating whether or not he admits the facts which the other side wishes to prove. Failure to do so means that the other side should then fix a directions hearing to establish a timetable for carrying out the experiments in the other side's presence. The timetable may also deal with any experiments that the other side may wish to carry out in response.

Because of the added expense and delay, parties are often reluctant to perform experiments Furthermore, even the most carefully designed experiments have a habit of going wrong, and failing to prove what they were designed to prove. But sometimes there is no other way to prove a certain point.

Further information

24–60   At any stage during proceedings, either party can request further information in order to clarify any matter in dispute. The court may order that that information be provided, within a specified time, if it is reasonably necessary for disposing fairly of the matter.

This can be especially useful in patent cases where the validity of the patent concerned has been put in issue. The patentee can, for example, be asked specific factual questions about a pleaded prior use or prior publication. Or the defendant could be asked details about his allegedly infringing product. But requests for information cannot be used to "fish" for information to construct a case.

Further directions hearing

24–61   There may be a further directions hearing, or "pre-trial review" a month or so before the trial is due to begin, at which directions for its conduct will be given. The parties will normally discuss and try to agree appropriate directions in advance, but the court will not be obliged to agree to the proposed directions.

The aim again is to dispose of the case in the most expeditious way. The judge may give directions relating to:

(a)   the service of further statements of case or further information;

(b)   further disclosure of documents;

(c)   further requests for admissions;

(d)   the holding of a meeting of experts for the purpose of producing a joint expert report on the state of the relevant art;

(e)   the exchanging of expert's reports;

(f)   the determination of a specific question as a preliminary issue.

Other matters may also be considered, such as whether a party should be ordered to provide a product sample for analysis or whether an independent scientific adviser should be appointed to assist the court. The appointment of an independent adviser is rare but a "primer" will often be prepared to explain to the court the technical background to the case. The Patents Court is well used to grasping technical issues: a primer, ideally agreed by both parties, and the evidence of both sides' experts are usually sufficient for the court.

Preparation for trial

24–62 As has already been said, the current patents judges have made considerable efforts to shorten the interval between the start of an action and trial. Most trials now take place within 12–18 months of the action starting. Moreover, an order for a "speedy trial", which allows a case to "jump the queue", can be made if special circumstances are shown.

To save time and money at the trial, judges in patent cases read the main documents in the action in advance. Therefore the claimant has to lodge all the documents in the form to be used at the trial at least seven days before the trial. At the same time, to assist the trial judge in his preparation for the hearing, the parties will submit their "skeleton arguments" and if possible an agreed reading guide, briefly setting out the issues and the best way of reading into the documents.

Trial

24–63 The trial takes place in open court before one judge. For even a fairly complex action, the hearing will usually last less than a week; but some very complex trials may take several weeks.

The claimant's case will be presented first by his advocate. After the opening speech, witnesses are called and oral evidence is given. Each witness may be cross-examined by the defendant's advocate.

After the claimant's evidence has been presented, it is the defendant's turn, following a similar procedure but usually without an opening speech. His advocate then sums up the defence, followed by a summing-up by the claimant's advocate.

Rather than being given immediately, judgment is usually held over or "reserved" to a later date in order to give the judge an opportunity to consider in full all of the evidence and arguments that have been put before him and to frame his decision.

If the judge decides in favour of the claimant, an injunction will almost always be granted if it has been requested.[77] But in suitable circumstances, it is possible for the injunction to be stayed pending an appeal by the defendant.[78] An order for delivery up of infringing articles held by the defendant may also be granted. Monetary remedies are discussed below.

Appeals[79]

24–64 An appeal lies to the Court of Appeal on questions of fact as well as matters of law.[80] However, the Court of Appeal does not generally interfere with the trial judge's findings of fact, particularly where they turn on which witnesses are to be believed or the weight to be attached to particular evidence.

A party wishing to appeal must obtain leave to appeal from the trial judge (failing which, the Court of Appeal itself),[81] and serve a notice of appeal within four weeks of

---

[77] See para. 24–75, below.

[78] See in particular *Minnesota Mining and Manufacturing Co. v. Johnson and Johnson Ltd* [1976] R.P.C. 671 (at page 676) and *Chiron v. Organon and Murex* (No. 10) [1995] F.S.R. 325 (at pages 339 to 344).

[79] CPR, schedules 1 and 2; and *Practice Direction (Court of Appeal: Leave to Appeal and Skeleton Arguments)* [1999] 1 W.L.R. 2 (the "1999 Appeals PD").

[80] On their face, the new rules appear to have removed the right of a party to appeal an interim order of a High Court judge, because the old RSC, Order 58, r. 6 is not reproduced in Schedules 1 or 2 (or anywhere else) of the CPR. However, the better view appears to be that this is not in fact the case, or at least that it was not intended. (See for example the discussion of this point in Blackstone's *Guide* at pp. 325 to 327.

[81] As stated in the 1999 Appeals PD, para. 2, the court below is often in the best position to decide whether a case is suitable for an appeal. If it (the lower court) is in any doubt as to whether an appeal would have a real prospect of success or involve a point of general principle (the basic test for allowing a party to take a point on appeal), it should adopt the safe course and refuse permissions to appeal, leaving the decision to the Court of Appeal if the party decides to take the matter further.

the judgment. If an injunction or other award (such as damages or delivery up) has been made, the defendant can apply for a stay of execution pending the hearing of the appeal.

The final stage of appeal on the law is to the House of Lords, sitting as a court, usually in a panel of five judges. Again, it is necessary for a party first to obtain leave to appeal. Sometimes the Court of Appeal itself grants leave, but more often the Court of Appeal leaves it to the Appeals Committee of the House of Lords to decide whether the issues in the case are of sufficient public importance to warrant a House of Lords decision. Very few appeals reach this stage.[82]

Costs

24–65 The costs of an action to enforce an intellectual property right (including the fees of solicitors, barristers and expert witnesses) do of course vary considerably. The scope of disclosure, the necessity for experiments, and the degree of admissions are all examples of factors which are indeterminate at the start. Sometimes the parties co-operate in narrowing down the dispute, but it is always possible to come up against an unreasonable opponent.

Where there is a serious dispute, of commercial importance, the costs of a typical action will normally be in the hundreds of thousands of pounds. That is only a general guide: simpler cases can cost less and, of course, particularly complex or bitterly fought actions can cost very much more.

24–66 Costs orders The court has a wide discretion in making suitable orders about legal costs but, in general, costs "follow the event": a party who loses will be ordered to pay the costs incurred by the party who wins. Exceptions to this include the example given earlier, where a party refuses unreasonably to admit facts which are later proved. Parties may also be penalised in costs for unreasonable conduct during the litigation (*e.g.* pleading numerous prior art references that are not necessary to support their case), even though they may be successful. Under the new rules, the court will also take into account whether the amount of costs incurred was proportionate to the value, complexity and importance of the case.

24–67 Costs assessment If the amount of costs to be paid cannot be agreed between the parties, it will be assessed by an authorised court officer or by a costs judge. (Occasionally, and increasingly under the new rules, the trial judge will award a fixed sum without the need for an assessment hearing.[83]) The asessment procedure involves the successful party setting out in detail the costs it seeks to recover from the other side, and trying to justify these in relation to the work done. The party who is to pay the other's costs will argue strongly that these are too high. It is therefore imperative that throughout the action detailed file notes and records of time spent are made and kept for use at this stage.

A winning party will very seldom recover his legal costs in full: typically he may get back about 60–70 per cent following the costs assessment procedure. Moreover there

---

[82] Two recent patent cases which have gone to the House of Lords are *Biogen v. Medeva* [1997] R.P.C. 1 and *Merrell Dow Pharmaceuticals v. H. N. Norton & Co.* [1996] R.P.C. 76.
[83] Examples of decisions on summary costs assessments include the following: *Insurance Corp of Channel Islands v. Royal Hotel Limited* [1998] Lloyd's Rep I.R. 151; *Banque Nationale de Paris v. (1) Montman Limited (2) Shapland Inc (3) Ramchand Hirand Melwani* (unreported, 1999); *R. v. Cardiff County Council Ex p. Brown* (unreported, 1999); *Hobin v. Douglas* (unreported, 1998); and *Re A Company* (No. 003089 of 1998).

is no compensation for the party's own management time spent handling the litigation. There may also be a risk that the loser has insufficient funds to meet the award against him. A party who sues an individual funded by Legal Aid should not expect to recover anything by the way of costs. (Legal Aid cannot be granted to a company, whatever its financial state.)

24–68 Security for costs[84] A defendant may in certain circumstances apply to the court for an order that the claimant should give appropriate security for the Defendant's costs of the action. The grounds for doing so include:

(a) that the claimant is ordinarily resident outside the jurisdiction[85]; or

(b) that there is reason to believe that the claimant will not be able to pay the defendant's costs should the defendant succeed, for example because it is insolvent; or

(c) that the claimant is a nominal claimant suing for the benefit of another and will be unable to pay the defendant's costs if ordered to do so.

This provision is a useful safeguard for a defendant who is worried about the claimant's ability or intention to pay if the claimant loses the action. It can also help to slow down the first steps in the action, which is often in the defendant's interests.

Security for costs is usually given stage by stage, not all at once for the whole action. Usually the manner of giving security (for example by a bank guarantee) and the amounts are agreed between the parties, but an application can be made to the court to resolve any differences.

Procedure — Patents County Court

Overview

24–69 As explained above, the idea of the PCC was to simplify and reduce the cost of patent actions, especially for the smaller litigant, and the PCC has indeed succeeded in providing access to justice in a new category of cases

Solicitors and patent attorneys (as well as barristers) have rights of audience before the PCC. The need to involve fewer professionals is one of the reasons cited for expecting the process to be cheaper than in the High Court. In practise, the change has not made a significant difference in complicated cases: the costs of preparation are still substantial and in most major cases specialist advocates will probably still be used as part of the legal team, as in the High Court.

The rules of procedure in the PCC now derive primarily from the new Civil Procedure Rules which govern both High Court and County Court litigation. The specific rules relating to patent litigation contained in the new Patents Court Practice Direction also apply to the PCC.

Transfer to PCC

24–70 There are provisions in the rules for the transfer of cases from the High Court to the PCC and vice versa. This is largely a discretionary decision but the financial position of each party is one important factor. Although there was a proposal to set

---

[84] Rules 25.12 to 25.15.
[85] Courts will generally treat parties within the E.U. or other Brussels or Lugano Convention countries as "within the jurisdiction".

an upper limit on the amount or value of the proceedings that the PCC could entertain, no limit has in fact been imposed.

One aim is to have shorter trials than in the Patents Court. This goal has sometimes been realised. On the other hand, some preliminary procedural matters have been dealt with less quickly than would be the case in the High Court. Another of the PCC's objectives is to encourage settlement of cases out of court. Under the new rules that will equally become an objective of the High Court.

In comparing the High Court and PCC, it is important to compare like with like. The typical case in the PCC should be simpler than a High Court case, so it would be surprising if it were not disposed of more cheaply and more quickly. In fact, the recent streamlining of patent procedures in the High Court and the more robust attitude of the judges there mean that comparable cases have actually been disposed of more efficiently in the High Court than in the PCC. This has led to a marked drop in the caseload of the PCC in the last few years.

Statements of case

24–71 Under the old PCC rules, parties were required to serve statements of case giving full particulars of their cases, including all facts, matters and arguments to be relied upon. This necessitated going into much more detail than a statement of claim in a High Court action. One had to explain, for example, which part of the alleged infringing product was covered by which claims, and how. A claim of invalidity of a patent for obviousness would have not only listed each piece of prior art relied upon, but also explained how each one related to the relevant parts of the patented invention.

The detail which was involved at this early stage meant that a claimant had to do a substantial amount of preparation work in advance of bringing the case, including talking to potential expert witnesses, and actually setting the case out in full. This tended to make the early part of litigation in the PCC more expensive than in the High Court. The rationale was that an early in-depth analysis of the arguments should prevent some (weaker) cases ever being brought. Or a defendant, presented with a strong and detailed case, might be persuaded to attempt to settle at an earlier stage.

Under the new rules, a similar "front loading" approach will in theory apply to High Court proceedings (though without the need to plead "arguments"). However, it is very doubtful whether High Court pleadings will in practice become as extensive as those in the PCC.

Subsequent procedure

24–72 There were significant differences between the procedure for PCC and High Court litigation under the old rules. For example, there was no automatic "discovery" procedure in the PCC; it was not possible for the parties to agree time extensions of more than 42 days without applying to the court; and there was a Preliminary Consideration hearing, in which the case was reviewed in detail by the judge, which took place after the exchange of "pleadings" (now known as Statements of Case). Under the new rules, High Court and PCC proceedings will be much more similar. In many ways, the old High Court rules have been changed to reflect PCC practice rather than vice versa. For example, the scope of discovery has been cut down; the freedom of parties to agree time extensions or timetables to suit themselves has been curtailed; and there will be much greater court intervention by way of Case Management Conferences and further directions hearings. There will also be a much greater emphasis on limiting costs and promoting settlements.

The present Patents Court Practice Direction provides that all Patents Court business within the High Court or PCC shall be subject to the more flexible but less streamlined "multi-track"[86] rules. These are intended for complex cases of considerable financial value which are likely to involve trials lasting several days. In due course it may be that the PCC will issue its own practice direction covering detailed procedure. It is logical that there should be some differences between High Court and PCC proceedings given that it is intended for smaller cases. For example, it may be that some aspects of the "fast-track" procedure provided for in the Civil Procedure Rules may be appropriate for the PCC. Fast-track proceedings follow tight timetables and may involve restrictions on the scope of disclosure and evidence. The fast track is intended for straightforward disputes of limited financial value which are likely to involve trials lasting no longer than a day.

Trial, judgment and appeals

29–73 Shorter cases in the PCC typically come to trial after six to 12 months. Some of the longer cases (say, one to two weeks or more) are taking rather longer to come to trial. At times, a small number of very big cases have clogged up the list and held up the shorter cases. On average, it has been taking a year to 18 months between the start of a longer case and trial.

The trial is where, in theory, there could be real savings in time and expenditure compared with High Court trials. PCC trials should be kept short, with much of the evidence and argument being in writing. There should be fewer professional advisers attending the trial. Judgment in the PCC, as in the High Court, may be either immediate or reserved, depending on the complexity of the case. Appeals lie direct to the Court of Appeal, provided that leave to appeal can be obtained either from the trial judge or the Court of Appeal itself.

Remedies and costs

24–74 The remedies available in the PCC are the same as those the High Court may grant (notably injunctions, delivering up of infringing goods, and damages or an account of profits).[87]

As in the High Court, the loser generally pays the winner's costs on an assessed basis. The judge can, if he chooses, make an order that the County Court's fixed rates should apply. This means limiting a party's recovery to a set amount per hour or day for the time spent by the patent agents, solicitors, barristers and experts. The limits are very low and can be waived by the judge if convinced that they are unrealistic in the context of the nature of the case concerned. In short, the winning party in the PCC may receive a much smaller contribution to his legal costs and the losing party may be less exposed.

E. REMEDIES

24–75 The ultimate objective of the party which has resorted to litigation is to be awarded some form of relief. If the advice set out earlier in this chapter has been followed, the litigant will have considered very carefully at the outset what its commer-

---

[86] Under the new CPR, cases are allocated to one of three categories or "tracks" (the small-claims track, the fast-track, and the multi-track) based to a large extent on the value of the case.
[87] See below at Section E.

cial objectives are, and which remedies available through litigation are most likely to help it achieve those objectives.

If your objective is to stop an infringing activity quickly, rather than to seek financial compensation, your best remedy will be an interim injunction. Even if financial compensation is your main objective, you may still need to seek urgent relief to prevent the defendant's assets disappearing before trial. Similarly, you may be concerned that if you give notice to the defendant of your intentions to sue him, relevant evidence may be destroyed. The forms of urgent relief you would be seeking in those circumstances are a freezing order and a search order respectively.[88]

The procedures you will need to follow, and the strategic issues you will need to have in mind, in seeking these forms of urgent relief are discussed at paragraph 24–84.

First, however, we consider the main remedies that are available after a full trial, namely an injunction to restrain the infringement (or in the case of breach of contract, to prevent any ongoing breach) and damages, or, as an alternative form of financial relief, an account of profits. Other remedies available are an order for delivery up or destruction of the infringing products and a declaration of validity and infringement,[89] but these are not discussed further here.

Injunctions

24–76 The following paragraphs consider two issues relating to injunctive relief. First, whether the claimant can request one as of right, and if not, what factors the courts take into account in deciding whether to award an injunction. Second, the scope of the injunction which should be awarded.

Available as of right?

24–77 Injunctions are equitable remedies. The court therefore always has a discretion whether to grant them. Where the court considers it inappropriate to grant an injunction it may award damages instead.[90] Nevertheless, when a defendant has been found to infringe a valid patent or other IP right, an injunction normally follows.

An example of a situation in which a court may not consider an injunction appropriate is where its grant is shown to be contrary to the public interest. This was raised in *Chiron v. Organon and Murex Ltd* (No. 10).[91] The patent in suit related to test kits for the Hepatitis C virus. The defendant argued that the grant of an injunction would be contrary to the public interest because it would prevent the public having access to its kits which it alleged were better than those of the plaintiffs. The court held that, in exercising its discretion, it should be guided by the considerations laid down in *Shelfer v. City of London Electric Lighting Company*.[92] Under those guidelines, an injunction should be refused and damages awarded instead where the following criteria are met:

(a) the injury to the claimant's legal rights is small;

(b) it is capable of being estimated in monetary terms;

(c) it can adequately be compensated by damages; and

(d) it would be oppressive to grant an injunction.

---

[88] These were called a "Mareva Order" and an "Anton Piller" order respectively before the CPR came into force.
[89] Section 61(1)(a)–(e).
[90] Section 50, of the Supreme Court Act 1981.
[91] [1995] F.S.R. 325.
[92] [1895] 1 Ch 287.

The court emphasised that the test is a flexible one. The courts have an overriding discretion and the interests of third parties and the public at large can be taken into
account.

Applying this to the facts of the case, and taking account of the need to protect the
monopoly rights of patentees, the court refused to limit the injunctions sought in that
case. It remains possible in theory, to avoid the grant of an injunction on public interest grounds, but only in very unusual circumstances.

Scope

24–78 Until very recently, this was not an issue. The form of injunction awarded to
a successful claimant in an action for infringement of an intellectual property right,
simply said that the defendant must "not infringe" the right. However, this can no
longer be taken as read. The courts may be reluctant to award an injunction in such
broad terms, and instead may seek to be more specific about the acts prohibited by the
injunction.

In *Coflexip v. Stolt Comex*,[93] it was held that the purpose of an injunction should be
to give effect to a judgment on liability, protecting the claimant from a continuation
of the infringements by the threatened activities of the defendant. But the injunction
also had to be fair to the defendant. While a broad injunction could meet the first
requirement, there was a risk that it did not meet the second. This was for three
reasons:

(i)   an injunction could only remedy a threatened infringement. There would be
       no threat where there was no indication that the defendant would continue the
       infringing acts after judgment. Further, it was common for the infringement
       to occupy only a small part of the successful claimant's monopoly, particu-
       larly in patent cases. So the injunction should not restrain the defendant from
       doing things that he had not threatened to do, might not have thought of
       doing, or might even be incapable of doing.

(ii)   the defendant should know what he could and could not do. So where the
       scope of the monopoly right was unclear, as in many patent cases, a broad
       injunction was likely to lead to uncertainty.

(iii) the defendant should not be at risk of being in contempt of court by doing
       something which, while different from the original infringement, might yet
       infringe. Here the question of infringement would have to be determined in
       contempt proceedings brought by the successful claimant. That was very
       unsatisfactory and could lead to the "bizarre" situation where an appellate
       court tried an infringement action as part of an appeal of a contempt decision.

He accepted that there were also arguments in favour of the traditional approach,
including the fact that the successful claimant might not be adequately protected without an injunction in general terms. However, he came down in favour of the narrow
form of injunction.

This approach has since been applied, albeit in more limited terms, in *Microsoft v.
Plato*[94] where the judge said that:

---

[93] [1999] F.S.R. 473.
[94] [1999] Masons C.L.R. 87; and July 15, 1999, CA; unreported.

"... it is well established on authority that, once the claimant has established any infringements of his rights at all, he is entitled as of right to an injunction in the usual wide form to restrain all future infringements ... My own view of the authorities cited ... is that, while they undoubtedly support [this] ... as a general proposition, they do not rule out the ability of the court, in an exceptional case, to grant relief in a narrower form. In most of the cases cited the court inferred, ... that the defendant would or might repeat the infringement if not restrained from doing so. In others the defendant could not be said to have acted as an honest trader would have acted. In each case the judge has a discretion ... to tailor the relief granted to match the wrong which has been committed or threatened."

Only time will tell whether this trend will continue, or whether these cases will prove simply to be the exceptions which prove the rule that injunctions should be in general terms.

Damages inquiries and accounts of profits

24–79 If the claimant succeeds in his claim for infringement, he is entitled to choose between either damages or an account of profits from the defendant, but not both. The claimant has to make this choice without knowing which will be the best financially. He should have some idea of his losses (compensated by damages) but he may have little idea what profits the defendant has made and would have to account for. However, the defendant can be ordered to disclose his accounting documents to enable the claimant to make an informed choice.

The assessment of damages or account of profits takes place at a separate hearing after judgment. At this hearing, each party puts forward its evidence on the level of compensation that it thinks should be payable by the defendant. The trial judge makes the final decision. This procedure may be deferred if the case is appealed.

Amount of damages

24–80 Recovery of damages in IP cases is governed by the same general rules as for other torts. A claimant can recover damages for loss which is foreseeable, caused by the tort and not excluded from recovery by public or social policy. The damages are assessed in such a way as to put the claimant in the position he would have been in had the infringement not taken place.

Lost sales are the main factor to take into account where the claimant himself manufactures and sells products under his patent, and these have to be proved by the claimant. A claimant who exploits a patent by the grant of licences can claim as damages the amount of the royalties that the infringer should have paid. To find an appropriate royalty rate the court may look at similar licences in the same field, if such comparables are available. A claimant who does not himself make and sell and so has suffered no lost sales, or a claimant who cannot prove lost sales, is at least entitled to a reasonable royalty on all infringements. This is assessed by the court. In appropriate cases, a claimant may also be able to recover damages in respect of so-called "convoy sales" — sales of products or services other than those covered by the patent but which are normally purchased with them. A parent of a wholly owned subsidiary may be able to recover for losses of the subsidiary resulting from diminution in the value of its shareholding or loss of dividends. This will not necessarily be exactly the same as the loss to the subsidiary.[95]

---

[95] Unlike in the U.S., there is relatively little English case law on damages. The leading recent case is *Gerber v. Lectra* [1997] R.P.C. 443, in which the Court of Appeal (reversing in part the first instance decision of Jacob J. [1995] R.P.C. 383) confirmed the principles on which damages should be assessed and addressed them in some detail.

Court's discretion

24–81 In patent cases the court has a discretion whether or not to award damages in four specific cases:

(i)   when a renewal fee for the patent has been paid late, for infringements during the period of delay[96];

(iv)   where the patent has been amended, the court must be satisfied that the specification of the patent as published was framed in good faith and with reasonable skill and knowledge[97];

(v)   where the patent is found to be only partially valid[98]; or

(vi)   in relation to infringements before grant of the patent but after publication of the application, if it would not have been reasonable, from a consideration of the application as published, to expect the acts sued upon to have infringed the final patent.[99]

Neither damages nor an account of profits will be awarded against a person who is an "innocent" infringer: this is someone who can prove that at the date of infringement he was not aware, and had no reasonable grounds for supposing, that the patent existed. This relates only to ignorance of the existence of the patent and has no application where an infringer is simply unaware that (or does not consider that) his actions amount to infringement.

The way to defeat most pleas of innocence by infringers is for the patentee, wherever possible, to mark his products and relevant literature with the patent number. (At the latest, the claimant's warning letter or Claim Form will provide a cut-off date for this exception.)

Accounts of profits

24–82 The claimant may ask that, instead of an inquiry into damages, he should get the sum equivalent to the profits derived by the defendant from the infringement. This leads to complicated calculations to allocate the profits derived from the infringement *per se*, isolating them from those derived from the general business of the defendant.

It is less common to seek an account of profits than an inquiry into damages. The complications and uncertainties involved were illustrated in the recent *Hoechst Celanese v. BP Chemicals* case.[1] The defendants had been found to infringe the claimant's patent relating to the final purifying stage of an acetic acid manufacturing process. The claimant claimed the defendants' total profits on sales of the acetic acid. The judge rejected this approach. He referred to the "tin whistle example much beloved of patent practitioners where a patent has claim 1 directed to the whistle and claim 15 directed to a battleship with a funnel to which the whistle is attached". He said that there was no doubt that only the profits attributable to the whistle should have to be accounted for. The court had to do its best to split the profits between the infringing and non-infringing parts. The overall result in this case was that the defendants only had to account for 0.3 per cent of their profits.

---

[96] Section 62(2), of the Patents Act 1977.
[97] Section 62(3), *ibid.*
[98] Section 63(2), *ibid.*
[99] Section 69(3), *ibid.*
[1] [1999] R.P.C 203.

The court has a general discretion as to whether to permit the claimant to obtain an account of profits. As with damages, the "innocent" infringer escapes.

Assignees/exclusive licencees

24–83 No compensation may be awarded to an assignee of a patent or an exclusive licensee in respect of the period between the transaction and its registration, if he fails to register his assignment or licence at the Patent Office within six months from the date of the transaction.[2] This is a sanction designed to ensure that the register is reliable.

Interlocutory or interim injunctions

General principles

24–84 In certain circumstances it is possible to obtain an "interlocutory" or interim injunction (that is, one which is granted before the matter is heard in full) to prevent infringement of a United Kingdom patent or other IP right at an early stage in the action. Interim injunctions are rarer in patent cases. This is particularly true in "modern" times, where cases can in appropriate circumstances reach full trial stage as quickly as they would in the past have reached the hearing of the application for an interim injunction. The possibility should, however, always be considered. Interim injunctions have often been refused on the basis that the patentee's monetary losses could be calculated, therefore damages would be an adequate remedy instead. However, there are a number of cases in which injunctions have been granted, for example, where the court has accepted that a defendant should not be allowed to establish a "bridgehead" by deliberately infringing a patent in the years prior to expiry of the patent, thereby using his infringing acts as a "springboard" from which to gain a competitive advantage.

If an interim injunction is granted, it often disposes of the action altogether. The defendant may give up or propose a reasonable settlement, having been prevented — for what is likely to be a significant length of time — from exploiting the product or process concerned.

24–85 Procedure The court has a general power to grant interim injunctions. An application should normally be made in writing, stating the order that is sought and the reasons. It should be supported by evidence.

There are two types of application:

- "on notice", where the respondent has been given advance notice and is present at the hearing; and

- "without notice", where the respondent has been given no notice and usually knows nothing of the application until the resulting court order (if granted) is served on him.

A claimant is usually given either a fixed date for the hearing, or a date from which it will "float". This means that the court will tell the claimant the time that the hearing will commence just the day before the hearing. A case "floating" from (say) June 10, will typically begin on June 10 or 11, but may be later if previous cases over-run.

---

[2] Sections 68 and 33 of the Patents Act 1977.

24–86 *Basis for granting interlocutory injunctions* The leading case which set down the principles involved is *American Cyanamid v. Ethicon*.[3] Recently, these principles were re-examined in *Series 5 Software v. Clarke*.[4] In *Series 5* it was emphasised that the grant of an interlocutory injunction is a matter of discretion and depends on all the facts of the case; there are no fixed rules as to when an injunction should or should not be granted.

24–87 *Delay* Any delay by the claimant may be fatal. Before reaching the *American Cyanamid* tests, a defendant can knock out the application if the claimant has delayed unduly in bringing his claim. It is difficult to argue that it is very important to get an early injunction if you have delayed in asking for one. The length of delay that will be unacceptable will vary from case to case. It depends very much on when the claimant first knew of the infringement or threatened infringement and on how the defendant's position has changed during this period. One of the first things to ask a client requesting urgent relief is how long he has known about the problem.

24–88 *Serious question to be tried* If there are no preliminary reasons for throwing the case out, the court will look to see if there is a serious question to be tried. It is therefore vital that the evidence in support of the application sets out the facts in full (and an outline of the arguments) on which the claimant will rely in the main action. To get an application started, it is not necessary to show a strong case: just an arguable one; but in reaching its decision the court may still take a preliminary look at the merits, as explained below.

24–89 *Damages an adequate remedy?* Once the court is satisfied that the legal case is arguable it will consider various factors, largely economic ones, conventionally starting by considering whether it really is critical for the claimant to have an injunction. If damages after a proper trial will be an adequate remedy for the claimant, and if the defendant will be able to pay them, interim relief is not appropriate and will be refused.

If damages in due course will not adequately compensate the claimant, the next question is whether damages will adequately compensate the defendant if an injunction were now granted and later (at trial) found not to have been justified. If they would, then an injunction will be granted and the applicant will be ordered to undertake to compensate the defendant for loss suffered, should the claimant then lose at trial (a "cross-undertaking in damages").

24–90 *Balance of convenience* If there is significant doubt about the adequacy of damages, the court goes on to the next test which is to consider the "balance of justice" between the parties (sometimes called the "balance of convenience"). This can include:

(a) whether it would cause greater hardship (financial or otherwise) to grant or refuse the injunction;

(b) the effect of the injunction on third parties, including customers of the parties;

(c) whether the status quo is more likely to be maintained by granting or refusing the injunction.

---

[3] [1975] A.C. 396.
[4] [1996] 1 All E.R. 853.

24–91 *Merits* Only if this last test fails clearly to point to the correctness of granting or refusing the injunction, will the court look at the relative strengths of each party's case. If the claimant appears to have a much stronger case and is likely to succeed at trial, he will probably get his injunction. If not, then (subject to appeal) he has reached the end of the road and must make do with an ordinary court action.

Part of the *American Cyanamid* decision suggested that the court should not take into account the relative strengths of each party's case unless there was "no credible dispute" about the facts which showed that "the strength of one party's case is disproportionate to that of the other". That issue was re-examined in *Series 5*. The judge said that there was no inflexible rule. On the contrary, he said that the relative strength of each party's case was a "major factor" which the court should take into account where a "clear view" could be reached on the basis of "credible evidence". However, if there is a conflict of evidence on a fundamental issue, such that the court cannot reach a clear view on the merits, the court will not normally attempt to resolve the conflict at the interlocutory stage, when cross-examination of witnesses is extremely rare.[5]

24–92 *Exceptions* There are some exceptions to the application of the *American Cyanamid* tests:

(a) where the hearing of the application is very likely to dispose of the entire proceedings because the infringements will have ceased by the time the case comes to full trial, the court should look carefully at the relative merits of the claim and defence (if it does not, then the alleged infringer could well be prejudiced);

(b) "without notice" applications for search orders and freezing injunctions require evidence of a very strong prima facie case, as well as other preconditions outside the scope of the *American Cyanamid* principles (see below);

(c) mandatory injunctions (orders requiring people to do things, as opposed to restraining orders) require proof of a strong likelihood that a similar injunction will be granted at the trial, and will not be granted if the facts relied upon are heavily contested. (These are uncommon in patent actions.)

24–93 *Effect of interlocutory injunction* If an interlocutory injunction is granted the Defendant will be bound for as long as the injunction lasts (usually until full trial or a further interim hearing). Breach of the injunction is contempt of court. If a "penal notice" is endorsed on the order and the order is then properly served, proceedings for breach can result in the defendant being fined or committed to prison.

Instead of contesting the hearing, the defendant may give an undertaking in the form of the injunction sought. The claimant usually requires the undertaking to be given to the court and included in a "consent order". Breach of the undertaking is then as serious as breach of an injunction.

24–94 *Appeal* It is possible to appeal to the Court of Appeal against a judge's decision on the grant or refusal of an interlocutory injunction, but appeals do not often

---

[5] This approach was later followed by the High Court in *CMI Centers for Medical Innovation GmbH v. Phytopharm PLC* [1999] F.S.R. 235. However, the Court of Appeal in *Greet and Anr v. Rilett* (March 11, 1999, unreported) stated that it was not appropriate to disregard the well-settled approach laid down by the House of Lords in the *American Cyanamid* case.

meet with success. Since the first instance judge has a great deal of discretion, the Court of Appeal will overturn the original judge's decision only if he was wrong in principle: for example, if he failed to consider some relevant factor, or gave too much weight to an irrelevant matter, or made an error of law. Such appeals can only be made at all with the permission of the court where it is a pure matter of discretion, thus permission may well be refused.

24–95 **"Without notice" applications — injunctions** Applications for injunctions may be made without notice "if it appears to the court that there are good reasons for not giving notice". If the claimant makes an application without giving notice, the evidence in support of the application must state the reasons why notice was not given.

Such injunctions normally last for a week or so only, that is, until an "on notice" hearing may be held. By then the defendant will have had time to serve evidence in response. The court will then consider whether the injunction is to continue, for example, up until trial.

Search orders

24–96 The search order, which used to be known as an Anton Piller order, is an extremely useful weapon against an infringer who, the claimant believes, may attempt to remove or destroy evidence of what he has been doing. It is a kind of "civil search warrant".

The application to court must be supported by detailed evidence setting out the facts, and in particular:

(a)   evidence of a very strong prima facie case for final relief at trial;

(b)   strong evidence that the defendant is in possession or control of property that will establish his liability to the claimant;

(c)   evidence that the claimant will be seriously damaged if the defendant destroys or conceals the property concerned; and

(d)   evidence of a real risk that the defendant will destroy or conceal the property in question if given notice of the application (based, for example, on previous misconduct in similar matters, or criminal intent).

These requirements were originally set out in *Anton Piller v. Manufacturing Processes,*[6] and additional requirements have been elaborated in subsequent cases.

Full and frank

24–97 Full and frank disclosure must be made by the claimant of all facts and matters that are relevant to the application, whether or not they are favourable to him. Failure to comply with this obligation will result in the order being discharged. Solicitors must take great care to ensure that they are getting the full story from their clients so as to comply with this duty.

---

[6] [1976] Ch 55.

Undertakings required

24–98 In addition to the normal cross-undertaking as to damages, the claimant or his solicitor usually has to give a number of undertakings, including the following:

(a) that the order and a copy of the evidence used to obtain it will be served by a solicitor;

(b) to tell the defendant of his right to legal advice and to give him reasonable time to obtain it;

(c) to explain the meaning and effect of the order in simple language;

(d) to make a detailed record of all property seized or inspected during the search; and

(e) to keep in safe custody anything seized as a result of the search.

In recent years the procedures have been tightened still further: for example, by having an independent solicitor[7] attend and report on the execution of the order.

Form of order

24–99 The order (if granted) will be detailed, specifying the documents or property that are being sought, the addresses of the premises to be searched and the times during which the search may be carried out. It will require the defendant to disclose the whereabouts of the items in question and sometimes to disclose the identity and addresses of his accomplices (for example, suppliers, distributors or retailers of infringing products). If the defendant does not comply with the order he will be in contempt of court.

Execution

24–100 Sometimes during the execution of these orders material is found which has not been made the subject of the order. The claimant cannot seize or use this material without returning to the court for permission to do so. It may be appropriate in some circumstances to telephone the court and ask for the order to be widened, subject to an undertaking to draft and file evidence in support.

In cases where search orders are appropriate one is often dealing with people of dubious credibility and character. In cases where there is reason to expect violence, it can be advisable for the claimant's solicitors to notify the police that the order is to be executed and ask them to be on hand in case there is trouble. The reason for their presence must be explained to the defendant.

It is also important to emphasise the penalties that can be imposed on a defendant who fails to comply with a search order. This was very clearly illustrated in a 1997 case, *Davy v. VAI Industries*.[8] In addition to the court's powers to impose fines, sequestration of assets and/or imprisonment of the defendant or its representatives, there is a real risk that failure to comply will discredit the defendant. This is what happened in the *Davy* case. At trial, the judge disbelieved much of the managing director's evidence, rejected testimony from other senior employees, and found that the claimant (then plaintiff) had, by reason of breaches of the Order (such as concealment and

---

[7] Referred to as the Supervising Solicitor, whose role is to monitor the execution of the order and report back to the court on the raid generally.

[8] [1999] 1 All E.R. 103, CA.

destruction of documents) only recovered a portion of its property. Such findings can undermine a defendant's credibility and burden it in any subsequent case on the merits.

On the other hand, the court will exercise caution in using these powers against a defendant it believes acted honestly in attempting to comply with the order, particularly when there were defects in the wording of the order and the way in which it had been served.[9]

Subsequent hearing

24–101 After execution of the order, perhaps a week later, an "on notice" hearing will be held. The defendant may file his own evidence in advance of the hearing and may apply to have the order varied or set aside. The action will then continue with service of Statements of Case in the ordinary way.

Freezing orders

24–102 Another form of "without notice" order is the freezing order. This used to be known as a *"Mareva Order"* after the *Mareva Compania Naviera v. International Bulk Carriers* case.[10] It freezes the assets and finances of the defendant to prevent them from being dissipated.

Such Orders are relevant in any case of a defendant with liability to pay damages who is likely to dissipate his assets or remove them from the jurisdiction, to avoid paying damages. Again, breach of the order constitutes a contempt of court.

Application for order

24–103 A "without notice" application is made to the court supported by evidence and a draft order for the relief sought. The order should specify the maximum amount of assets affected and must identify the defendant's assets as specifically as possible. It will generally apply only to assets within the jurisdiction of the court. The order must also provide for liberty to apply for variation or discharge of the order at a subsequent "on notice" hearing and will reserve the costs to be determined at the end of the action.

The claimant's evidence must disclose a cause of action and at least a good arguable case. It must also explain the reason for believing that the defendant is likely to dissipate or remove his assets from the jurisdiction. As in the case of a search order application, full and frank disclosure must be made of all relevant matters.

The claimant will be required to undertake to inform any other persons affected by the order of their right to apply for directions or for a variation of the order. He must also agree to meet any expenses incurred by third parties in complying with the injunc-

---

[9] *Adam Phones v. Goldschmidt*, Chancery Division, Jacob J. July 9, 1999 (unreported). This was a case in which a so-called "doorstep piller" had been awarded, allowing the claimant to serve the order requiring the defendant to produce materials at his doorstep, but not actually to enter his premises. The order referred to computer disks, but is was not self-evident from the order that the defendants were to copy the programs to disk, hand over the disks and delete the programs from their computers. The defendant had made an honest mistake in not copying all of the software because he thought the claimant already had copies. So, although there was authority that even this placed him in contempt, the judge did not consider that committal proceedings were justified. The judge was clearly influenced in this by the fact that there were errors in the way the order had been drawn up and served — on a Saturday (when the defendant would have great difficulty getting legal advice) and without a supervising solicitor present.

[10] [1975] 2 Lloyd's Rep. 509, CA.

tion; for example, a bank's costs for searching through its records for accounts held in the name of the defendant and stopping payments out of those accounts.

CONCLUSION

24–104   This Chapter has considered how to enforce intellectual property rights and the remedies that might be available. Section A emphasised the importance of choosing the most appropriate dispute resolution mechanism for your particular circumstances — identifying the merits and disadvantages of litigation, arbitration and ADR. It outlined the various forms of dispute resolution and the pros and cons of each and concluded that litigation is not the only option; other dispute-resolution mechanisms should be considered at the outset, not least in light of the new Civil Procedure Rules in the United Kingdom.

The various strategic issues that might come into play at each stage of the litigation process were then considered in Section B. These are of utmost importance, particularly where there are parallel disputes in more than one country. The key issues are to establish your commercial objectives and then determine which country's laws and procedures are likely to result in these objectives being achieved most cost effectively.

It may not always be possible to implement the desired strategy; jurisdictional issues may constrain the decision-making process. As was explained in Section C, determining which country's courts have jurisdiction is not at all straightforward, particularly where questions of validity *and* infringement of an intellectual property right are at issue. Clarification may become available in the future from either the European Court of Justice or as a result of legislative reform. But for the foreseeable future, the situation is likely to remain complex and unclear.

The Chapter then considered in Section D the procedures that must be followed in English litigation, with particular reference to the major changes introduced by the new Civil Procedure Rules, and, finally, Section E reviewed the remedies that may be available to a successful party, both in the short-term and following a full trial.

The dispute-resolution process is not an easy one. To be successful one has to achieve a delicate balance of strategic, legal and other issues. The author hopes that the comments in this Chapter will provide a useful guide for those who become involved in this complex area.

INDEX

THE COMPANION CD

Instructions for Use

Introduction

These notes are provided for guidance only. They should be read and interpreted in the context of your own computer system and operational procedures. It is assumed that you have a basic knowledge of WINDOWS. However, if there is any problem please contact our help line on 020 7393 7266 who will be happy to help you.

CD Format and Contents

To run this CD you need at least:
- IBM compatible PC with Pentium processor
- 8MB RAM
- CD-ROM drive
- Microsoft Windows 95

The CD contains data files of relevant statutory material (see contents list on CD). It does not contain software or commentary.

Please refer to the "**Contents**" document for the numbering of the files.

Installation

The following instructions make the assumption that you will copy the data files to a single directory on your hard disk (e.g. C:\Working with Technology).

Please follow the instructions below for WordPerfect 9 and above

Using the materials with Microsoft Word

N.B. for other versions of Word, and other Windows word processors in general, the instructions will be similar, but if you are not sure refer to the documentation that came with your word processor.

To open a Working with Technology document in Word, select **File, Open** from the menu. Highlight the Working with Technology directory in the **Directories** list box. Select the desired document, e.g. "21." from the list and press **OK**.

LICENCE AGREEMENT

Definitions

1. The following terms will have the following meanings:
"The PUBLISHERS" means Sweet & Maxwell of 100 Avenue Road, London NW3 3PF (which expression shall, where the context admits, include the PUBLISHERS' assigns or successors in business as the case may be) of the other part on behalf of Thomson Books Limited of Cheriton House, North Way, Andover SP10 5BE.
"The LICENSEE" means the purchaser of the title containing the Licensed Material.
"Licenced Material" means the data included on the disk;
"Licence" means a single user licence;
"Computer" means an IBM-PC compatible computer.

Grant of Licence; Back-up Copies

2.(1) The PUBLISHERS hereby grant to the LICENSEE, a non-exclusive, non-transferable licence to use the Licensed Material in accordance with these terms and conditions.

(2) The LICENSEE may install the Licensed Material for use on one computer only at any one time.

(3) The LICENSEE may make one back-up copy of the Licensed Material only, to be kept in the LICENSEE's control and possession.

Proprietary Rights

3.(1) All rights not expressly granted herein are reserved.

(2) The Licensed Material is not sold to the LICENSEE who shall not acquire any right, title or interest in the Licensed Material or in the media upon which the Licensed Material is supplied.

(3) The LICENSEE shall not erase remove, deface or cover any trademark, copyright notice, guarantee or other statement on any media containing the Licensed Material.

(4) The LICENSEE shall only use the Licensed Material in the normal course of its business and shall not use the Licensed Material for the purpose of operating a bureau or similar service or any online service whatsoever.

(5) Permission is hereby granted to LICENSEES who are members of the legal profession (which expression does not include individuals or organisations engaged in the supply of services to the legal profession) to reproduce, transmit and store small quantities of text for the purpose of enabling them to provide legal advice to or to draft documents or conduct proceedings on behalf of their clients.

(6) The LICENSEE shall not sublicence the Licensed Material to others and this Licence Agreement may not be transferred, sublicensed, assigned or otherwise disposed of in whole or in part.

(7) The LICENSEE shall inform the PUBLISHERS on becoming aware of any unauthorised use of the Licensed Material.

Warranties

4.(1) The PUBLISHERS warrant that they have obtained all necessary rights to grant this licence.

(2) Whilst reasonable care is taken to ensure the accuracy and completeness of the Licensed Material supplied, the PUBLISHERS make no representations or warranties, express or implied, that the Licensed Material is free from errors or omissions.

(3) The Licensed Material is supplied to the LICENSEE on an "as is" basis and has not been supplied to meet the LICENSEE's individual requirements. It is the sole responsibility of the LICENSEE to satisfy itself prior to entering this Licence Agreement that the Licensed Material will meet the LICENSEE's requirements and be compatible with the LICENSEE's hardware/software configuration. No failure of any part of the Licensed Material to be suitable for the LICENSEE's requirements will give rise to any claim against the PUBLISHERS.

(4) In the event of any material inherent defects in the physical media on which the licensed material may be supplied, other than caused by accident abuse or misuse by the LICENSEE, the PUBLISHERS will replace the defective original media free of charge provided it is returned to the place of purchase within 90 days of the purchase date. The PUBLISHERS' entire liability and the LICENSEE's exclusive remedy shall be the replacement of such defective media.

(5) Whilst all reasonable care has been taken to exclude computer viruses, no warranty is made that the Licensed Material is virus free. The LICENSEE shall be responsible to ensure that no virus is introduced to any computer or network and shall not hold the PUBLISHERS responsible.

(6) The warranties set out herein are exclusive of and in lieu of all other conditions and warranties, either express or implied, statutory or otherwise.

(7) All other conditions and warranties, either express or implied, statutory or otherwise, which relate to the condition and fitness for any purpose of the Licensed Material are hereby excluded and the PUBLISHERS shall not be liable in contract or in tort for any loss of any kind suffered by reason of any defect in the Licensed Material (whether or not caused by the negligence of the PUBLISHERS).

Limitation of Liability and Indemnity

5.(1) The LICENSEE shall accept sole responsibility for and the PUBLISHERS shall not be liable for the use of the Licensed Material by the LICENSEE, its agents and employees and the LICENSEE shall hold the PUBLISHERS harmless and fully indemnified against any claims, costs, damages, loss and liabilities arising out of any such use.

(2) The PUBLISHERS shall not be liable for any indirect or consequential loss suffered by the LICENSEE (including without limitation loss of profits, goodwill or data) in connection with the Licensed Material howsoever arising.

(3) The PUBLISHERS will have no liability whatsoever for any liability of the LICENSEE to any third party which might arise.

(4) The LICENSEE hereby agrees that
(a) the LICENSEE is best placed to foresee and evaluate any loss that might be suffered in connection with this Licence Agreement;
(b) that the cost of supply of the Licensed Material has been calculated on the basis of the limitations and exclusions contained herein; and
(c) the LICENSEE will effect such insurance as is suitable having regard to the LICENSEE's circumstances.

(5) The aggregate maximum liability of the PUBLISHERS in respect of any direct loss or any other loss (to the extent that such loss is not excluded by this Licence Agreement or otherwise) whether such a claim arises is contract or tort shall not exceed a sum equal to that paid as the price for the title containing the Licensed Material.

Termination

6.(1) In the event of any breach of this Agreement including any violation of any copyright in the Licensed Material, whether held by the PUBLISHERS or others in the Licensed Material, the Licence Agreement shall automatically terminate immediately, without notice and without prejudice to any claim which the PUBLISHERS may have either for moneys due and/or damages and/or otherwise.

(2) Clauses 3 to 5 shall survive the termination for whatsoever reason of this Licence Agreement.

(3) In the event of termination of this Licence Agreement the LICENSEE will remove the Licensed Material.

Miscellaneous

7.(1) Any delay or forbearance by the PUBLISHERS in enforcing any provisions of this Licence Agreement shall not be construed as a waiver of such provision or an agreement thereafter not to enforce the said provision.

(2) This Licence Agreement shall be governed by the laws of England and Wales. If any difference shall arise between the Parties touching the meaning of this Licence Agreement or the rights and liabilities of the parties thereto, the same shall be referred to arbitration in accordance with the provisions of the Arbitration Act 1996, or any amending or substituting statute for the time being in force.